Functional Soft Matter II

Masahiro Funahashi
Editor

Functional Soft Matter II

Analytical Methods and Topics

Springer

Editor
Masahiro Funahashi
Department of Chemical Science
and Engineering
Kobe University
Kobe, Kagawa, Japan

ISBN 978-981-92-0099-3 ISBN 978-981-92-0100-6 (eBook)
https://doi.org/10.1007/978-981-92-0100-6

This Springer imprint is published by the registered company Springer Nature Singapore Pte Ltd.
The registered company address is: 152 Beach Road, #21-01/04 Gateway East, Singapore 189721, Singapore

Preface

In traditional materials science, substances with rigid structures have been investigated for a long time. Metals and ceramics are typical examples. These rigid materials have been mainly utilized for structural materials used in machinery, architecture, containers, and so on. On the other hand, human beings have utilized soft materials, such as textiles, fibers, and leathers. Most of them have been obtained from natural resources.

The assumption of macromolecules by Staudinger in 1920 and the subsequent invention of nylon by Carothers in 1935 brought industrial production of synthetic polymers, and many kinds of synthetic plastics, resins, and fibers are available for practical usage now.

Not only structural materials, but soft materials with electronic, optical, and other functions (functional soft matter) have been studied these 50 years. The observation of the electronic conduction in violanthrone crystals in 1950 by Akamatsu and Inokuchi was an epoch-making discovery. Synthesis of polyacetylene films with high electric conductivity in 1977 by Shirakawa, MacDiarmid, and Heeger extended electronics to soft matter.

In addition to polymers, the display application of liquid crystals has remarkably grown since the announcement of the dynamic scattering display by Heilmeier of RCA in 1968. In the liquid crystal displays, optical anisotropy coupled with the mobility of liquid crystal molecules creates a light-shuttering function.

Now, soft materials with optoelectronic functions are extensively studied for applications to flexible displays, wearable sensors, and printed devices. For example, organic transistors and sensors are integrated on flexible polymer substrates by the printing method. Many materials are under study, while some are applied for practical usage. The significance of functional soft materials is just increasing for the creation of new technologies. A few review articles on this topic have been published in top journals. However, textbooks covering basic science, applications, and characterization methods have been very limited so far. This textbook covers the basic concepts of functional soft matter, characterization methods of nanostructures and physical properties of soft matter, functions of soft matter, and device applications, including photoreactive liquid crystals, liquid crystalline semiconductors, photoalignment polymers, circularly polarized luminescent materials, ionic liquids, conjugated polymers, and so on.

This book consists of two volumes: *Functional Soft Matter I: Basic Principles*, and *Functional Soft Matter II: Analytical Methods and Topics. Functional Soft Matter II* comprises two parts. In this volume, *Functional Soft Matter II*, Part I, is related to analytical methods for structural, optical, and electronic properties of functional soft matter. Scanning microscope, diffraction methods, and thermal analysis are basic techniques for structural characterization of soft matter. Characterization methods of electronic, optical, electrochemical, and interfacial properties are significant for device applications of functional soft matter. Not only experimental technique but also molecular simulation is also an effective method for the design of functional soft matter. This part should be helpful for graduate students and young researchers carrying out their researches actually. The authors and the editor hope that readers will study the chapters with the scientific basis, which the readers mastered in the first volume. Part II includes recent research topics in the field of functional soft matter science. This part covers representative device applications and scientific progress. Graduate students can obtain insights into the recent trend of functional soft matter science and recognize various aspects of functional soft matter by reading this part. The content described in this part should inspire the research activities of the readers. The authors and editor hope that the reader will obtain great insights into functional soft matter from this volume, as well as the first volume, *Functional soft matter I: Basic Principles.*

The editor expresses my sincere gratitude to all authors for their valuable contributions to Functional Soft Matter—Electronic and Optical Functions. The editor hopes this book will serve as a valuable resource for advanced studies in functional materials fields.

Kobe, Japan Masahiro Funahashi

Contents

Analytical Methods

Scanning Probe Microscopy for the Investigation of Organic Materials

Shinobu Uemura

1 Introduction

In the development of organic materials, molecular arrangements in assembled structures have a significant effect on properties such as electronic conduction. Consequently, the observation of surface morphology of films and molecular arrangements in the assembled structures is an essencial and inseparable part of the research process. In soft matters, structural observation is typically conducted using optical microscopes and electron microscopes. However, scanning probe microscopy (SPM) is also frequently employed because it enables precise control of the measurement environment.

SPM is a technique that uses probes to detect the physical properties of samples and convert them into images. Therefore, the probes, substrates, and measurement environments used vary depending on the property to be detected. SPM tequniques range from scanning tunneling microscopy (STM), which utilizes the tunneling effect to enable observation of molecules and atoms, to atomic force microscopy (AFM), which has become a prevalent measurement technique. Other SPMs include friction force microscopy (FFM) and Kelvin probe force microscopy (KFM) techniques, which image friction force and surface potential, respectively. Additionally, SPMs employ special detection probes to detect features beyond surface morphology.

In this chapter, the imaging and physical property detection of organic materials using SPM is introduced.

S. Uemura (✉)
Faculty of Engineering and Design, Kagawa University, Takamatsu, Japan
e-mail: uemura.shinobu@kagawa-u.ac.jp

M. Funahashi (ed.), *Functional Soft Matter II*,
https://doi.org/10.1007/978-981-92-0100-6_1

2 SPM Techniques

2.1 STM: Tip, Substrate, and Conditions

As mentioned above, STM is a technique using the tunneling current that flows when a bias voltage is applied between a probe and a substrate separated by a small gap (approximately 1 nm) [1]. In STM, an atomically sharp conductive wire, such as platinum-iridium or tungsten, is typically used as a tip (probe), and an atomically flat conductive substrate, such as metal single crystal facets and HOPG, is commonly used. Since the tunneling current decays exponentially with increasing tip-substrate distance, STM provides extremely high resolution. In principle, over 90% of the tunneling current flowing between the atom of the tip (closest to the substrate) and the substrate is detected by the tip, resulting in a resolution of tens of picometers. Therefore, since the tunneling current is also affected due to molecules present between the tip and substrate, to visualize self-assembled molecular layers and supramolecular structures [2, 3].

STM provides atomic resolution imaging, and therefore, STM investigations of isolated molecules under ultra-high vacuum (UHV) and at low temperatures, which the molecular motion is suppressed, have been reported frequently. Under UHV, there are no adsorbed gas molecules or organic contaminants from the atmosphere, allowing the use of various metal single crystal surfaces, such as those with catalytic properties. Under such conditions, manipulation of molecules using a tip can also be performed, as demonstrated a molecular motor [4], and a nanocar [5]. Furthermore, temperature control is relatively easy in UHV, and chemical reactions have also been achieved [6]. For instance, Grill et al. reported that brominated porphyrin molecules can be covalently bonded by the thermally induced debromination, constructing mesh polymer structures, so-called covalent organic frameworks (COFs), on a substrate. Other typical condensation reactions, such as esterification and imine reactions, at various interfaces for the preparation of 2D-COFs have also been reported [7, 8].

Practically, STM can observe molecules and atoms at medium-substrate interfaces, such as in air or low-polarity solvents, where current leakage through the medium between the probe and substrate is eliminated. However, observing the self-assembled structures of organic molecules adsorbed on a substrate in air is relatively difficult. This is presumably because organic molecules without chemical bonds to the substrate can be physically moved by the tip scanning over the molecules, because of the narrow distance between the tip and molecules. In contrast, at the interface between a substrate and an organic solution in which the target organic molecules are appropriately dissolved, molecular assemblies can be observed when several conditions are satisfied (Fig. 1) [9]. These conditions include sufficient interactions between the traget molecules and the substrate, preferential partition of the molecules to the interface rather than remaining in solution, and alignment of the molecules with the substrate lattice. The partition of the molecules to the solid–liquid interface also competes with the organic solvent molecules. Therefore, the

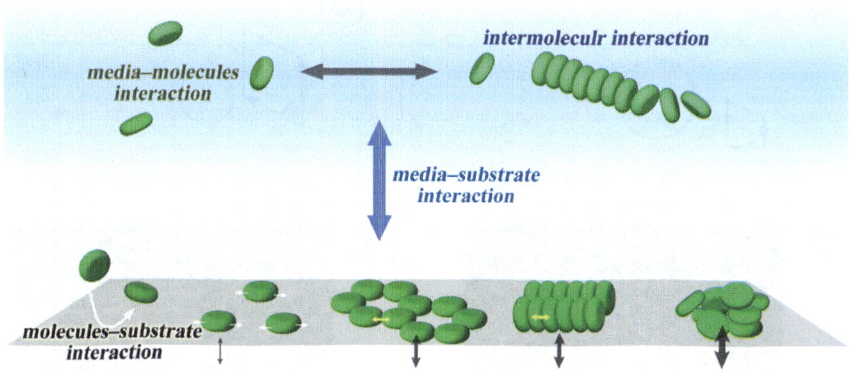

Fig. 1 Schematic illustration of the interaction between solute molecules, solvents, and substrates at solution–solid interface

equilibrium balance among molecule-molecule, molecule-solvent, and molecule-substrate interactions is extremely important. Under such conditions, the molecules are less likely to move toward the solution, and even if the molecules are removed from the substrate by the tip, they may be adsorbed onto the substrate again. Conversely, when the interaction between the molecules and the substrate is too weak, the molecules do not adsorb onto the substrate and cannot be visible. When the interaction is too strong, the molecules cannot rearrange into ordered structures and may adsorb excessively, often forming multi layers. From the above, molecules observed by STM at solution–substrate interfaces usually form ordered structures.

Relatively low-volatility solvents such as 1-phenyloctane or trichlorobenzene are commonly used as the solvent for STM investigation. Trimesic acid (TMA) is a typical molecule forming multiple hydrogen-bonding network structures, and STM observations have been conducted under various environmental conditions [10, 11]. Fatty acids, in which TMA is relatively dissolved, are used as a solvent for STM investigation at solution–solid interfaces [12]. TMA forms "chicken-wire" structures (honeycomb network) with lower density at the interface of long-chain fatty acids and HOPG, while "flower" structures with higher density are formed at the interface of short-chain fatty acids and HOPG (Fig. 2). The structural polymorphism is due to the porous structures (low density) and the partition of TMA and the fatty acids between the solution and the solid, which indicates thermodynamic control. Since porous structures are generally thermodynamically unfavorable, multiple hydrogen bondings are essential for stabilizing such structures. Additionally, the reversible hydrogen bonding allows structural transitions (polymorphism) depending on the concentration and solvent. Furthermore, to form thermodynamically stable structures, it has been also reported that lamellar structures are formed by hydrogen bonds between TMA and long-chain fatty acids as solvents [13], and that host–guest structures were formed by putting guest molecules such as coronene [14] and C_{60} [15] into the pores of TMA network structures. Thus, thermodynamically unstable porous structures can

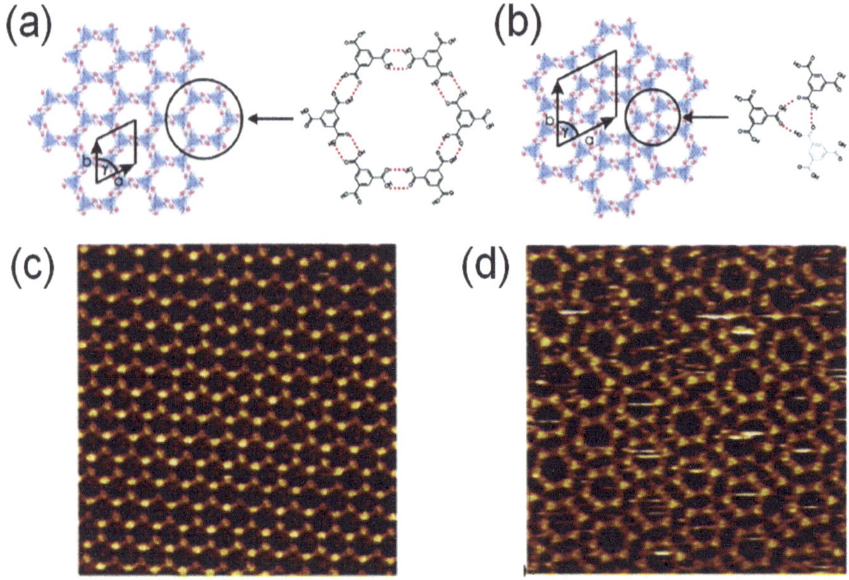

Fig. 2 Structure models of **a** "chicken-wire" structure and **b** "flower" structure of TMA. Typical STM images of TMA at the interface between HOPG and a saturated solution of fatty acids. **c** In heptanoic acid and **d** in pentanoic acid. Adapted with permission from Ref. [12]. Copyright 2025 American Chemical Society

be formed through TMA and/or molecules with multiple hydrogen-bonding sites, and guest molecules or molecules that interact with the network-forming molecules can fill the pores or rearrange into other ordered structures.

When leakage current from the tip to the solution is prevented, water can also be used as a solvent for STM. For instance, when the tip is insulated with Apeizon wax or nail polish, drying it with the tip pointing upward, the coated tip with only a few atoms exposed can be prepared. Using the coated tip, an STM investigation in an electrochemical environment is also possible (electrochemical (EC) STM). Under electrochemical potential control in the electrolyte solution, potential changes of the substrate induce the ion concentration changes at the interfacial double layer [9]. This phenomenon can be utilized to control the interfacial concentration of organic molecules. For instance, iodine adlayers on metal single crystal surfaces, such as Au(111) and Pt(111), are known to form chemically inert surfaces and to control the adsorption and desorption of iodine adlayers by the electrochemical potential. By utilizing such properties of iodine, it is easy to create surfaces with different adsorption strengths. Figure 3 shows fullerene C_{60} adlayers onto various metal substrates by a wet process [16]. On iodine-modified Au(111) (I/Au(111)), it was difficult to observe C_{60} adlayers, whereas the iodine on Au(111) could be observed. C_{60} can be easily replaced from iodine adlayers by electrochemical potential control, and C_{60} epitaxially grown adlayers onto Au(111) with low defects can be constructed,

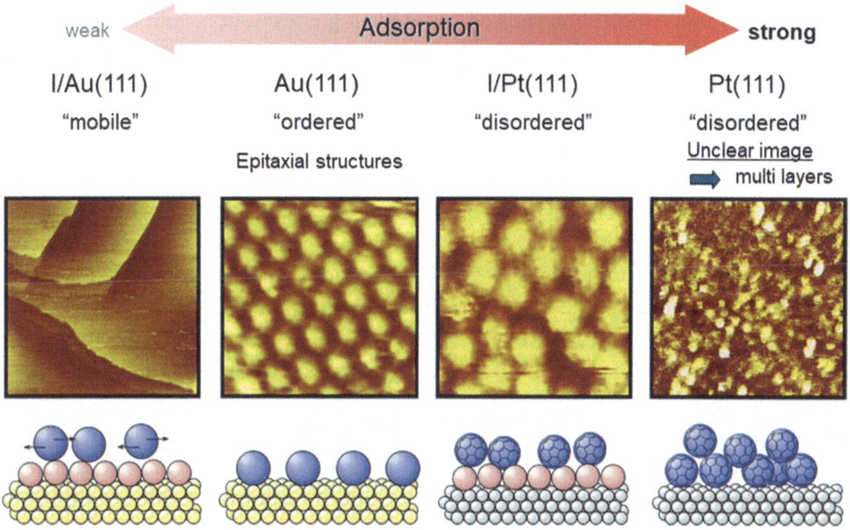

Fig. 3 EC-STM images and the schematic illustrations of C_{60} adlayers on various substrates at room temperature in an aqueous solution. Reprinted (adapted) with permission from Ref. [16]. Copyright 2025 American Chemical Society

comparable to those on Au(111) obtained by vapor deposition. Additionally, excessive adsorption of C_{60} (C_{60} multilayers) on Pt(111) was observed, whereas random but mostly single adlayers of C_{60} on I/Pt(111) were observed at the aqueous solution–solid interface. These results indicate that the adsorption strength of the substrates is in the order of Pt(111) > I/Pt(111) > Au(111) > I/Au(111). This is a good example of the relationship between the effect of substrate and the morphology of the organic molecular adlayer observed by STM.

2.2 *AFM: Cantilever, Substrates, and Conditions*

AFM visualizes surface topography by detecting the vertical deflection or movement of a cantilever caused by the interactions between the tip and the sample, known as atomic force. Since atomic forces act between all materials, the cantilever, substrates, and samples can range from insulators to conductors. In principle, AFM observation is suitable for soft matters, which often consist of large nonconductive structures such as aggregates and polymers. However, AFM investigations strongly depend on the measurement mode. In contact mode, which enables high resolution imaging, the cantilever effectively contacts with the sample surface, causing physical scratching or deformation of soft samples, even if the cantilever with a relatively small spring constant (or stiffness) is used. Therefore, measuring aggregates or thin films of organic molecules by contact mode AFM is challenging.

To observe organic molecular aggregates with high resolution, amplitude modulation mode (AM-AFM, called "tapping mode"), where the cantilever oscillates and intermittently taps the sample surface during observing, is preferred.

The imaging resolution of AFM, such as AM mode, is typically limited to several nanometers for crystalline ordered structures due to the finite radius of curvature of the cantilever tip. When nanoscale materials such as isolated nanoparticles are observed, the image is broadened by tip convolution, and the apparent size is detected as being larger than the actual size. For instance, zinc chlorin derivatives have been shown to align onto HOPG through π-π interactions between zinc chlorin groups and HOPG, as well as van der Waals interactions between long alkyl chains and HOPG (Fig. 4) [17]. In AM-AFM observation, two types of ordered structures depending on the number of alkyl chains were found: linearly ordered structures with 4.8 nm in width and hexagonally packed cyclic structures with 7.6 nm in the center-to-center distance. However, individual molecules were not resolved by AM-AFM. Conversely, in STM observations at the solution-solid interfaces, although the periodicities of the ordered structures were approximately consistent with those oberved by AFM, individual molecules in the ordered structures can be revealed by the differences in the imaging contrast, which reflect the electron density of the functional groups. The linear structures were assigned to lamellar structures composed of interdigitated zinc chlorin derivatives, and the hexagonal structures consisted of six derivatives. Thus, it can be seen that the ordered structures were observed by AM-AFM without molecular resolution under ambient conditions.

In AM-AFM, phase images representing phase shifts in the cantilever oscillation can be detected simultaneously. Phase imaging emphasizes the edges of thin films with different heights and visualizes differences in physical properties such as friction and elasticity. Surface metal–organic frameworks (SURMOFs) [18], which are composed of a porphyrin derivative with phenyl carboxyl groups (H_2TCPP) and copper ions, were reported to rearrange into large nanosheets on HOPG upon exposure to acetic acid vapor due to thermodynamic equilibrium [19]. AFM height images of SURMOFs after the vapor treatment reveal uniform height nanosheets, however the phase images often show areas with different contrasts in the same nanosheet (Fig. 5). This contrast difference is attributed to the difference in crystalline orientation of SURMOFs and the sensitivity of the AM-AFM phase imaging to anisotropy to the scan direction. Additionally, the reconstruction of crystalline self-assembled structures formed by weak hydrogen bonds is occasionally observed under ambient conditions (Ostwald ripening). However, the reconstruction of structures with relatively strong interactions, such as metal coordination, is unlikely to occur under ambient conditions. In this case, the presence of acetic acid, which can promote a thermodynamic equilibrium state, likely facilitated structural reconstruction.

STM investigations of organic molecules at high temperatures in air are generally unsuitable because of the active molecular motion, although some exceptions have been reported [20]. In contrast, AFM for imaging relatively macroscopic structures can also be used for observations at relatively high temperatures. In the research for polymer crystals [21] and liquid crystals [22], AFM investigations have been

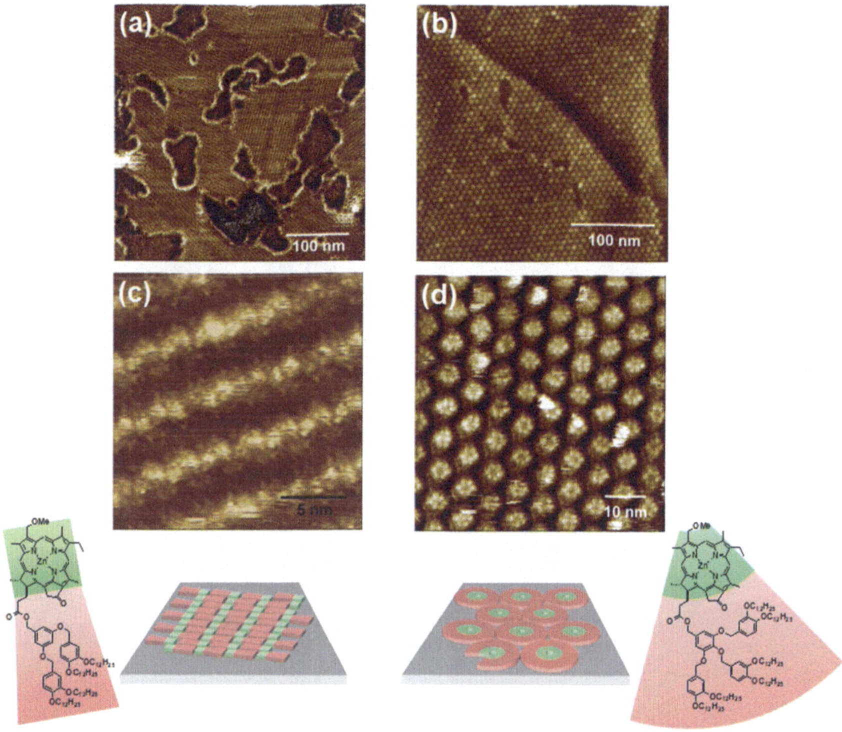

Fig. 4 AM-AFM images (**a**, **b**) and STM images (**c**, **d**) of zinc chlorin derivatives at the solution–HOPG interface at room temperature. Reprinted (adapted) with permission from Ref. [17]. Copyright 2025 John Wiley and Sons

occasionally conducted at temperatures around phase transitions to observe crystal growth processes and structural rearrangement in each phase.

Generally, under solutions or at lower temperature, AM-AFM based on cantilever oscillation cannot achieve sufficient resolution due to the viscous resistance from the solution or the stiffness of the cantilever, although AFM can observe any sample. Recently, it was reported that non-contact AFM operating in frequency modulation mode (FM-AFM) enables to observe molecular internal structures under UHV at extremely low temperatures, as well as in the solution. Especially, in the solution, video-rate observations of proteins and three-dimensional mapping of adsorbed water have been achieved by FM-AFM.

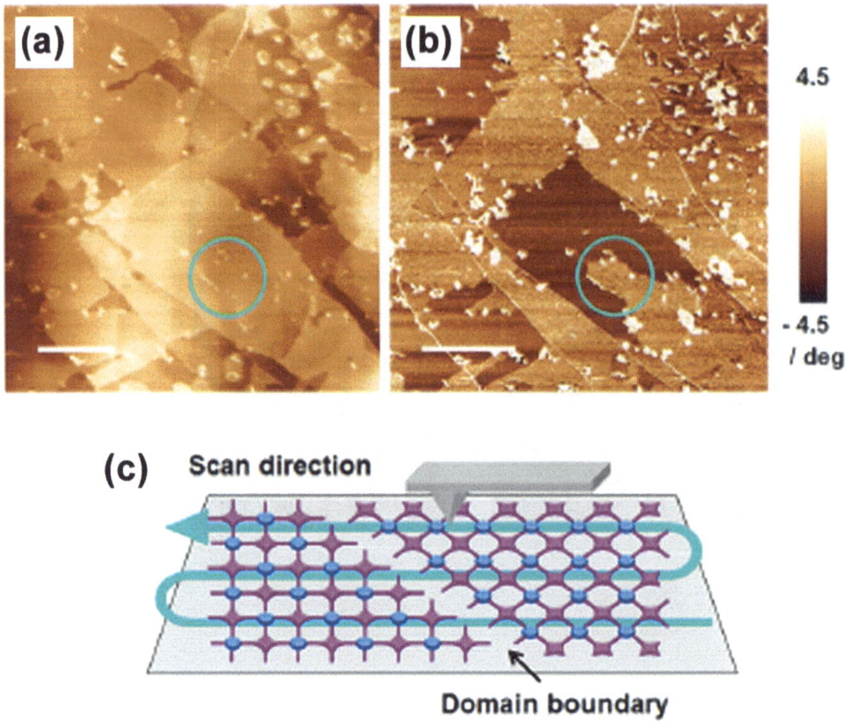

Fig. 5 AM-AFM **a** height image and **b** phase image of H_2TCPP–Cu MOFs after the vapor treatment. **c** Schematic illustration of discrimination of crystalline domains in MOF nanosheets by phase imaging. Reprinted (adapted) with permission from Ref. [19]. Copyright 2025 John Wiley and Sons

3 Measurements of Surface Properties with SPM

3.1 Conductivity

In the development of nanodevices, measurement of the properties of assembled structures and/or aggregates is sufficient; however, measurements at the single-molecule level would provide clearer insights into effects such as orientation. SPM uses a sharp probe to scan a sample surface and detect its topography, and by utilizing this probe, it is possible to detect properties such as electrical conductivity and mechanical properties.

Since STM typically uses tunnel current, differences in the electronic states of molecules are imaged. For instance, even though graphite surfaces are atomicaly flat, it occasionally exhibits misaligned layered structures, and long-period contrast (moiré patterns) are observed. Additionally, as shown in Fig. 4c, d, aromatic groups with high electronic density appear bright, while alkyl chain regions appear dark, forming a lamellar structure that can be clearly distinguished. Furthermore, scanning tunneling spectroscopy (STS) is a method for quantitatively evaluating the electronic structure of molecules. Even for molecules arranged at the organic solvent–HOPG interface, it is possible to determine the energy distribution of surface electron state density of states [23].

Conversely, while the conductivity of thin films and/or self-assembled structures of organic materials is measured preferably by AFM, commercial conductive AFMs operate in contact mode, which makes it difficult to measure the conductivity without damage to the structures. Therefore, KFM is often used for evaluating the electrical property measurements of organic materials [24]. KFM is a technique that applies a voltage between an oscillating conductive cantilever and a sample (or substrate) and the contact potential difference is quantitatively measured through the electrostatic interactions . Like AM-AFM, KFM involves an oscillating cantilever, making it a measurement method that causes minimal physical damage to the sample. Since surface potential, work function, and electrostatic forces can be evaluated, the conductivity of battery materials and conductive polymers has been measured by KFM. The films of poly-γ-methyl-L-glutamic acid (PMLG) with a helical structure were prepared using the layer-by-layer (LbL) method via an ester–amide exchange reaction [25]. On the surface of the films, ester groups or free amino groups existed alternately. While the height images show fiber-like features on the film surface, the surface potential images of KFM reveal completely different contrasts at the same height (Fig. 6). This indicates that the introducing reaction of amine has not been fully completed. In other words, it demonstrates that the presence of amine groups on the surface of insulator PMLG can be observed using KFM.

3.2 *Mechanical Properties of Soft Matters*

In AFM, a cantilever tip touching the samples leads to measuring mechanical properties of the samples, such as viscoelasticity and friction. However, the forces acting between the tip and the sample are not only the targeted mechanical properties, such as viscosity; other interactions, including electrostatic forces, van der Waals forces, and surface forces, also come into play. Therefore, generally three methods are used: force curves, which measure force by pressing the cantilever and measuring its deflection; phase shift measurements in AM mode; and AM-FM mode. By selecting an appropriate measurement mode, it is possible to measure elasticity, viscoelasticity, and other properties of materials such as rubber [26], cells [27], and the curing process of resins [28], among others.

Frictional force between films on substrates and cantilevers can be measured by the deflection of a cantilever mentioned above. Commercial friction force microscopy, or lateral force microscopy, can measure the friction between thin films and cantilevers; however, it is difficult to measure the adhesion and friction force of a single molecule (including polymers) to substrates. Recently, graphene nanoribbons have been synthesized from aromatic compounds on a substrate, and their physical properties have been investigated [29]. By positioning the edge of a graphene nanoribbon

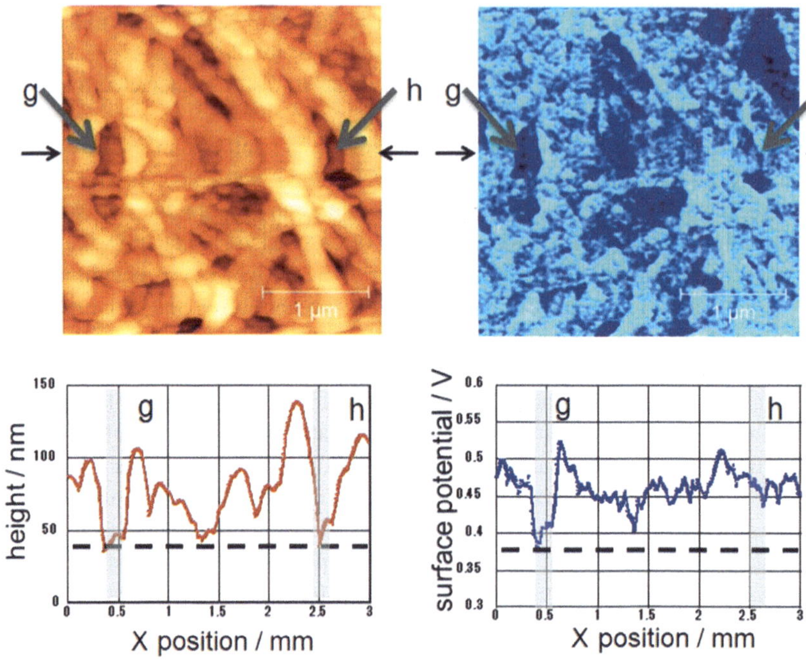

Fig. 6 (Left) A typical height image and (right) a surface potential image of PMLG LbL films on ITO glass. Reprinted (adapted) with permission from Ref. [24]. Copyright 2025 American Chemical Society

by the tip of an AFM cantilever and then lifting it, the cantilever bends, and it has been possible to successfully measure the frictional force between the graphene nanoribbon and the substrate (Fig. 7). It has also been confirmed that there is no significant change even if the number of units in the graphene nanoribbon is different.

4 Conclusion and Outlook

SPM is a useful tool for visualizing structures with molecular resolution in various environments. However, the observed structures can vary depending on the sample (solvent, concentration, film thickness, viscosity, etc.), probe (cantilever), and substrate. In particular, the softness or viscosity of the medium or sample can influence the structures observed by SPM. Therefore, optimal conditions for visualizing target molecules or assembled structures can be determined through trial and error.

Additionally, with the growing interest in 2D materials such as graphene and COFs, SPM is increasing being used as a tool for in situ observation of structures. In the future, SPM is expected to be used to measure the properties of materials, such as 2D-COFs, created in situ, and determine their potential for direct application in

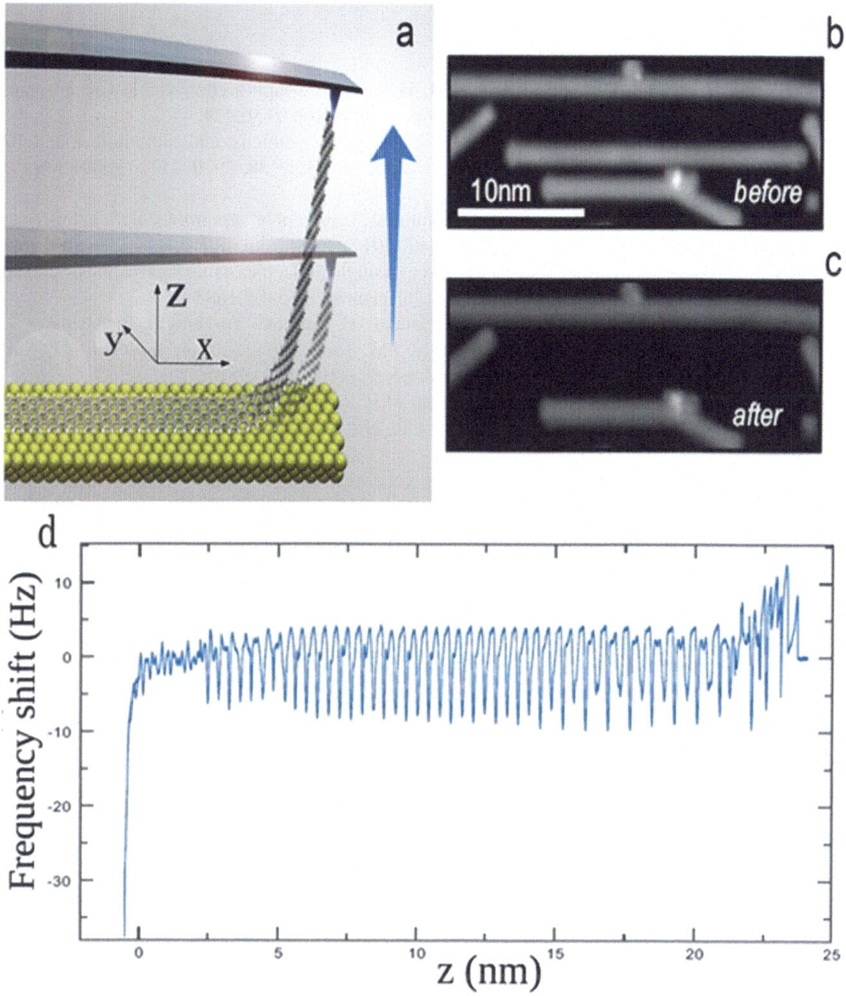

Fig. 7 **a** Schematic illustration of the graphene nanoribbon detachment by the tip. **b**, **c** STM images of graphene nanoribbons before and after the detachment of single graphene nanoribbon. **d** Frequency shift trace recorded as graphene nanoribbon lifted up until detachment. Reprinted with permission from Ref. [29]. Copyright 2025 American Chemical Society

devices. Further development of new SPM-based measurement methods is highly anticipated.

References

1. G. Binnig, H. Rohrer, C. Gerber, E. Weibel, Tunneling through a controllable vacuum gap. Appl. Phys. Lett. **40**, 178–180 (1981). https://doi.org/10.1063/1.92999
2. T. Kudernac, S. Lei, J.A.A.W. Elemans, S. de Feyter, Two-dimensional supramolecular self-assembly: nanoporous networks on surfaces. Chem. Soc. Rev. **38**, 2520–2542 (2009). https://doi.org/10.1039/B708902N
3. K.S. Mali, N. Pearce, S. de Feyter, N.R. Champness, Frontiers of supramolecular chemistry at solid surfaces. Chem. Soc. Rev. **46**, 2520–2542 (2017). https://doi.org/10.1039/C7CS00113D
4. J.K. Gimzewski, C. Joachim, Nanoscale science of single molecules using local probes. Science **283**, 1683–1688 (1999). https://doi.org/10.1126/science.283.5408.1683
5. Y. Shirai, A.J. Osgood, Y. Zhao, Y. Yao, L. Saudan, H. Yang, C. Yu-Hung, L.B. Alemany, T. Sasaki, J.F. Morin, J.M. Guerrero, K.F. Kelly, J.M. Tour, Surface-rolling molecules. J. Am. Chem. Soc. **128**, 4854–4864 (2006). https://doi.org/10.1021/ja058514r
6. L. Grill, M. Dyer, L. Lafferentz, M. Persson, M.P. Peters, S. Hecht, Nano-architectures by covalent assembly of molecular building blocks. Nat. Nanotech. **2**, 687–691 (2007). https://doi.org/10.1038/nnano.2007.346
7. S. Clair, Sylvain, D.G. de Oteyza, Controlling a chemical coupling reaction on a surface: Tools and strategies for on-surface synthesis, Chem. Rev. **119**, 4717-4776 (2019). https://doi.org/10.1021/acs.chemrev.8b00601
8. J. Sakamoto, J. van Heijst, O. Lukin, A.D. Schlüter, Two-dimensional polymers: just a dream of synthetic chemists? Angew. Chem. Int. Ed. **48**, 993–1167 (2009). https://doi.org/10.1002/anie.200801863
9. S. Uemura, R. Tanoue, N. Yilmaz, A. Ohira, M. Kunitake, Molecular dynamics in two-dimensional supramolecular systems observed by STM. Materials **3**, 4252–4276 (2010). https://doi.org/10.3390/ma3084252
10. Y. Ishikawa, A. Ohira, M. Sakata, C. Hirayama, M. Kunitake, A two-dimensional molecular network structure of trimesic acid prepared by adsorption-induced self-organization. Chem. Commun. 2652–2653 (2002). https://doi.org/10.1039/B207556C
11. M. Lackinger, W.M. Heckt, Carboxylic acids: versatile building blocks and mediators for two-dimensional supramolecular self-assembly. Langmuir **25**, 11307–11321 (2009). https://doi.org/10.1021/la900785f
12. M. Lackinger, S. Griessl, W.M. Heckl, M. Hietschold, G.W. Flynn, Self-assembly of trimesic acid at the liquid−solid interface—a study of solvent-induced polymorphism. Langmuir **21**, 4984–4988 (2005). https://doi.org/10.1021/la0467640
13. K.G. Nath, O. Ivasenko, J.M. MacLeod, J.A. Miwa, J.D. Wuest, A. Nanci, S.F. Perepichka, F. Rosei, Crystal engineering in two dimensions: an approach to molecular nanopatterning. J. Phys. Chem. C **111**, 16996–17007 (2007). https://doi.org/10.1021/jp0762774
14. S.J.H. Griessl, M. Lackinger, F. Jamitzky, T. Markert, M. Hietschold, W.M. Heckl, Incorporation and manipulation of coronene in an organic template structure. Langmuir **20**, 9403–9407 (2004). https://doi.org/10.1021/la049441c
15. S.J.H. Griessl, M. Lackinger, F. Jamitzky, T. Markert, M. Hietschold, W.M. Heckl, Room-temperature scanning tunneling microscopy manipulation of single C_{60} molecules at the liquid−solid interface: playing nanosoccer. J. Phys. Chem. B **108**, 11556–11560 (2004). https://doi.org/10.1021/jp049521p
16. S. Uemura, M. Sakata, C. Hirayama, M. Kunitake, Fullerene adlayers on various single-crystal metal surfaces prepared by transfer from L films. Langmuir **20**, 9198–9201 (2004). https://doi.org/10.1021/la048982z
17. S. Uemura, S. Sengupta, F. Würthner, Cyclic self-assembled structures of chlorophyll dyes on HOPG by the dendron wedge effect. Angew. Chem. Int. Ed. **48**, 7825–7828 (2009). https://doi.org/10.1002/anie.200902801
18. S. Motoyama, R. Makiura, O. Sakata, H. Kitagawa, Highly crystalline nanofilm by layering of porphyrin metal−organic framework sheets. J. Am. Chem. Soc. **133**, 5640–5643 (2011). https://doi.org/10.1021/ja110720f

19. K. Sakata, S. Kashiyama, G. Matsuo, S. Uemura, N. Kimizuka, M. Kunitake, Growth of two-dimensional metal-organic framework nanosheet crystals on graphite substrates by thermal equilibrium treatment in acetic acid vapor. ChemNanoMat **1**, 259–263 (2015). https://doi.org/10.1002/cnma.201500062
20. R. Gutzler, T. Sirtl, J.F. Dienstmaier, K. Mahata, W.M. Heckl, M. Schmittel, M. Lackinger, Reversible phase transitions in self-assembled monolayers at the liquid−solid interface: Temperature-controlled opening and closing of nanopores. J. Am. Chem. Soc. **132**, 5084–5090 (2010). https://doi.org/10.1021/ja908919r
21. A.-K. Löhmann, T. Henze, T. Thurn-Albrecht, Direct observation of prefreezing at the interface melt–solid in polymer crystallization. PNAS **111**, 17368–17372 (2014). https://doi.org/10.1073/pnas.1408492111
22. S. Sengupta, S. Uemura, S. Patwardhan, V. Huber, F.C. Grozema, L.D.A. Siebbeles, U. Baumeister, F. Würthner, Columnar mesophases based on zinc chlorophyll derivatives functionalized with peripheral dendron wedges. Chem. Eur. J. **17**, 5300–5310 (2011). https://doi.org/10.1002/chem.201002659
23. F. Jäckel, M.D. Watson, K. Müllen, J.P. Rabe, Prototypical single-molecule chemical-field-effect transistor with nanometer-sized gates. Phys. Rev. Lett. **92**, 188303 (2004). https://doi.org/10.1103/PhysRevLett.92.188303
24. A. Liscio, V. Palermo, P. Samorì, Nanoscale quantitative measurement of the potential of charged nanostructures by electrostatic and Kelvin probe force microscopy: Unraveling electronic processes in complex materials. Acc. Chem. Res. **43**, 541–550 (2010). https://doi.org/10.1021/ar900247p
25. R. Higuchi, M. Hirano, M. Ashaduzzaman, N. Yilmaz, T. Sumino, D. Kodama, S. Chiba, S. Uemura, K. Nishiyama, A. Ohira, M. Fujiki, M. Kunitake, Construction and characterization of molecular nonwoven fabrics consisting of cross-linked poly(γ-methyl-L-glutamate). Langmuir **29**, 7478–7487 (2013). https://doi.org/10.1021/la3045576
26. T. Igarashi, S. Fujinami, T. Nishi, N. Asao, K. Nakajima, Nanorheological mapping of rubbers by atomic force microscopy. Macromolecules **46**, 1916–1922 (2013). https://doi.org/10.1021/ma302616a
27. Y. Fujii, W.C. Koizumi, T. Imai, M. Yokobori, T. Matsuo, K. Oka, K. Hotta, T. Okajima, Spatiotemporal dynamics of single cell stiffness in the early developing ascidian chordate embryo. Commun. Biol. **4**, 341 (2021). https://doi.org/10.1038/s42003-021-01869-w
28. H.K. Nguyen, M. Aoki, X. Liang, S. Yamamoto, K. Tanaka, K. Nakajima, Local mechanical properties of heterogeneous nanostructures developed in a cured epoxy network: Implications for innovative adhesion technology. ACS Appl. Nano Mater. **4**, 12188–12196 (2021). https://doi.org/10.1021/acsanm.1c02692
29. L. Gigli, S. Kawai, R. Guerra, N. Manini, R. Pawlak, X. Feng, K. Müllen, P. Ruffieux, R. Fasel, E. Tosatti, E. Meyer, A. Vanossi, Detachment dynamics of graphene nanoribbons on Gold. ACS Nano **13**, 689–697 (2019). https://doi.org/10.1021/acsnano.8b07894

Diffraction Methods

Mitsuo Hara

1 Introduction

Photoelectric and ion-conduction functions are closely related to the macroscopic arrangements of molecules and molecular aggregates, as well as their arrangements near interfaces. Therefore, understanding the internal structures of materials is crucial for enhancing device performance. Among various structural-analysis methods, those based on diffraction (X-ray, electron, and neutron) are among the most powerful tools for analyzing the details of internal structures, thereby enabling the morphologies, periodicities, and heterogeneities of self-assembled structures, atomic arrangements, and molecular structures to be elucidated. The information provided by these analysis techniques is vital when designing and evaluating functional materials, improving existing materials, and optimizing structures according to function.

The characteristics of X-ray, electron, and neutron diffraction methods are listed in Table 1. Given that differences in radiation wavelengths translate into different interactions with matter, these methods should be chosen according to the analysis purpose. Specifically, unlike X-rays and electrons, which are scattered owing to interactions with electrons, neutrons are scattered through interactions with atomic nuclei (Fig. 1a) [1], which enables the detection of light elements and the discrimination of different isotopes. Figure 1b illustrates the different strengths of X-ray element and neutron element interactions [2]. In view of the abundance of literature describing the basic principles of each diffraction method, the present chapter does not duplicate this information but rather provides an overview of the specific information that can be obtained by applying these methods to functional soft matter (e.g., liquid crystals, polymers, and colloids).

M. Hara (✉)
Program in Advanced Materials Science, Faculty of Engineering and Design, Kagawa University, Takamatsu, Kagawa, Japan
e-mail: hara.mitsuo@kagawa-u.ac.jp

M. Funahashi (ed.), *Functional Soft Matter II*,
https://doi.org/10.1007/978-981-92-0100-6_2

Table 1 Comparing the features of X-ray, electron, and neutron diffraction methods

	X-ray diffraction	Electron diffraction	Neutron diffraction
Radiation used	X-rays	Electrons	Neutrons
Wavelength	0.01–10 nm	0.001–0.1 nm	0.01–1 nm
Energy	Low to medium (10–100 keV)	High (tens to hundreds of keV)	Low (meV to keV)
Interaction target	Electrons	Electrons	Nuclei
Sample state	Solids, powders, films, liquids	Ultrathin films, particles	Solids, powders, films, liquids
Type of equipment	X-ray generators (X-ray tubes, synchrotron)	Electron microscopes (TEM, SEM)	Neutron sources (research reactors, accelerators)
Main applications	Crystal structure analysis, phase identification, grain size measurement	High-resolution structure analysis, nanoscale structure analysis	Crystal structure analysis, phase identification, magnetic material analysis
Advantages	Nondestructive structural analysis possible	High-resolution fine structure analysis	Complex structure analysis including light elements, isotope discrimination

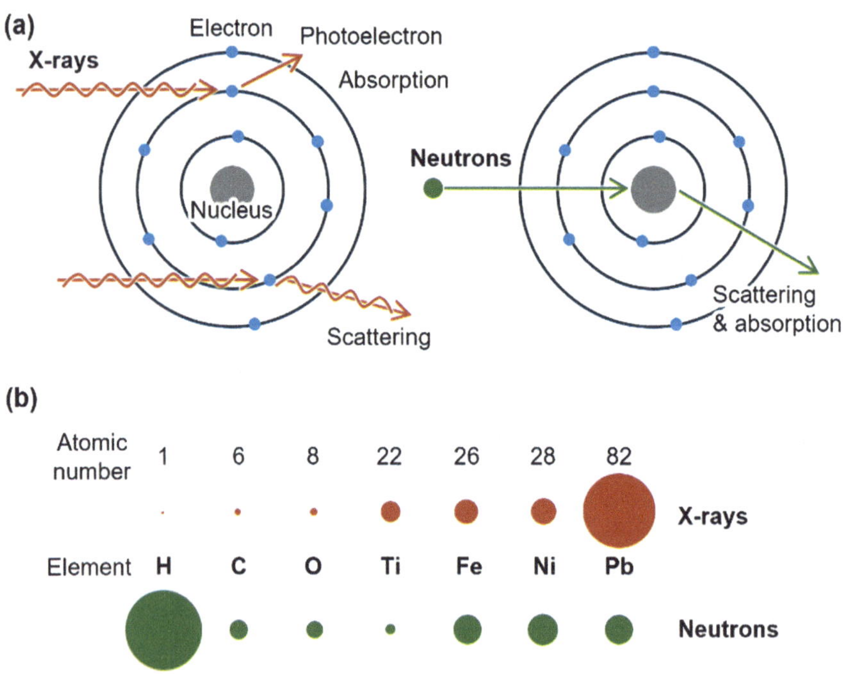

Fig. 1 **a** Interactions between X-rays/neutrons and matter. **b** Comparing neutron and X-ray scattering cross-sections

2 Bulks

Liquid crystals are soft materials that form self-assembled nanoperiodic structures and exhibit molecular cooperativity [3]. Therefore, when mechanical stimuli, electromagnetic waves, dielectric anisotropy, or surface free energy affect liquid–crystal molecules at the substrate interface or near the free surface, these influences easily propagate within the liquid–crystal phase. By leveraging this property, one can uniformly align the self-assembled structure of a liquid crystal over a large area [4], as exemplified by liquid–crystal displays. Additionally, considerable research effort has been directed toward applying the self-assembled structures of functional liquid–crystal molecules, such as those exhibiting photoelectric and ion-conduction functions, as carrier-transport materials in devices. Numerous studies have focused on the correlation between the types and orientations of liquid–crystal phases and their functions and have highlighted the importance of accurately determining these phases and their orientations for enhancing device performance.

X-ray diffraction (XRD) methods are commonly used to evaluate the self-assembled morphologies of liquid crystals, thereby facilitating rigorous identification of the liquid–crystal phase when high-order peaks are observed in the XRD profile, as described in detail by Ohta [5].

The study of liquid–crystal materials exhibiting electron conduction is an example of research focusing on the correlation between liquid–crystal phases and their functions. Hanna and Funahashi synthesized 2-phenylnaphthalene derivatives and reported that charge mobility increased as the liquid–crystal phases became more ordered (smectic A (SmA) → smectic B (SmB) → smectic E (SmE)) [6]. Similar trends have been observed for terthiophene derivatives [7], following the order of 10^{-4} to 10^{-3} $cm^2\ V^{-1}\ s^{-1}$ (SmA, SmC) < 10^{-3} to 10^{-2} $cm^2\ V^{-1}\ s^{-1}$ (SmB, SmF) < 10^{-2} $cm^2\ V^{-1}\ s^{-1}$ (SmE, SmG). The rate of intermolecular charge transfer depends on the orderliness of the molecular arrangement, even for the same molecule, which highlights the importance of understanding the morphologies of molecular aggregates during measurement. Hanna and Iino reported that films fabricated at temperatures compatible with the formation of liquid–crystal phases feature elevated uniformity [8], noting that Ph-BTBT-10 showed very high mobilities (2 and 10 $cm^2\ V^{-1}\ s^{-1}$ its SmE and polycrystalline phases, respectively), which are comparable to those observed for oxide semiconductors (IGZO) and highlight its usefulness as an organic semiconductor film for transistors [9].

The morphology of a liquid–crystal phase is also crucial for the functional expression of ionically conductive liquid–crystal materials. Kato et al. synthesized ammonium salt-containing thermotropic liquid crystals that form hexagonal columnar (Col_h) and bicontinuous cubic (Cub_{bi}) phases, depending on the temperature and demonstrated that the ion conductivity of the latter phase exceeded that of the former phase (Fig. 2) [10, 11]. This behavior was ascribed to the presence of continuous ion channels in the Cub_{bi} phase, as opposed to interrupted ion channels between the domains of the Col_h phase due to macroscopic nonalignment. Zhang et al. synthesized liquid–crystal molecules containing sodium sulfonate and captured the behavior of

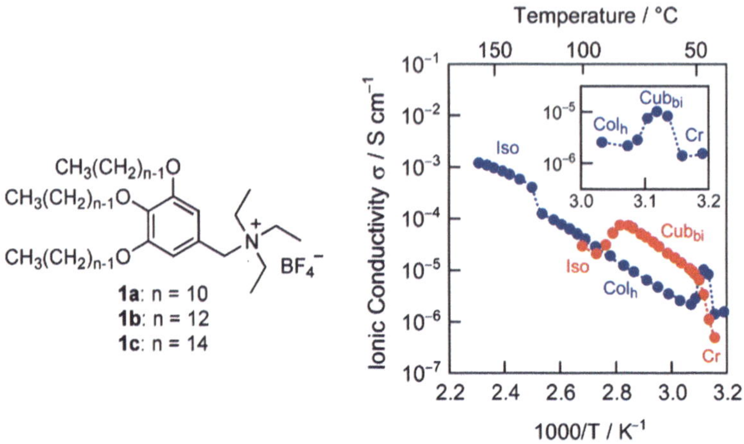

Fig. 2 Ionic conductivities of **1a** (red) and **1b** (blue) as functions of temperature. Adapted from Ref. [10] with layout modified with permission of the ACS. Copyright 2007 American Chemical Society

the Col_h-to-Cub_{bi} phase transition under humidity control using in situ XRD methods [12]. This phase transition was induced by the selective absorption of water into ion nanochannels, resulting in an increase in the volume fraction of the hydrophilic regions. The ion conductivity of the Cub_{bi} phase exceeded that of the Col_h phase by three orders of magnitude. Moreover, the Cub_{bi} phase not only featured continuously connected ion nanochannels but also exhibited two water-diffusion modes and showed transport properties equivalent to those of polymeric ion conductors [13].

Organic–inorganic mesostructured materials and mesoporous metal oxides can be synthesized using sol–gel reactions involving metal alkoxides in the presence of surfactants or amphiphilic block copolymers and have been effectively structurally characterized using XRD methods. Various composite-material morphologies have been evaluated by XRD methods [14, 15]; for example, Seki et al. reported that chromonic liquid crystals can be fixed in self-assembled structure in solution through sol–gel reactions with amorphous silica [16]. The combination of this composite approach with liquid–crystal film alignment enabled the synthesis of optical films that exhibit absorption anisotropies [17].

Lyotropic liquid crystals, which self-assemble in solution, can also be evaluated by XRD methods. The large thicknesses of glass bottles or tubes used for storing or reacting samples result in X-rays being absorbed by the container material, which prevents the sample inside the container from being evaluated by XRD methods. This problem can be solved using glass tubes with thin (0.01-mm-thick) walls through which X-rays can pass and are typically used for single-crystal X-ray analyses. The container material (e.g., soda-lime, Lindemann, borosilicate, or quartz glass) affects X-ray transmission and the background, which highlights the importance of selecting containers suitable for the targeted 2θ range. Polyimide tubes are also effective [18]. The use of these tubes is not limited to evaluating lyotropic liquid crystals but also include characterizing nanoparticle, colloid, or gel particle size, shape, and aggregation state using XRD, small-angle X-ray scattering (SAXS), and neutron diffraction methods [19–28]. In addition to container wall thickness and material, the X-ray absorption rate of the sample also needs to be considered. One is advised to reduce the outer diameter of the tube and thin the sample when elements with high X-ray absorption cross-sections (e.g., halogens and heavy metals) are present.

3 Thin Films

One can fully utilize the functions of molecules or molecular aggregates by uniformly aligning self-assembled structures over large areas. While local self-assembled structures can be evaluated using atomic force or scanning electron microscopy, evaluating the relationships between the structures/arrangements observed in these local regions and macroscopic regions is difficult using microscopy-based techniques. In such cases, X-ray or neutron diffraction methods are effective, with grazing incidence X-ray diffraction (GI-XRD) methods a particularly powerful tool (Fig. 3).

When X-rays are incident on a film at a near-critical angle, one can separate the information provided by the in-plane and out-of-plane directions. As a result, we can discuss the orientation in the macroscopic region. GI-XRD methods require highly parallel (small-divergence-angle) high-brightness X-rays to be incident on the sample. Until the end of the twentieth century, such analyses were mainly conducted at synchrotron radiation facilities. However, toward the end of the twentieth century, the performance of X-ray focusing mirrors improved, enabling the use of highly parallel high-brightness X-rays in laboratory-scale X-ray diffractometers. As a result, GI-XRD methods have been increasingly performed using laboratory equipment

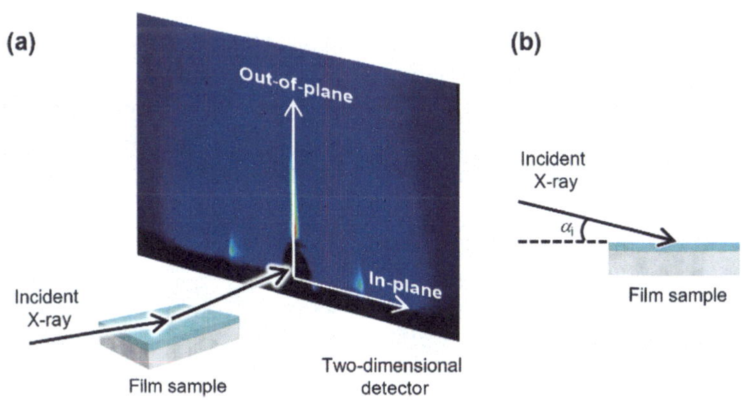

Fig. 3 **a** Illustrating a GI-XRD methods viewed from an oblique angle. **b** Side-view diagram. The symbol α_i indicates the X-ray incidence angle

since the beginning of the twenty-first century [29]. For detailed measurement principles, the readers are referred to existing papers and reference books [30–34]. Here, we provide some examples of characterizing functional materials by GI-XRD methods.

π-Conjugated polymers with condensed π-electron systems, such as polythiophene and polyfluorene derivatives, exhibit photoelectric properties. The introduction of branched or long side chains to improve solubility enables these polymers to be solution—processed into films, and self-assembled structures can be aligned by sublayer or mechanical shearing. Therefore, π-conjugated polymers are attractive materials for enhancing device performance. Knaapila et al. showed through GI-XRD methods that polyfluorene derivatives with suitable molecular weights form nematic phases, which lead to the good alignment of the main chains [35, 36]. Osaka et al. investigated not only side-chain length but also copolymers with units that feature different side-chain lengths and branched side chains and discussed the structures and alignments of self-assembled π-conjugated polymers determined by GI-XRD measurements (Fig. 4) [37–40].

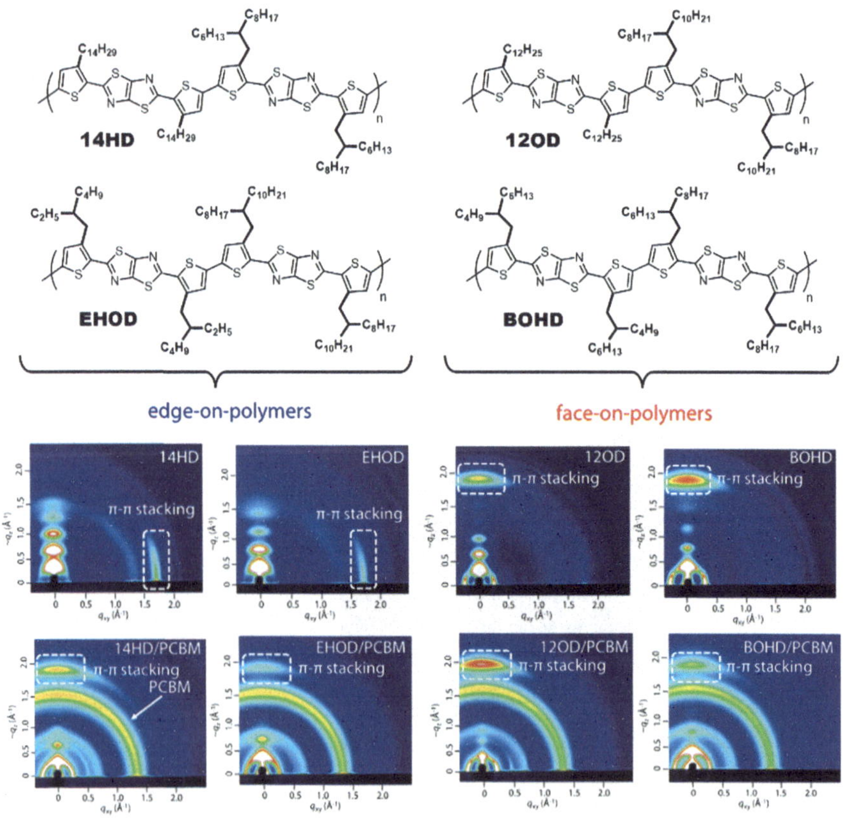

Fig. 4 Chemical structures of π-conjugated polymers with different side chains and their GI-XRD images. Adapted from Ref. [40]. Copyright 2018 American Chemical Society

The mesoporous metal oxides mentioned in Sect. 2 can be applied to solar cell electrodes, battery separators, and molecular sieves, among others, by perpendicularly aligning the mesopores relative to the substrate surface [41–46]. Such materials are referred to as vertically aligned mesoporous materials and are often analyzed by GI-XRD methods. When mesoporous metal-oxide films are prepared within spatial templates, such as the grooves created by anodized porous alumina [47, 48] or lithography [49], the mesopores align themselves along the inner walls of the grooves perpendicularly to the support surface. Vertical alignment can also be achieved using external (e.g., magnetic and electric) fields [50–53]. Various methods of preparing vertically aligned mesoporous materials have been developed and evaluated using GI-XRD methods [54–65].

4 In Situ GI-XRD Methods

GI-XRD measurements should ideally be performed under conditions identical to those used during device operation to correlate device performance with the morphology, order, and arrangement of a self-assembled structure. Nagao et al. subjected thin films of polyimide derivatives with alkylsulfonic acid side chains to GI-XRD methods in humidity-controlled environments, which revealed that lamellar hydrophilic–hydrophobic phase-separated structures become more ordered with increasing humidity and form nanoscale water-molecule channels [66]. These authors also showed that a polyimide derivative with a molecular weight of 2.6×10^5 exhibited larger domains and proton conductivities that were two orders of magnitude higher than that with a molecular weight of 1.3×10^4 [67]. GI-XRD methods were also effectively used when the morphologies of lyotropic liquid crystals used in solutions were manipulated by adjusting the humidity [68–70]; moreover, it can also be used to evaluate ionomer self-assembly behavior [71–74]. GI-XRD methods revealed that moderate humidity led to lamellar self-assembly when applied to an ionic polysiloxane thin film (Fig. 5) [75]. Lowering the neutralization degree prevented self-assembly, which indicates that self-assembly requires ionic-group aggregation. Interestingly, diluting the ionic-group concentration by copolymerization improved the self-assembly orderliness. The results of ionomer evaluation in humid environments suggest the possibility of constructing new academic theories.

Nagano and Seki et al. synthesized block copolymers with photoisomerizable side chains and observed the photoalignment of microphase-separated structures under linearly polarized light using time-resolved experiments at synchrotron facilities. The authors found that scattering from the liquid–crystal phase and the formation of microphase-separated structures occur almost synchronously, indicative of cooperative photoalignment [76]. Furthermore, the domains split into ones that are submicron in size, rotated, and subsequently fused during the photoalignment process [77, 78].

Changes in the self-assembly of photoresponsive molecules under light irradiation are generally evaluated using thin films to enable light to penetrate deep into the sample, as exemplified by the work of Nagano and Seki et al. [76–78].

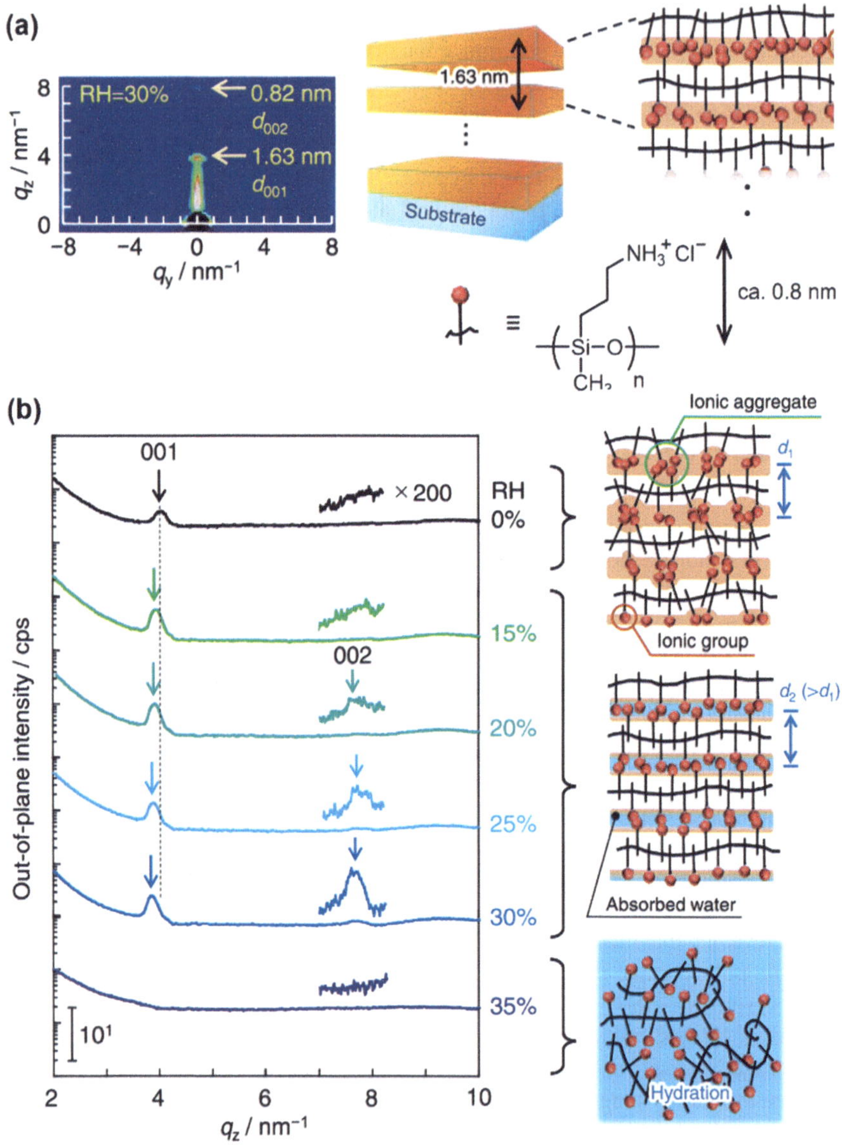

Fig. 5 **a** GI-SAXS image and structural model of an ionic linear polysiloxane. **b** Out-of-plane intensity profiles from humidity-controlled GI-SAXS methods involving ionic linear polysiloxane films. Adapted from Ref. [75] with layout modified with permission of the ACS. Copyright 2022 American Chemical Society

However, film formation is often accompanied by dewetting at relatively low sample molecular weights, resulting in aggregation that hinders light penetration and alignment. Conversely, the use of thick samples to alleviate dewetting prevents light from reaching the deep interior, thereby hindering the ability to accurately evaluate light-irradiation effects. Consequently, light-irradiation experiments are performed by considering the trade-off between film stability and light transmittance for samples with relatively low molecular weights. To solve this problem, one can assemble a cell with a gap of several microns and fill it with the desired sample to prevent dewetting while enabling light to reach the sample interior. By utilizing the GI-XRD stage and introducing X-rays into the cell through the cell gap, one can observe diffraction from the sample, and XRD patterns of the sample can be obtained during photoisomerization when irradiated with light from the top of the cell. Yamakado and Maeda et al. used this method to elucidate the crystal–crystal phase transitions of ion-pair aggregates using XRD methods [79].

XRD methods can also be performed on a liquid–crystal cell. Events previously difficult to evaluate using XRD methods can be assessed by assembling a cell using a glass substrate coated with a polymeric liquid–crystal PAz film and injected with low-molecular-weight 5CB liquid crystals. Hara and Seki et al. found that a highly ordered smectic liquid–crystal phase formed near the PAz film, where the PAz film and 5CB molecules mixed (Fig. 6) [80]. This type of ordering in bulk systems has been reported previously [81]. While the ordering of low-molecular-weight liquid crystals at alignment-film interfaces has long been known from optical experiments [82], this is the first example confirmed by XRD methods.

Cell assembly is not the only way of preventing dewetting. Saito et al. wrapped thin films of photoresponsive molecules with polyvinyl alcohol to enable changes in molecular aggregation to be evaluated under light irradiation while preventing dewetting, which revealed that a columnar liquid–crystal phase with anthracene dimers as the core melted when irradiated with UV light [83]. Furthermore, the authors demonstrated the application of this columnar liquid crystal-phase as a light-melt adhesive.

5 Tender X-Rays

XRD methods typically rely on Cu K_α radiation ($\lambda = 0.154$ nm), and similar wavelength ranges (energy = 8.6–12.4 keV) are used at synchrotron radiation facilities. Recently, X-rays in the weaker energy range of 1–5 keV, referred to as "tender X-rays", have been increasingly utilized despite their lower energies and, hence, lower penetration ability. Specifically, tender X-rays are used in GI-XRD methods. When Cu Kα radiation is incident on organic thin films, even at angles minimally wider than the critical angle for total reflection, X-rays penetrate deep into the film and reach the substrate. In contrast, low-energy X-rays penetrate the film gradually even beyond the critical angle for total reflection and do not reach the deepest part of the

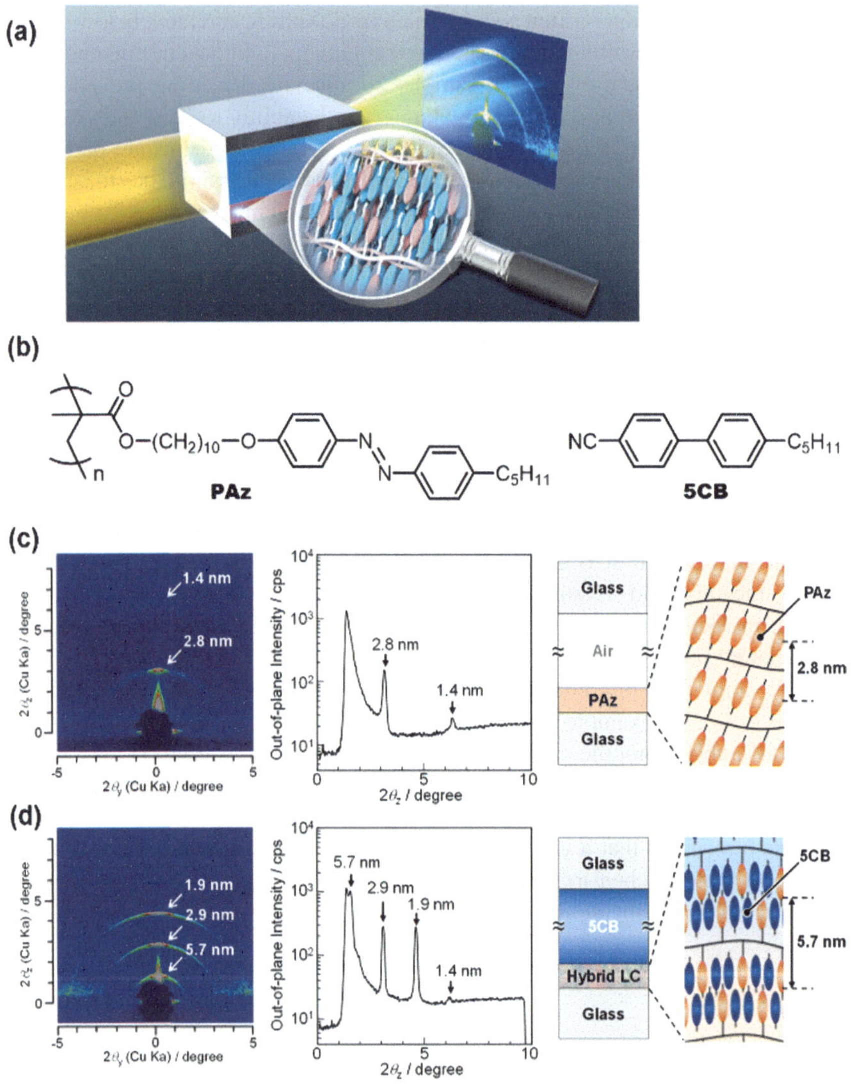

Fig. 6 **a** Schematic showing how self-assembly in a sandwiched cell is evaluated using X-ray diffraction methods. **b** Chemical structures of PAz and 5CB. **c**, **d** Two-dimensional diffraction images, out-of-plane intensity profiles, and structural models **c** before and **d** after the injection of 5CB into the cells. Adapted from Ref. [80] with layout modified with permission of the ACS. Copyright 2023 American Chemical Society

film all at once (Fig. 7a) [34, 84]. Therefore, tender X-rays allow the depth-direction analysis of thin films with a resolution of 10–100 nm.

Yamamoto et al. used tender X-rays to evaluate the lamellar microphase-separated structures of block-copolymer thin films [85] and revealed that the domain spacing near the film surface was larger than that in the film interior. This behavior was attributed to the higher mobility and faster relaxation of polymer chains near the surface compared to those in the film interior. Nagano and Seki et al. applied tender X-rays to thin films of side-chain-type polymer liquid crystals and analyzed mesogen alignment in the depth direction, which revealed that films $\leq$ 15 nm thick exhibited planar alignment, while thicker films were characterized by the coexistence of planar alignment at the substrate interface and homeotropic alignment at the surface (Fig. 7b) [86].

6 Ultrafast-Time-Resolved Electron Diffraction

Electrons are more sensitive for detecting light elements compared to X-rays. Recently, a high-speed time-resolved electron diffraction method [87–89] was developed that enabled structural dynamics in chemical reactions to be analyzed. Hada and Saito et al. observed mesogen movement in liquid–crystal molecules using electron diffraction by irradiating a π-extended cyclooctatetraene derivative with ultraviolet light. They discovered that the saddle-shaped molecular skeleton deformed into a planar shape on a timescale of about one trillionth of a second (1 ps) when irradiated with light, resulting in changes to the surrounding liquid–crystal structure [90]. This observation represents the first example of directly observed excited-state aromaticity emerging on a timescale of about 1–100 ps upon irradiation with light. No method for directly structurally analyzing the rapid dynamic behavior of a molecular skeleton buried in the carbon chain of a liquid crystal composed solely of light elements had previously existed. Hada and Kato et al. directly observed the rearrangement of azobenzene molecular assemblies induced by the Weigert effect with a response time of one ten-billionth of a second by irradiating azobenzene molecules that form the smectic B phase with linearly polarized light (Fig. 8) [91].

7 Summary

Soft-material self-assembly can be evaluated using diffraction methods. This chapter introduced various examples of diffraction experiments that were applied to bulk samples ranging from powders to liquids and colloids. GI-XRD methods were shown to be a powerful tool for evaluating the arrangements of self-assembled structures; these structures become uniformly aligned over a macroscopic area, which effectively enables the manifestation of functions. In situ GI-XRD methods in humidity-controlled or light-irradiated environments provide a wealth of information, and

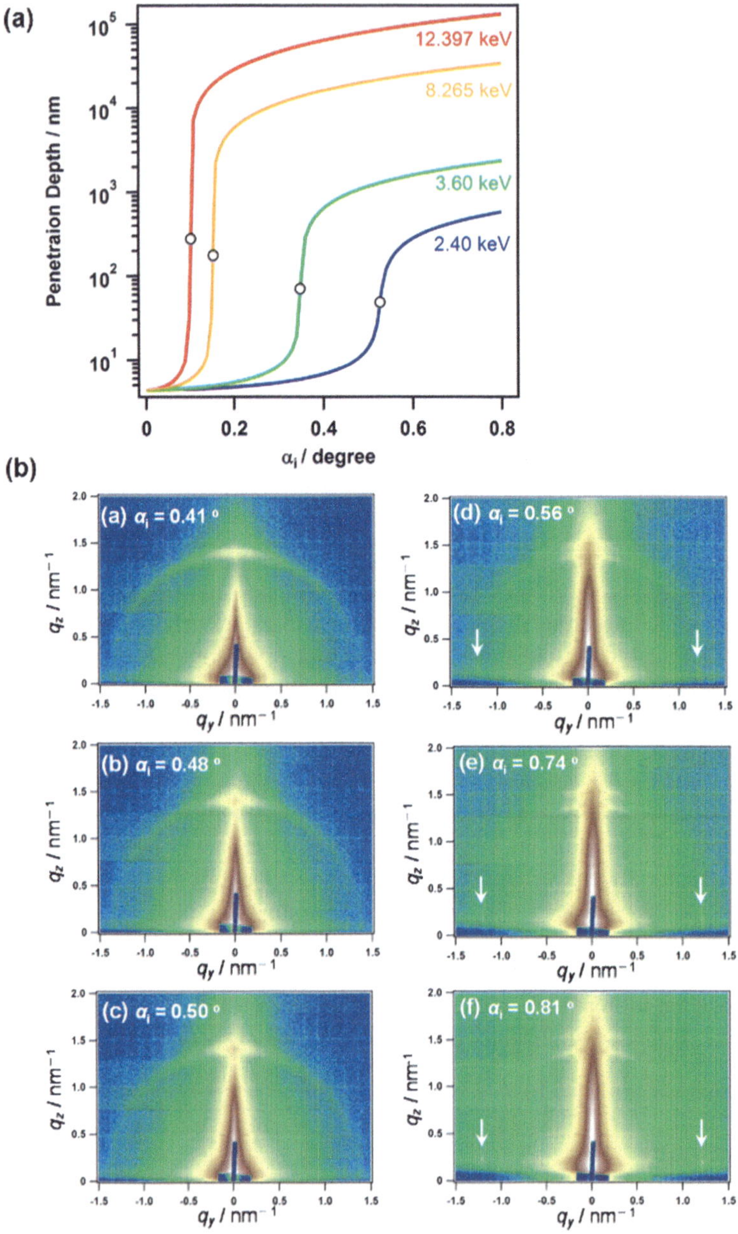

Fig. 7 **a** Calculated penetration depths of X-rays in polystyrene as functions of incidence angle (α_i). Symbols show the critical angle for each energy. Adapted from Ref. [34]. Copyright 2015 American Chemical Society. **b** GI-SAXS images of 30-nm-thick poly(cyanobiphenyl acrylate) derivative acquired using synchrotron tender X-rays ($\lambda = 0.539$ nm, 2.30 keV). Adapted from Ref. [86]. Copyright 2016 American Chemical Society

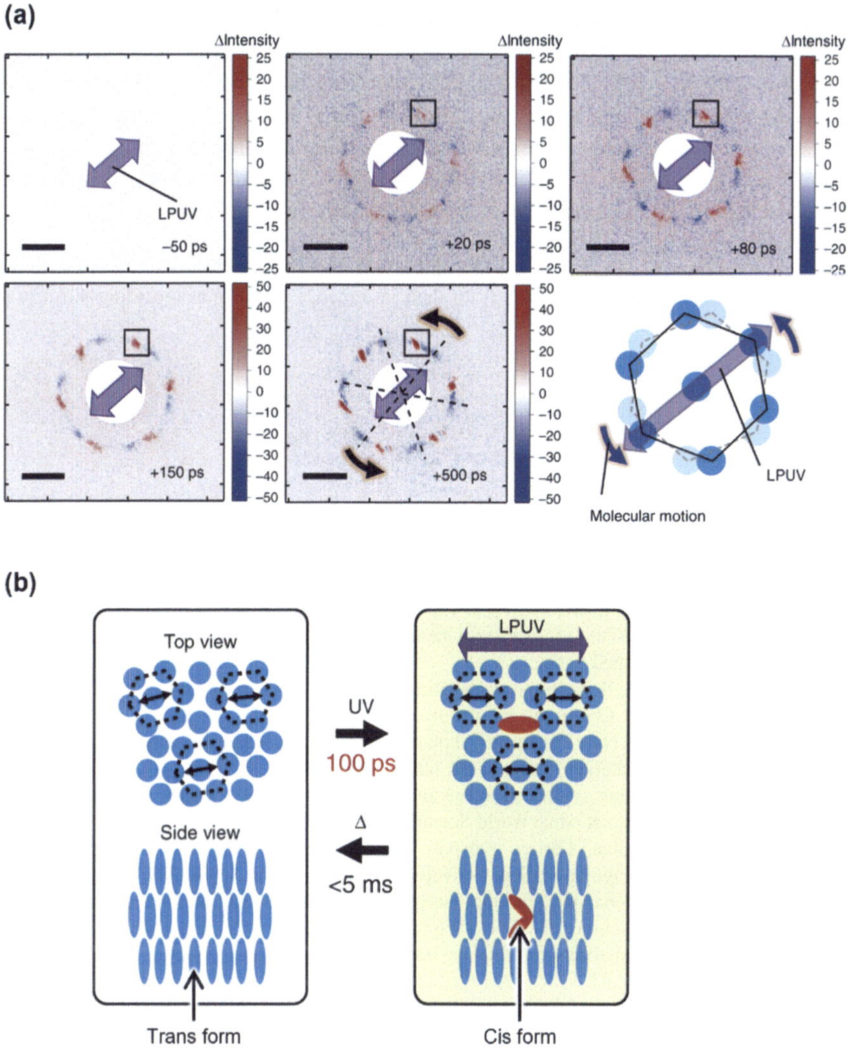

Fig. 8 **a** Snapshots of differential ultrafast time-resolved electron diffraction images before and after excitation with linearly polarized UV (LPUV) light. **b** Schematic depicting ultrafast molecular orientation in an azobenzene liquid–crystal assembly. Adapted from Ref. [91] with layout modified with permission under Creative Commons Attribution 4.0 International License (http://creativecommons.org/licenses/by/4.0/)

one can also evaluate photoresponsive soft materials while suppressing dewetting through the use of GI-XRD attachments. Examples of depth-direction analysis by purposefully varying the wavelength (energy) of the incident X-rays and liquid–crystal phase behavior with a time resolution of 1 ps by ultrafast electron diffraction were also provided. Thus, diffraction experiments are shown to be effective for nondestructively evaluating periodic structures over a wide area. However, one should not overly rely on diffraction methods alone but rather aim to combine them with other approaches, such as microscopic observations, as well as UV–visible and infrared spectroscopy, evaluating wettabilities and mechanical properties, thermal analyses, and mobility measurements. Understanding the strengths and weaknesses of each method and evaluating materials in a multifaceted manner is crucial for comprehensively understanding their structures and properties.

Acknowledgements The author thanks Mr. Tatsuo Hikage (Nagoya University) and Dr. Shintaro Nakagawa (The University of Tokyo) for their informative discussions regarding diffraction methods.

References

1. L. Gao, S. Han, H. Ni, J. Zhu, L. Wang, S. Gao, Y. Wang, D. Huang, Y. Zhao, R. Zou, Application of neutron imaging in observing various states of matter inside lithium batteries. Natl. Sci. Rev. **10**, nwad238 (2023)
2. S.F. Parker, P.J. Baker, R. McGreevy, A vision for the future of neutron scattering and muon spectroscopy in the 2050s. ACS Phys. Chem Au (2024). https://doi.org/10.1021/acsphyschemau.4c00026. published online
3. T. Kato, N. Mizoshita, K. Kishimoto, Functional liquid-crystalline assemblies: self-organized soft materials. Angew. Chem. Int. Ed. **45**, 38–68 (2006)
4. T. Seki, S. Nagano, M. Hara, Versatility of photoalignment techniques: from nematics to a wide range of functional materials. Polymer **54**, 6053–6072 (2013)
5. T. Ohta, Crystals: X-ray crystal structure analysis. *Chapter 1 in Physics and Chemistry of Molecular Assemblies*, 1st edn. (World Scientific, 2020), pp. 1−40
6. M. Funahashi, J. Hanna, Carrier transport in calamitic mesophases of liquid crystalline photoconductor 2-phenylnaphthalene derivatives. Mol. Cryst. Liq. Cryst. **331**, 509–516 (1999)
7. M. Funahashi, J. Hanna, High ambipolar carrier mobility in self-organizing terthiophene derivative. Appl. Phys. Lett. **76**, 2574–2576 (2000)
8. H. Iino, J. Hanna, Availability of liquid crystallinity in solution processing for polycrystalline thin films. Adv. Mater. **23**, 1748–1751 (2011)
9. H. Iino, T. Usui, J. Hanna, Liquid crystals for organic thin-film transistors. Nat. Commun. **6**, 6828 (2015)
10. T. Ichikawa, M. Yoshio, A. Hamasaki, T. Mukai, H. Ohno, T. Kato, Self-organization of room-temperature ionic liquids exhibiting liquid-crystalline bicontinuous cubic phases: formation of nano-ion channel networks. J. Am. Chem. Soc. **129**, 10662–10663 (2007)
11. T. Ichikawa, M. Yoshio, A. Hamasaki, J. Kagimoto, H. Ohno, T. Kato, 3D interconnected ionic nano-channels formed in polymer films: self-organization and polymerization of thermotropic bicontinuous cubic liquid crystals. J. Am. Chem. Soc. **133**, 2163–2169 (2011)
12. H. Zhang, L. Li, M. Möller, X. Zhu, J.J. Hernandez Reuda, M. Tosenthal, D.A. Ivanov, From channel-forming ionic liquid crystals exhibiting humidity-induced phase transitions to nanostructured ion-conducting polymer membranes. Adv. Mater. **25**, 3543−3548 (2013).
13. Y. Chen, M.D. Lingwood, M. Goswami, B.E. Kidd, J.J. Hernandez, M. Rosenthal, D.A. Ivanov, J. Perlich, H. Zhang, X. Zhu, M. Möller, L.A. Madsen, Humidity-modulated phase control and nanoscopic transport in supramolecular assemblies. J. Phys. Chem. B **118**, 3207–3217 (2014)
14. P.F.W. Simon, R. Ulrich, H.W. Spiess, U. Wiesner, Block copolymer-ceramic hybrid materials from organically modified ceramic precursors. Chem. Mater. **13**, 3464–3486 (2001)

15. Y. Wan, D. Zhao, On the controllable soft-templating approach to mesoporous silicates. Chem. Rev. **107**, 2821–2860 (2007)
16. M. Hara, S. Nagano, N. Mizoshita, T. Seki, Chromonic/silica nanohybrids: synthesis and macroscopic alignment. Langmuir **23**, 12350–12355 (2007)
17. M. Hara, S. Nagano, N. Kawatsuki, T. Seki, Photoalignment and patterning of a chromonic-silica nanohybrid on photocrosslinkable polymer thin films. J. Mater. Chem. **18**, 3259–3263 (2008)
18. https://www.pro-wave.co.jp/
19. T. Li, A.J. Senesi, B. Lee, Small angle X-ray scattering for nanoparticle research. Chem. Rev. **116**, 11128–11180 (2016)
20. S. Mourdikoudis, R.M. Pallares, N.T.K. Thanh, Characterization techniques for nanoparticles: comparison and complementarity upon studying nanoparticle properties. Nanoscale **10**, 12871–12934 (2018)
21. M.M. Modena, B. Rühle, T.P. Burg, S. Wuttke, Nanoparticle characterization: what to measure? Adv. Mater. **31**, 1901556 (2019)
22. E. Paineau, A.M. Philippe, K. Antonova, I. Bihannic, P. Davidson, I. Dozov, J.C.P. Gabriel, M. Impéror-Clerc, P. Levitz, F. Meneau, L.J. Michot, Liquid-crystalline properties of aqueous suspensions of natural clay nanosheets. Liq. Cryst. Rev. **1**, 110–126 (2013)
23. L. Wang, T. Sasaki, Titanium oxide nanosheets: graphene analogues with versatile functionalities. Chem. Rev. **114**, 9455–9486 (2014)
24. R. Ma, T. Sasaki, Two-dimensional oxide and hydroxide nanosheets: controllable high-quality exfoliation, molecular assembly, and exploration of functionality. Acc. Chem. Res. **48**, 136–143 (2015)
25. P. Davidson, C. Penisson, D. Constantin, J.-C.P. Gabriel, Isotropic, nematic, and lamellar phases in colloidal suspensions of nanosheets. Proc. Natl. Acad. Sci. U.S.A. **115**, 6662–6667 (2018)
26. T. Nakato, Y. Nono, E. Mouri, M. Nakata, Panoscopic organization of anisotropic colloidal structures from photofunctional inorganic nanosheet liquid crystals. Phys. Chem. Chem. Phys. **16**, 955–962 (2014)
27. N. Miyamoto, M. Shintate, S. Ikeda, Y. Hoshida, Y. Yamauchi, R. Motokawa, M. Annaka, Liquid crystalline inorganic nanosheets for facile synthesis of polymer hydrogels with anisotropies in structure, optical property, swelling/deswelling, and ion transport/fixation. Chem. Commun. **49**, 1082–1084 (2013)
28. T. Inadomi, S. Ikeda, Y. Okumura, H. Kikuchi, N. Miyamoto, Photo-induced anomalous deformation of poly(*N*-isopropylacrylamide) gel hybridized with an inorganic nanosheet liquid crystal aligned by electric field. Macromol. Rapid Commun. **35**, 1741–1746 (2014)
29. M. Komura, K. Watanabe, T. Iyoda, T. Yamada, H. Yoshida, Y. Iwasaki, Laboratory-GISAXS measurements of block copolymer films with highly ordered and normally oriented nanocylinders. Chem. Lett. **38**, 408–409 (2009)
30. B. Lee, I. Park, J. Yoon, S. Park, J. Kim, K.-W. Kim, T. Chang, M. Ree, Structural analysis of block copolymer thin films with grazing incidence small-angle X-ray scattering. Macromolecules **38**, 4311–4323 (2005)
31. M. Ree, Probing the self-assembled nanostructures of functional polymers with synchrotron grazing incidence X-ray scattering. Macromol. Rapid Commun. **35**, 930–959 (2014)
32. G.J.B. Voss, E.A.C. Panduro, A. Midttveit, J.B. Fløystad, K. Høydalsvik, A. Gibaud, D.W. Breiby, M. Rønning, Mesostructured alumina as powders and thin films. J. Mater. Chem. A **2**, 9727–9735 (2014)
33. Z. Jiang, GIXSGUI: a MATLAB toolbox for grazing-incidence X-ray scattering data visualization and reduction, and indexing of buried three-dimensional periodic nanostructured films. J. Appl. Crystallogr. **48**, 917–926 (2015)
34. I. Saito, T. Miyazaki, K. Yamamoto, Depth-resolved structure analysis of cylindrical microdomain in block copolymer thin film by grazing-incidence small-angle X-ray scattering utilizing low-energy X-rays. Macromolecules **48**, 8190–8196 (2015)
35. M. Knaapila, R. Stepanyan, B.P. Lyons, M. Torkkeli, T.P.A. Hase, R. Serimaa, R. Güntner, O.H. Seeck, U. Scherf, A.P. Monkman, The influence of the molecular weight on the thermotropic

alignment and self-organized structure formation of branched side chain hairy-rod polyfluorene in thin films. Macromolecules **38**, 2744–2753 (2005)
36. M. Knaapila, R. Stepanyan, B.P. Lyons, M. Torkkeli, A.P. Monkman, Towards general guidelines for aligned, nanoscale assemblies of hairy-rod polyfluorene. Adv. Funct. Mater. **16**, 599–609 (2006)
37. I. Osaka, T. Kakara, N. Takemura, T. Koganezawa, K. Takimiya, Naphthodithiophene-naphthobisthiadiazole copolymers for solar cells: alkylation drives the polymer backbone flat and promotes efficiency. J. Am. Chem. Soc. **135**, 8834–8837 (2013)
38. I. Osaka, M. Saito, T. Koganezawa, K. Takimiya, Thiophene-thiazolothiazole copolymers: significant impact of side chain composition on backbone orientation and solar cell performances. Adv. Mater. **26**, 331–338 (2014)
39. V. Vohra, K. Kawashima, T. Kakara, T. Koganezawa, I. Osaka, K. Takimiya, H. Murata, Efficient inverted polymer solar cells employing favourable molecular orientation. Nat. Photonics **9**, 403–408 (2015)
40. M. Saito, T. Koganezawa, I. Osaka, Correlation between distribution of polymer orientation and cell structure in organic photovoltaics. ACS Appl. Mater. Interfaces **10**, 32420–32425 (2018)
41. A. Yamaguchi, F. Uejo, T. Yoda, T. Uchida, Y. Tanamura, T. Yamashita, N. Teramae, Self-assembly of a silica-surfactant nanocomposite in a porous alumina membrane. Nat. Mater. **3**, 337–341 (2004)
42. H.K.M. Tanaka, Y. Yamauchi, T. Kurihara, T. Sakka, K. Kuroda, A.P. Millis Jr., Exploration of a standing mesochannel system with antimatter/matter atomic probes. Adv. Mater. **20**, 4728–4733 (2008)
43. M.-L. Lin, C.-C. Huang, M.-Y. Lo, C.-Y. Mou, Well-ordered mesoporous carbon thin film with perpendicular channels: application to direct methanol fuel cell. J. Phys. Chem. C **112**, 867–873 (2008)
44. S. Tominaka, C.-W. Wu, K. Kuroda, T. Osaka, Electrochemical analysis of perpendicular mesoporous Pt electrode filled with pure water for clarifying the active region in fuel cell catalyst layers. J. Power. Sources **195**, 2236–2240 (2010)
45. W. Weng, T. Higuchi, M. Suzuki, T. Fukuoka, T. Shimomura, M. Ono, L. Radhakrishnan, H. Wang, N. Suzuki, H. Oveisi, Y. Yamauchi, A high-speed passive matrix electrochromic display using a mesoporous TiO_2 electrode with vertical porosity. Angew. Chem. Int. Ed. **49**, 3956–3959 (2010)
46. X. Jiang, A. Ishizumi, N. Suzuki, M. Naito, Y. Yamauchi, Vertically-oriented conjugated polymer arrays in mesoporous alumina via simple drop-casting and appearance of anisotropic photoluminescence. Chem. Commun. **48**, 549–551 (2012)
47. B. Platschek, N. Petkov, D. Himsl, S. Zimdars, Z. Li, R. Köhn, T. Bein, Vertical columnar block-copolymer-templated mesoporous silica via confined phase transformation. J. Am. Chem. Soc. **130**, 17362–17371 (2008)
48. Y. Yamauchi, T. Nagaura, S. Inoue, Oriented growth of small mesochannels utilizing a porous anodic alumina substrate: preparation of continuous film with standing mesochannels. Chem. Asian J. **4**, 1059–1063 (2009)
49. C.-W. Wu, T. Ohsuna, T. Edura, K. Kuroda, Orientational control of hexagonally packed silica mesochannels in lithographically designed confined nanospaces. Angew. Chem. Int. Ed. **46**, 5364–5368 (2007)
50. S.H. Tolbert, A. Firouzi, G.D. Stucky, B.F. Chmelka, Magnetic field alignment of ordered silicate-surfactant composites and mesoporous silica. Science **278**, 264–268 (1997)
51. Y. Yamauchi, M. Sawada, A. Sugiyama, T. Osaka, Y. Sakka, K. Kuroda, Magnetically induced orientation of mesochannels in 2D-hexagonal mesoporous silica films. J. Mater. Chem. **16**, 3693–3700 (2006)
52. A. Walcarius, E. Sibottier, M. Etienne, J. Ghanbaja, Electrochemically assisted self-assembly of mesoporous silica thin films. Nat. Mater. **6**, 602–608 (2007)
53. N. Vilá, J. Ghanbaja, E. Aubert, A. Walcarius, Electrochemically assisted generation of highly ordered azide-functionalized mesoporous silica for oriented hybrid films. Angew. Chem. Int. Ed. **53**, 2945–2950 (2014)

54. Y. Zhang, J. Wang, Evaporation-induced alignment of cylindrical mesopores in TiO_2 thin films. J. Am. Ceram. Soc. **93**, 365–369 (2010)
55. F. Shan, X. Lu, Q. Zhang, J. Wu, Y. Wang, F. Bian, Q. Lu, Z. Fei, P.J. Dyson, A facile approach for controlling the orientation of one-dimensional mesochannels in mesoporous titania films. J. Am. Chem. Soc. **134**, 20238–20241 (2012)
56. Z. Teng, G. Zheng, Y. Dou, W. Li, C.-Y. Mou, X. Zhang, A.M. Asiri, D. Zhao, Highly ordered mesoporous silica films with perpendicular mesochannels by a simple Stöber-solution growth approach. Angew. Chem. Int. Ed. **51**, 2173–2177 (2012)
57. C.-W. Wu, T. Ohsuna, M. Kuwabara, K. Kuroda, Formation of highly ordered mesoporous titania films consisting of crystalline nanopillars with inverse mesospace by structural transformation. J. Am. Chem. Soc. **128**, 4544–4545 (2006)
58. C.-W. Koh, U.-H. Lee, J.-K. Song, H.-R. Lee, M.-H. Kim, M. Suh, Y.-U. Kwon, Mesoporous titania thin film with highly ordered and fully accessible vertical pores and crystalline walls. Chem. Asian J. **3**, 862–867 (2008)
59. Y.-S. Ko, C.-W. Koh, U.-H. Lee, Y.-U. Kwon, Synthesis of mesoporous Titania thin films with vertical pore channels and thick and crystalline walls. Microporous Mesoporous Mater. **145**, 141–145 (2011)
60. H. Oveisi, X. Jiang, M. Imura, Y. Nemoto, Y. Sakamoto, Y. Yamauchi, A mesoporous γ-alumina film with vertical mesoporosity: the unusual conversion from a *Im3m* mesostructure to vertically oriented γ-alumina nanowires. Angew. Chem. Int. Ed. **50**, 7410–7413 (2011)
61. V.R. Koganti, D. Dunphy, V. Gowrishankar, M.D. McGehee, X. Li, J. Wang, S.E. Rankin, Generalized coating route to silica and Titania films with orthogonally tilted cylindrical nanopore arrays. Nano Lett. **6**, 2567–2570 (2006)
62. M. Hara, S. Nagano, T. Seki, π-π interaction-induced vertical alignment of silica mesochannels templated by a discotic lyotropic liquid crystal. J. Am. Chem. Soc. **132**, 13654–13656 (2010)
63. M. Hara, S. Nagano, T. Seki, Spontaneous formation of vertically aligned lamellae in thin films of block copolymer-silica hybrid material. Bull. Chem. Soc. Jpn **86**, 1151–1157 (2013)
64. C. Ma, L. Han, Z. Jiang, Z. Huang, J. Feng, Y. Yao, S. Che, Growth of mesoporous silica film with vertical channels on substrate using Gemini surfactants. Chem. Mater. **23**, 3583–3586 (2011)
65. A. Mikó, A.L. Demirel, M. Somer, Vertically oriented hexagonal mesoporous zirconia thin films by block copolymer templating. J. Mater. Chem. **22**, 3705–3707 (2012)
66. K. Krishnan, H. Iwatsuki, M. Hara, S. Nagano, Y. Nagao, Proton conductivity enhancement in oriented, sulfonated polyimide thin films. J. Mater. Chem. A **2**, 6895–6903 (2014)
67. K. Krishnan, H. Iwatsuki, M. Hara, S. Nagano, Y. Nagao, Influence of molecular weight on molecular ordering and proton transport in organized sulfonated polyimide thin films. J. Phys. Chem. C **119**, 21767–21774 (2015)
68. S. Hara, H. Wada, A. Shimojima, K. Kuroda, Formation of nanogrooves with sub-5 nm periodicity using local silicification at the interspace between a Si substrate and lyotropic liquid crystals. ACS Nano **11**, 5160–5166 (2017)
69. M. Hara, T. Orito, S. Nagano, T. Seki, Humidity-responsive phase transition and on-demand UV-curing in a hygroscopic polysiloxane-surfactant nanohybrid film. Chem. Commun. **54**, 1457–1460 (2018)
70. M. Hara, N. Wakitani, A. Kodama, S. Nagano, T. Seki, Hierarchical photocomposition of heteronanostructures in a surfactant–polysiloxane hybrid film toward next-generation nanolithography. ACS Appl. Polym. Mater. **2**, 2284–2290 (2020)
71. S. Imai, M. Arakawa, Y. Nakanishi, M. Takenaka, H. Aoki, M. Ouchi, T. Terashima, Water-assisted microphase separation of cationic random copolymers into sub-5 nm lamellar materials and thin films. Macromolecules **55**, 9113–9125 (2022)
72. R. Sujita, S. Imai, M. Ouchi, H. Aoki, T. Terashima, Microphase separation of cationic homopolymers bearing alkyl ammonium salts into sub-4 nm lamellar materials with water intercalation channels. Macromolecules **56**, 9738–9749 (2023)
73. T. Ikami, H. Aoki, T. Terashima, Lamellar microphase separation and phase transition of hydrogen-bonding/crystalline statistical copolymers: amide functionalization at the interface. ACS Macro Lett. **13**, 446–452 (2024)

74. R. Sujita, H. Aoki, M. Takenaka, M. Ouchi, T. Terashima, Universal access to water-compatible and nanostructured materials via the self-assembly of cationic alternating copolymers. ACS Macro Lett. **13**, 747–753 (2024)
75. M. Hara, A. Kodama, S. Washiyama, Y. Fujii, S. Nagano, T. Seki, Humidity-induced self-assembled nanostructures via ion aggregation in ionic linear polysiloxanes. Macromolecules **55**, 4313–4319 (2022)
76. S. Nagano, Y. Koizuka, T. Murase, M. Sano, Y. Shinohara, Y. Amemiya, T. Seki, Synergy effect on morphology switching: real-time observation of photo-orientation of microphase separation in a block copolymer. Angew. Chem. Int. Ed. **51**, 5884–5888 (2012)
77. M. Sano, S. Nakamura, M. Hara, S. Nagano, Y. Shinohara, Y. Amemiya, T. Seki, Pathways toward photoinduced alignment switching in liquid crystalline block copolymer films. Macromolecules **47**, 7178–7186 (2014)
78. M. Sano, M. Hara, S. Nagano, Y. Shinohara, Y. Amemiya, T. Seki, New aspects for the hierarchical cooperative motions in photoalignment process of liquid crystalline block copolymer films. Macromolecules **48**, 2217–2223 (2015)
79. R. Yamakado, M. Hara, S. Nagano, T. Seki, H. Maeda, Photo-responsive soft ionic crystals: ion-pairing assemblies of azobenzene carboxylates. Chem. Eur. J. **23**, 9244–9248 (2017)
80. C. Kawakami, M. Hara, S. Nagano, T. Seki, Induction of highly ordered liquid crystalline phase of an azobenzene side chain polymer by contact with 4′-pentyl-4-cyanobiphenyl: an in situ study. Langmuir **39**, 619–626 (2023)
81. S. Sugisawa, Y. Tabe, Induced smectic phases of stoichiometric liquid crystal mixtures. Soft Matter **12**, 3103–3109 (2016)
82. K. Miyano, Wall-induced pretransitional birefringence: a new tool to study boundary aligning forces in liquid crystals. Phys. Rev. Lett. **43**, 51–54 (1979)
83. S. Saito, S. Nobusue, E. Tsuzaka, C. Yuan, C. Mori, M. Hara, T. Seki, C. Camacho, S. Irle, S. Yamaguchi, Light-melt adhesive based on dynamic carbon frameworks in a columnar liquid-crystal phase. Nat. Commun. **7**, 12094 (2016)
84. H. Okuda, K. Takeshita, S. Ochiai, S. Sakurai, Y. Kitajima, Near-surface relaxation structure of annealed block copolymer film on Si substrates examined by grazing-incidence small-angle scattering utilizing soft X-rays. J. Appl. Crystallogr. **44**, 380–384 (2011)
85. I. Saito, D. Shimada, M. Aikawa, T. Miyazaki, K. Shimokita, H. Takagi, K. Yamamoto, Orientation and relaxation behaviors of lamellar microdomains of poly(methyl methacrylate)-b-poly(n-butyl acrylate) thin films as revealed by grazing-incidence small-angle X-ray scattering. Polym. J. **48**, 399–406 (2016)
86. D. Tanaka, T. Mizuno, M. Hara, S. Nagano, I. Saito, K. Yamamoto, T. Seki, Evaluations of mesogen orientation in thin films of polyacrylate with cyanobiphenyl side chain. Langmuir **32**, 3737–3745 (2016)
87. A.H. Zewail, Four-dimensional electron microscopy. Science **328**, 187–193 (2010)
88. A.A. Ischenko, P.M. Weber, R.J.D. Miller, Capturing chemistry in action with electrons: realization of atomically resolved reaction dynamics. Chem. Rev. **117**, 11066–11124 (2017)
89. H. Jean-Ruel, M. Gao, M.A. Kochman, C. Lu, L.C. Liu, R.R. Cooney, C.A. Morrison, R.J.D. Miller, Ring-closing reaction in diarylethene captured by femtosecond electron crystallography. J. Phys. Chem. B **117**, 15894–15902 (2013)
90. M. Hada, S. Saito, S. Tanaka, R. Sato, M. Yoshimura, K. Mouri, K. Matsuo, S. Yamaguchi, M. Hara, Y. Hayashi, F. Röhricht, R. Herges, Y. Shigeta, K. Onda, R.J.D. Miller, Structural monitoring of the onset of excited-state aromaticity in a liquid crystal phase. J. Am. Chem. Soc. **139**, 15792–15800 (2017)
91. M. Hada, D. Yamaguchi, T. Ishikawa, T. Sawa, K. Tsuruta, K. Ishikawa, S. Koshihara, Y. Hayashi, T. Kato, Ultrafast isomerization-induced cooperative motions to higher molecular orientation in smectic liquid-crystalline azobenzene molecules. Nat. Commun. **10**, 4159 (2019)

Thermal Analysis

Kazushi Hayashi and Kyosuke Isoda

1 General of Thermal Analysis

Thermal analysis is a technique for measuring the physical properties of a substance as a function of temperature while continuously changing the temperature of the substance. When a substance undergoes a phase transition or chemical reaction, thermal changes such as exothermal and endothermal changes are observed. Also, when impurities are present in a pure substance, a rise in boiling point (b.p.) or a drop in freezing point occurs. For example, the calorific value change, melting point (m.p.), and b.p. of a chemical reaction can be measured to study catalytic reactions, solid-phase reactions, and state diagrams. The summary of representative thermal analyses is shown in Table 1.

K. Hayashi · K. Isoda (✉)
Organic Materials Chemistry Group, Sagami Chemical Research Institute (SCRI), Ayase, Kanagawa, Japan
e-mail: isoda032@toyo.jp

K. Isoda
Department of Applied Chemistry, Faculty of Science and Engineering, Toyo University, Kawagoe, Saitama, Japan

M. Funahashi (ed.), *Functional Soft Matter II*,
https://doi.org/10.1007/978-981-92-0100-6_3

Table 1 Representative thermal analyses

Techniques	Parameter observed	Instruments used
Thermogravimetric analysis (TG)	Weight of sample	Thermal balance
Differential thermal analysis (DTA)	Temperature difference between sample and standard	TGA
Thermogravimetric differential thermal analysis (TG–DTA)	Weight of sample and temperature difference between sample and standard	Thermal balance and differential thermal analyzer
Differential scanning calorimetry (DSC)	Heat supplied to the sample or standard	DSC

2 Thermal Analytical Methods

2.1 Thermogravimetry (TG), Differential Thermal Analysis (DTA), and Combined Thermogravimetry–Differential Thermal Analysis (TG–DTA)

Thermogravimetry (TG) is a technique that can measure the weight of a sample while increasing temperature according to a programmed heating method. The fundamental principle of TG involves detecting weight changes in a sample positioned on a thermobalance, monitored through alterations in the balance equilibrium. These weight changes are quantified based on the amount of current required for balance adjustment.

TG measurements can provide valuable insights into various processes, including (1) pyrolysis reactions to measure crystal water content and dehydration, (2) pyrolysis and solid-phase reactions forming volatile substances (e.g., decomposition of metal salts and complexes), (3) reactions involving solids and gases (e.g., high-temperature oxidation of metals), (4) determination of m.p. and b.p., and (5) analysis of solid mixtures through processes such as pyrolysis, evaporation, and sublimation.

Differential thermal analysis (DTA) is a technique used to measure temperature differences between a sample and a reference material while increasing temperature (Fig. 1). When the sample experiences a thermal change like a phase transition or chemical reaction, it alters the temperature difference with the reference material, which is recorded as a peak in the DTA curve. DTA measurements allow us to investigate various phenomena including (1) crystal and secondary transitions, (2) phase transformations like melting and evaporation, and (5) decomposition, redox, and solid phase reactions.

The combination of TG and DTA techniques, thermogravimetric differential thermal analysis (TG–DTA), is a powerful tool to simultaneously assess weight change and calorific value change (Fig. 1). Integrating TG and DTA curve patterns offers a more precise understanding of thermal phenomena for a sample compared to either technique alone.

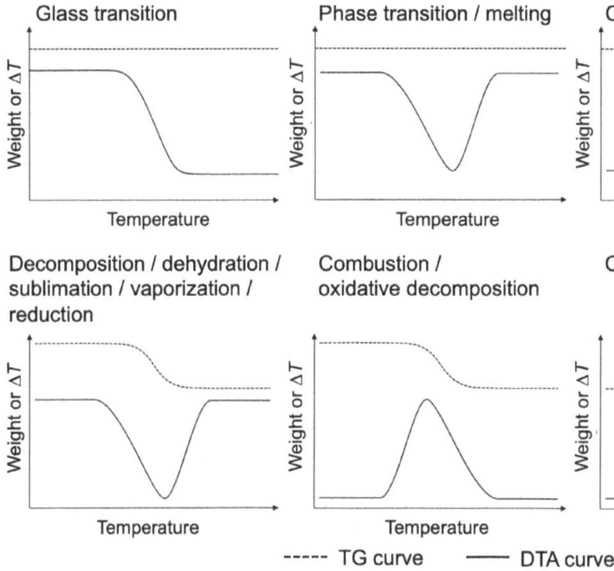

Fig. 1 Typical TG and DTA curves observed upon various processes

Examples of measured data obtained from TG–DTA analysis are shown in Fig. 2. Figure 2a shows a characteristic TG curve of sublimation process for naphthalene, in which the weight decreases during the heating process. At the same time, DTA analysis shows a downward convex derived from endothermic peak, indicating melting (lower) and sublimation (later). Figure 2b shows results of TG and DSC curves involving desorption process in porous coordination polymer of **$\{[Co(4\text{-}pmna)_2(H_2O)_2](NO_3)_2{\cdot}2CH_3OH\}_n$** [1]. Upon heating process, obvious decrease is observed in the temperature range of 50–90 °C with endothermal peak, which correspond to two MeOH and two coordinated H_2O. Also, the temperature range of 170 °C indicates the melt of **$\{[Co(4\text{-}pmna)_2](NO_3)_2\}_n$** and gradual decomposition to form an amorphous phase with the loss of 4-pmna ligands.

2.2 *Differential Scanning Calorimetry (DSC)*

DSC measurement is a technique used to measure the energy difference between a sample and a reference material as temperature varies. Unlike other methods, DSC method involves a single temperature-controlled heat source to transfer heat to both the sample and reference material. Also, DSC measurements can be conducted multiple times as long as the sample does not decompose. The temperature difference between the sample and reference material correlates with the heat energy per unit time exchanged between them. Specific heat can be determined from the DSC curve,

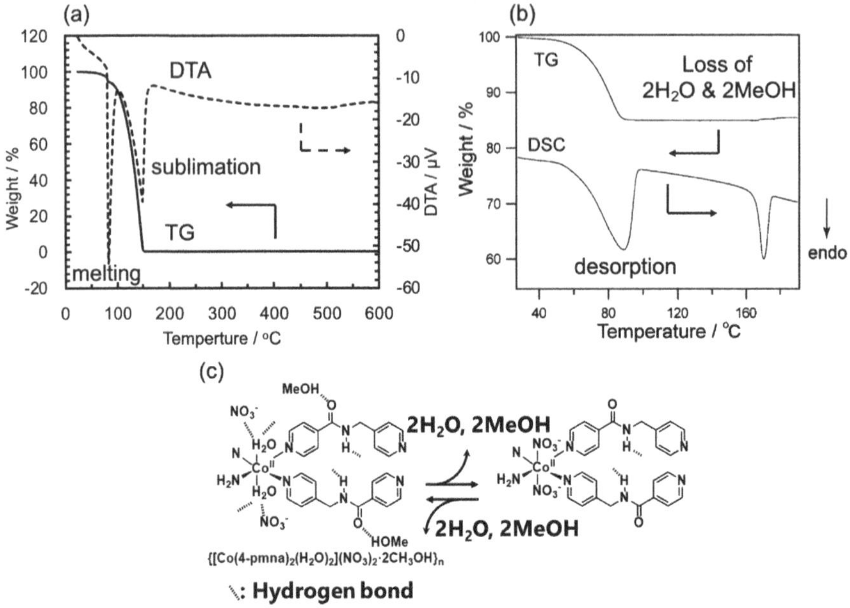

Fig. 2 **a** TG–DTA curves of naphthalene; **b** TG and DSC curves of **{[Co(4-pmna)$_2$(H$_2$O)$_2$](NO$_3$)$_2$·2CH$_3$OH}$_n$**; **c** predicted mechanism for desorption of H_2O and MeOH of **{[Co(4-pmna)$_2$(H$_2$O)$_2$](NO$_3$)$_2$·2CH$_3$OH}$_n$**. Copyright 2006 American Chemical Society [1]

while endothermic and exothermic quantities can be calculated from peak areas. This enables DSC to (1) quantify specific heat, reaction heat, and heat of transition, (2) assess crystallinity and purity, and (3) measure reaction and crystallization rates. Observable events include glass transition, crystallization, melting, hardening, cross-linking, oxidation, and decomposition. In melting process, a downward convex due to endothermic peak is observed, while in crystallization process, an upward convex due to exothermic peak is evident (Fig. 3a). These peaks allow determination of m.p. and crystallization temperature, of which peak area provides the enthalpy change (ΔH) of materials measured. On the other hand, Fig. 3b shows how to determine the glass transition temperature (T_g). Generally, T_g is defined as the middle temperature between extrapolated glass transition start temperature (T_{sg}) and extrapolated glass transition end temperature (T_{eg}). It is defined that T_{sg} and T_{eg} are the crosspoint between the baseline before or after phase transition and the tangent at the inflection point, respectively.

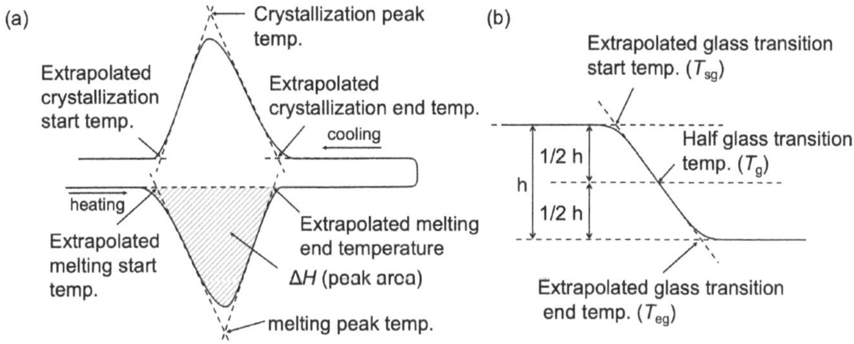

Fig. 3 Typical DSC curves observed upon **a** melting and crystallization, **b** glass transition

2.3 DSC of Organic Materials

2.3.1 Liquid-Crystalline Materials

DSC measurements enable the continuous monitoring of temperature during the phase transitions of liquid-crystalline molecules, by using only a small amount of sample (typically a few milligrams). This technique is adept at measuring the subtle endothermal and exothermal heats associated with phase transitions. Moreover, DSC facilitates the precise observation and recording of each phase transition and sequence. When coupled with polarized optical microscopic (POM) observations and X-ray diffraction (XRD) measurements, DSC measurement can thoroughly identify distinct liquid-crystalline phases within specific temperature ranges.

Liquid-crystalline molecules exhibit diverse shapes, with rod-shaped and disk-shaped molecules being prominent examples (Fig. 4). These molecules can show multiple liquid-crystalline phases, each characterized by subtle differences in molecular motion and aggregation states. The transitions between adjacent phases involve changes in both enthalpy and entropy. Notably, the transition enthalpy is intricately linked to alterations in molecular ordering and arrangement accompanying the phase transition. Larger changes in ordering generally correspond to higher transition enthalpies.

2.3.2 Rodlike Molecules

In this section, an initial topic is rod-shaped molecules as an example, because it is easy to visualize the aggregated structure of molecules. Here, we focus on the exothermic peak observed during the cooling process after melting to isotropic (Iso) phase for liquid-crystalline molecules.

The cooling process from Iso phase tends to form a nematic liquid-crystalline phase (nematic, N), in which only molecular orientation is ordered (Fig. 5). The

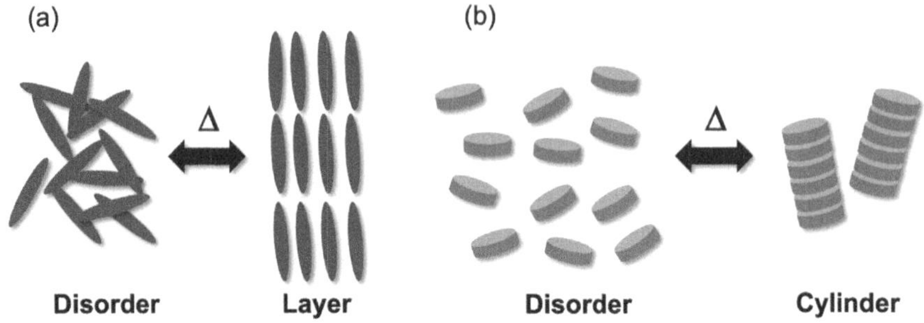

Fig. 4 Schematic illustrations of typical molecular assembling for rodlike and disclike molecules

transition enthalpy (ΔH) is generally reported to be less than 1 kJ mol^{-1} [2]. In the further cooling process, a smectic liquid-crystalline phase (smectic, Sm) is observed on the low-temperature region of the N phase, in which the molecules are arranged in a layered structure. For instance, in the smectic A (SmA) phase, the molecules in each layer are parallel to the layer normal (Fig. 5). In the lower temperature region to SmA phase, the smectic C (SmC) phase appears, in which the molecules are tilted at a certain angle to the layer normal. Both SmA and SmC phases are not ordered within the layers, whereas there are also many Sm phases with molecular order (not shown here). Values of ΔH associated with SmA to SmC phases are relatively small, < 0.042–2.8 kJ mol^{-1}, suggesting that there is no dramatic change in molecular order. On the other hand, values of ΔH for N to SmA or SmC phases are < 0.21–4.6 kJ mol^{-1} and < 0.70–9.6 kJ mol^{-1}, respectively, and tend to be larger than for the Iso-N phase transition. Furthermore, the value of ΔH from the N to Sm phase tends to be larger (< 2.9–12 kJ mol^{-1}) because the molecules become oriented from random to form an oriented state.

Next, some examples of rod-shaped molecules are shown below. The molecule drawn here (Fig. 6) is a liquid-crystalline molecule based on an anthraquinone framework, **AQ12** [3], which forms three crystalline phases (Crystal, Cr) during the heating process, and upon further heating, a phase transition from Cr_3 to SmC is observed at 132 °C ($\Delta H = 20$ kJ mol^{-1}). Further increasing the temperature, a phase transition from SmC to N is observed at 184 °C and from N to Iso at 223 °C (both with $\Delta H =$ ca. 2 kJ mol^{-1}) is very small compared to that of Cr_3 to SmC phase transition. These results suggest that the magnitude of the ΔH depends on the difference in molecular ordering in the phases before and after the phase transition. Likewise, for **AQ14** with different alkyl lengths, the phase transition temperature and transition enthalpy show a similar phase sequence to those of **AQ12** [3].

Next, we focus on the phase sequence of liquid-crystalline organic semiconductor **Ph-BTBT-10** (Fig. 7) [4], an asymmetric rod-shaped molecule consisting of a benzothienobenzothiophene (**BTBT**) framework tethering with Ph group and alkyl chain at terminals. **Ph-BTBT-10** undergoes a phase transition from Iso to SmA phases at around 225 °C during the cooling process. Upon further cooling, **Ph-BTBT-10** shows a phase transition from SmA to SmE phases at around 210 °C.

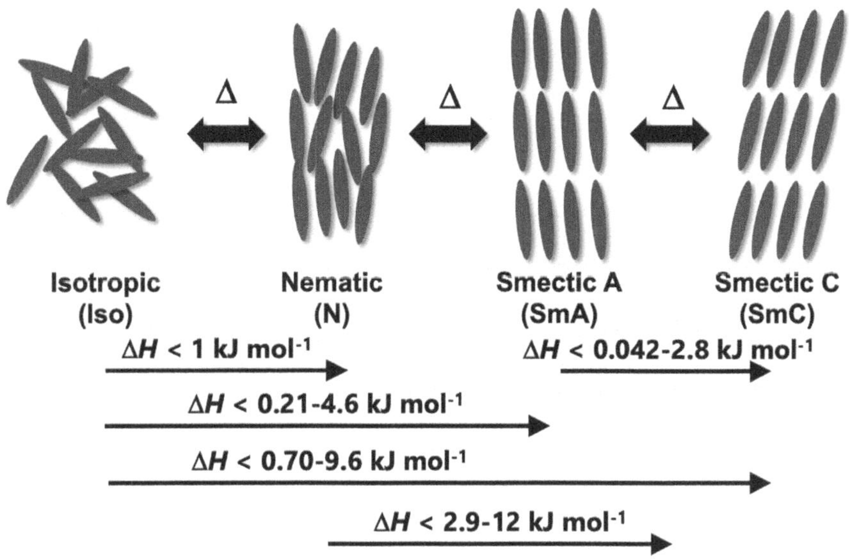

Fig. 5 Phase transition of rodlike molecules

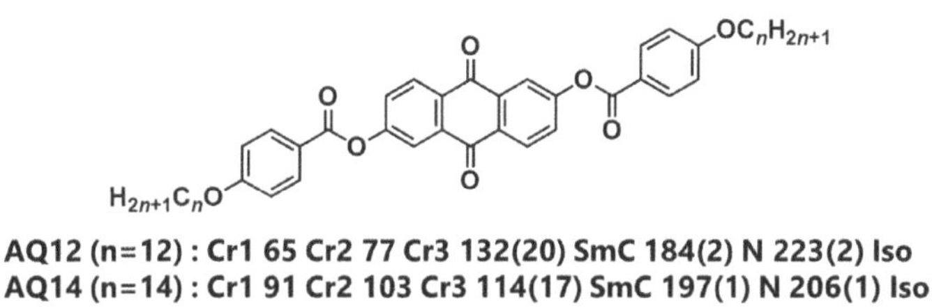

Fig. 6 Transition temperatures (°C) and enthalpies of transition (kJ mol^{-1}, in parentheses) determined by DSC on second heating (first heating reflects recrystallization conditions) at 5 °C min^{-1}. Cr: crystalline; N: nematic; SmC: smectic C; Iso: isotropic

This phase transition involves remarked changes in molecular ordering from SmA with little intermolecular order to SmE phase forming a highly ordered rectangular structure (Fig. 4), resulting in an obvious exothermic peak. To provide higher charge-transporting properties in liquid-crystalline organic semiconductors, it is important to form a highly ordered structure by controlling intermolecular interactions. Therefore, higher-order SmE phase exhibits higher charge-transporting properties (> 10 cm^2/Vs) than SmA phase. As a result, the phase sequence information obtained from DSC measurements is important for the design and application of liquid-crystalline organic semiconductors.

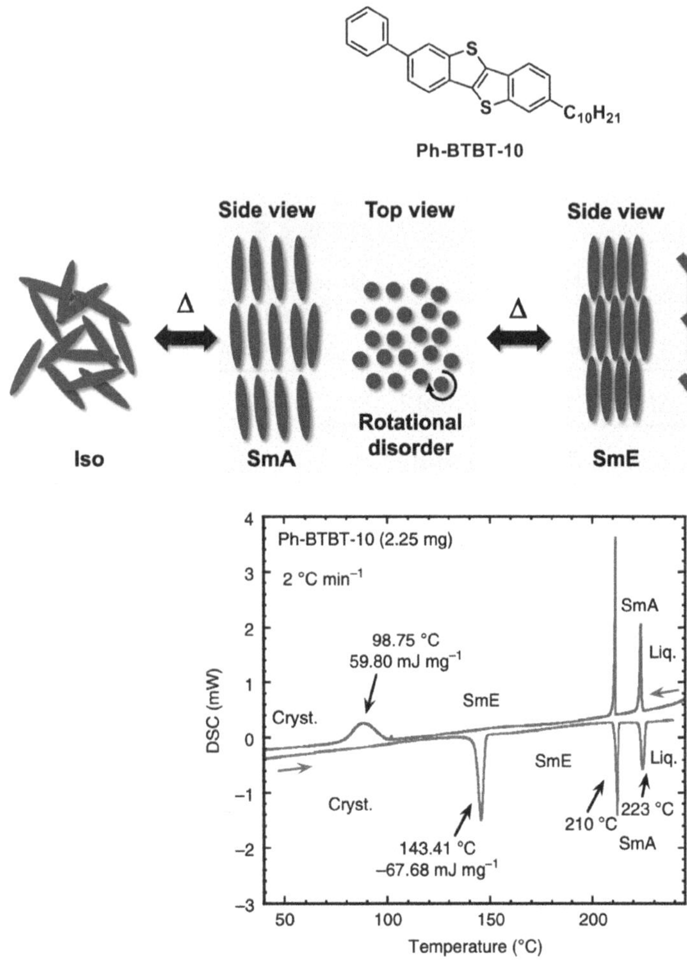

Fig. 7 Molecular structure of **Ph-BTBT-10** (top); molecular ordering in SmA and SmE phases (middle); DSC curves measured at 2 °C/min. Copyright 2015 Springer Nature [4]

2.3.3 Disclike Molecules

A disklike molecule is a typical shape and building block for the formation of the liquid-crystalline phase (Fig. 8). As for general molecular design, multiple alkyl chains are incorporated into the core of disclike molecules. The typical molecular orientation of disklike molecules is a columnar structure, in which molecules self-assembled into a one-dimensional stacked structure. For example, liquid-crystalline molecules based on a discotic triphenylene framework stack to form a columnar liquid crystal (columnar, Col) phase. The column with a hexagonal lattice is categorized

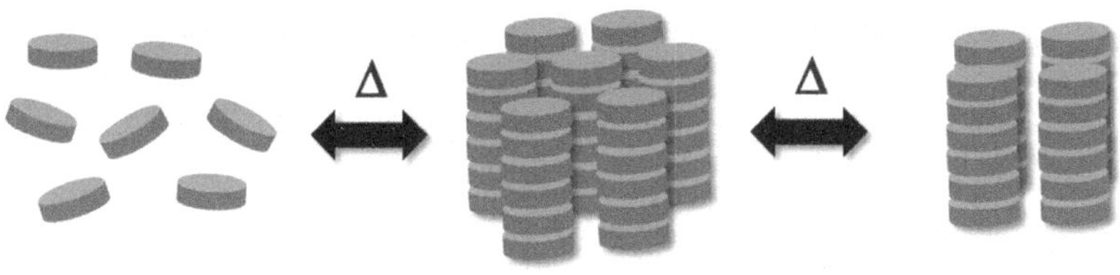

Fig. 8 Schematic illustrations of Col$_h$ and Col$_t$ phases formed by disclike molecules

into a hexagonal columnar (hexagonal columnar, Col$_h$) phase, whereas the one with a tetragonal lattice is called a tetragonal columnar (tetragonal columnar, Col$_t$) phase.

TP-X*n* [5], the most representative discotic molecule, **TP-O4** has an Iso-Col$_h$ phase transition temperature around 150 °C and exhibits the Col$_h$ phase over a wide temperature range (Fig. 9). **TP-X*n*** tends to be a gradual decrease in transition temperature derived from the Iso–Col$_h$ phases as the alkyl chains are elongated. This is due to the elongation of the alkyl chains increases the mobility and the volume, which weakens the intermolecular interactions between triphenylene cores, making it difficult to form a column structure.

For **TP-O5**, the ΔH value observed during Iso–Col$_h$ phase transition has been reported to be 8.20 kJ mol^{-1}, suggesting a significant change in molecular order [6]. On the other hand, **TP-Ph-5**, in which multiple alkynylphenyl moieties are introduced into triphenylene, underwent a phase transition from Iso to discotic nematic (N$_D$) phases in the cooling process (Fig. 10) [7]. This corresponding phase transition ΔH is about 0.2 kJ mol^{-1}, which is considerably smaller than the value due to the Iso–Col$_h$ phase transition for **TP-O5**. These results suggest that the intermolecular order in the N$_D$ phase is smaller than in the Col phase.

TP derivatives have long been recognized for their utility as liquid-crystalline organic semiconductors, transporting charge carriers along the one-dimensional axis due to the stacking of molecules. Higher carrier-transporting properties hinge on the precise ordering of molecules within the column. In a previous works, Adam et al. reported the synthesis of **TP-S6** [8], where oxygen atoms of **TP-O6** [9] are substituted with sulfur atoms, successfully inducing a highly ordered hexagonal columnar (Col$_{ho}$) phase (Fig. 9). Notably, **TP-S6** molecules adopt a structure exhibiting increased intermolecular order within the column, which is clearly distinct from the disordered hexagonal columnar (Col$_{hd}$) phase with loose ordering in the column [8]. Consequently, the hole-transporting properties of **TP-S6** in the Col$_{ho}$ phase can reach up to 0.1 cm^{2}V^{-1} s^{-1}, which were 100 times larger than those of **TP-O5** (0.001 cm^{2}V^{-1} s^{-1}) in the Col$_{hd}$ phase. These results are an important example of how the novel TP derivative design successfully improves carrier-transporting properties.

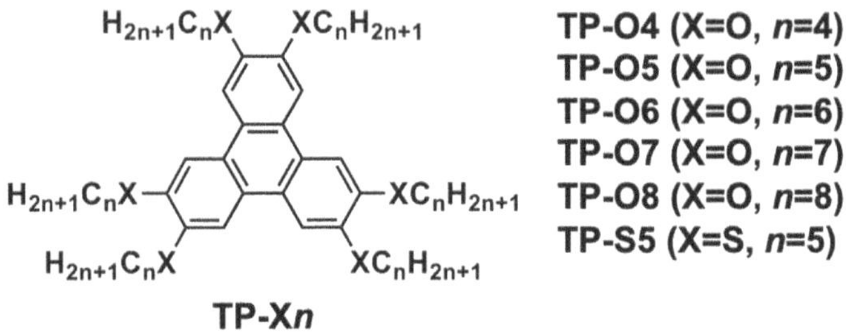

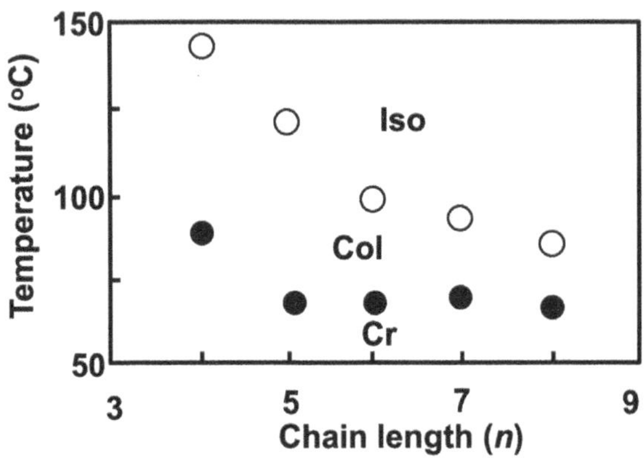

Fig. 9 Molecular structures of **TP-X*n*** and **TP-S5** (top); chain-length dependence for the formation of Col phases of **TP-X*n***

2.3.4 Polymer

The thermal analysis of polymeric materials plays a crucial role in assessing heat resistance-related properties, encompassing parameters like decomposition temperature and combustion temperature in TG measurement. Additionally, in DSC measurements, key aspects such as melting point (T_m), crystallization temperature (T_c), and glass transition temperature (T_g) can be measured. Generally, polymeric materials exhibit a combination of crystalline and amorphous regions, where molecules are regularly arranged in crystalline region and exist in a disordered state in amorphous region. Among them, crystalline polymers have both crystalline and amorphous regions, resulting in T_m derived from the crystalline phase (Cr-Iso) and T_g originating from the amorphous phase (Fig. 11). Conversely, amorphous polymers solely tend to possess T_g. These temperatures vary depending on the structure of the molecule and its density and orientation in the solid state. T_m and T_g of common polymers are summarized in Table 2.

Polyethylene (PE), with its main chain comprising methylene units, can exist in two distinct states, namely low-density PE (LDPE) and high-density PE (HDPE),

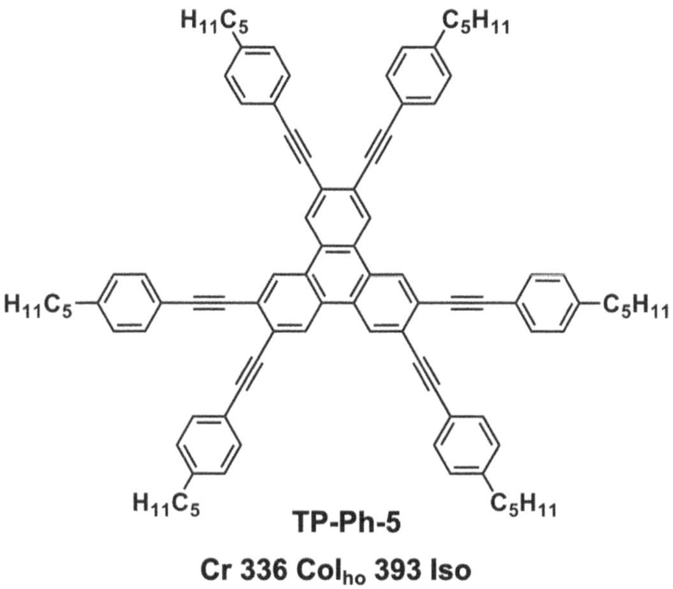

Fig. 10 Molecular structure of **TP-Ph-5**

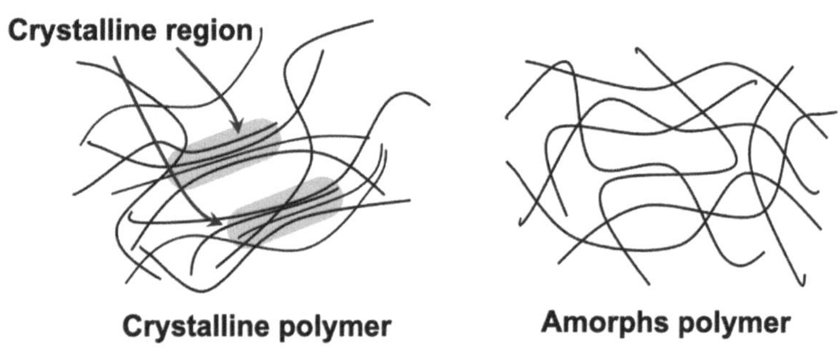

Fig. 11 Schematic illustrations of polymer chain alignment of crystalline (left) and amorphous polymers

each characterized by a different T_m. Polyoxymethylene (POM), in which methylene moieties replaced with oxygen atoms in PE, significantly enhances its T_m to 175 °C higher than PE. Moreover, the incorporation of an aromatic terephthalate ester in polyethylene terephthalate (PET), partially replacing methylene moieties, enables a remarkable increment in both T_g and T_m. This enhancement is attributed to the increased intermolecular interaction between the main chains. Furthermore, polyimide (PI), featuring an aromatic moiety in the main chain, exhibits heightened intermolecular interactions, surpassing PET in T_m and demonstrating excellent thermal stability.

Table 2 T_g and T_m of various polymers

		T_g / °C	T_m / °C
Polyethylene (PE) LDPE	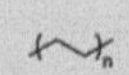	-120∼-20	110∼125 (LDPE) 125∼135 (HDPE)
Polypropyrene (PP)	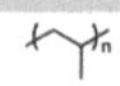	-20∼0	165∼175
Polyoxymethylene (POM)		-60∼-50	170∼180
Polyethylene terephtharate (PET)	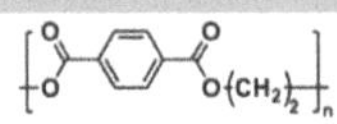	69	256
Polystyrene (PS)	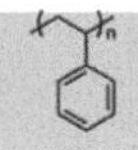	90∼100	-
Polymethyl methacrylate (PMMA)	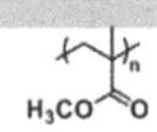	100	-
Polyimide (Kapton)	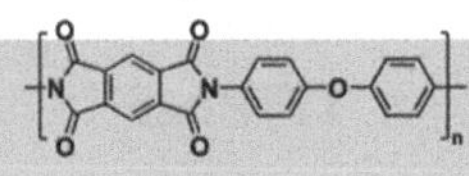	∼400	-

During the heating process, PET pellet exhibits T_g at approximately 74 °C (Fig. 12) [10]. Subsequent heating reveals a crystallization-derived peak around T_c of 160 °C. Upon further heating, a distinct T_m is observed at approximately 257 °C. In comparison, T_g and T_m values for the film state are lower to those of the pellet. However, it is evident that the PET pellet exhibits a higher T_c, indicating a higher level of crystallinity than the quenched PET film.

Thermophysical properties of polymeric materials are influenced by the tacticity of the polymer main chain. Poly(methyl methacrylate) (**PMMA**), known for its transparency and amorphous nature, possesses a chiral central carbon atom in the main chain (Fig. 13). By controlling the orientation of the central carbon, three distinct types of tacticity can be prepared: "Syndiotactic," "Isotactic," and "Atactic" (Fig. 10). These variations in tacticity have shown different T_g despite slight differences in chain configurations (Fig. 13) [11, 12].

Thermophysical properties of polymer blends consisting of amorphous **PMMA** and polystyrene (**PS**) are discussed [13]. Both amorphous **PS** and **PMMA** possess T_g of 102.6 and 108.26 °C, respectively (Table 3). A subtle change in peaks derived from their respective T_g values was observed in the polymer blend containing **PS**:**PMMA** = 1:1 (wt%) (**PS/PMMA** blend (1/1)). Mathur et al. conducted T_g measurements by adjusting the **PS** and **PMMA** ratio, revealing that the phase interactions between **PS** and **PMMA** are physical rather than chemical properties. These interactions depend on changes in morphology accompanied by compositional changes [14]. Conversely, the block copolymer **PS-*b*-PMMA** exhibits an increase in T_g^1 attributed to **PS** and a significant increase in T_g^2 due to **PMMA**. Thus, blending and copolymerization of a diverse range for polymers offer a valuable approach to controlling polymer

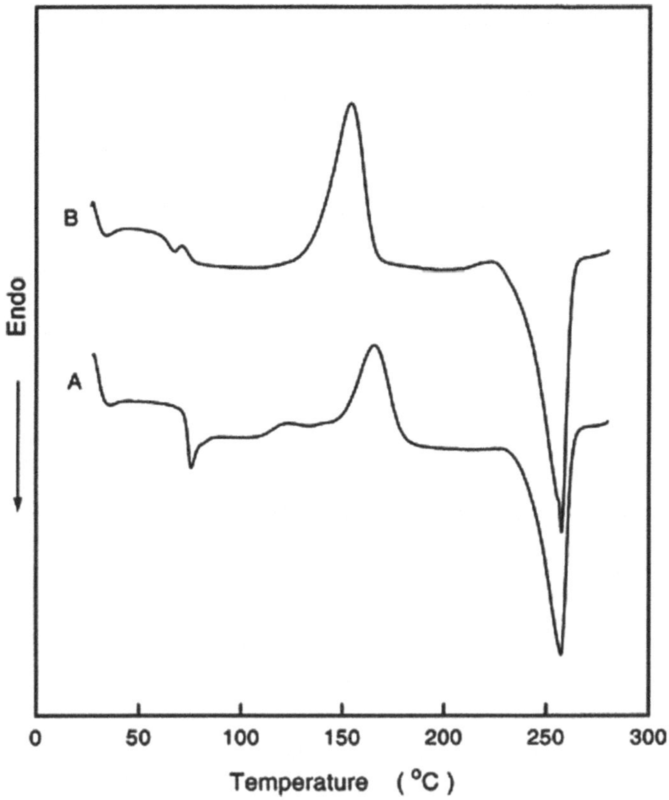

Fig. 12 DSC thermograms of (A) the "as-received" PET pellet and (B) the PET film (quenched in liquid nitrogen). Copyright 1999 Elsevier [10]

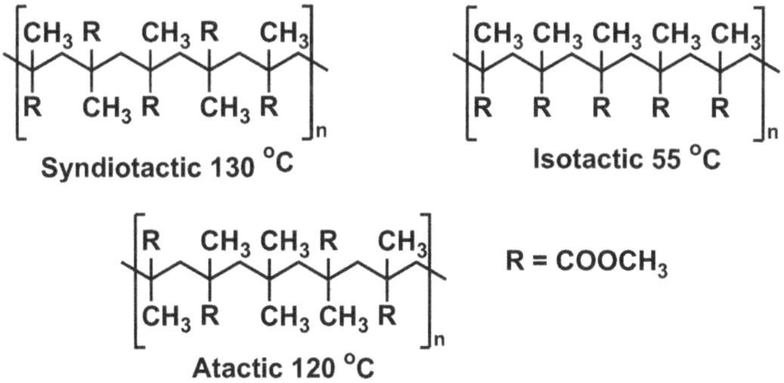

Fig. 13 Tacticity-dependent T_g for syndiotactic, isotactic, and atactic **PMMA**

Table 3 T_g of PS, PMMA, PS/PMMA blend (1/1) (wt%), and PS-*b*-PMMA (1/1) (wt%)

	*M*w (kg/mol)	*M*n (kg/mol)	T_g^1	T_g^2
PS	329.1	133.1	102.60	
PMMA	79.3	43.7		108.26
PS/PMMA blend (1/1)			100.50	110.89
PS-*b*-PMMA (1/1)	170.9	162.4	107.27	132.04

properties, providing opportunities for the exploration of new functional materials in the future.

References

1. K. Uemura, K. Saito, S. Kitagawa, H. Kita, Hydrogen-bonded porous coordination polymers: structural transformation, sorption properties, and particle size from kinetic studies. J. Am. Chem. Soc. **128**, 16122–16130 (2006)
2. S. Singh, Phase transitions in liquid crystals. Phys. Rep. **324**, 107–269 (2000)
3. K. Isoda, T. Yasuda, T. Kato, Dipole-driven self-assembly of redox-active mesogenic tetracyanoanthraquinodimethanes. J. Mater. Chem. **18**, 4522–4528 (2008)
4. H. Iino, T. Usui, J.-I. Hanna, Liquid crystals for organic thin-film transistors. Nat. Commun. **6**, 6828 (2015)
5. D.S.S. Rao, S.K. Prasad, R.K. Bamezai, S. Kumar, High pressure investigations on the phase behaviour of discotic liquid crystals. Mol. Cryst. Liq. Cryst. **397**, 143–159 (2003)
6. W. Kranig et al., Phase behavior of discotic liquid crystalline polymers and related model compounds. Adv. Mater. **2**, 36–40 (1990)
7. K. Praefcke, B. Kohne, D. Singer, Hexaalkynyltriphenylene: a new type of nematic-discotic hydrocarbon. Angew. Chem. Int. Ed. Engl. **29**, 177–179 (1990)
8. D. Adam et al., Fast photoconduction in the highly ordered columnar phase of a discotic liquid crystal. Nature **371**, 141–143 (1994)
9. D. Adam et al., Transient photoconductivity in a discotic liquid crystal. Phys. Rev. Lett. **70**, 457–460 (1993)
10. Z. Zhong, S. Zheng, Y. Mi, High-pressure DSC study of thermal transitions of a poly(ethylene terephthalate)/carbon dioxide system. Polymer **40**, 3829–3834 (1999)
11. C. Duval-Terrié, L. Lebrun, Polymerization and characterization of PMMA. Polymer chemistry laboratory experiments for undergraduate students. J. Chem. Edu., **83**, 443 (2006)
12. L. Chang, E.M. Woo, Tacticity effects on glass transition and phase behavior in binary blends of poly(methyl methacrylate)s of three different configurations. Polym. Chem. **1**, 198–202 (2010)
13. C. Chuai, K. Almdal, J. Lyngaae-Jørgensen, Thermal behavior and properties of polystyrene/poly(methyl methacrylate) blends. J. Appl. Polym. Sci. **91**, 609–620 (2004)
14. V. Mathur, K. Sharma, Thermal response of polystyrene/poly methyl methacrylate (PS/PMMA) polymeric blends. Heat Mass Transf. **52**, 2901–2911 (2016)

Electronic Properties of Soft Matter

Hiroaki Iino and Masahiro Funahashi

1 Conductivity and Mobility in Organic Materials

A voltage is applied to organic materials such as soft matter, generating an electrical current. The current density is defined as the amount of charge carriers (holes and electrons) that pass-through a cross-section per unit area and per second. The current is determined by the density and velocity of the charge carriers. In other words, as the density and the velocity of charge carriers increase, the conductivity also increases. The velocity of charge carriers is usually proportional to the electric field, and the coefficient of proportionality is mobility. Therefore, mobility is the charge carrier velocity per unit electric field, and mobility can be calculated by dividing the charge carrier velocity by the applied electric field.

Mobility is a very important parameter for charge carrier transport materials, and there are two categories for evaluation methods of mobility. One method is to measure the charge carrier velocity directly and then calculate the mobility. The other method is to indirectly evaluate the mobility by dividing conductivity by carrier density. This indirect evaluation method is easier than the direct evaluation method because the conductivity is easy for evaluation from current–voltage characteristics and conductivity is a multiplication of charge carrier density and mobility, and then mobility is obtained from dividing conductivity by charge carrier density. The former direct measurement includes the time-of-flight method, charge carrier extraction by linearly increasing voltage measurement in the metal-insulator-semiconductor

H. Iino (✉)
Laboratory for Future Interdisciplinary Research of Science and Technology, Institute of Integrated Research, Institute of Science Tokyo, Yokohama, Japan
e-mail: iino@first.iir.isct.ac.jp

M. Funahashi
Department of Chemical Science and Engineering, Graduate School of Engineering, Kobe University, Kobe, Hyogo, Japan
e-mail: funahashi.masahiro@phoenix.kobe-u.ac.jp

M. Funahashi (ed.), *Functional Soft Matter II*,
https://doi.org/10.1007/978-981-92-0100-6_4

Table 1 Evaluation methods for charge carrier mobility

Categories	Methods	Requirement of contact
Direct measurement	Time-of-flight	Blocking contact
	MIS-CELIV	–
	Impedance spectroscopy	Ohmic contact
Indirect measurement	SCLC	Ohmic contact
	Hall measurement	Ohmic contact
	PR-TRMC	–
	FET measurement	Ohmic contact

(MIS-CELIV) method, impedance measurement method, and so on, as shown in Table 1 [1–7]. On the other hand, the latter indirect methods include space charge limited current (SCLC) method, Hall measurement, pulse-radiolysis time-resolved microwave conductivity (PR-TRMC) method, field effect transistor (FET) measurement, and so on, as shown in Table 1 [8–13]. Due to the measurement principle, it is very important for the latter indirect measurement to accurately evaluate charge carrier density as well as conductivity measurement.

In this chapter, the principles of the various measurement methods for mobility will be introduced and the time-of-flight method, which is particularly appropriate for the evaluation of soft matter, will be discussed in detail.

2 Contact Property between Organic Materials and Electrodes

When the mobility is evaluated, the electrical contact property between organic materials and electrodes is important. For the time-of-flight method, the blocking contact is required because the small dark current is needed for the measurement of transient photocurrent. On the other hand, in the evaluation of mobility from conductivity, it is necessary to determine the exact conductivity, so good contact property without injection barrier, i.e., ohmic contact, between the organic semiconductor and electrode is very important as shown in Table 1.

Considering the energy level for electrons, the difference between the work function of the electrode and the highest occupied molecular orbital (HOMO) and lowest unoccupied molecular orbital (LUMO) levels of the organic material will act as a contact resistance as an injection barrier from the electrode. In the time-of-fight method described above, the contact property must be blocking with a large injection barrier. On the other hand, in the method evaluating mobility from conductivity, if there is a barrier between the electrode and the semiconductor material, the observed current is influenced by the barrier, and calculated conductivity does not often indicate a true value. This makes it difficult to determine the correct mobility from a current measurement. This is a critical problem, and it is necessary to select an electrode with a smaller injection barrier to form ohmic contact so that the conductivity can be correctly evaluated. This is required for various measurement methods such as SCLC, Hall, and FET measurements. A four-terminal measurement is also effective to accurately measure a conductivity. This method measures a passing current through the two outer terminals of the four terminals as well as the potential difference between the two inner terminals, excluding contact resistance that appears at the two outer terminals.

In the PR-TRMC method described below, it is not necessary to consider any contact properties because electrodes are not used. Soft matter such as liquid crystals and polymers have a tendency to form high potential barrier at electrodes. Time-of-flight and PR-TRMC methods are often used for carrier transport properties of liquid crystals and supramolecular assemblies because they need no Ohmic contacts.

3 Time-of-Flight Method

The time-of-flight method, which can directly measure charge carrier velocity and estimate mobility, is one of the most appropriate methods for evaluating the mobility of soft mater including liquid crystals. The time-of-flight method is an evaluation technique that directly measures the charge carrier transport properties by the transient photocurrent, which is the displacement current that flows through an external circuit when charge carriers are transported from the top electrode to the counter electrode. While the charge carriers are being transported in the sample, the photocurrent flows, and when the charge carriers reach the counter electrode, the photocurrent stops flowing, so the transporting time (transit time), τ_T, can be measured from the time of kink point of the transient photocurrent. The charge carrier velocity can be directly determined to divide the distance between the electrodes, which is usually the sample thickness, by the transit time. Thus, the mobility can be directly measured to divide charge carrier velocity by the electric field. The mobility μ is calculated by following Eq. (1).

$$\mu = \frac{d^2}{\tau_T V} \tag{1}$$

In here, d is the sample thickness, V is the applied voltage.

Conventional equipment of time-of-flight method is pulse laser for the excitation of molecule for charge carrier generation, power supply unit for applied voltage, and digital oscilloscope as shown in Fig. 1.

The features of time-of-flight method are as follows:

1. The mobility as a function of the electric field can be directly determined.
2. By changing the polarity of applied voltage, the positive and negative charge carrier transport can be evaluated without changing the sample.

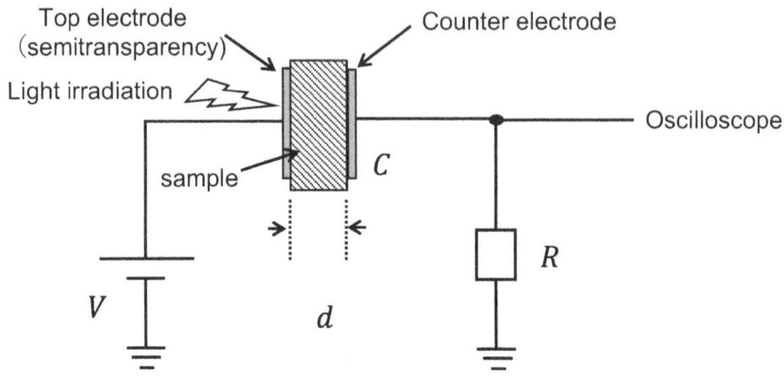

Fig. 1 Electrical circuit of time-of-flight method

3. The distribution of trapping sites and energy levels in the bulk can also be evaluated from the shape of the transient photocurrent [14, 15].
4. By integrating the transient photocurrents, the amount of collected charge can be evaluated. If all the charge carriers reach the counter electrode, the amount of collected charge corresponds to the amount of photocarriers generated, and the charge carrier generation efficiency can be evaluated to divide by the number of incident photons [16]. If no carriers reach the counter electrode, it is possible to evaluate how many carriers are annihilated in the sample, carrier lifetime, and so on [17].

3.1 Requirements of Time-of-Flight Method

(i) Generation Position of Charge Carriers.

In conventional time-of-flight method, it is necessary to generate charge carriers in the form of sheets near the top electrode by light irradiation. Therefore, the photoexcitation wavelength at which the excitation light is absorbed by the organic semiconductor material is chosen, so that charge carriers are generated as close as possible to near the top electrode. In general, the penetration depth is defined as the depth at which light intensity decreases to 1/e of the initial intensity. The sample thickness should be set 10 times or more of this penetration depth. Under the condition, the distribution of the generated charge carriers can be sheet-like.

In intrinsic photocarrier generation, organic molecules are excited by light and then an electric field generates hole and electron carriers at a ratio of 1:1, respectively. Charge carriers of the same polarity as the bias on the top electrode are drifted to the counter electrode, while charge carriers of an opposite polarity are quickly absorbed by the top electrode. For example, when the irradiating electrode is positively biased, the transported carriers to the counter electrode are positive charge carriers and negative charge carriers are quickly absorbed to the top electrode. Movement of negative charge carriers from the generated place to the top electrode is observed as an initial spike in the transient photocurrent, but it has little effect on the transit time because they do not contribute to the photocurrent immediately. It should be recalled that the photocurrent does not appear for the charge carrier that has reached the electrode, but for the charge carrier traveling in the sample.

The light absorption coefficient is 10^5 cm^{-1} at most, and the penetration depth is 0.1 μm, while the required thickness is ten times that, so a sample with a thickness of about 1 μm is required. Furthermore, if the material has a low light absorption coefficient of 10^4 cm^{-1}, a thickness of about 10 μm is required. In such cases, charge carrier generating layers (CGL) are also used. For example, dyes such as rhodamine 6G can be used as a CGL by forming about 10 nm on the electrode by vacuum deposition and irradiating it by excitation light with the wavelengths that match the absorption band of rhodamine 6G [18]. ITO electrodes can also be used like CGL for hole, whereby electrons are accepted from the excited state of the organic material, generating holes. ITO, as a transparent electrode, is very accessible due to the need for light irradiation although the effect of interface trapping is strong on the ITO electrode [19]. Semi-transparent Al electrodes, having a thickness of 20–30 nm, can be used. In this case, a native oxide film is formed on the surface of Al electrode, which acts as a blocking layer and can be used as an electrode with lower dark current and less interface trapping.

The use of a laser with a pulse width sufficiently shorter than the transit time is required. A flash light source can be also used instead of a pulse laser. It is important to irradiate the pulsed light within a time shorter than the transit time. In general, a duration time of an excitation light shorter than 10 ns is sufficient if the transit time is more than 1 μs.

(ii) Appropriate Photocarrier Generation.

Time-of-flight method is evaluated by assuming that the internal electric field is constant. The internal electric field applied inside the sample is formed by the charge, Q_0, on external electrode of the parallel plate, which is $Q_0 = CV$, in here, C is capacitance and V is applied voltage. Therefore, when a large charge of the same magnitude as $Q_0 = CV$ is generated inside, the electric field between the irradiation electrode and the location of the charge carrier sheet decreases and the electric field becomes stronger between the location of the charge carrier sheet and the counter electrode, resulting in a transient photocurrent curve with a peak, which is controlled by space charge limited. Therefore, the amount of generated charge carriers should be kept below 10% of Q_0, preferably below 2%. If the amount of charge carrier generation is greater than 10% of Q_0, the amount of light exposure should be reduced using a neutral density (ND) filter. If the concentration of photogenerated charge carriers is enough to fill the trap site, the trap site is filled and the mobility unaffected by the trap can be determined. However, as mentioned above, if the amount of light irradiation is increased too much in order to fill the trap sites, the internal electric field will not be constant and the correct mobility cannot be estimated by the time-of-flight method.

(iii) Blocking Contact.

Since the time-of-flight method measures photocurrent, if the dark current is large, the photocurrent cannot be measured. Therefore, the contact between the electrode and the organic semiconductor must be blocking. Furthermore, it is a problem that the distribution of injected charge carriers and intrinsic charge carriers from the electrode results in the formation of space charges and an internal electric field that is not constant. The time required for such a change in distribution is called dielectric relaxation time, τ_{rel}, which is estimated by the following Eq. (2).

$$\tau_{\text{rel}} = \frac{\varepsilon_r \varepsilon_0}{\sigma} \tag{2}$$

In here, ε_r is the relative dielectric constant, ε_0 is the permittivity of vacuum, and σ is the conductivity. It is determined by the electrical conductivity and relative dielectric constant, and if conductivity is smaller than 10^{-13} S cm^{-1}, the relaxation time is 1 s or longer. For general organic materials, this is not a major problem because they often have a higher resistance than this.

(iv) Resistance of Measurement Circuit.

The transient photocurrent curve is obtained with a digital oscilloscope from the displacement current in the external circuit. Therefore, it is necessary to put a resistance, *R*, in series with the sample to convert the displacement current into a voltage as shown in Fig. 1. This resistance should be smaller than the resistance of the sample and the input impedance of the digital oscilloscope. Furthermore, to measure the charge carrier transport inside the sample with the displacement current outside, the time constant of *RC* must be smaller than the time range in which the transit time appears. In here, *R* is the resistance to be measured and *C* is the total capacitance of the measurement circuit and sample. In the previous section, since the photocarrier generation must be smaller than $Q_0 = CV$, the amount of charge carriers cannot be increased. On the other hand, since the photocurrent is often small, a resistance value should be large enough, not to be limited by the time constant of the *RC* of the circuit. Otherwise, a current amplifier should be used. When looking at transient photocurrent curve for various time domains, it is useful to divide the evaluation into several measurement domains with different values of *R*.

3.2 Actual Transient Photocurrent: Non-Dispersive and Dispersive Transports

When charge carriers are transported to the counter electrode without being trapped, the charge carrier packet spreads out with time due to diffusion in general. Without diffusion effect, the distribution of charge carriers is kept in the sheet as shown in Fig. 2a and the resulting transient photocurrent curves are rectangular in shape as

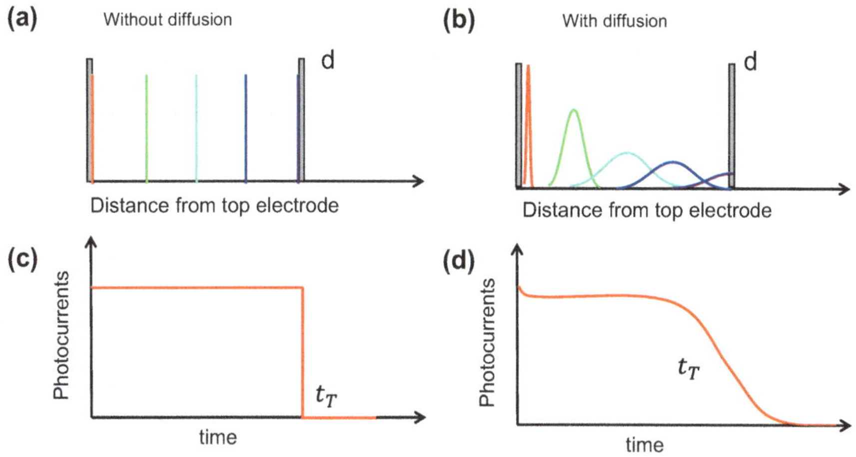

Fig. 2 Distribution of charge carrier in sample **a** without and **b** with diffusion effects. Transient photocurrent curves **c** without and **d** with diffusion effects

own in Fig. 2c. With diffusion of charge carriers, the distribution of the charge carrier is shown in Fig. 2b, a transient photocurrent curve with rounded corners is obtained as shown in Fig. 2d. Since the effect of diffusion increases with transit time, the mobility is estimated slightly higher if the kink point is estimated as the transit time.

In the same sample thickness, apparent transit time estimated from the kink point is shorter than the original transit time in a low electric field because of diffusion effect. This causes the mobility to increase at low electric fields where the transit time is longer. In order to evaluate the mobility without such diffusion effects, the transit time can be obtained by evaluating from the time of the transient photocurrent value that is half of the current value of the plateau [20].

These transient photocurrent curves in which the transit time is visible in a linear plot is called a non-dispersive transport property.

In the cases of materials having many trap sites and wide density of state, especially in amorphous materials, the distribution of the charge carrier is often anomalously dispersed as shown in Fig. 3a and it is difficult to find clear plateaus and transit times in a linear plot as shown in Fig. 3b. In such cases, the transit time can be estimated from its kink point in a double logarithm plot as shown in Fig. 3c [21]. Thus, the time-of-flight method can evaluate not only the mobility but also the trap sites and density of state that inhibits charge carrier transport.

In actual transient photocurrent curves, it is often unclear whether it is a kink point corresponding to a transit time. As a way to distinguish real kink points from those caused by the other reasons, it is very useful to evaluate how the kink point shifts with respect to the electric field. Furthermore, the shape of the exponential decay, where the charge carriers annihilate at a constant, results in transient photocurrent curves that look like a kink point in a double logarithm plot. It is also important to confirm

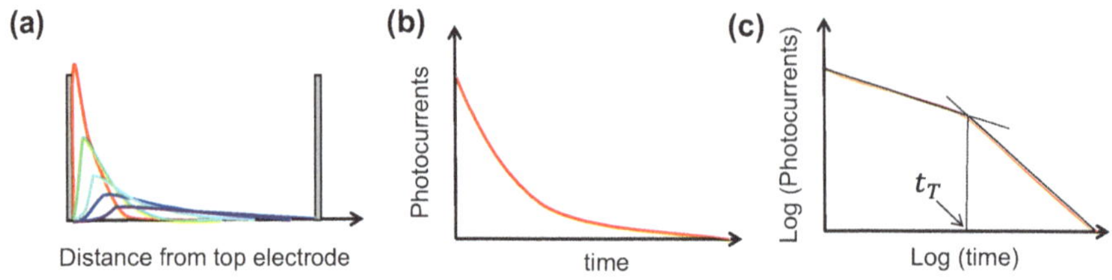

Fig. 3 Transient photocurrents in sample with dispersive transport. **a** The distribution of charge carrier. Transient photocurrent curves in sample with dispersive transport in **b** liner plot, and **c** double logarithm plot

that transient photocurrent curves are not a straight line by plotting the logarithm of the current versus time linearly.

3.3 Compatibility of Time-of-Flight Method with Soft Matter

As mentioned above, a thickness of 10 μm is often required when the time-of-flight method to generate photocarriers through absorption by its own sample is used. In the case of liquid crystalline materials, such thickness samples can be easily formed by a liquid crystal cell structure which is fabricated by bonding two substrates with an adhesive. Of course, in liquid crystal materials, appropriate orientation control is necessary because of their strong anisotropy property. For smectic phases consisting of rod-like molecules, the liquid crystal molecules should align parallel to the electrode surface and the smectic layers orient perpendicular to the surface because an efficient charge carrier transport occurs within the smectic layer [4]. In contrast, disk-like molecules form columnar stacks in a typical columnar phase, resulting in preferential charge carrier transport along the columnar axis. Stronger and well-defined photocurrent signals should be obtained in homeotropically aligned columnar liquid crystal samples [3].

Charge carrier transport includes not only electronic conduction by electrons and hole hopping transport but also ionic conduction, in which the ionized molecules themselves are transported. In a system containing carrier-trapping impurities, not only electronic but also ionic conduction are sometimes observed. In some materials, transition from an electronic transport to an ionic conduction was observed with an increase in the impurity concentration [22, 23].

Another major advantage of the time-of-flight method is that it allows direct evaluation of charge carrier velocity and mobility at each electric field. Charge carrier transport in soft matter, including amorphous materials and liquid crystals, often follows a Gaussian disorder model [24]. In this model of charge carrier transport, mobility is known to depend on the electric field, and the time-of-flight method is very useful because it can evaluate the mobility in each electric field and temperature.

So the evaluation of time-of-flight method can reveal the distribution width of the density of states from the electric field and temperature dependence of mobility.

4 MIS-CELIV Method

The MIS-CELIV method [5, 25, 26] accumulates charge carriers at the semiconductor-insulator interface of a metal/insulator/semiconductor (MIS) structure, then allows the accumulated charge carrier to drift in the bulk and reads the transient current curves as shown in Fig. 4a. First, charge carriers are injected from the electrode on the semiconductor side to accumulate at the semiconductor/insulator interface. If the carrier to be measured is a positive charge carrier, the electrode on the insulator is negatively biased. If the carrier to be measured is a negative charge carrier, the electrode on the insulator is positively biased. Although ohmic contact is desirable at the semiconductor/electrode interface during accumulation, even if a little contact resistance exists, accumulation can be achieved over time. Then, by increasing the voltage linearly, current flows under transient space-charge-limited current conditions as shown in Fig. 4b. At this time, by linearly increasing the voltage, the displacement current component corresponding to the capacitance component is constant, so the transient space-charge limiting current corresponding to carrier transport can be easily isolated and evaluated. The current value obtained by linearly increasing the voltage without accumulation is j_0, and the mobility can be obtained from the transit time t_{tr} by Eq. (3). t_{tr} is expressed by Eq. (4) and t_{2j_0} is obtained from the time of transient current curve to reach twice the current value j_0 as shown in Fig. 4c.

$$\mu = \frac{2d_s^2}{At_{tr}^2}\left(1 + \frac{\varepsilon_s d_i}{\varepsilon_i d_s}\right) \tag{3}$$

$$t_{tr} = \frac{4}{\pi} t_{2j_0} \tag{4}$$

In here, μ is the mobility, A is the rate of increasing applied voltage, ε_s is the dielectric constant of the organic semiconductor film, d_s is the thickness of the organic semiconductor film, ε_i is the dielectric constant of the insulator, and d_i is the thickness of the insulator.

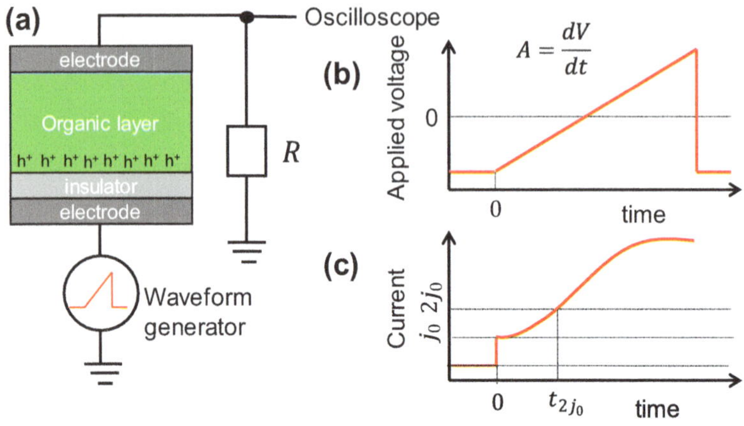

Fig. 4 MIS-CELIV method. **a** Device structure and electrical circuit. Schematic illustration of **b** applied triangle voltage and **c** observed transient current

5 Impedance Spectroscopy Method

In impedance measurements, a small sinusoidal voltage signal is applied to a sandwich-type sample and the frequency response of the amplitude and phase of the current signal is evaluated [6, 7]. An Ohmic contact is required for the determination of a charge carrier mobility in an organic semiconductor device by the impedance spectroscopy. The injected charge carriers through the Ohmic contact indicate different behaviors above and below the critical frequency. In the high frequency region, the injected charge carriers do not arrive at the counter electrode within one period, and they do not reach a thermal equilibrium state. The capacitance of the device equals the geometrical capacitance of the device. In the low frequency region, the injected carriers are able to arrive at the counter electrode within one period, resulting in distribution of the carriers over the device. This distribution of the carriers reduces the capacitance (Fig. 5a). The traveling time and mobility can be estimated from the reciprocal frequency of the peak seen in the change in susceptance, $\Delta B = \omega(C - C_{\text{geo}})$ due to traveling time effects, as shown in Fig. 5b. In here, ω is the angular frequency, C is the measured capacitance, and C_{geo} is the geometric capacitance of sample.

These measurements do not require a film thickness of more than a few micrometers, which is required under conditions such as penetration depth in a time-of-flight method, and have the advantage of being able to evaluate thin-film samples of about 100 nm, which are used in OLEDs and other applications. Unlike a conventional impedance measurement, one electrode is required to be an Ohmic contact.

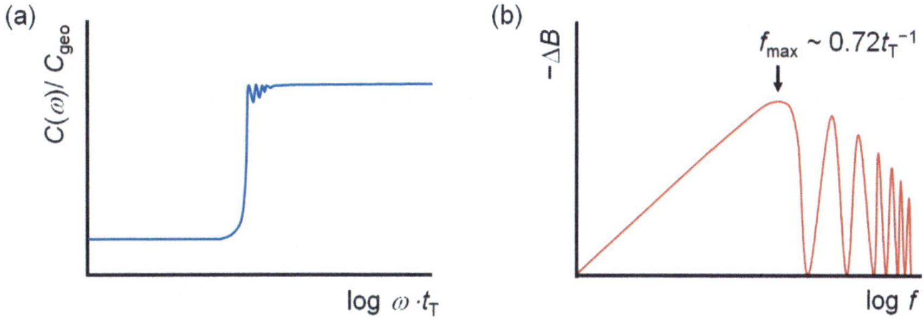

Fig. 5 Schematic illustration for frequency-dependence of **a** C/C_{geo} and **b** ΔB

6 Space Charge Limited Current Method

In a device in which an Ohmic contact between an electrode and a semiconductor layer, charge carriers are injected from the electrode to the semiconductor layer to form space charge when the carrier mobility of the semiconductor is relatively low. The electric field in the semiconductor layer is modulated by the space charge, as shown in Fig. 6a. When the injected charge carrier q is equal to or greater than the charge $Q_0 = CV$ stored outside forming the electric field, the electric field near the electrode becomes zero. In such a state, current is limited by the carrier mobility, the sample thickness, and the dielectric constant (space charge limited current). This phenomenon can be solved analytically, and the current value is determined by the following Eq. (5) [8].

$$J = \frac{9}{8}\mu\varepsilon_r\varepsilon_0\left(\frac{V^2}{d^3}\right) \tag{5}$$

where μ is mobility, ε_r is relative dielectric constant, ε_0 is permittivity of vacuum, V is voltage, and d is film thickness. From this equation, the space charge limiting current depends on V^2 and is inversely proportional to d^3. Interestingly, this equation does not depend on the carrier density, but is defined by the relative permittivity, mobility, voltage, and thickness. Therefore, if the relative permittivity of the material is known, it is possible to estimate the mobility.

Since this measurement method is evaluated by examining the current-voltage characteristics, it is an indirect way of estimating mobility, taking into account the carrier density from the conductivity. As shown in Fig. 6b, a double logarithmic plot of a current-voltage curve is used for the evaluation of the carrier mobility. The carrier mobility is estimated from the current values in the region where the current is proportional to V^2. When the carrier traps are present in the semiconductor layer, the current-voltage curve is modulated by trap-filling. In order to observe the space charge limited current, Ohmic contacts are required. When high work function electrodes such as Au were used, an Ohmic contact is sometimes formed for HOMO level of the semiconductor layer, giving a hole mobility. In order to obtain an electron

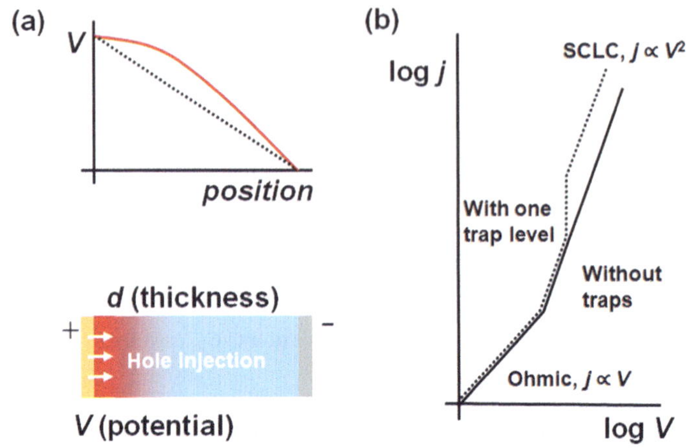

Fig. 6 **a** Schematic illustration of hole injection from an Ohmic contact anode in a hole-only device and a potential distribution in the device. **b** Typical double logarithmic plots of the *j*–*V* curve without traps and with one trap level

mobility, an ohmic contact is sometimes formed for LUMO level, using low work function metals such as Ag/Mg and Ca.

7 Hall Measurement

Hall measurement is a method of estimating majority carrier concentration or mobility from the Hall's effect. Hall voltage appears when moving electrons are subjected to Lorentz force under the influence of a magnetic field, creating a difference in charge density. It is possible to estimate the carrier density, polarity of the majority carriers, and Hall mobility simultaneously. In general, the Hall measurement is difficult for intrinsic organic semiconductors with low carrier densities because of the low current densities. In order to increase the carrier density, the Hall measurement is combined with a FET method, in which carrier density can be remarkably increased by a gate bias application [9].

In Hall measurement, the electrons must have momentum. However, since charge carriers in hopping conduction do not have momentum, the Hall voltage does not appear. On the other hand, in organic crystalline materials, some of them are band-conducting. For example, in a single-crystal thin film of rubrene, the mobility evaluated from Hall measurement agrees with the mobility measured by other method, but in a polycrystalline thin film of pentacene with a mobility of 1 cm^2/Vs, the mobility evaluated by the Hall measurement is smaller than the mobility determined by other method. From this point of view, the Hall measurements are used as a method to check whether band conduction is present or not in organic semiconductors [9].

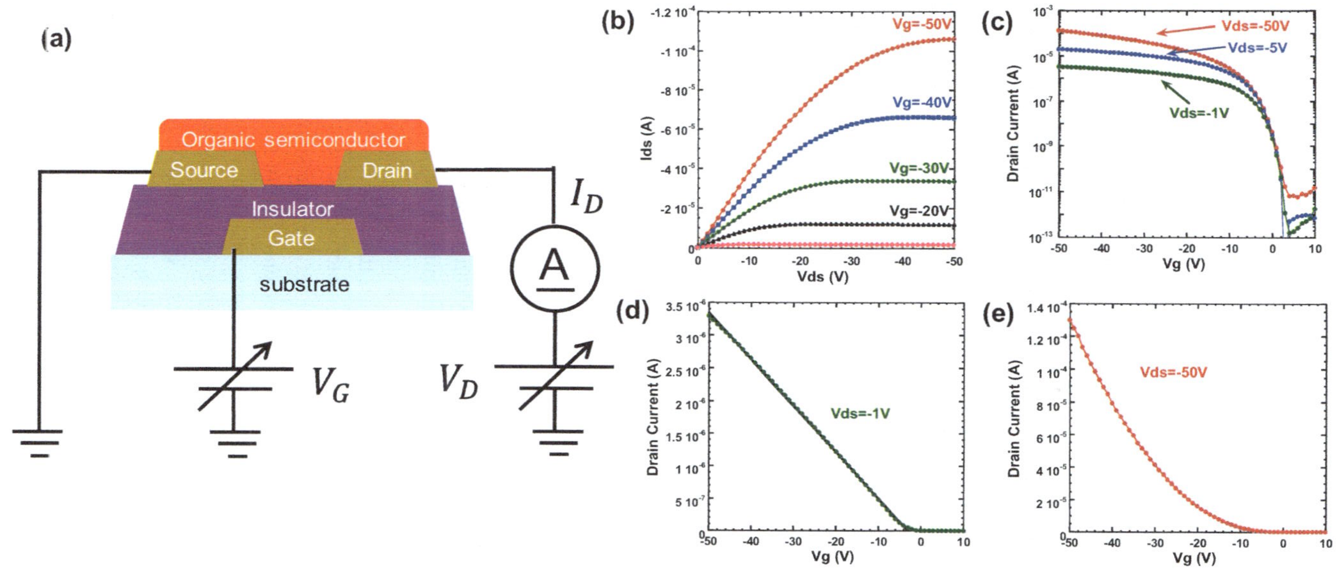

Fig. 7 Field effect transistor (FET) method. **a** Schematic illustration of structure of FET and measurement circuit. **b** Output characteristics. Transfer characteristics in **c** logarithm plot of drain current, **d** linear region, and **e** saturation region

8 Pulse-Radiolysis Time-Resolved Microwave Conductivity (PR-TRMC)

Microwave is a kind of electromagnetic waves and the frequency is several GHz. When the microwave is illuminated on a sample containing mobile charges, the microwave is absorbed into the sample by oscillation of the charges. The microwave absorption is evaluated as the conductivity of the sample. The charge carriers are generated by electron beam irradiation from a Van-de-Graaff accelerator or pulse laser illumination. This method provides the maximum mobility of the material, to exclude trapping and de-trapping processes of charge carriers. Any electrodes are not required in this method because the measured quantity is the absorption of microwave. Therefore, the measurement can be conducted, independent of morphologies of the samples, including powders, liquids, crystals, and so on. However, the sign of charge carriers, i.e., positive or negative charge carriers, could not be determined by this method. It is very important to evaluate an exact charge carrier density in order to obtain an accurate carrier mobility [10–12]. Soft mater such as liquid crystals and supramolecular assemblies sometimes exhibits bad contact with the electrode. Time-resolved microwave conductively is an effective method for soft matter.

9 Field Effect Transistor Measurement

A field-effect transistor (FET) is an electronic device that works as a switching element as shown in Fig. 7c. From these transistor characteristics, it is possible to evaluate FET mobility. Basically, the mobility is obtained by dividing the charge carrier density from the conductivity. In the case of field-effect transistors, it is possible to define the amount of charge that can be stored in the channel, $Q = C_i V_G$, by applying a gate voltage V_G, and the mobility can be estimated. In practice, mobility can be estimated by evaluating the increase in current when the gate voltage V_G is changed.

The transistor has the structure as shown in the Fig. 7a. Applying gate voltage causes carriers to accumulate at the organic semiconductor/gate insulator interface. Applying drain voltage causes current to flow between the source and drain, but the magnitude of the gate voltage changes the concentration of the accumulated charge carriers and the current value between the source and drain. When the drain voltage is small in relation to the gate voltage, charge carriers accumulate on the entire interface on the gate insulator between the source and drain electrodes. But when the gate voltage becomes the same level as the drain voltage, the electric field from the gate electrode becomes small near the drain electrode, resulting in a depletion state (high resistance state) as charge carrier accumulation becomes small, and the electric field between the source and drain electrodes is applied only to the high-resistance state, resulting in saturation of the current value. The current saturates because the electric field between the source and drain only hangs on the high resistance state as shown in Fig. 7b. The area where the drain voltage is small becomes a linear region, and the area where the drain voltage is high becomes a saturated region, each showing current–voltage characteristics according to the following Eqs. (6) and (7), respectively [13].

$$I_D = \frac{W}{L}\mu C_i\left(V_G - V_{th} - \frac{V_D}{2}\right)V_D \tag{6}$$

$$I_D = \frac{W}{2L}\mu C_i(V_G - V_{th})^2 \tag{7}$$

where I_D is drain current, W and L are channel width and length, respectively, μ is the mobility, C_i is the gate insulator capacitance per unit area, V_{th} is threshold voltage, V_D is drain voltage.

The mobility is obtained from the slope of the transfer characteristic in linear region, which plots the source-drain current against the drain voltage as shown in Fig. 7d. In the saturation region where the drain voltage is high, the current value increases by V_G^2 as shown in Fig. 7e, so the mobility is calculated by checking the linearity of the slope by making a rooted plot.

The value obtained by this method, also called FET mobility, is the transport at the gate dielectric interface and is strongly affected by traps at the gate dielectric interface. In addition, because the evaluation is based on the conductivity between the source and drain, if the contact resistance between the source-drain electrode and the organic semiconductor is large, it is either significantly smaller or the linearity is poor, making it difficult to accurately measure the mobility.

10 Summary

In this chapter, we indicate the measurement methods for evaluating charge carrier transport properties and mobility appropriate for soft matter are presented. There are two categories of measurement: direct measurement of charge carrier transport and indirect measurement. The direct measurement method is to evaluate transit time when charge carriers reach counter electrode and mobility from the transit time such as time-of-flight method. On the other hand, an indirect measurement method is to divide conductivity by the charge carrier density in order to obtain the mobility. In this case, ohmic contact is required and charge carrier density must be accurately measured. In generally, it is difficult to realize the ohmic contact between the soft matter and electrode, therefore, time-of-flight and PR-TRMC methods were mainly used. On the other hand, the crystalline thin films of liquid crystalline materials are easier to realize ohmic contact, so mobility is evaluated by the field effect transistor structure. For soft matter such as liquid crystals, the time-of-flight method is a relatively suitable method because thin films with a thickness of several micrometers can be easily formed in the liquid crystal cell structure.

References

1. W.E. Spear, Drift mobility techniques for the study of electrical transport properties in insulating solids. J. Non-Cryst. Solid. **1**, 197–214 (1968). https://doi.org/10.1016/0022-3093(69)90001-5
2. M. Funahashi, Time-of-flight method for determining the drift mobility in organic semiconductors, in *Organic Semiconductors for Optoelectronics*, ed. by H. Naito (Wiley, 2021) (Chapter 6)
3. D. Adam, F. Closs, T. Frey, D. Funhoff, D. Haarer, P. Schuhmacher, K. Siemensmeyer, Transient photoconductivity in a discotic liquid crystal. Phys. Rev. Lett. **70**, 457–460 (1993). https://doi.org/10.1103/PhysRevLett.70.457
4. M. Funahashi, J. Hanna, Fast hole transport in a new Calamitic liquid crystal of 2-(4′-heptyloxyphenyl)-6-dodecylthiobenzothiazole. Phys. Rev. Lett. **78**, 2184–3187 (1997). https://doi.org/10.1103/PhysRevLett.78.2184
5. G. Juška, N. Nekrašas, K. Genevičius, Investigation of charge carriers transport from extraction current transients of injected charge carriers. J. Non-Cryst. Solids **358**, 748–750 (2012). https://doi.org/10.1016/j.jnoncrysol.2011.12.016
6. H.C.F. Martens, J.N. Huiberts, P.W.M. Blom, Simultaneous measurement of electron and hole mobilities in polymer light-emitting diodes. Appl. Phys. Lett. **77**, 1852–1854 (2000). https://doi.org/10.1063/1.1311599
7. K. Takagi, H. Naito, Characterization of transport properties of organic semiconductors using impedance spectroscopy, in *Organic Semiconductors for Optoelectronics*, ed. by H. Naito (Wiley, 2021) (Chapter 5)
8. P.W.M. Blom, M.J.M. de Jong, J.J.M. Vleggaar, Electron and hole transport in poly(p-phenylene vinylene) devices. Appl. Phys. Lett. **68**, 3308–3310 (1996). https://doi.org/10.1063/1.116583
9. V. Podzorov, E. Menard, J.A. Rogers, M.E. Gershenson, Hall effect in the accumulation layers on the surface of organic semiconductors. Phys. Rev. Lett. **95**, 226601 (2005). https://doi.org/10.1103/PhysRevLett.95.226601
10. A.M. van de Craats, J.M. Warman, M.P. de Haas, D. Adam, J. Simmerer, D. Haarer, P. Schuhmacher, The mobility of charge carriers in all four phases of the columnar discotic material hexakis(hexylthio)triphenylene: combined TOF and PR-TRMC results. Adv. Mater. **8**, 823–826 (1996). https://doi.org/10.1002/adma.19960081012
11. A. Saeki, S. Seki, T. Sunagawa, K. Ushida, S. Tagawa, Charge carrier dynamics in polythiophene films studied by in-situ measurement of flash-photolysis time-resolved microwave conductivity (FP-TRMC) and transient optical spectroscopy (TOS). Philos. Mag. **86**, 1261–1276 (2006). https://doi.org/10.1080/14786430500380159
12. A. Saeki, Microwave and terahertz spectroscopy, in *Organic Semiconductors for Optoelectronics*, ed. by H. Naito (Wiley, 2021) (Chapter 7)
13. G. Horowitz, Organic field-effect transistors. Adv. Mater. **10**, 365–377 (1998). https://doi.org/10.1002/(SICI)1521-4095(199803)10:5<365::AIDADMA365>3.0.CO;2-U
14. T. Nagase, H. Naito, Localized-state distributions in molecularly doped polymers determined from time-of-flight transient photocurrent. J. Appl. Phys. **88**, 252–259 (2000). https://doi.org/10.1063/1.373649
15. A. Ohno, J. Hanna, D.H. Dunlap, Analysis of trap distribution using time-of-flight spectroscopy. Jpn. J. Appl. Phys. **47**, 1079–1088 (2008). https://doi.org/10.1143/JJAP.47.1079
16. R.R. Chance, C.L. Braun, Temperature dependence of intrinsic carrier generation in anthracene single crystals. J. Chem. Phys. **64**, 3573–3581 (1976). https://doi.org/10.1063/1.432707
17. J.M. Marshall, Measurement of carrier lifetime in materials exhibiting anomalous dispersion. J. Non-Cryst. Solids **198-200**, 182–185 (1996). https://doi.org/10.1016/0022-3093(96)80021-4
18. C. Im, H. Bässler, H. Rost, H.H. Hörhold, Hole transport in polyphenylenevinylene-ether under bulk photoexcitation and sensitized injection. J. Chem. Phys. **113**, 3802–3807 (2000). https://doi.org/10.1063/1.1287657
19. H. Zhang, J. Hanna, Electrode effect on photohole generation in smectic phenylnaphthalene liquid crystalline photoconductor. J. Appl. Phys. **88**, 270–277 (2000). https://doi.org/10.1063/1.373652

20. A. Hirao, T. Tsukamoto, H. Nishizawa, Analysis of nondispersive time-of-flight transients. Phys. Rev. B **59**, 12991–12995 (1999). https://doi.org/10.1103/PhysRevB.59.12991
21. H. Scher, E.W. Montroll, Anomalous transit-time dispersion in amorphous solids. Phys. Rev. B **12**, 2445–2477 (1975). https://doi.org/10.1103/PhysRevB.12.2455
22. M. Funahashi, J. Hanna, Impurity effect on charge carrier transport in smectic liquid crystals. Chem. Phys. Lett. **397**, 319–323 (2004). https://doi.org/10.1016/j.cplett.2004.08.125
23. K. Tokunaga, Y. Takayashiki, H. Iino, J. Hanna, Electronic conduction in nematic phase of small molecules. Phys. Rev. B **79**, 033201 (2009). https://doi.org/10.1103/PhysRevB.79.033201
24. H. Bässler, Charge transport in disordered organic photoconductors a Monte Carlo simulation study, phys. Stat. Sol. (b) **175**, 15–56 (1993). https://doi.org/10.1002/pssb.2221750102
25. A. Armin, G. Juška, M. Ullah, M. Velusamy, P.L. Burn, P. Meredith, A. Pivrikas, Balanced carrier mobilities: balanced carrier mobilities: not a necessary condition for high-efficiency thin organic solar cells as determined by MIS-CELIV. Adv. Energy Mater. **4**, 1300954 (2014). https://doi.org/10.1002/aenm.201300954
26. C. Katagiri, T. Yoshida, M.S. White, C. Yumusak, N.S. Sariciftci, K. Nakayama, Application of MIS-CELIV technique to measure hole mobility of hole-transport material for organic light-emitting diodes. AIP Adv. **8**, 105001 (2018). https://doi.org/10.1063/1.5045711

Optical Properties of Anisotropic and Chiral Media

Ryotaro Ozaki

1 Introduction

Soft matter, such as polymers and liquid crystals, have optical anisotropy that appears due to the molecular shape and orientation. Optical anisotropy is the property caused by a refractive index that varies with direction. Polymers and liquid crystals can have two refractive indices, one in the direction of the major axis of the molecules and the other in the direction of the minor axis of the molecules. Optically anisotropic materials are important for liquid crystal displays (LCDs), polarization filters, and optical elements. However, it is difficult to confirm optical anisotropy with the human eye, but when we observe optically anisotropic materials using a polarizer, the beautiful colors that appear due to optical anisotropy can be seen. Figure 1a shows a transparent plastic spoon photographed with a commercial digital camera. In this picture, the background is an LCD monitor that displays the color white. That is, we observe the light passing through the transparent spoon from a linearly polarized light source, because the LCD monitor has a polarizer on its top surface. There is no visual change from the image of the transparent spoon that we normally see because the conventional digital camera cannot detect the polarization of the light. On the other hand, Fig. 1b is a photograph taken with a polarizer in front of the digital camera. The transparent plastic spoon is colored in a rainbow, resulting in a beautiful picture. To explain the mechanism of the color change, an understanding of refractive index anisotropy and polarization is necessary. Figure 1c, d shows polarization microscope images of nematic and smectic liquid crystals, respectively. These polarization microscope images also show colors resulting from optical anisotropy caused by molecular orientation. In this chapter, we discuss the basics of optical properties in anisotropic media for the measurement and analysis of soft matter.

R. Ozaki (✉)
Graduate School of Science and Engineering, Ehime University, Matsuyama, Ehime, Japan
e-mail: ozaki.ryotaro.mx@ehime-u.ac.jp

M. Funahashi (ed.), *Functional Soft Matter II*,
https://doi.org/10.1007/978-981-92-0100-6_5

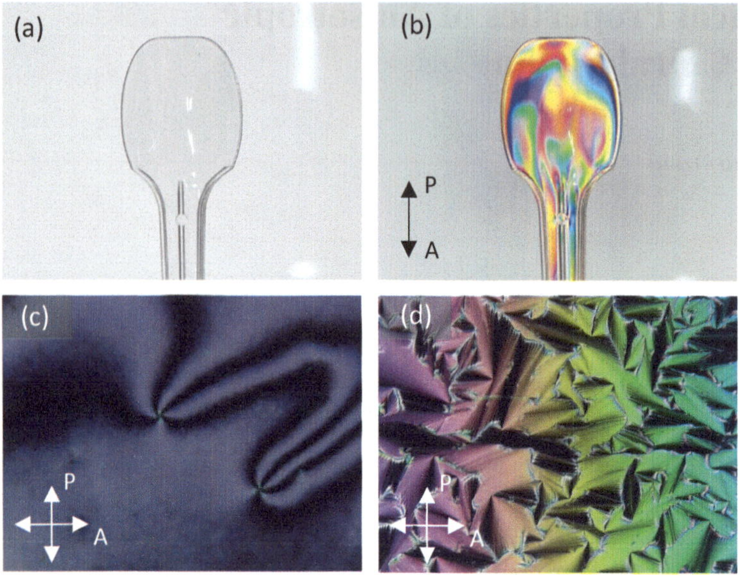

Fig. 1 **a** Transparent plastic spoon, **b** Transparent plastic spoon under parallel polarized light, **c** Nematic liquid crystal under cross polarized light, and **d** Smectic liquid crystal under cross polarized light

2 Polarization of Light

Light polarization is important for understanding the optical properties of anisotropic media. This section provides an introduction to the fundamentals of polarization. Figure 2a–c shows horizontal, vertical, and 45° linear polarizations, respectively, where the light travels along the z-axis. In this coordinate system, the horizontally polarized light corresponds to the E_x component of the electric field, while the vertically polarized light corresponds to the E_y component. Figure 2d, e shows right and left circular polarizations. The light propagates from $z = 0$ to $z = 8$ μm, and the observer is located at $z > 8$ μm. In optics, the direction of circular polarization is determined by the rotation as it appears to the observer, as indicated by the black arrow. Therefore, in optics, the light in Fig. 2d, e is defined as right and left circular polarizations, respectively [1]. However, the definition in optics is different from that in materials science and quantum mechanics [2]. The definition of circular polarization in materials science and quantum mechanics is opposite to that in optics, as shown by the red arrow. In soft materials such as liquid crystals and polymers, researchers study materials based on materials science and use optics to measure their properties, but it should be noted that the definition of circular polarization is opposite in the two fields of materials science and optics. In this chapter, we use the definition of optics to align with the definitions found in textbooks on optics or crystallography.

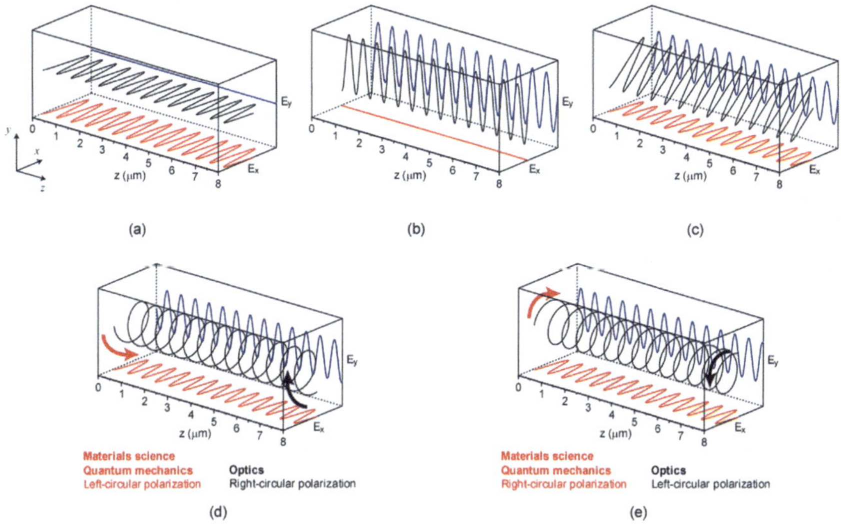

Fig. 2 Electric field profiles and their E_x and E_y components of light polarizations: **a** Horizontal polarization, **b** Vertical polarization, **c** 45° linear polarization, **d** Right circular polarization, and **e** Left circular polarization. Note that the two fields of materials science and optics have opposite definitions of circular polarization

On the other hand, the 45° linear, right circular, and left circular polarizations show different polarization states although the amplitudes of the E_x and E_y components are the same. This difference in polarization is determined by the phase difference between E_x and E_y. Figure 3a–c shows two-dimensional plots of E_x and E_y for 45° linear, right circular and left circular polarizations. In 45° linear polarization, E_x and E_y are in phase. However, in right circularly polarized light, E_x leads E_y by 90° and in left circularly polarized light, E_x lags E_y by 90°. This phase difference gives the state of polarization.

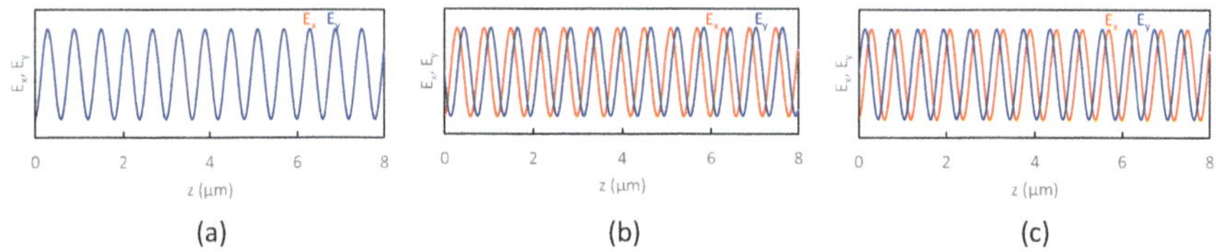

Fig. 3 E_x and E_y components of **a**45° linear polarization, **b**Right circular polarization, and **c** Left circular polarization

3 Optical Anisotropy

A medium whose refractive index is the same for all directions, such as glass or air, is called an isotropic medium. In contrast, polymers and liquid crystals oriented in a certain direction have two refractive indices depending on the molecular orientation. Such a medium is called an anisotropic medium. Figure 4a, b illustrates the index ellipsoids for isotropic and anisotropic media, respectively. The refractive index in the isotropic medium is equal in all directions, while the refractive index in the anisotropic medium has an extraordinary refractive index n_e in the y-axis direction and an ordinary refractive index n_o in the x- and z-axis directions.

When polarized light propagates through an anisotropic medium, the velocity of the light depends on the polarization direction. In Fig. 5a, b, the major axis of the index ellipsoid is oriented toward the y-axis. In this case, the velocities of E_x and E_y propagating in the z-direction are determined by n_o and n_e, respectively. The velocity of E_x is faster than that of E_y because n_o is smaller than n_e. Figure 5c shows that the velocity of E_y propagating in the z-direction is determined by the effective refractive index n', when the major axis of the index ellipsoid is inclined in the yz-plane by θ from the y-axis. In this case, E_x is not affected by the inclination and the velocity of E_x is determined by n_o. This effective refractive index n' is given by the following equation:

$$n' = \frac{n_o n_e}{\sqrt{n_e^2 \sin^2\theta + n_o^2 \cos^2\theta}} \tag{1}$$

The effective refractive index is derived when the polar angle θ is the sole variable undergoing change. It can also be calculated when the azimuthal angle ϕ is varied, either independently or in combination with θ. However, since the dielectric tensor

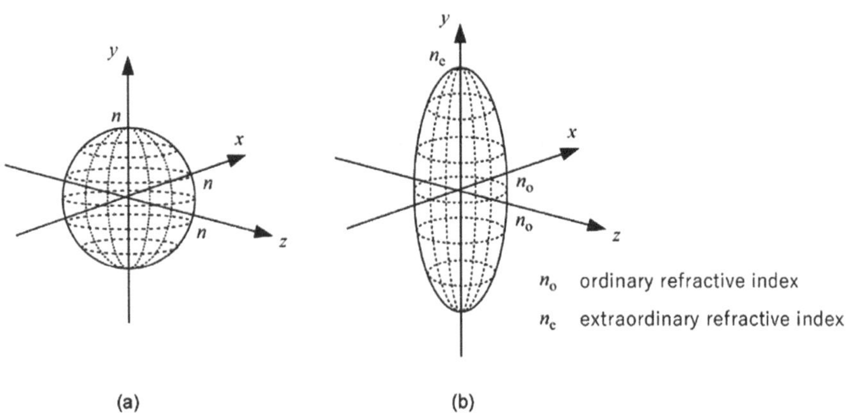

Fig. 4 Index ellipsoids for **a** isotropic and **b**anisotropic media

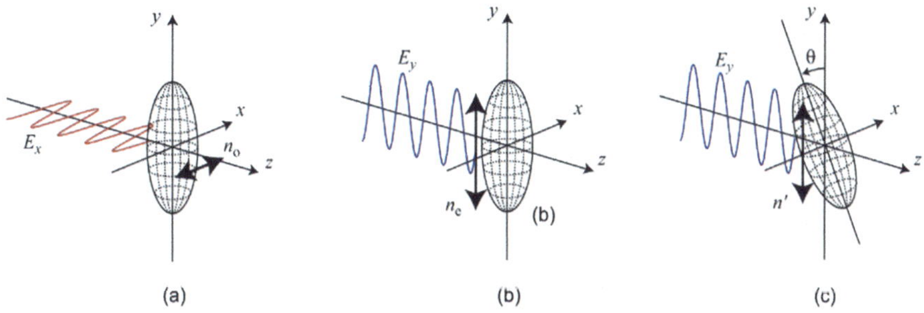

Fig. 5 Refractive index for polarized light in anisotropic medium: **a** Refractive index for E_x, **b** Refractive index for E_y, and **c**Effective refractive index for E_y when the director is tilted by θ

must be considered, a detailed explanation is not provided here. For further details regarding the dielectric tensor, please refer to a technical book on optics [3, 4].

4 Light Propagation in Anisotropic Medium

The case of a system with its major axis oriented along the y-axis, as shown in Fig. 6, is examined. In this system, 45° linearly polarized light is incident. The 45° linearly polarized light consists of E_x and E_y. The propagation velocity of E_x is c/n_o, and that of E_y is c/n_e, where c is the speed of light. Figure 7 shows the electric fields of E_x and E_y in the anisotropic medium with $n_e = 1.6$ and $n_o = 1.5$. The background is an isotropic medium with $n = 1.5$, and an incident light with 45° linear polarization enters the anisotropic medium placed from $z = 0$ to 6 μm. Before the light enters the anisotropic medium, E_x and E_y are in phase. In contrast, in the material, the phases of E_x and E_y are not the same because the velocity of E_x is faster than that of E_y. As a result, the polarization gradually changes from 45° linear polarization to right circular polarization, – 45° linear polarization, left circular polarization, and back to 45° linear polarization. For better understanding, Fig. 8 shows the electric field E_x, E_y and their superposition in a 3D view.

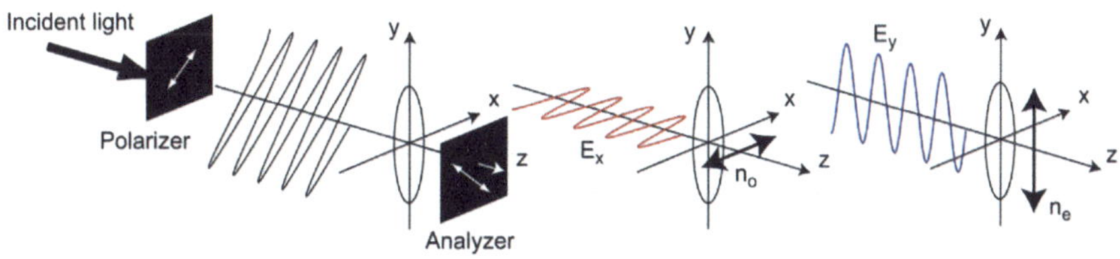

Fig. 6 Light propagation in anisotropic medium where the direction of polarization is at a 45° angle to the molecular director. The propagation velocities of E_x and E_y are c/n_o and c/n_e, respectively

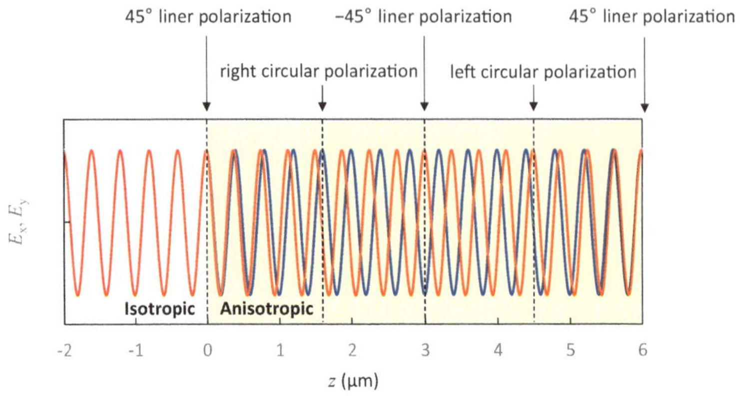

Fig. 7 E_x and E_y in the anisotropic medium with $n_e = 1.6$ and $n_o = 1.5$. λ is 600 nm in the air. E_x and E_y in the isotropic medium are in phase, while the phases of E_x and E_y are not the same as each other because the velocity of E_x is faster than that of E_y

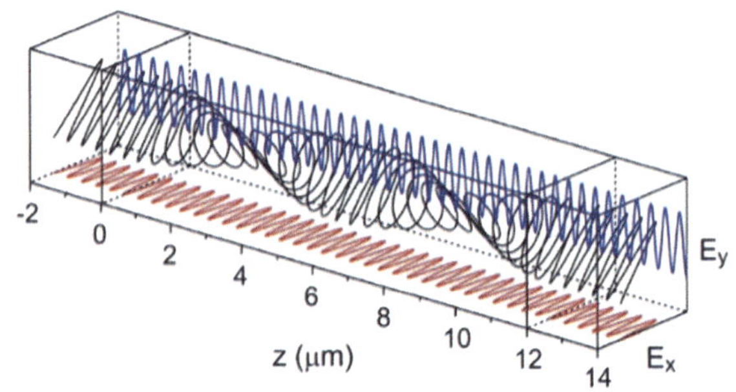

Fig. 8 Electric field profile of the E_x and E_y in the anisotropic medium with $n_e = 1.6$ and $n_o = 1.5$. The polarization gradually changes from 45° linear polarization to right circular polarization, then to − 45° linear polarization, left circular polarization, and back to 45° linear polarization

The transmittance of the light passing through the anisotropic medium obtained with an analyzer set at an angle of − 45°, as shown in Fig. 6, is given by

$$T = \frac{1}{2}\sin^2 2\phi \sin^2\left(\frac{\pi \Delta nd}{\lambda}\right) \quad (2)$$

where ϕ is an angle from the y-axis in the xy-plane, d is the thickness of the anisotropic medium, and λ is the wavelength of the incident light. The maximum transmittance is 50%, as 50% of the light is absorbed by the polarizer before entering the anisotropic medium. Figure 9 shows the transmittance for $\lambda = 600$ nm as a function of the thickness d of the anisotropic medium. The transmittance varies with increasing d. At $d = 3$ μm, the transmittance is at its maximum, as the polarization is aligned with the polarization angle of the analyzer. In contrast, at $d = 6$ μm, the transmittance is

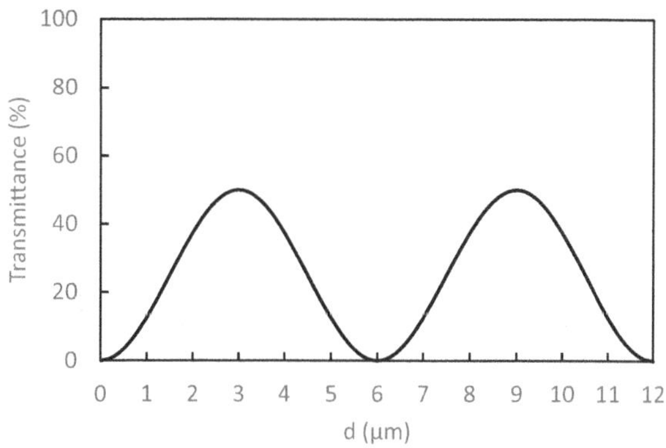

Fig. 9 Transmittance of light passing through the anisotropic medium of $\Delta n = 0.1$ as shown in Fig. 6, as a function of the thickness d of the anisotropic medium

at its minimum, as the polarization is perpendicular to the polarization angle of the analyzer.

As can be seen from the equation, the transmittance also depends on wavelength and retardation factor Δnd. Figure 10 shows transmission spectra of the anisotropic medium for $\Delta nd = 0.4$, 0.6, and 1.2. The transmission spectra exhibit peaks at different wavelengths. The peaks at different wavelengths result in the anisotropic medium appearing to have a different color to the observer. To clearly show the relation between wavelength and color, the color bar is displayed at the top of the spectrum. The spectra for $\Delta nd = 0.4$, 0.6, and 1.2 have a peak in the red, blue, and green regions, respectively. The inset shows the color of the spectrum, which is obtained through color calculation using the CIE color space. The colors of the anisotropic media with $\Delta nd = 0.6$ and 1.2 are close to the peak color because the transmittance outside the peak is relatively weak. The color for $\Delta nd = 0.4$ is orange due to the broad transmission spectrum around red and yellow regions. Consequently, the color of the anisotropic medium for polarized light is determined by the retardation factor Δnd. Figure 11 shows the dependence of the color of an anisotropic medium on Δnd. When polarized light passes through an anisotropic medium and the transmitted light is observed with a polarizer, different colors are seen from the sample depending on Δnd. In Fig. 1, we can see that the various colors are due to Δnd, which depends on the molecular orientation. In other words, if we understand the relationship between polarization and retardation color, we can determine the molecular orientation from the observed image. The transparent spoon shown in Fig. 1 exhibits the beautiful colors corresponding to the optical anisotropy due to the molecular orientation along the shape of the mold. Both nematic and smectic liquid crystals also show colors corresponding to their molecular orientation.

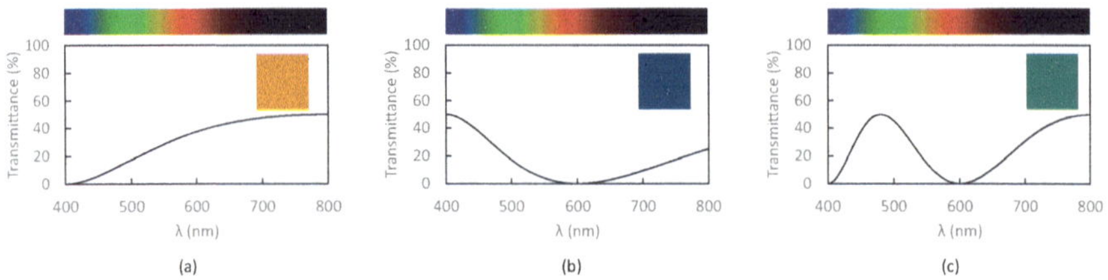

Fig. 10 Transmission spectra of the anisotropic medium for **a** $\Delta nd = 0.4$, **b** $\Delta nd = 0.6$, and **c** $\Delta nd = 1.2$. The inset shows the color of the spectrum which is obtained through color calculation using the CIE color space

Fig. 11 Color bar of the retardation Δnd

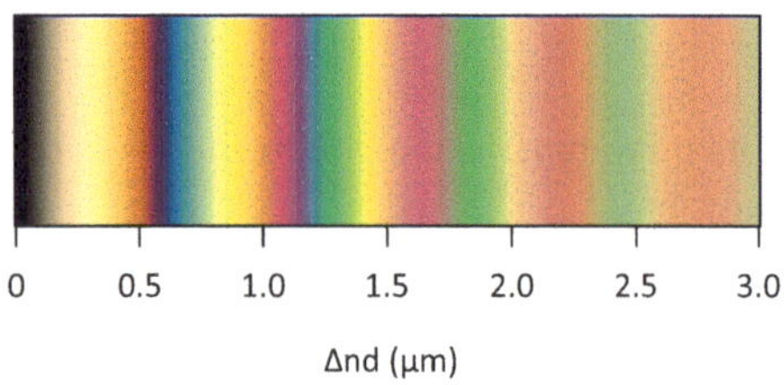

5 Light Absorption in a Material

Most materials exhibit light absorption in certain frequency bands. This section discusses the light absorption and the complex refractive index n^*. When light is absorbed in a material, the electric field in the material is attenuated exponentially. Figure 12 shows the light attenuation in media with different absorption coefficients. The refractive index of an absorbing material can be expressed as a complex number $n + i\kappa$, where the speed of light is determined by the real part n and the absorption of light depends on the imaginary part κ.

The electric field of the light propagating in an absorbing material along the z-axis is expressed as

$$E(z) = \exp\left[i\left(\frac{2\pi n^*}{\lambda}z - \omega t\right)\right] \tag{3}$$

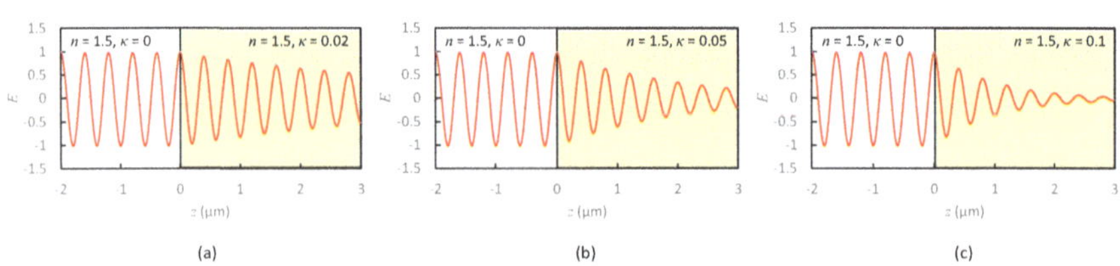

Fig. 12 Light attenuation in lossy media with different absorption coefficients: **a** $n = 1.5$ and $\kappa = 0.02$, **b** $n = 1.5$ and $\kappa = 0.05$, and **c** $n = 1.5$ and $\kappa = 0.1$

By substituting $n^* = n + i\kappa$ into the equation, $E(z)$ can be written as

$$E(z) = \exp\left\{i\left[\frac{2\pi(n + i\kappa)}{\lambda}z - \omega t\right]\right\}$$
$$= \exp\left(-\frac{2\pi\kappa}{\lambda}z\right)\exp\left[i\left(\frac{2\pi n}{\lambda}z - \omega t\right)\right] \tag{4}$$

Now we consider only the real part of this equation. We obtain

$$E(z) = \exp\left(-\frac{2\pi\kappa}{\lambda}z\right)\cos\left(\frac{2\pi n}{\lambda}z - \omega t\right) \tag{5}$$

This equation represents that the light wave expressed as a cosine function is exponentially attenuated. Thus, the introduction of the complex refractive index n^* enables the reproduction of the exponential attenuation of the electric field. Figure 13 shows the electric field profile in an absorbing material for $n = 1.5$ and $\kappa = 0.1$ at a wavelength of 600 nm. The broken, solid, and red lines represent a cosine curve, an exponential curve, and the attenuated electric field in the material, respectively. The extinction coefficient κ determines the strength of absorption.

In general, n and κ are frequency dependent. In particular, n changes significantly near the absorption peak wavelength. Figure 14 shows the frequency dispersion of n and κ for Rhodamine 6G, which has an absorption band from 450 to 600 nm [5]. The absorbance A is defined by

$$A = \log_{10}\left(\frac{I_0}{I}\right) = \log_{10}\left(\frac{1}{T}\right) \tag{6}$$

where I_0 and I are the intensities of incident and transmitted light. The calculated transmission spectra of Rhodamine 6G aqueous solution for different concentrations are shown in Fig. 15a. In this calculation, the reflection at the interface of a cell

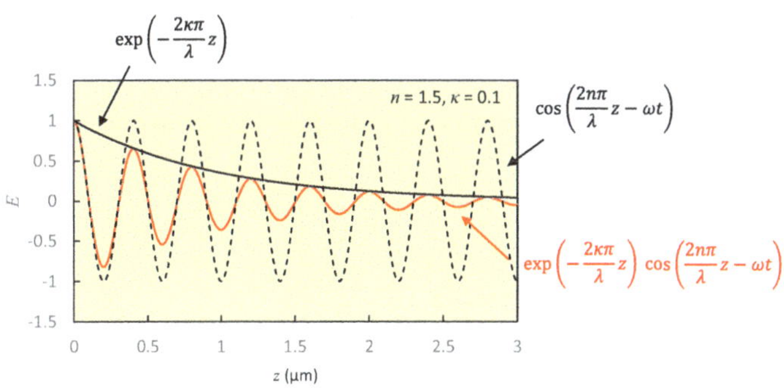

Fig. 13 Electric field profile in an absorbing material for $n = 1.5$ and $\kappa = 0.1$ at a wavelength of 600 nm

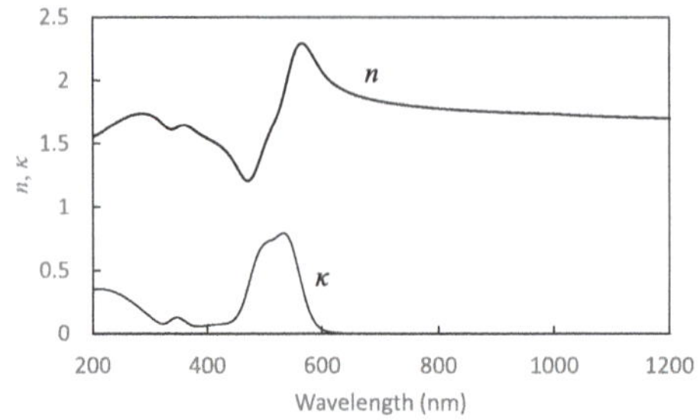

Fig. 14 Frequency dispersion of n and κ for Rhodamine 6G [5]

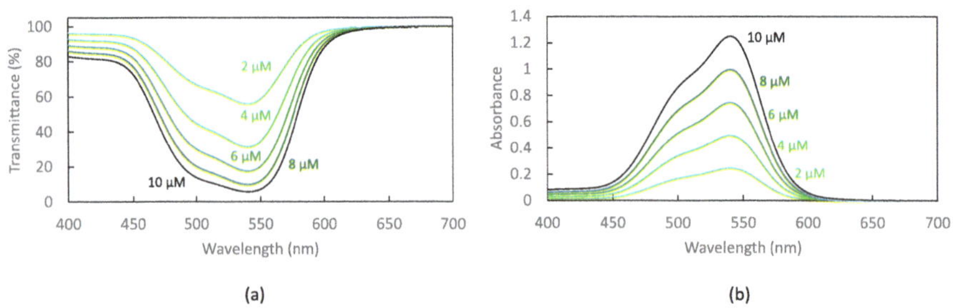

Fig. 15 **a** Calculated transmission spectra and **b** absorbance of Rhodamine 6G aqueous solution at different concentrations

is ignored for simplicity, and it is assumed that the transmission spectrum can be obtained by $T = [\exp(-2\kappa\pi d/\lambda)]^2$, where $d = 1$ cm. The transmission spectra reveal a significant decrease in transmittance as the concentration increases, particularly in the wavelength regions where absorption peaks are observed. However, it is notable that the transmission spectra do not show a constant decrease between the concentrations, indicating a non-linear relationship between concentration and transmittance reduction. In contrast, the absorbance spectra exhibit a constant increase in absorption peaks with increasing concentrations, reflecting a linear relationship between concentration and absorbance. This linearity in absorbance makes it a more reliable and straightforward parameter for understanding the absorption characteristics, as it directly correlates with the concentration, providing clearer insights into its optical behavior.

6 Cholesteric Liquid Crystals

Cholesteric liquid crystals are materials in which the liquid crystal molecules have a helical structure, as shown in Fig. 16a. The spatial modulation of the refractive index caused by the helical structure results in multiple reflections for light inside cholesteric liquid crystals. When these multiple reflections are in phase, the

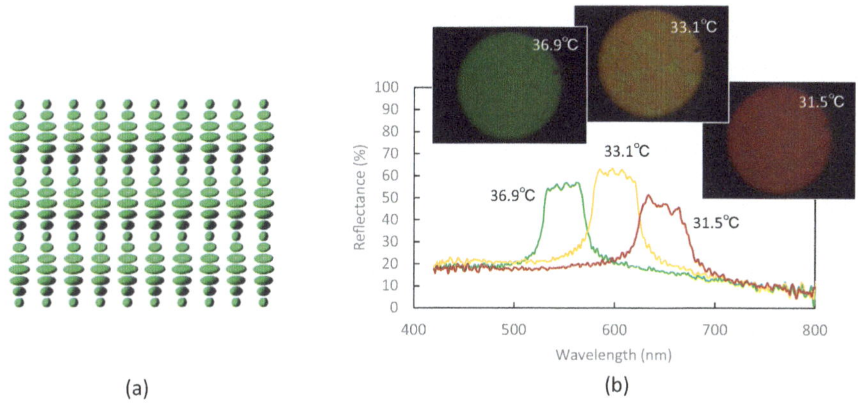

Fig. 16 **a** Schematics of cholesteric liquid crystals. **b** Temperature dependence of the selective reflection of a cholesteric liquid crystal

cholesteric liquid crystal exhibits strong light reflection, which is referred to as selective reflection. The helical pitch of the cholesteric liquid crystal is the length of 360° rotation of the molecular director. When the helical pitch is approximately equal to the wavelength of visible light, the cholesteric liquid crystal appears colored because visible light is reflected. The selective reflection is one of Bragg reflections. In contrast to the typical Bragg reflection, the reflected light is circularly polarized. Furthermore, the wavelength of the selective reflection of the cholesteric liquid crystal can be controlled by varying the pitch in response to changes in temperature and electric field. This feature is difficult to achieve with dielectric multilayers of inorganic materials, which creates a variety of active applications. Such applications include display devices, laser, thermometers, and others.

Figure 16b shows the temperature dependence of the selective reflection of a cholesteric liquid crystal. At 31.5 °C, the spectrum has a reflection band from 620 to 680 nm. This wavelength region, which corresponds to red, can be confirmed from the microscopic image. As the pitch of the cholesteric liquid crystal is shortened with increasing temperature, the selective reflection shifts to shorter wavelengths. The helical pitch of this liquid crystal shortens with increasing temperature, but conversely, for some liquid crystals, the helical pitch lengthens with increasing temperature [6, 7].

The helical pitch can be controlled by electric or magnetic fields. Figure 17a shows the calculated director profiles of the cholesteric liquid crystal when an electric field is applied perpendicular to the helix. The director profile at $E = 0$ is a sinusoidal function, and then the directors gradually align parallel to the electric field with increasing electric field. The pitch elongation caused by the application of an electric field distorts the helix, and it is no longer a sinusoidal function. Above the critical electric field E_c, the directors align completely parallel to the electric field, and the pitch length becomes infinite. The relationship between the electric field and the helical pitch is shown in Fig. 17b. The helical pitch does not show a linear

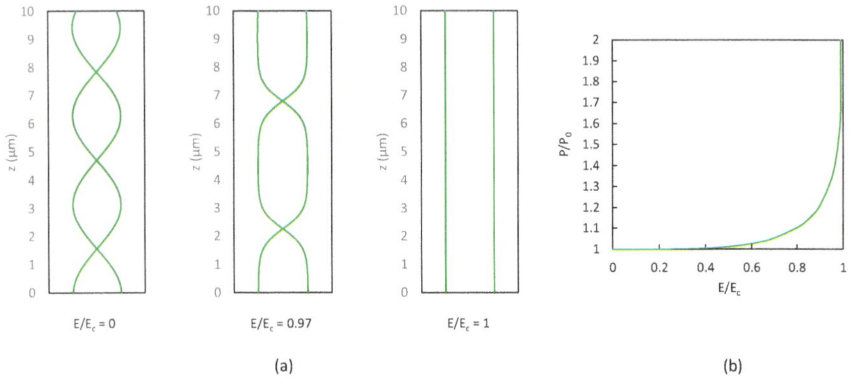

Fig. 17 **a** Director profiles of the cholesteric liquid crystal when an electric field is applied perpendicular to the helix. **b** Electric field dependence of helical pitch elongation in cholesteric liquid crystal

dependence on the electric field; instead, it elongates rapidly starting from near the threshold electric field.

The optical properties, such as reflection and transmission, of cholesteric liquid crystals are commonly analyzed numerically using Berreman's 4 × 4 matrix method or the finite-difference time-domain (FDTD) method [8–14]. Both methods solve Maxwell's equations and can also calculate light behavior in complicated systems and the angle of incidence dependence. On the other hand, the reflection and transmission spectra of a cholesteric liquid crystal for normal incidence can be calculated analytically [4]. The analytical calculation procedure is as follows.

The average refractive index of n_e and n_o and its wave vector k are written as

$$n_{avg} = \sqrt{\frac{n_e^2 + n_o^2}{2}} \tag{7}$$

and

$$k = \frac{2\pi n_{avg}}{\lambda} \tag{8}$$

The reflectance R of the cholesteric liquid crystal with a thickness of d and a helical pitch p for normal incidence is given by

$$R = \left|r^2\right| = \begin{cases} \dfrac{\kappa^2 \sinh^2 sd}{s^2 \cosh^2 sd + \left(\frac{\Delta k}{2}\right)^2 \sinh^2 sd} & \text{for} - \kappa < \frac{\Delta k}{2} < \kappa \\ \dfrac{\kappa^2 \sin^2 sd}{\left(\frac{\Delta k}{2}\right)^2 + \kappa^2 \cos^2 sd} & \text{for} \left(\frac{\Delta k}{2}\right)^2 > \kappa^2 \end{cases} \tag{9}$$

where q, Δk, κ, and s are

$$q = \frac{2\pi}{p} \tag{10}$$

$$\Delta k = 2k - 2q \tag{11}$$

$$\kappa = \frac{\pi\left(n_e^2 - n_o^2\right)}{2\lambda n_{\text{avg}}} \tag{12}$$

$$s^2 = \kappa^2 - \left(\frac{\Delta k}{2}\right)^2 \tag{13}$$

Figure 18 shows the analytically derived reflection spectra of cholesteric liquid crystals for $\Delta n = 0.1$, 0.2, and 0.3, where the other parameters are as follows: $d =$ 4 μm and $p = 400$ nm. They have a reflection band caused by selective reflection. The bandwidth becomes wider as Δn increases. The long and short wavelength edges of the reflection band for an infinitely long cholesteric liquid crystal are defined as $n_e p$ and $n_o p$, respectively. The short wavelength edges for $\Delta n = 0.1$, 0.2, and 0.3 are close to 600 nm, but not exactly 600 nm. This is because the wavelength edges of $\lambda_o = n_o p$ and $\lambda_e = n_e p$ are obtained from the ideal state of infinite length. Therefore, they are slightly different from $n_o p$ and $n_e p$ in the real system. The reflectivity of the selective reflection increases with increasing Δn. This is because the selective reflection is due to multiple reflections inside the cholesteric liquid crystal, and each individual reflection inside becomes strong for the larger Δn. The fringing of wavelengths outside the reflection band is due to interference in the cell. It might not be observed in experiments due to the non-uniformity of the cell thickness, which eliminates the interference fringes.

Figure 19 shows the thickness dependence of the reflection spectra of the cholesteric liquid crystal for $n_o = 1.5$ and $n_e = 1.7$. The reflectivity is weaker and the reflection band is wider at thinner thicknesses. With increasing thickness, the reflectivity increases and the wavelength edges of the selective reflection approach $n_o p$ and $n_e p$. The width of the interference fringe decreases as the thickness of the cell increases.

The reflection color of cholesteric liquid crystals changes depending on the incident angle. Figure 20 shows a schematic illustration of the light path for oblique incidence. The dependence of the central wavelength of the selective reflection from

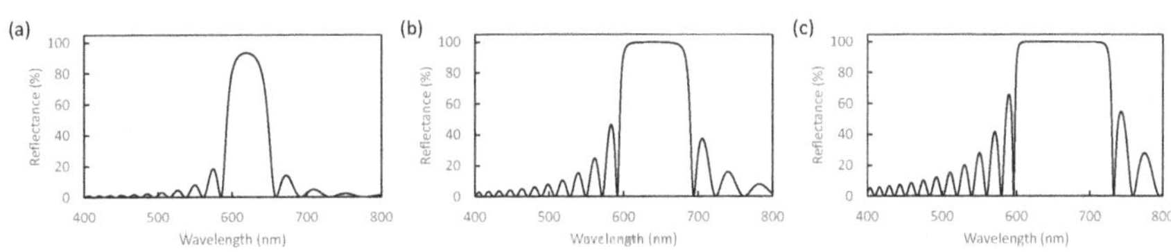

Fig. 18 Reflection spectra of cholesteric liquid crystals for **a** $\Delta n = 0.1$, **b** $\Delta n = 0.2$, and **c** $\Delta n = 0.3$, where $d = 4$ μm and $p = 400$ nm

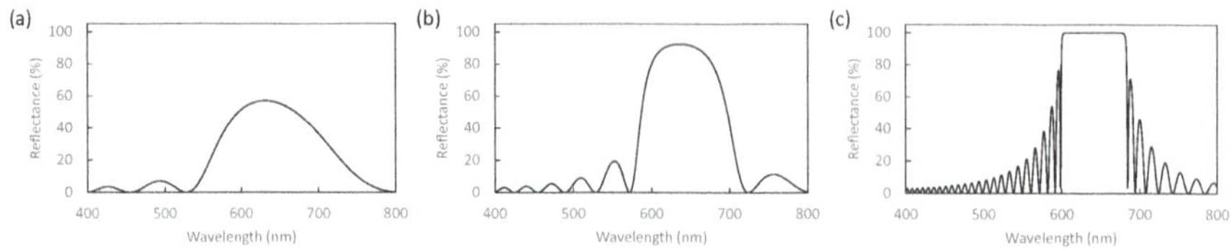

Fig. 19 Thickness dependence of reflection spectra of cholesteric liquid crystal with $n_o = 1.5$ and $n_e = 1.7$: **a** $d = 1.0\ \mu m$, **b** $d = 2.0\ \mu m$, and **c** $d = 8.0\ \mu m$

Fig. 20 Schematics of selective reflection of a cholesteric liquid crystal at oblique incidence

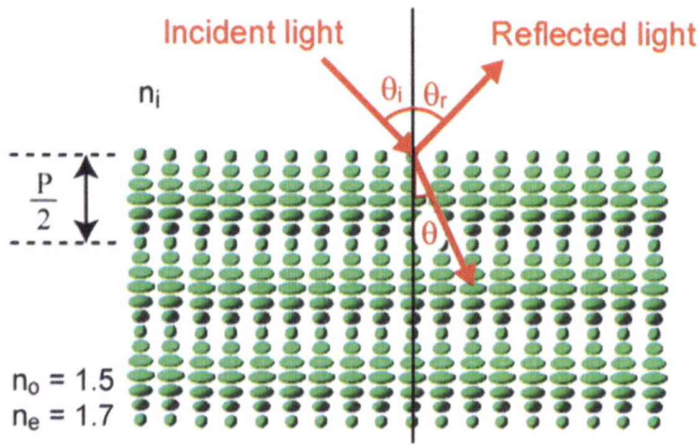

cholesteric liquid crystals is given by

$$\lambda_c = n_{avg} p \cos\theta \tag{14}$$

Note that θ is the angle in the cholesteric liquid crystal, not the angle of incidence from the air.

$$\theta = \sin^{-1}\left(\frac{n_i \sin\theta_i}{n_{avg}}\right) \tag{15}$$

Although the exact solution of the selective reflection for normal incidence exists, numerical methods are commonly used to calculate the spectrum of the selective reflection for oblique incidence. The dependence of the central wavelength of the selective reflection on the angle of incidence can be calculated from Eq. 14 [15]. For practical use, however, the dependence of the bandwidth on the angle of incidence is also of interest. The dependence of the reflection bandwidth on the angle of incidence can also be determined by the geometrical-optics approximation [16]. Figure 21 shows the incident angle dependence of the transmission spectrum of a cholesteric liquid crystal calculated by the 4×4 matrix method. Light with 45° linear polarization enters the cholesteric liquid crystal at an angle of incidence θ_i. The selective reflection shifts toward shorter wavelengths with increasing incident angle. It is well known that the selective reflection of cholesteric liquid crystals has a reflectance of 50%,

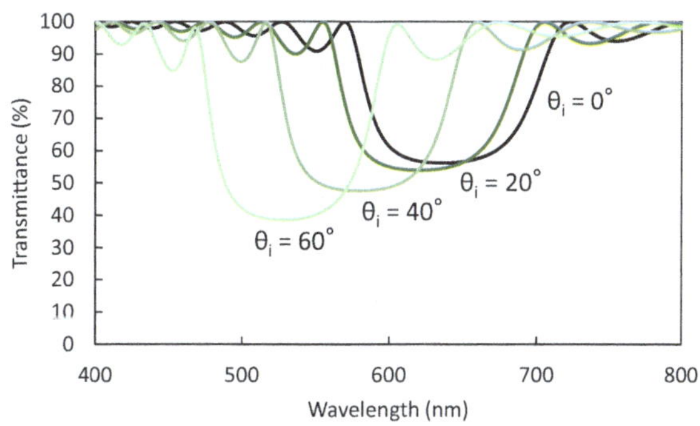

Fig. 21 Incident angle dependence of the transmission spectrum of a cholesteric liquid crystal calculated by the 4 × 4 matrix method. The calculation parameters are as follows: $n_o = 1.5$, $n_e = 1.7$, $p = 400$ nm, and $d = 2$ μm

but only for normal incidence. When the incident angle is oblique, the reflectance changes and also depends on the polarization [17, 18].

In the selective reflection of cholesteric liquid crystals, standing waves occur at the wavelength edges of the selective reflection $\lambda_o = n_o p$ and $\lambda_e = n_e p$. At these wavelengths, an electric field enhancement occurs in the material and is used for laser applications [19–21]. To understand the optical properties of the standing wave at the wavelength edge of the selective reflection, we discuss the electric field profile in the cholesteric liquid crystal. Figure 22 shows the schematic illustrations of the cholesteric liquid crystal and the electric field at $\lambda_o = n_o p$ and $\lambda_e = n_e p$, respectively. In Fig. 22a, the arrows represent electric fields of light in the cholesteric liquid crystal parallel to the major molecular axis. In this case, the light in the cholesteric liquid crystal is affected only by n_e, and the spatial length of the light is equal to the helical pitch. Since a wavelength is given by multiplying the refractive index by the spatial length of the light, we can easily understand the reason why the longer wavelength edge of the selective reflection is represented as $n_e p$. In contrast, the electric fields in Fig. 22b are perpendicular to the major molecular axis. Here, the spatial length of the light is also equal to the helical pitch, but the light in the cholesteric liquid crystal is affected only by n_o. Therefore, the shorter wavelength edge of the selective reflection is expressed as $n_o p$.

A propagating wave moves through a material as shown in Fig. 23a, while the electric field of a standing wave oscillates in the direction indicated in Fig. 23b. At wavelengths λ_e and λ_o, as shown in Fig. 22, the electric fields are always aligned parallel and perpendicular to the long axis of the molecules, respectively. This means that the electric fields at λ_e and λ_o do not propagate as shown in Fig. 23a, but instead form standing waves, as shown in Fig. 23b. Since the group velocity of the standing wave is close to zero, the incident light is confined within the cholesteric liquid crystal for an extended time. This confinement enhances the electric field, enabling applications in laser.

For a better understanding of the selective reflection, Fig. 24 shows the calculated transmission spectrum of a cholesteric liquid crystal and the electric field profiles in the cholesteric liquid crystal. These are calculated using Berreman's 4 × 4 matrix

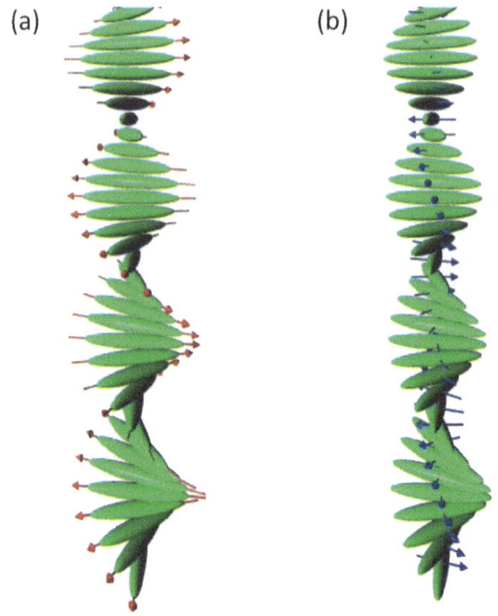

Fig. 22 Schematic of cholesteric liquid crystal molecules and directions of electric field oscillation at the wavelength edges at **a** $\lambda_e = n_e p$ and **b** $\lambda_o = n_o p$

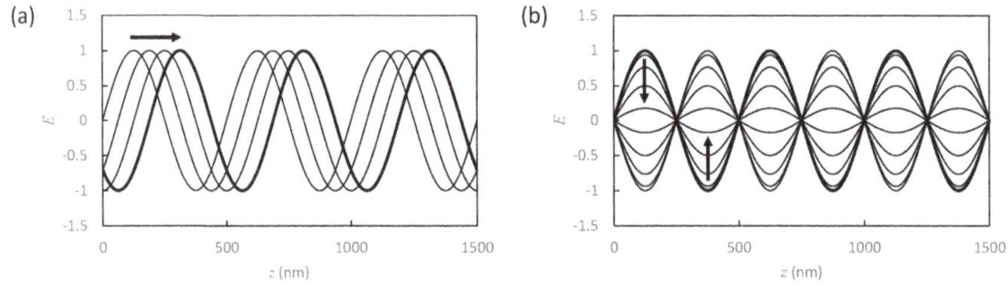

Fig. 23 Difference in oscillation between **a** propagating and **b** standing waves. In the propagating wave, the peak positions move to the positive direction, while in the standing wave, the peak positions do not move

method. This method has been widely used to determine the transmission and reflection properties of liquid crystals. Recently, a technique for calculating internal electric fields using the 4×4 matrix method has been developed [10]. Figure 24a shows the calculated transmission spectrum of cholesteric liquid crystal with a right-handed helix. The calculation parameters are as follows: $n_e = 1.7$, $n_o = 1.5$, $p = 400$ nm, $d = 4$ μm, and incident light has linear polarization. Figure 24b shows the electric field profile when incident light with right circular polarization and a wavelength of 640 nm enters. Since 640 nm corresponds to the central wavelength of the selective reflection, the incident light is reflected and decays exponentially in the cholesteric liquid crystal. Conversely, Fig. 24c shows the electric field profile for incident light with left circular polarization at the same wavelength. Here, the light passes through the cholesteric liquid crystal, as its polarization is opposite to the helix. In Fig. 24d, the electric field profile is shown for incident light with right circular polarization and a wavelength of 500 nm. This wavelength is outside the range of selective reflection, allowing the light to pass through the sample. Figure 24e shows the electric field profile for incident light with right circular polarization and a wavelength of 590 nm. In this case, the electric field is enhanced at the center of the cholesteric liquid crystal due to the formation of a standing wave.

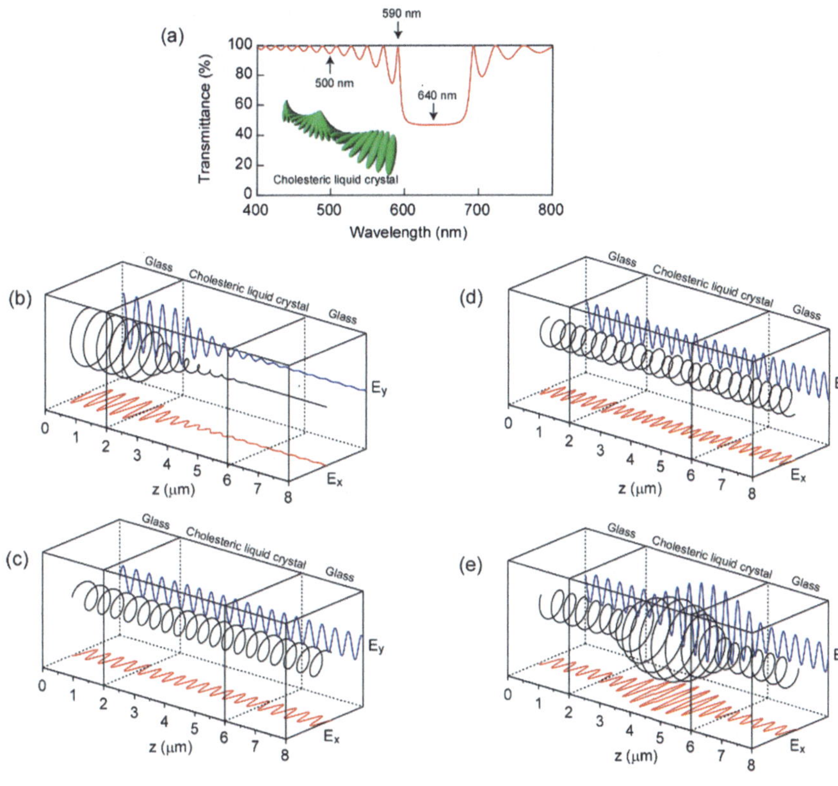

Fig. 24 **a** Transmittance of cholesteric liquid crystal. The calculation parameters are as follows: $n_o = 1.5$, $n_e = 1.7$, $p = 400$ nm, $d = 4$ μm, and the incident light is linearly polarized. Electric field profiles in the cholesteric liquid crystal: **b** right circular polarized light at $\lambda = 640$ nm, **c** left circular polarized light at $\lambda = 640$ nm, **d** right circular polarized light at $\lambda = 500$ nm, and **e** right circular polarized light at $\lambda = 590$ nm

7 Summary

This chapter serves as a detailed guide on the optical properties of anisotropic materials and chiral media. It begins by explaining optical anisotropy and its significance in various optical applications. Key concepts of light polarization are introduced, illustrating how different polarization states affect light propagation. The chapter then explores the refractive index anisotropy, explaining how the direction-dependent refractive indices influence light velocity and polarization in anisotropic media. The discussion extends to light absorption, detailing how the complex refractive index accounts for the attenuation of light in absorbing materials. Cholesteric liquid crystals are examined for their distinctive selective reflection properties, with explanations on how their helical structure affects light interaction. The chapter is supported by figures and diagrams that elucidate the theoretical concepts, making it a valuable resource for understanding the optical behavior of anisotropic materials.

References

1. E. Hecht, *Optics*, 5th edn. (Pearson, 2016)
2. R.P. Feynman, R.B. Leighton, M. Sands, *The Feynman Lectures On Physics* (Addison Wesley, 1963)
3. D.-K. Yang, S.-T. Wu, *Fundamentals of Liquid Crystal Devices* (Wiley, 2006)
4. P. Yeh, C. Gu, *Optics of Liquid Crystal Displays*, 2nd edn. (Wiley, 2009)
5. S. Seçkin, P. Singh, A. Jaiswal, T.A.F. König, Super-radiant SERS enhancement by plasmonic particle gratings. ACS Appl. Mater. Interfaces **15**, 43124–43134 (2023). https://doi.org/10.1021/acsami.3c07532
6. R.S. Pindak, C.-C. Huang, J.T. Ho, Divergence of cholesteric pitch near a smectic-*A* transition. Phys. Rev. Lett. **32**, 43–45 (1974). https://doi.org/10.1103/PhysRevLett.32.43
7. H. Kimura, M. Hosino, H. Nakano, Temperature dependent pitch in cholesteric phase. J. Phys. Colloques **40**, C3-174–C3-177 (1979). https://doi.org/10.1051/jphyscol:1979335
8. D.W. Berreman, T.J. Scheffer, Bragg Reflection of Light from Single-Domain Cholesteric Liquid-Crystal Films. Phys. Rev. Lett. **25**, 577–581 (1970). https://doi.org/10.1103/PhysRevLett.25.577
9. D.W. Berreman, Optics in stratified and anisotropic media: 4×4-matrix formulation. J. Opt. Soc. Am. **62**, 502–510 (1972). https://doi.org/10.1364/JOSA.62.000502
10. R. Ozaki, K. Nakahiro, Electric field analysis in chiral liquid crystals by Berreman's 4 × 4 matrix method. Jpn. J. Appl. Phys. **61**, 061006 (2022). https://doi.org/10.35848/1347-4065/ac65c9
11. K. Yee, Numerical solution of initial boundary value problems involving maxwell's equations in isotropic media. IEEE Trans. Antennas Propagat. **14**, 302–307 (1966). https://doi.org/10.1109/TAP.1966.1138693
12. A. Taflove, S.C. Hagness, *Computational Electromagnetics: The Finite-Difference Time-Domain Method*, 3rd edn (Artech House on Demand, 2005)
13. C.M. Titus, P.J. Bos, J.R. Kelly, E.C. Gartland, Comparison of analytical calculations to finite-difference time-domain simulations of one-dimensional spatially varying anisotropic liquid crystal structures. Jpn. J. Appl. Phys. **38**, 1488–1494 (1999). https://doi.org/10.1143/JJAP.38.1488
14. R. Ozaki, Y. Matsuhisa, H. Yoshida, K. Yoshino, M. Ozaki, Optical properties and electric field enhancement in cholesteric liquid crystal containing different periodicities. J. Appl. Phys. **100**, 023102 (2006). https://doi.org/10.1063/1.2215124

15. R. Dreher, G. Meier, A. Saupe, Selective reflection by cholesteric liquid crystals. Mol. Cryst. Liq. Cryst. **13**, 17–26 (1971). https://doi.org/10.1080/15421407108083534
16. R. Ozaki, Simple model for estimating band edge wavelengths of selective reflection from cholesteric liquid crystals for oblique incidence. Phys. Rev. E **100**, 012708 (2019). https://doi.org/10.1103/PhysRevE.100.012708
17. H. Takezoe, Y. Ouchi, M. Hara, A. Fukuda, E. Kuze, Experimental studies on reflection spectra in monodomain cholesteric liquid crystal cells: total reflection, subsidiary oscillation and its beat or swell structure. Jpn. J. Appl. Phys. **22**, 1080–1091 (1983). https://doi.org/10.1143/JJAP.22.1080
18. C. Oldano, E. Miraldi, P.T. Valabrega, Comparison between measured and calculated reflectance spectra from a monodomain cholesteric liquid crystal. Jpn. J. Appl. Phys. **23**, 802–809 (1984). https://doi.org/10.1143/JJAP.23.802
19. V.I. Kopp, B. Fan, H.K.M. Vithana, A.Z. Genack, Low-threshold lasing at the edge of a photonic stop band in cholesteric liquid crystals. Opt. Lett. **23**, 1707–1709 (1998). https://doi.org/10.1364/OL.23.001707
20. H. Finkelmann, S.T. Kim, A. Muæoz, P.P. Muhoray, B. Taheri, Tunable mirrorless lasing in cholesteric liquid crystalline elastomers. Adv. Mater. **13**, 1069–1072 (2001)
21. T. Matsui, R. Ozaki, K. Funamoto, M. Ozaki, K. Yoshino, Flexible mirrorless laser based on a free-standing film of photopolymerized cholesteric liquid crystal. Appl. Phys. Lett. **81**, 3741–3743 (2002). https://doi.org/10.1063/1.1522498

Molecular Dynamics Simulation

Go Watanabe and Yoshiki Ishii

1 Introduction

It is well known that atoms and molecules are the fundamental units of matter. Various experimental techniques—such as electron microscopy, atomic force microscopy, and X-ray diffraction—are available to visualize these building blocks. However, for functional soft materials, challenges remain in sample preparation, atomic species identification, and the observation of internal structures and the dynamics of atoms. To address these challenges at the mesoscopic scale, molecular simulation is considered a powerful and effective approach. While much of the interesting physical phenomena, properties, and functions of functional soft materials at the mesoscopic and macroscopic scales are well understood, the relation between the detailed structure of the constituent molecules and the macroscopic structure and properties of their assembled structure remains a challenge. To address this issue, it is essential to understand the structure and dynamics of the molecules in the system at the mesoscopic scale, and MD simulation is one of the most appropriate molecular simulation methods. The most common aims of MD simulation are to predict static and dynamical properties or to confirm theoretical predictions. Below, we focus on some important techniques for carrying out MD simulation and analyze its results to understand the physical properties of the system.

To perform MD simulations effectively, the simulation models and conditions must be appropriate for capturing the structure and dynamics of the target system. Setting and tuning force field parameters is especially important for functional soft materials, although it can be technically challenging and is not well documented for all material types. Therefore, the first part of this chapter introduces

G. Watanabe (✉) · Y. Ishii
Department of Data Science, School of Frontier Engineering, Kitasato University, Sagamihara, Kanagawa, Japan
e-mail: go0325@kitasato-u.ac.jp

M. Funahashi (ed.), *Functional Soft Matter II*,
https://doi.org/10.1007/978-981-92-0100-6_6

methods for accurately setting force field parameters, including both basic knowledge and advanced techniques. Subsequently, we discuss several physical parameters that should be calculated to investigate the detailed structure and dynamics of the constituent molecules. These parameters form the foundation for understanding static and dynamic properties, and additional parameters may be required depending on the specific goals of the MD simulation.

2 Force Field

In computational science, classical mechanical models used to describe molecular interactions are defined by a set of potential functions and their parameters [1–3], collectively referred to as a molecular force field. In the 1990s, Jorgensen et al. developed the OPLS force field for organic compounds such as water, alkanes, and alcohols [4, 5]. This force field has since been refined through various molecular simulations of aqueous and organic solutions. Despite employing relatively simple rules for assigning point charges and van der Waals parameters, the OPLS force field maintains robust parameter validity. Lopes-Pádua et al. extended this model to ionic liquid systems, enabling the estimation of transport coefficients in electrolyte solutions [6].

The AMBER and CHARMM force fields are also widely used in biomolecular simulations [7, 8], offering high transferability across diverse molecular structures, including peptides, proteins, base oils, and polymers. Zannoni et al. constructed both united-atom and all-atom models for liquid-crystalline compounds based on the AMBER force field to investigate phase transitions and their microscopic mechanisms [9, 10]. The general AMBER force field (GAFF) further enhances parameterization by optimizing atomic charges through quantum chemical calculations in isolated systems [11, 12]. Additionally, the DREIDING force field, which models intramolecular interactions using universal harmonic oscillator parameters, remains a valuable tool when specific chemical groups are unsupported by other force fields [13]. These advancements have significantly broadened the scope of molecular simulations for functional soft materials.

In MD simulation, the internal energy of a system is expressed as the sum of potential and kinetic energies as Φ and K, respectively. These energies are critical for determining thermodynamic quantities such as free energy via the partition function in a statistical ensemble. Within the molecular force field framework, the potential interaction energy Φ is generally composed of intramolecular and intermolecular interaction terms:

$$\begin{aligned}\Phi &= \Phi^{\text{intra}} + \Phi^{\text{inter}} \\ &= \Phi^{\text{intra}}_{\text{bond}} + \Phi^{\text{intra}}_{\text{angle}} + \Phi^{\text{intra}}_{\text{dihedral}} + \Phi^{\text{intra}}_{\text{nonbond}} + \Phi^{\text{inter}}_{\text{nonbond}} \end{aligned} \tag{1}$$

where Φ^{intra} and Φ^{inter} represent the intramolecular and intermolecular energies, respectively. The intramolecular interaction energy Φ^{intra} consists of the harmonic potentials for stretching and bending motions of covalent bonds as $\Phi^{\mathrm{intra}}_{\mathrm{bond}}$ and $\Phi^{\mathrm{intra}}_{\mathrm{angle}}$. Furthermore, the torsional motion of dihedral angle within a molecule is also considered with the periodic cosine-based function for organic molecules in $\Phi^{\mathrm{intra}}_{\mathrm{dihedral}}$. These parameters are finely tuned to reproduce potential energies and normal-mode frequencies obtained from quantum chemical calculations of isolated systems. It is important to note that harmonic oscillator models cannot accommodate bond breaking. While this limitation is acceptable for simulating organic compounds without chemical reactions, it becomes unsuitable for systems involving covalent bond recombination, such as proton transfer or photoactive compounds. In such cases, reactive force fields, machine learning/deep learning-based force fields, and first-principles calculations using density functional theory (DFT) are more appropriate [14–16].

The nonbonding interactions included in the intramolecular and intermolecular terms as $\Phi^{\mathrm{intra}}_{\mathrm{nonbond}}$ and $\Phi^{\mathrm{inter}}_{\mathrm{nonbond}}$ are often designed with the electrostatic and Lennard–Jones (LJ) interactions with the function form as follows:

$$\Phi_{\mathrm{nonbond}} = \sum_{i>j}\sum_{a,b}\left[\frac{q_{ia}q_{jb}}{4\pi\varepsilon_0 r_{iajb}} + 4\varepsilon_{iajb}\left\{\left(\frac{\sigma_{iajb}}{r_{iajb}}\right)^{12} - \left(\frac{\sigma_{iajb}}{r_{iajb}}\right)^{6}\right\}\right] \tag{2}$$

where r_{iajb} is the interatomic distance between two atoms a and b in molecule i and molecule j, respectively. q_{ia} represents the atomic charge of atom a in molecule i, and σ_{iajb} and ϵ_{iajb} mean the size and energy parameters of LJ interaction between atoms a and b. These LJ parameters of σ_{iajb} and ε_{iajb} for unlike atoms are defined as the arithmetic mean for σ_{ia} and σ_{jb}, and the geometric means for ε_{ia} and ε_{jb} in the AMBER manner (Lorentz-Berthelot rule), and the geometric means for the both parameters in the OPLS manner. In this function, the electrostatic interactions between molecules can be defined as the point-charge resolution, and the van der Waals terms are expressed with the 12-power and 6-power functions for repulsive and attractive interactions, respectively. However, from the perspectives of theoretical chemistry, the Buckingham function of an exponential repulsive term would be a rational choice rather than the LJ function to express the van der Waals interactions of atoms [17]. For example, various force fields for inorganic compounds have been designed with the form of exponential functions [18]. Nevertheless, intermolecular interactions for organic molecules can be known empirically to be softly expressed with the conformational degrees-of-freedom by change of their molecular structures even within the framework of LJ potentials. The computational cost of LJ potential further advantages with the stability of numerical algorithm. Therefore, classical simplified treatment for the molecular interactions is a powerful choice to evaluate physicochemical properties of self-assembled soft matters [19].

However, there is a theoretical limitation in the parameterization of molecular force fields based on isolated systems. For instance, in condensed phases, surrounding molecules perturb the electrostatic interactions of any given molecule, leading to electron density redistribution. This many-body interaction necessitates reoptimization of force field parameters originally designed for isolated systems [2]; otherwise, the accuracy of MD simulations may deteriorate.

These changes in electrostatic properties can be interpreted as effects of polarization and charge transfer, which are particularly pronounced in electrolyte systems. A computational approach to address this challenge is the use of polarization force fields [20, 21]. Various classical mechanical treatments—such as induced multipole moments, dynamic charge transfer, charge fluctuations within atoms-in-molecules, and spring-bonded polarizable atoms—can enhance the accuracy of electrostatic interactions and the reliability of computed physical properties [20–23].

Borodin et al. and Yethiraj et al. have developed polarization force fields that account for the effects of induced moments in ionic liquid compounds [21, 24, 25]. Nevertheless, the theoretical treatment of electrostatic interactions significantly increases the computational cost of classical MD simulations. Moreover, a larger number of force field parameters are required to accurately represent refined intermolecular interactions.

As a result, these methods may limit the applicability of molecular force fields to only nanoscopic systems and simple compounds. Therefore, the rational design of intermolecular interactions remains a critical challenge in efforts to simulate nanoscopic and mesoscopic self-assembled soft materials using large-scale MD simulations [19].

Instead of explicit considerations for polarization and charge-transfer effects, the uniform charge scaling method is often employed as an empirical approach to adjust electrostatic interactions in electrolyte systems. This technique also serves as an alternative strategy for modifying force fields in soft matter simulations. Maginn et al., Pfaendtner et al., and Acevedo et al. recommend a scaling factor of 0.80 for ionic liquids, which significantly improves the accuracy of ion transport coefficients calculated via MD simulations [26–28]. Vega et al. similarly propose modified parameters for inorganic salts dissolved in aqueous solutions, using a scaling factor of 0.85.

However, uniform charge scaling does not account for electron density redistribution when transitioning from isolated systems to condensed phases. The determination of scaling factors remains empirical and compound-specific, which may compromise the realism of MD simulations [29].

To address this limitation, we recently proposed a self-consistent modeling method that integrates classical MD and DFT calculations to update nonpolarizable force fields [30, 31]. This method iteratively adjusts atomic charges in a classical MD system using reference electron density distributions obtained from DFT. The atomic charges typically converge within three to four cycles of the self-consistent scheme, and the resulting nonpolarizable force fields have demonstrated improved accuracy in predicting energetic and transport properties of ionic liquids, particularly as candidates for battery electrolytes [30, 32, 33].

In this methodology, electrostatic properties are optimized for the condensed phase of soft materials rather than isolated molecules (Fig. 1a and b). Consequently, this scheme functions as an extension of conventional force fields for simulating

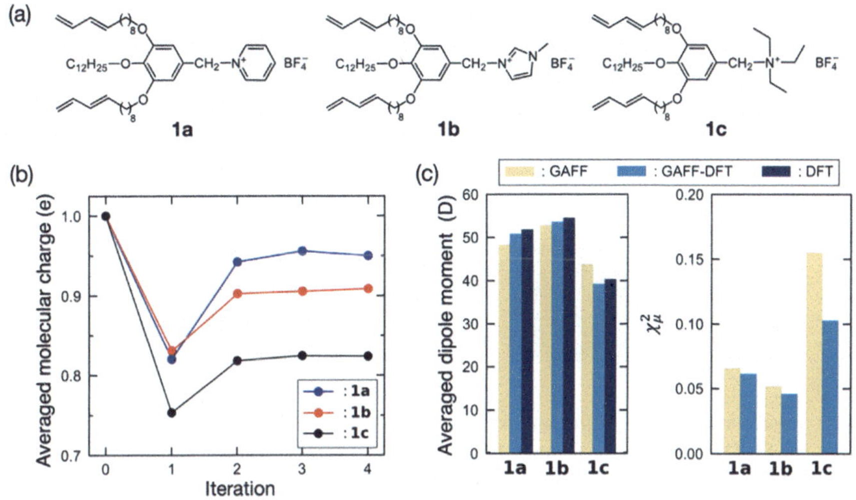

Fig. 1 **a** Molecular structures of ionic liquid-crystalline (LC) compounds **1a**, **1b**, and **1c**. **b** Convergence of the molecular charges of self-assembled ionic LCs obtained in the self-consistent scheme with MD simulations and DFT calculations. **c** Dipole moments for the cation of ionic LC compounds obtained with the force fields before and after the self-consistent modeling scheme, that is, original GAFF and GAFF-DFT. **d** Deviations of dipole moments obtained from the referenced DFT calculations. Adapted with permission from [34]. Copyright 2022 American Chemical Society

soft matter systems. Using this approach, we developed nonpolarizable force fields for self-assembled ionic liquid crystals [34, 35]. The atomic charges derived in the condensed phase enable accurate modeling of electrostatic interactions and dipole moments, as validated by DFT calculations (Fig. 1c).

Furthermore, large-scale MD simulations using the optimized force field successfully reproduced self-assembled nanostructures featuring 1D and 3D interconnected nanochannels observed in columnar and bicontinuous phases, respectively (Fig. 2) [31, 34]. These results are consistent with experimental X-ray diffraction measurements [35]. Theoretical models at all-atom resolution thus allow us to investigate nanoscopic behaviors of water and ion transport within hydrophilic nanochannels and their interfaces in self-assembled mesostructures of various soft materials [19, 36].

3 Order Parameter

When performing MD simulations of soft material systems, the orientational order of molecules is a key microscopic property that correlates with macroscopic properties measured in experiments. To confirm whether the simulated system exhibits a liquid crystal (LC) phase—a representative state of functional soft materials—it

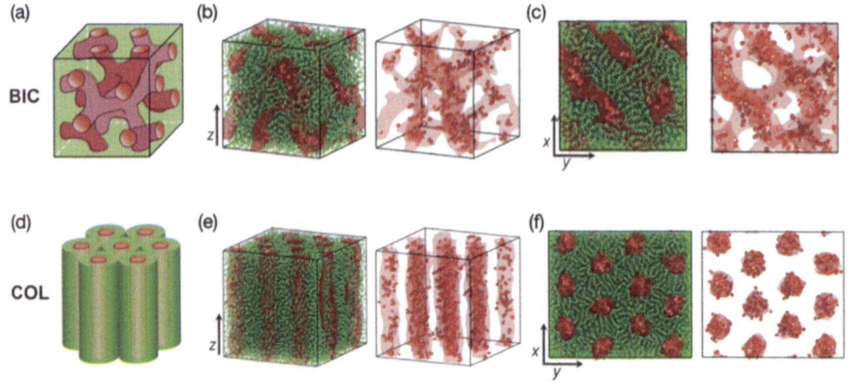

Fig. 2 **a** and **d** Schematic figures of self-assembled bicontinuous cubic and columnar LC structures. **b**, **c**, **e**, and **f** MD simulation snapshots of **1c** at 0.50 of the molar fraction of water species. Adapted with permission from [32]. Copyright 2021 American Association for the Advancement of Science

is a common practice to analyze the orientational scalar order parameter, defined as follows:

$$\langle P_2 \rangle = \frac{1}{2}\left\langle 3(\boldsymbol{u}_i \cdot \boldsymbol{n})^2 - 1 \right\rangle_i \tag{3}$$

where $\boldsymbol{u}_i$ is the long axis unit vector of molecule i, and $\boldsymbol{n}$ is the director. The long axis vector is typically defined as the vector connecting the terminal atoms of the rigid central core, or as the principal axis corresponding to the minimum moment of inertia. The scalar order parameter $\langle P_2 \rangle$ represents the average degree of molecular alignment with respect to the director. In all-atom MD simulation, P_2 can be estimated as the largest positive eigenvalue of the order parameter tensor $\boldsymbol{Q}$, defined as follows:

$$Q_{\alpha\beta} = \frac{1}{N}\sum_{i=1}^{N} \frac{1}{2}\left(3u_{i\alpha}u_{i\beta} - \delta_{\alpha\beta}\right) \tag{4}$$

where N is the total number of constituent molecules, subscripts α and β denote the Cartesian coordinates x, y, and z, and $\boldsymbol{u}_i$ is the normalized long axis vector of molecule i. Since $\langle P_2 \rangle$ can be computed at each time step of the MD simulation, its time evolution can be tracked to determine when the system reaches equilibrium (Fig. 3). A value of $\langle P_2 \rangle = 1$ indicates perfect alignment of all molecules in the same direction, while $\langle P_2 \rangle = 0$ corresponds to a completely random molecular orientation.

The temperature dependence of $\langle P_2 \rangle$ reveals the nematic–isotropic transition temperature T_{NI}, at which $\langle P_2 \rangle$ typically falls within the range of 0.3–0.4 [9]. Comparing the temperature-dependent behavior of $\langle P_2 \rangle$ between simulation and experiment serves as a primary benchmark for validating the simulation model and methodology. Based on pioneering work by Zannoni et al. [10], which successfully

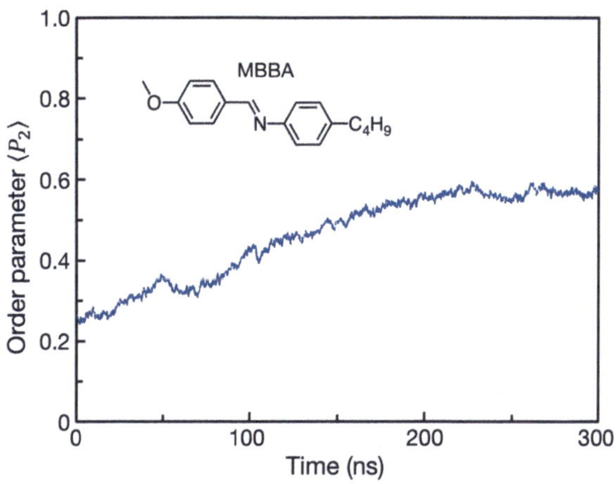

Fig. 3 Time dependence of the scalar order parameter $\langle P_2 \rangle$ for a nematic LC system composed of 2016 MBBA molecules at 300 K. The inset shows the molecular structure of MBBA

captured the nematic–isotropic phase transition of nCB, all-atom MD simulations have since demonstrated similar transitions in various liquid crystals [37, 38] (Fig. 4).

However, because the order parameter $\langle P_2 \rangle$ does not account for the head-to-tail orientation of molecules, it cannot be used to determine whether molecular dipoles tend to align in a specific direction—a feature characteristic of ferroelectric ordering. In such cases, the polar order parameter $\langle P_1 \rangle$ should be calculated using the following expression:

$$\langle P_1 \rangle = \langle \boldsymbol{u}_i \cdot \boldsymbol{n} \rangle_i = \frac{1}{N} \sum_{i=1}^{N} \cos \theta_i \tag{5}$$

where θ_i is the angle between the long axis unit vector $\boldsymbol{u}_i$ of molecule i and the director $\boldsymbol{n}$. A value of $\langle P_1 \rangle = 1$ indicates that all the molecules are perfectly aligned in the same head-to-tail direction, while $\langle P_1 \rangle = 0$ corresponds to random orientation of molecular dipoles.

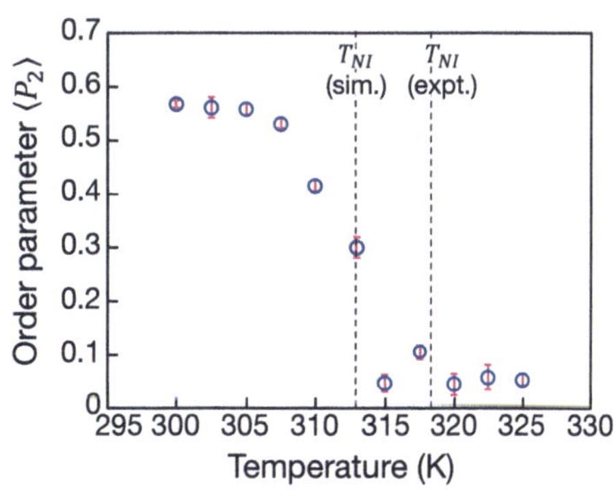

Fig. 4 Temperature dependence of the order parameter $\langle P_2 \rangle$ for a nematic LC system consisting of 2016 MBBA molecules

The alignment of specific molecules—such as liquid crystal (LC) molecules and other functional organic compounds—on surfaces or interfaces can be evaluated by analyzing the molecular orientational order parameter with respect to the surface [39–41] (Fig. 5). This type of order parameter is defined by the following equation:

$$S = \frac{1}{N}\sum_{i=1}^{N}\frac{1}{2}\left(3\cos^2\theta_i - 1\right) \tag{6}$$

where θ_i is the angle between the surface normal and a reference vector of molecule i, such as the long axis vector or the principal axis of inertia. The value of S reflects the average orientation of molecules relative to the surface:

- $S = -0.5$ indicates that all the molecules are aligned parallel to the surface,
- $S = 1$ indicates that all the molecules are oriented perpendicular to the surface,
- $S = 0$ suggests random molecular orientation.

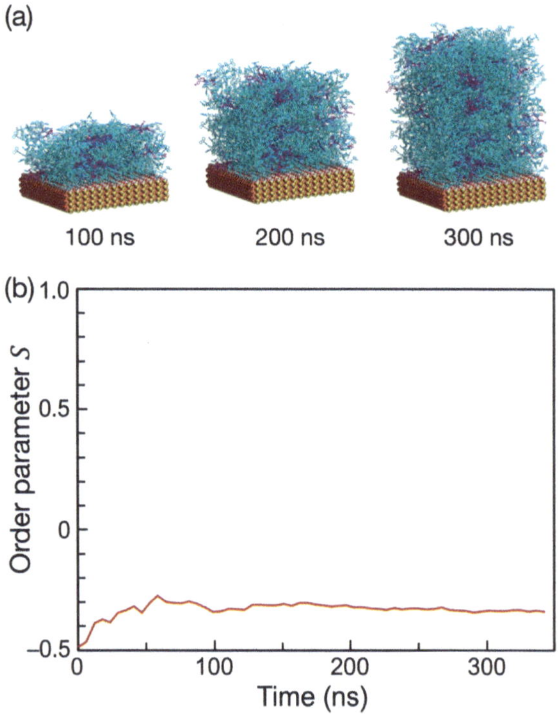

Fig. 5 **a** MD simulation snapshots of the vacuum-deposited blends of donor–acceptor π-conjugated macrocycle molecules (guest) and host CBP molecules on the SiO_2 substrate after 100 ns, 200 ns, and 300 ns at 300 K. **b** Time dependence of the orientational order parameter S averaged for guest molecules. Its time-averaged value is approximately −0.34 and it indicates that guest molecules exhibited a horizontal molecular orientation on the substrate. Adapted with permission from [41]. Copyright 2021 Wiley-VDH

4 Density Profiles

Similar to the order parameter, the density profile is a crucial physical property for quantitatively understanding the alignment and spatial distribution of molecules within an assembled structure. Density profiles can refer to various types of density, including number density, mass density, and electron density. Unless otherwise specified, the term typically refers to either the total number density or mass density, and both will be treated equivalently in the following discussion.

Number and mass density profiles describe how atoms or molecules—and their total mass—are distributed along a specific spatial direction, such as the layer normal. These profiles are useful for analyzing phase separation in complex systems composed of multiple components, microsegregated or patterned structures, and the morphology of blended films [41].

In membrane systems such as lipid bilayers, all-atom MD simulations have been successfully performed. To validate the simulated structure, the density profile of lipid molecules along the layer normal is typically calculated and compared with experimental data [42]. The density profile perpendicular to the bilayer often exhibits periodic features, reflecting the distribution of the head and tail groups of lipid molecules.

Bulk smectic liquid crystals (LCs), which possess layered structures, also require density profile analysis. The density profile in such systems is commonly modeled using the following expression:

$$\rho(z) = \rho_0 \left\{ 1 + \eta \sum_l \cos\left(\frac{2\pi l z}{d} \right) \right\} \tag{7}$$

where η is the amplitude of the density wave, d is the layer spacing, and l is the layer index. If the density profile along a given direction exhibits cosine periodicity, the system is considered to exhibit smectic layer ordering [43–46]. While MD simulation snapshots can visually suggest the presence of layered structures, the density profile provides quantitative confirmation of such ordering.

For systems consisting of a monolayer or a thin film at the water interface, hierarchical density profiles of water molecules—from the interface to the bulk—can be obtained using all-atom MD simulations. This approach is particularly valuable because experimental techniques such as X-ray or neutron scattering often face challenges in resolving such fine interfacial structures. Analyzing the density profiles of water molecules provides insights into their dynamics and hydrogen-bonding networks at interfaces with solid surfaces [46], hydrophobic layers [47], and organic molecular assemblies.

The electron density profile, which quantifies the number of electrons per unit volume as a function of position, can also be estimated from MD simulations. However, classical MD treats each atom as a point mass with a fixed partial charge and does not explicitly model electrons. As a result, it cannot reproduce a true electron density profile derived from the quantum mechanical distribution of electron clouds.

Nevertheless, under a simplified approximation—where a fixed number of electrons is assigned to each atom—an approximate electron density profile can be constructed. While this method does not take quantum effects into account, it can still be used to compare with electron densities obtained from X-ray crystallography and to validate molecular structures derived from MD simulations.

In the case of smectic liquid crystals, where LC molecules adopt multiple conformations within layered structures, combining electron density profiles from MD simulations with X-ray crystallographic data has successfully revealed detailed orientational arrangements of the constituent molecules [48] (Fig. 6).

5 Diffusion Properties

When the material being simulated is a pure liquid and soft matter, the molecules usually have diffusivity because they are not in a solid state. Then, to confirm the fluidity of the system, the self-diffusion coefficient of the molecules needs to be discussed, and it can be obtained from the all-atom MD simulation results. Depending on the system, the value of the diffusion coefficient of the molecules can be investigated experimentally, and its comparison with that obtained from the MD simulation result shows its validation. While MD simulation does not always give us the accurate diffusion coefficients corresponding to the experiments, the dynamic behavior of the molecule at different environments, such as temperature, surrounding molecules, and the external field applied, can be analyzed.

The diffusion coefficient is related to the mean square displacement (MSD), $\langle [\boldsymbol{r}(t) - \boldsymbol{r}(0)]^2 \rangle$, defined as follows:

$$D = \lim_{t \to \infty} \frac{\langle [\boldsymbol{r}_i(t+s) - \boldsymbol{r}_i(s)]^2 \rangle}{2nt} \tag{8}$$

where n is the dimensionality of the system, $\boldsymbol{r}_i(t)$ is the position of atom i at time t, and the brackets denote averaging over particles and time origins s. This relation, known as Einstein's relation, is strictly valid in the long-time limit.

In practice, MSD is calculated by averaging over all particles in the system to reduce statistical noise, often over multiple time origins during the equilibrium phase. The diffusion coefficient is then obtained from the slope a of the linear region of the MSD versus time plot:

$$D = \frac{a}{2n} \tag{9}$$

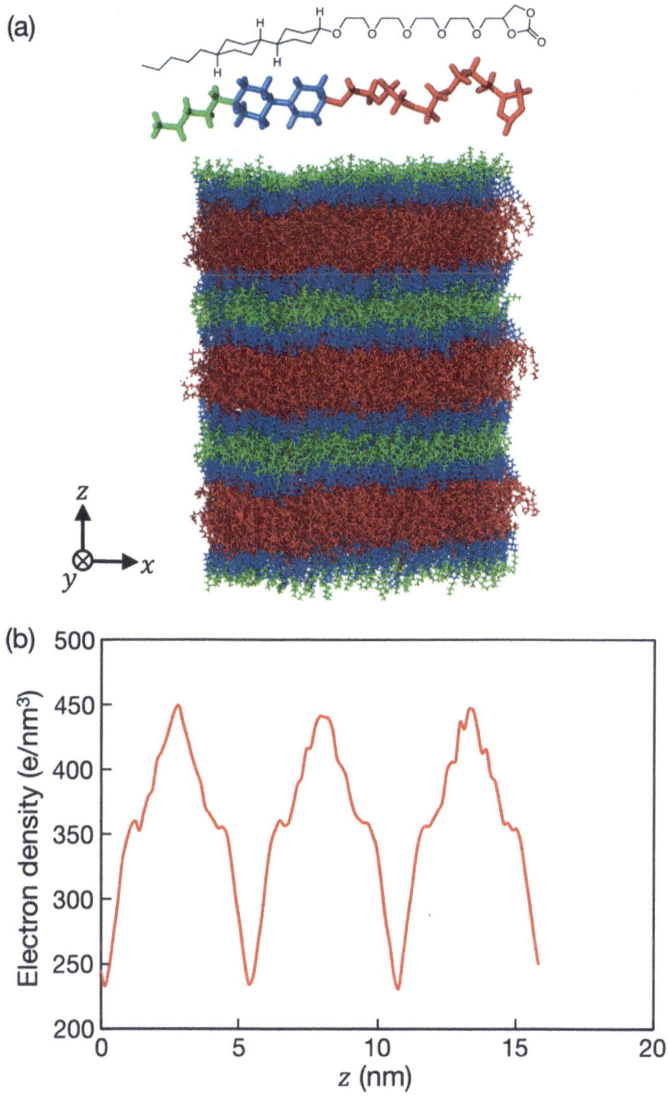

Fig. 6 **a** Molecular structure of the simulated LC molecule and MD simulation snapshot of the SmA LC phase exhibited by the molecule. **b** Electron density profile of the SmA LC phase. Adapted with permission from [48]. Copyright 2023 Wiley-VDH

The slope is determined by fitting a straight line to the linear portion of the MSD curve, typically excluding both the initial short-time region and the long-time region where deviations may occur. In fluid systems such as liquids and liquid crystals, MSD increases steadily over time. By contrast, in solid systems, MSD tends to oscillate around a constant value, which indicates restricted molecular motion.

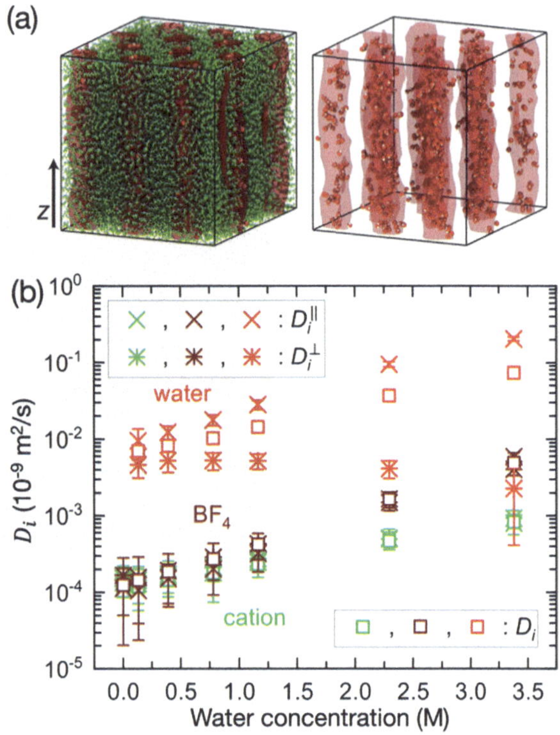

Fig. 7 **a** MD simulation snapshots for the columnar LC structure formed by the ionic LC molecule. **b** Three-dimensional diffusion coefficients and those parallel and perpendicular to the column of water, cation, and anion molecules. Adapted with permission from [31]. Copyright 2021 American Association for the Advancement of Science

To investigate anisotropic diffusion, MSD can be analyzed along specific directions or planes. In liquid crystals, where molecules exhibit orientational order, the ratio of diffusion coefficients parallel and perpendicular to the director is often calculated to identify phase transitions, such as those between nematic and isotropic phases [29, 44, 46, 49] (Fig. 7).

6 Radial Distribution Functions

The radial distribution function (RDF), denoted as $g(r)$, is a valuable tool for describing the structural organization of a system. While molecules in a crystalline phase exhibit clear positional order, those in liquid or amorphous phases lack such long-range order. RDF quantifies the probability of finding an atom within a spherical shell of thickness v at a distance r from a reference atom (Fig. 8).

The volume of the shell is given by:

$$V = \frac{4}{3}\pi(r + \Delta r)^3 - \frac{4}{3}\pi r^3$$

Fig. 8 Schematic figure of spherical shell of thickness Δr at a distance r from atom i

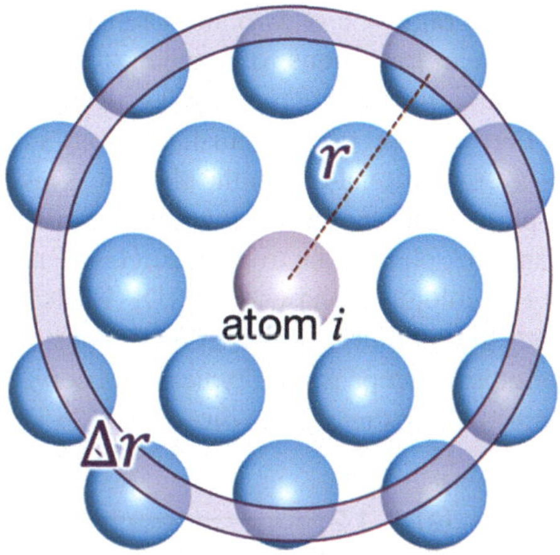

$$
\begin{aligned}
&= 4\pi r^2 \Delta r + 4\pi r \Delta r^2 + \frac{3}{4}\pi \Delta r^3 \\
&\approx 4\pi r^2 \Delta r
\end{aligned} \tag{10}
$$

Assuming the number density of atoms is ρ, the number of atoms within the shell surrounding atom i can be calculated as follows:

$$\Delta n_i(r) = \rho \cdot 4\pi r^2 \Delta r \cdot g(r) \tag{11}$$

Since ρ is N/V where N and V is the total number of atoms and the total volume of the system, respectively, and Δn_i can be averaged over all the atoms and the simulation time at the equilibrium, the RDF $g(r)$ is expressed by:

$$g(r) = \frac{1}{4\pi r^2 \Delta r} \frac{V}{N} \left[\frac{1}{N} \sum_{i=1}^{N} \Delta n_i(r) \right] \tag{12}$$

In crystalline systems, the RDF exhibits an infinite series of sharp peaks, with their spacing and intensities reflecting the characteristic lattice structure. In contrast, the RDF of a liquid displays a continuous curve with only a few short-range peaks, which gradually converge to a constant value while oscillating at longer distances. This behavior indicates the absence of long-range positional order in liquid systems.

When the simulated system lacks fluidity, as determined by diffusion coefficient analysis, the RDF can help distinguish whether the system is crystalline or glassy. The RDF in glassy systems does not exhibit the discrete, periodic peaks observed in crystals.

RDF analysis is also useful for characterizing liquid crystal (LC) phases, which can be classified according to their degree of positional order. At temperatures where the system exhibits a nematic LC phase, the RDF appears liquid-like, typically showing only one or two short-range peaks (Fig. 9a) [50, 51]. In contrast, when the system transitions to a smectic LC phase, multiple peaks emerge in the RDF, indicating long-range positional order and the presence of layered structures.

Furthermore, to distinguish between SmA and SmB phases, which differ in their in-layer positional order, the sharpness and height of peaks in the two-dimensional in-plane RDF should be examined. SmB phases exhibit sharper and higher peaks than SmA phases due to the hexagonal ordering of molecules [48] (Fig. 9b).

For columnar phases, the RDF is also effective in identifying positional order along the column axis. In the one-dimensional RDF calculated parallel to the column, periodic sharp peaks are typically observed, indicating that the molecules are well aligned along the column direction [52–54].

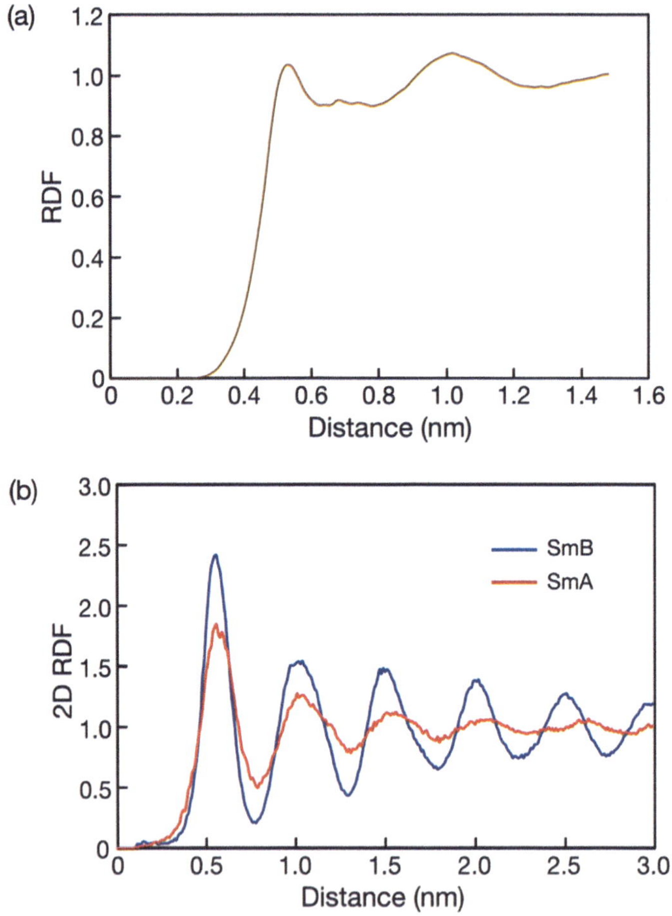

Fig. 9 **a** RDF for the nematic LC phase formed by MBBA molecules. **b** Two-dimensional RDFs in the smectic layer of SmA (red line) and SmB (blue line). Adapted with permission from [49] and [48]. Copyright 2018 American Chemical Society and 2023 Wiley-VDH

7 Orientational Correlation Functions

To understand how the alignment or orientation of rigid molecules changes depending on the distance, it is useful to analyze the orientational correlation function (OCF). The OCF indicates the degree to which the orientation of a single molecule is correlated with the orientation of another molecule at a distance r and is provided by the following equation:

$$g_2(r) = \left\langle \frac{1}{2}\left(3\cos^2\theta_{ij}(r) - 1\right) \right\rangle \tag{13}$$

where θ_{ij} is the angle between the orientation vectors of reference molecule i and molecule j, located within a spherical shell of thickness Δr at a distance r and the brackets indicate ensemble averaging over all pairs of molecules i and j. The orientation vector is typically defined using the long axis or principal inertia axis of each molecule, disregarding head-to-tail directionality.

An OCF value of 1 indicates perfect orientational correlation between molecules at distance r, while a value of 0 suggests no correlation. By plotting $g_2(r)$ as a function of distance, one can assess how rapidly orientational correlations decay with increasing separation (Fig. 10).

It is also possible to calculate the dipole correlation function in the system, defined as follows:

$$g_1(r) = \left\langle \cos\theta_{ij}(r) \right\rangle \tag{14}$$

Plotting $g_1(r)$ as a function of distance reveals how dipole–dipole correlations decay with increasing separation. This analysis is useful for determining whether

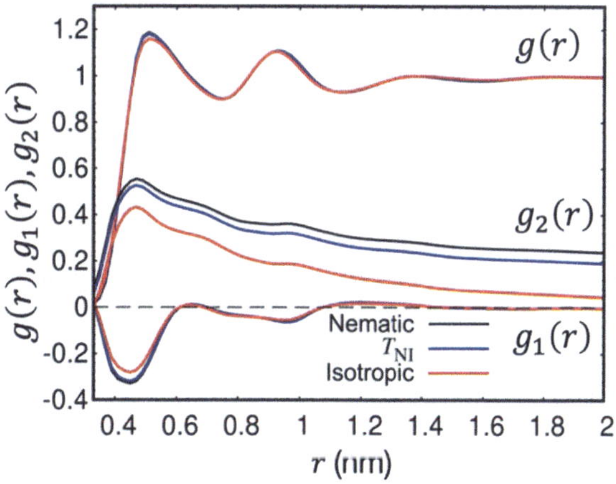

Fig. 10 Radial distribution function $g(r)$ and orientational correlation function $g_1(r)$ and $g_2(r)$ for 5CB. Adapted with permission from [55]. Copyright 2025 AIP Publishing

the system exhibits global or local polarity, and for understanding how dipole–dipole interactions influence molecular alignment within the assembled structure [10, 45, 54, 55].

8 B-factors

The B-factor, also known as the temperature factor or Debye–Waller factor, was originally introduced in X-ray crystallography. In all-atom MD simulations, it is commonly used to evaluate how much each atom in a crystal structure or molecular assembly fluctuates from its average position during the equilibration phase.

The B-factor for atom i is calculated using the following formula:

$$B_i = \frac{8}{3}\pi^2\Delta_i^2 \tag{15}$$

where Δ_i is the root mean square fluctuation (RMSF) of atom i. The RMSF is estimated as follows:

$$\Delta_i = \sqrt{\frac{1}{T}\sum_{j=1}^{T}\left|\boldsymbol{r}_i(t_j) - \bar{\boldsymbol{r}}_i\right|^2} \tag{16}$$

Here, T is the number of time steps, $\boldsymbol{r}_i(t_j)$ is the position of atom i, and $\bar{\boldsymbol{r}}_i$ is the average position of atom i over the simulation period. This analysis is performed during the equilibrium phase of the MD simulation.

For functional soft materials, particularly organic crystals, the B-factor is a key metric for comparing the thermal stability of different crystal structures of organic molecular derivatives or analogues. It can also be used to assess the stability of molecular assemblies under varying conditions, such as temperature, molecular alignment (or arrangement), and structural form.

By mapping B-factor values onto molecular structures using a color-coded scheme, one can visually identify which regions of a molecule are rigid and which are flexible [56–58] (Fig. 11).

Although B-factors are included in crystal structure data obtained from experimental techniques such as X-ray crystallography, they reflect a combination of dynamic fluctuations and static disorder. Importantly, these fluctuations do not always correspond to molecular behavior at room temperature or under practical conditions. Therefore, for crystalline or molecular assembly structures of organic molecules, it is essential to perform all-atom MD simulations and analyze the B-factors derived from the simulation results to obtain more realistic insights into molecular flexibility (Fig. 11).

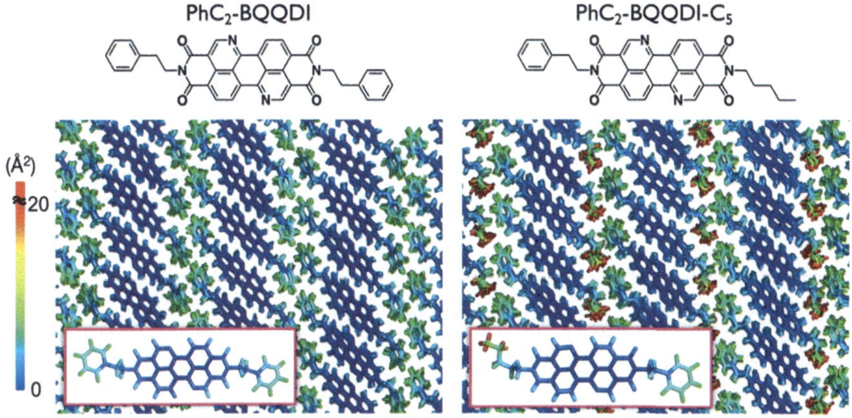

Fig. 11 MD simulation snapshots of two organic semiconductors, PhC_2-BQQDI and PhC_2-BQQDI-C_5, color-coded by the B-factor (Å) values obtained from the trajectories after reaching the equilibrium. The magnitude of B-factors is represented by the color-coded scale bar ranging from blue (small value) to red (large value). Adapted with permission from [56]. Copyright 2023 Wiley–VCH

9 Concluding Remarks

With the continuous advancement of computational power and high-performance computing technologies, the temporal and spatial scales, accessible to all-atom MD simulations, are steadily expanding. However, simulating phenomena on the millimeter or millisecond scale remains beyond current capabilities.

Since no universal methodology can be applied to all mesoscale systems, it is essential to carefully design simulation models and conditions that are appropriate for the specific system under investigation. To validate the results obtained from MD simulations, a variety of physical parameters—such as molecular arrangement, positional order, and dynamic behavior—should be analyzed. These parameters must be selected based on the phase or state of the system and the specific objectives of the simulation.

References

1. M.P. Allen, D.J. Tildesley, *Computer Simulation of Liquids, Oxford Science* (1989). https://doi.org/10.1007/BF00646086
2. A.J. Stone, *The Theory of Intermolecular Forces* (Oxford University Press, 1996). https://doi.org/10.1093/acprof:oso/9780199672394.001.0001
3. D. Frenkel, B. Smit, *Understanding Molecular Simulation*, 2nd edn. (Academic Press, 2002). https://doi.org/10.1016/B978-0-12-267351-1.X5000-7

4. W.L. Jorgensen, J.D. Madura, C.J. Swenson, Optimized intermolecular potential functions for liquid hydrocarbons. J. Am. Chem. Soc. **106**, 6638–6646 (1984). https://doi.org/10.1021/ja00334a030
5. W.L. Jorgensen, D.S. Maxwell, J. Tirado-Rives, Development and testing of the OPLS all-atom force field on conformational energetics and properties of organic liquids. J. Am. Chem. Soc. **118**, 11225–11236 (1996). https://doi.org/10.1021/ja9621760
6. J.N. Canongia Lopes, J. Deschamps, A.A.H. Pádua, Modeling ionic liquids using a systematic all-atom force field. J. Phys. Chem. B **108**, 2038–2047 (2004). https://doi.org/10.1021/jp0362133
7. C.I. Bayly, K.M. Merz, D.M. Ferguson, W.D. Cornell, T. Fox, J.W. Caldwell, P.A. Kollman, P. Cieplak, I.R. Gould, D.C. Spellmeyer, A second generation force field for the simulation of proteins, nucleic acids, and organic molecules. J. Am. Chem. Soc. **117**, 5179–5197 (1995). https://doi.org/10.1021/ja00124a002
8. A.D. MacKerell, D. Bashford, M. Bellott, R.L. Dunbrack, J.D. Evanseck, M.J. Field, S. Fischer, J. Gao, H. Guo, S. Ha, D. Joseph-McCarthy, L. Kuchnir, K. Kuczera, F.T.K. Lau, C. Mattos, S. Michnick, T. Ngo, D.T. Nguyen, B. Prodhom, W.E. Reiher, B. Roux, M. Schlenkrich, J.C. Smith, R. Stote, J. Straub, M. Watanabe, J. Wiórkiewicz-Kuczera, D. Yin, M. Karplus, All-atom empirical potential for molecular modeling and dynamics studies of proteins. J. Phys. Chem. B **102**, 3586–3616 (1998). https://doi.org/10.1021/jp973084f9
9. R. Berardi, L. Muccioli, C. Zannoni, Can nematic transitions be predicted by atomistic simulations? A computational study of the odd-even effect. ChemPhysChem **5**, 104–111 (2004). https://doi.org/10.1002/cphc.200300908
10. G. Tiberio, L. Muccioli, R. Berardi, C. Zannoni, Towards in silico liquid crystals. Realistic transition temperatures and physical properties for n-cyanobiphenyls via molecular dynamics simulations. ChemPhysChem **10**, 125–136 (2009). https://doi.org/10.1002/cphc.200800231
11. J. Wang, R.M. Wolf, J.W. Caldwell, P.A. Kollman, D.A. Case, Development and testing of a general amber force field. J. Comput. Chem. **25**, 1157–1174 (2004). https://doi.org/10.1002/jcc.20035
12. D.A. Case, H.M. Aktulga, K. Belfon, D.S. Cerutti, G.A. Cisneros, V.W.D. Cruzeiro, N. Forouzesh, T.J. Giese, A.W. Götz, H. Gohlke, S. Izadi, K. Kasavajhala, M.C. Kaymak, E. King, T. Kurtzman, T.-S. Lee, P. Li, J. Liu, T. Luchko, R. Luo, M. Manathunga, M.R. Machado, H.M. Nguyen, K.A. O'Hearn, A.V. Onufriev, F. Pan, S. Pantano, R. Qi, A. Rahnamoun, A. Risheh, S. Schott-Verdugo, A. Shajan, J. Swails, J. Wang, H. Wei, X. Wu, Y. Wu, S. Zhang, S. Zhao, Q. Zhu, T.E. Cheatham, D.R. Roe, A. Roitberg, C. Simmerling, D.M. York, M.C. Nagan, K.M. Merz, AmberTools. J. Chem. Inf. Model. **63**, 6183–6191 (2023). https://doi.org/10.1021/acs.jcim.3c01153
13. S.L. Mayo, B.D. Olafson, W.A. Goddard, DREIDING: a generic force field for molecular simulations. J. Phys. Chem. **94**, 8897–8909 (1990). https://doi.org/10.1021/j100389a010
14. T.P. Senftle, S. Hong, M.M. Islam, S.B. Kylasa, Y. Zheng, Y.K. Shin, C. Junkermeier, R. Engel-Herbert, M.J. Janik, H.M. Aktulga, T. Verstraelen, A. Grama, A.C.T. van Duin, The ReaxFF reactive force-field: development, applications and future directions. Npj Comput. Mater. **2**, 15011 (2016). https://doi.org/10.1038/npjcompumats
15. O.T. Unke, S. Chmiela, H.E. Sauceda, M. Gastegger, I. Poltavsky, K.T. Schütt, A. Tkatchenko, K.R. Müller, Machine learning force fields. Chem. Rev. **121**, 10142–10186 (2021). https://doi.org/10.1021/acs.chemrev.0c01111
16. S. Takamoto, C. Shinagawa, D. Motoki, K. Nakago, W. Li, I. Kurata, T. Watanabe, Y. Yayama, H. Iriguchi, Y. Asano, T. Onodera, T. Ishii, T. Kudo, H. Ono, R. Sawada, R. Ishitani, M. Ong, T. Yamaguchi, T. Kataoka, A. Hayashi, N. Charoenphakdee, T. Ibuka, Towards universal neural network potential for material discovery applicable to arbitrary combination of 45 elements. Nat. Commun. **13**, 1–6 (2022). https://doi.org/10.1038/s41467-022-30687-9
17. J.N. Israelachvili, *Intermolecular and Surface Forces*, 3rd edn. (Academic Press, 2010). https://doi.org/10.1016/C2009-0-21560-1
18. C. Massobrio, J. Du, M. Bernasconi, P.S. Salmon, *Molecular Dynamics Simulations of Disordered Materials* (Springer International Publishing, 2015). https://doi.org/10.1007/978-3-319-15675-0

19. K. Jeong, S. Jeong, S. Lee, C.Y. Son, Predictive molecular models for charged materials systems: from energy materials to biomacromolecules. Adv. Mater. **35**, 1–29 (2023). https://doi.org/10.1002/adma.202204272
20. M. Salanne, Simulations of room temperature ionic liquids: from polarizable to coarse-grained force fields. Phys. Chem. Chem. Phys. **17**, 14270–14279 (2015). https://doi.org/10.1039/C4CP05550K
21. D. Bedrov, J.-P. Piquemal, O. Borodin, A.D. MacKerell, B. Roux, C. Schröder, Molecular dynamics simulations of ionic liquids and electrolytes using polarizable force fields. Chem. Rev. **119**, 7940–7995 (2019). https://doi.org/10.1021/acs.chemrev.8b00763
22. H. Nakano, T. Yamamoto, S. Kato, A wave-function based approach for polarizable charge model: systematic comparison of polarization effects on protic, aprotic, and ionic liquids. J. Chem. Phys. **132**, 2038–2047 (2010). https://doi.org/10.1063/1.3298873
23. K. Goloviznina, Z. Gong, A.A.H. Padua, The CL&Pol polarizable force field for the simulation of ionic liquids and eutectic solvents. WIREs Comput. Mol. Sci. **12**, 1–16 (2022). https://doi.org/10.1002/wcms.1572
24. O. Borodin, Polarizable force field development and molecular dynamics simulations of ionic liquids. J. Phys. Chem. B **113**, 11463–11478 (2009). https://doi.org/10.1021/jp905220k
25. E. Choi, J.G. McDaniel, J.R. Schmidt, A. Yethiraj, First-principles, physically motivated force field for the ionic liquid [BMIM][BF4]. J. Phys. Chem. Lett. **5**, 2670–2674 (2014). https://doi.org/10.1021/jz5010945
26. Y. Zhang, E.J. Maginn, A simple AIMD approach to derive atomic charges for condensed phase simulation of ionic liquids. J. Phys. Chem. B **116**, 10036–10048 (2012). https://doi.org/10.1021/jp3037999
27. K.G. Sprenger, V.W. Jaeger, J. Pfaendtner, The general AMBER force field (GAFF) can accurately predict thermodynamic and transport properties of many ionic liquids. J. Phys. Chem. B **119**, 5882–5895 (2015). https://doi.org/10.1021/acs.jpcb.5b00689
28. B. Doherty, X. Zhong, S. Gathiaka, B. Li, O. Acevedo, Revisiting OPLS force field parameters for ionic liquid simulations. J. Chem. Theory Comput. **13**, 6131–6135 (2017). https://doi.org/10.1021/acs.jctc.7b00520
29. K. Cui, A. Yethiraj, J.R. Schmidt, Influence of charge scaling on the solvation properties of ionic liquid solutions. J. Phys. Chem. B **123**, 9222–9229 (2019). https://doi.org/10.1021/acs.jpcb.9b08033
30. Y. Ishii, N. Matubayasi, Self-consistent scheme combining MD and order-N DFT methods: an improved set of nonpolarizable force fields for ionic liquids. J. Chem. Theory Comput. **16**, 651–665 (2020). https://doi.org/10.1021/acs.jctc.9b00793
31. Y. Ishii, N. Matubayasi, G. Watanabe, T. Kato, H. Washizu, Molecular insights on confined water in the nanochannels of self-assembled ionic liquid crystal. Sci. Adv. **7**, 1–15 (2021). https://doi.org/10.1126/sciadv.abf0669
32. L. Hakim, Y. Ishii, K. Matsumoto, R. Hagiwara, K. Ohara, Y. Umebayashi, N. Matubayasi, Transport properties of ionic liquid and sodium salt mixtures for sodium-ion battery electrolytes from molecular dynamics simulation with a self-consistent atomic charge determination. J. Phys. Chem. B **124**, 7291–7305 (2020). https://doi.org/10.1021/acs.jpcb.0c04078
33. L. Hakim, Y. Ishii, N. Matubayasi, Spatial-decomposition analysis of electrical conductivity in mixtures of ionic liquid and sodium salt for sodium-ion battery electrolytes. J. Phys. Chem. B **125**, 3374–3385 (2021). https://doi.org/10.1021/acs.jpcb.1c00372
34. Y. Ishii, N. Matubayasi, H. Washizu, Nonpolarizable force fields through the self-consistent modeling scheme with MD and DFT methods: from ionic liquids to self-assembled ionic liquid crystals. J. Phys. Chem. B **126**, 4611–4622 (2022). https://doi.org/10.1021/acs.jpcb.2c02782
35. T. Sakamoto, T. Ogawa, H. Nada, K. Nakatsuji, M. Mitani, B. Soberats, K. Kawata, M. Yoshio, H. Tomioka, T. Sasaki, M. Kimura, M. Henmi, T. Kato, Development of nanostructured water treatment membranes based on thermotropic liquid crystals: molecular design of sub-nanoporous materials. Adv. Sci. **5**, 1 9 (2018). https://doi.org/10.1002/advs.201700405
36. T. Kato, J. Uchida, Y. Ishii, G. Watanabe, Aquatic functional liquid crystals: design, functionalization, and molecular simulation. Adv. Sci. **11**, 1–13 (2024). https://doi.org/10.1002/advs.202306529

37. F. Chami, M.R. Wilson, V.S. Oganesyan, Molecular dynamics and EPR spectroscopic studies of 8CB liquid crystal. Soft Matter **8**, 6823–6833 (2012). https://doi.org/10.1039/c2sm25429h
38. N.J. Boyd, M.R. Wilson, Validating an optimized GAFF force field for liquid crystals: T NI predictions for bent-core mesogens and the first atomistic predictions of a dark conglomerate phase. Phys. Chem. Chem. Phys. **20**, 1485–1496 (2018). https://doi.org/10.1039/C7CP07496DL
39. Y.-F. Mo, C.-L. Yang, Y.-F. Xing, M.-S. Wang, X.-G. Ma, L.-Z. Wang, Effects of silica surface on the ordered orientation of polyethylene: a molecular dynamics study. Appl. Surf. Sci. **311**, 273–278 (2014). https://doi.org/10.1016/j.apsusc.2014.05.055
40. C. Tonnelé, M. Stroet, B. Caron, A.J. Clulow, R.C.R. Nagiri, A.K. Malde, P.L. Burn, I.R. Gentle, A.E. Mark, B.J. Powell, Elucidating the spatial arrangement of emitter molecules in organic light-emitting diode films. Angew. Chem. Int. Ed. **56**, 8402–8406 (2017). https://doi.org/10.1002/anie.201610727
41. S. Shikita, G. Watanabe, D. Kanouchi, J. Saito, T. Yasuda, Alternating donor-acceptor π-conjugated macrocycle exhibiting efficient thermally activated delayed fluorescence and spontaneous horizontal molecular orientation. Adv. Photonics Res. **2**, 2100021 (2021). https://doi.org/10.1002/adpr.202100021
42. S. Moradi, A. Nowroozi, M. Shahlaei, Shedding light on the structural properties of lipid bilayers using molecular dynamics simulation: a review study. RSC Adv. **9**, 4644–4658 (2019). https://doi.org/10.1039/C8RA08441F
43. A. Pizzirusso, M. Savini, L. Muccioli, C. Zannoni, An atomistic simulation of the liquid-crystalline phases of sexithiophene. J. Mater. Chem. **21**, 125–133 (2011). https://doi.org/10.1039/c0jm01284j
44. M.F. Palermo, A. Pizzirusso, L. Muccioli, C. Zannoni, An atomistic description of the nematic and smectic phases of 4-n-octyl-4′ cyanobiphenyl (8CB). J. Chem. Phys. **138**, 204901 (2013). https://doi.org/10.1063/1.4804270
45. W. Cao, B. Senthilkumar, V. Causin, V.P. Swamy, Y. Wang, G. Saielli, Influence of the ion size on the stability of the smectic phase of ionic liquid crystals. Soft Matter **16**, 411–420 (2020). https://doi.org/10.1039/c9sm02115a
46. M. Salanne, P.A. Madden, Polarization effects in ionic liquids. Langmuir **26**, 16896 (2010). https://doi.org/10.1021/la000009e
47. H. Fan, D.E. Resasco, A. Striolo, Amphiphilic silica nanoparticles at the decane-water interface: insights from atomistic simulations. Langmuir **27**, 5264–5274 (2011). https://doi.org/10.1021/la200428r
48. K. Hamaguchi, H. Lu, S. Okamura, S. Kajiyama, J. Uchida, S. Sato, G. Watanabe, Y. Ishii, H. Washizu, G. Ungar, T. Kato, Reentrant 2D nanostructured liquid crystals by competition between molecular packing and conformation: potential design for multistep switching of ionic conductivity. ChemPhysChem **24**, 1–11 (2023). https://doi.org/10.1002/cphc.202200927
49. J. Yoshida, S. Tamura, K. Hoshino, H. Yuge, H. Sato, A. Yamazaki, S. Yoneda, G. Watanabe, Comprehensive understanding of host- and guest-dependent helix inversion in chiral nematic liquid crystals: experimental and molecular dynamics simulation study. J. Phys. Chem. B **122**, 10615–10626 (2018). https://doi.org/10.1021/acs.jpcb.8b07653
50. G. Watanabe, J. Saito, Y. Fujita, Y. Tabe, Molecular dynamics simulation study of two-dimensional diffusion behavior in smectic liquid crystalline monolayers. J. Phys. Soc. Japan **82**, 084603 (2013). https://doi.org/10.7566/JPSJ.82.084603
51. N.J. Boyd, M.R. Wilson, Optimization of the GAFF force field to describe liquid crystal molecules: the path to a dramatic improvement in transition temperature predictions. Phys. Chem. Chem. Phys. **17**, 24851–24865 (2015). https://doi.org/10.1039/c5cp03702f
52. V. Marcon, T. Vehoff, J. Kirkpatrick, C. Jeong, D.Y. Yoon, K. Kremer, D. Andrienko, Columnar mesophases of hexabenzocoronene derivatives I. Phase transitions. J. Chem. Phys. **129**, 094505 (2008). https://doi.org/10.1063/1.2969763
53. V. Marcon, J. Kirkpatrick, W. Pisula, D. Andrienko, Supramolecular structure of perylene tetracarboxdiimides. Phys. Status Solidi Basic Res. **245**, 820–824 (2008). https://doi.org/10.1002/pssb.200743436

54. J. Peláez, M. Wilson, Molecular orientational and dipolar correlation in the liquid crystal mixture E7: a molecular dynamics simulation study at a fully atomistic level. Phys. Chem. Chem. Phys. **9**, 2968–2975 (2007). https://doi.org/10.1039/b614422e
55. S. Ogita, Y. Ishii, G. Watanabe, H. Washizu, K. Kim, N. Matubayasi, Atomistic analysis of nematic phase transition in 4-cyano-4′-n-alkyl biphenyl liquid crystals: sampling for the first-order phase transition and the free-energy decomposition. J. Chem. Phys. **162**, (2025). https://doi.org/10.1063/5.0242416
56. C.P. Yu, S. Kumagai, M. Tsutsumi, T. Kurosawa, H. Ishii, G. Watanabe, D. Hashizume, H. Sugiura, Y. Tani, T. Ise, T. Watanabe, H. Sato, J. Takeya, T. Okamoto, Asymmetrically functionalized electron-deficient π-conjugated system for printed single-crystalline organic electronics. Adv. Sci. **10**, 1–8 (2023). https://doi.org/10.1002/advs.202207440
57. S. Sato, B. Dhara, D. He, D. Miyajima, G. Watanabe, Molecular dynamics simulation to predict assembly structures of bowl-shaped π-conjugated molecules. Chem. Commun. **61**, 4951–4954 (2025). https://doi.org/10.1039/D4CC06482H
58. K. Nitta, Y. Shoji, T. Fukushima, G. Watanabe, Molecular dynamics insights into orientation and hexagonal ordering of tripodal triptycenes on solid surfaces. Nanoscale Horiz. **11**(4), 999–1005 (2026).https://doi.org/10.1039/D5NH00837A

Interfacial Analysis of Organic Devices

Keishiro Tahara and Tomofumi Kadoya

1 Introduction

Improving the performance of organic devices is essential for next-generation electronics [1, 2]. Organic devices have numerous advantages, including lightweightness, flexibility, affinity for the human body, the use of low-cost materials, and simple manufacturing processes. These advantages are primarily due to the use of organic materials as device components. The first proposition in the field of organic electronics involved replacing inorganic materials with organic ones. This proposition was partially resolved by the accumulation of knowledge and technology in this field. Furthermore, the features of organic devices that are superior to those of their inorganic counterparts have been established.

Whether inorganic or organic, no substance can manifest any electronic device function as a single component. All devices are fabricated by laminating multiple components to achieve the desired operation. Representative examples of organic devices include organic field effect transistors (OFETs), organic light emitting diodes (OLEDs), and organic photovoltaics (OPVs), the device configurations of which are shown in Fig. 1. Organic materials play major roles in the operation of an organic device. Charge carriers are transported across the channels formed by organic semiconductors (OSCs) in an OFET, three organic layers are often sandwiched to produce electroluminescence in an OLED, while organic donor and acceptor layers are used for photoelectric conversion in an OPV.

To induce an electric field, each device contains electrodes made of Si, Au, Ag, Al, Ca, or indium tin oxide (ITO), with most electrodes composed of inorganic materials and metal oxides. Although these inorganic components can be replaced by organic

K. Tahara (✉)
Faculty of Engineering and Design, Kagawa University, Takamatsu, Kagawa, Japan
e-mail: tahara.keishiro@kagawa-u.ac.jp

T. Kadoya
Department of Chemistry, Konan University, Kobe, Hyogo, Japan

M. Funahashi (ed.), *Functional Soft Matter II*,
https://doi.org/10.1007/978-981-92-0100-6_7

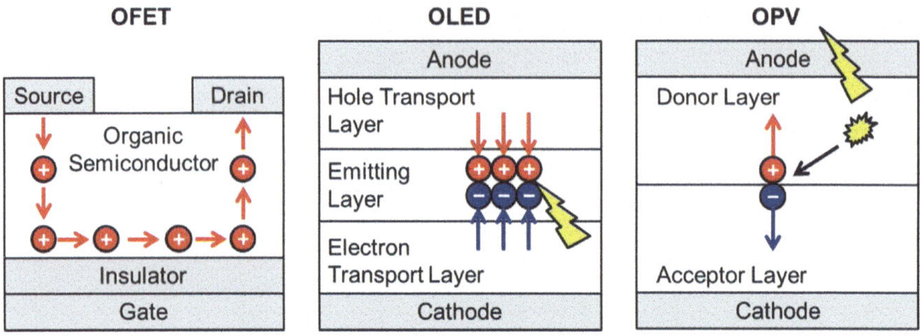

Fig. 1 Typical configurations and operations of organic devices. Inorganic components are shaded light gray

materials in the future, most current organic devices contain interfaces between organic and inorganic components. In particular, the process of charge injection from the metal electrode to the OSC is important for both OFETs and OLEDs.

In general, charges and excitons are localized in OSCs because they are covalent or noncovalent assemblies of organic constituents with low dielectric constants. Consequently, important processes proceed in the region very close to the interface (within a few nanometers) during device operation [3]. The device configuration of an OFET consists of an insulator film stacked with a gate electrode. A thermal oxide film of SiO_2 is widely used as the gate insulator film and is formed on the Si wafer surface by oxidizing agents. Charge transport occurs at the interface between the OSC layer and the gate insulator film (Fig. 1). Injected holes and electrons are transferred to the emitting layer in an OLED where they recombine, while charge separation and recombination occur at the interface between the donor and acceptor layers in an OPV. Hence, controlling interfacial structures and features is critical for attaining an improved understanding of elementary processes and enhancing the performance of organic devices.

Hereafter, we focus on the interfaces between OFET materials [4, 5]. OFETs are classified into four types of configurations based on the relative replacement of electrodes with OSC layers (Fig. 2). A gate electrode is located below an OSC in the bottom-gate type and above it in the top-gate type. The source and drain electrodes are located above the OSC in the top-contact type and below it in the bottom-contact type. Accordingly, there are two types of OSC-containing interface: electrode/OSC, where charge injection proceeds from the electrode into the semiconductor (I in Fig. 2), and insulator/OSC, where charge carriers are transported in the OSC (II in Fig. 2). The structures and features of these interfaces largely depend on the combination of materials, methods for forming the OSC thin film and constructing the interface, and device configuration. Differences in interfacial structures and features significantly impact OFET device performance. Hence, interfaces need to be precisely characterized to clarify the structure–function relationship.

In Sect. 2, we discuss methods used to analyze two types of OFET interface containing small-molecule OSCs (I and II in Fig. 2). First, we provide a general

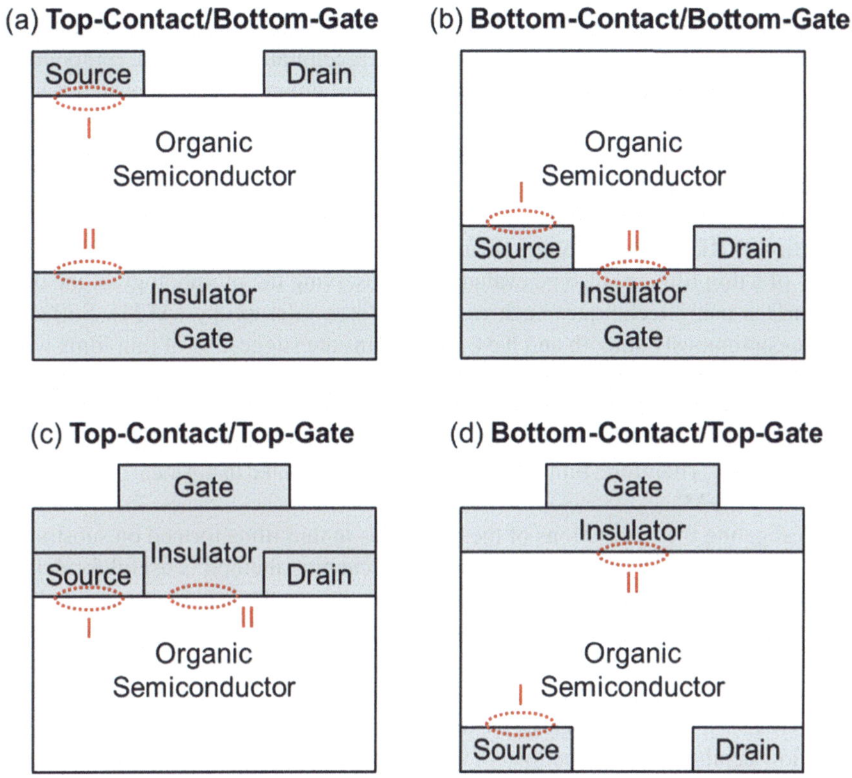

Fig. 2 Typical OFET configurations

statement regarding the analytical methods, after which we briefly introduce photoelectron spectroscopy, which has not been mentioned much in other chapters. In Sect. 3, we discuss the details of relatively recent interface examples.

2 Methods for Analyzing Interfaces Between Organic Transistor Materials

2.1 General Statement

Forming thin films of highly crystalline OSCs is essential for the fabrication of OFET devices. The quality and crystallinity of a thin film are highly dependent on the film forming method. Two types of method are used to produce thin film OSCs: dry and wet. The dry method involves growing a thin film in the gas phase. For example, vacuum deposition has the advantage of producing thin films with high crystallinity;

however, the film formation rate is slow, and the introduction and maintenance of the apparatus are costly. The wet method involves the precipitation of organic compounds from a solution, as exemplified by spin-coating and dip-coating. The resultant film crystallinity is inferior to that obtained by the dry method; however, films are formed faster. The crystallinity of a thin film largely depends on the type and properties of the substrate (electrode or insulating film).

The crystallinity of a thin film produced on a substrate can be evaluated by X-ray diffraction (XRD) analysis. Sharp diffraction peaks indicate high crystallinity. The quality of a thin film can also be evaluated by observing the morphology of the thin film surface using techniques such as atomic force microscopy (AFM). Surfaces that are continuously smooth and have large grains are suggestive of thin films with few defects. The principles of XRD and scanning probe microscopy are provided in other chapters. In the conventional θ–2θ XRD scanning method, diffraction data are acquired from the crystal plane parallel to the substrate surface, providing information on the bulk film. Therefore, thin films should also be evaluated using local information obtained by AFM.

Investigating the orientations of the molecules in thin films formed on substrates is also important. The conventionalθ–2θ XRD scanning method is useful for determining the preferred orientation of the crystal-growth process on the substrate, molecular orientation in the thin film, and interplanar spacing. Recently, XRD has been used to evaluate the in-plane structures of organic thin films. The angle of the incident X-rays relative to the sample is fixed at approximately 1° or less in grazing incidence XRD (GIXRD), while the angle of the detector is changed. Because X-rays penetrate to shallow depths, the contributions of the organic thin film to the diffraction pattern are larger than those of the substrate. In addition, AFM observations of the external shapes of the microcrystals on the substrate provide information on the molecular orientation of the crystalline film. Crystal habits are often exposed when crystals are grown at low degrees of supersaturation; consequently, molecular orientation can be inferred from the external forms of the microcrystals. In addition, the height of a molecular step on a flat crystal plane can be measured. Hence, analyzing the cross-sectional profile enables the spacing between constituent molecules to be quantified. Again, the combined use of XRD and AFM is useful for obtaining more precise insight into molecular orientation.

A fundamental problem in investigating the interface between an organic thin film and a substrate is that the interface is not exposed to the surface, but is covered and buried with organic matter and substrate which are continuous from the interface. Therefore, the actual interface must be examined in a manner that avoids physical contact. However, the actual interface is subject to the influence of the covering process of the substrate surface with the organic matter and the corresponding temporary properties. The vacuum deposition method involves sublimation, adsorption, the aggregation of small organic molecules, and crystal growth. Interface formation is extensions of these processes. Hence, precise analyses of these elementary processes are important for obtaining a true picture of the interface. In particular, a thorough investigation of adsorption and aggregation is advantageous for understanding the interactions between the substrate surface and small organic molecules. AFM can effectively determine the interplanar spacings in ultrathin films comprising a few molecular layers and specific crystal grains. In addition, an experimental environment for high-intensity light using synchrotron radiation has recently been developed, leading to the advanced acquisition of in-plane diffraction patterns using X-ray irradiation at minute angles close to the critical angle for total reflection. This advanced measurement technique enables the evaluation of the in-plane structure of monolayers with a thickness of less than a few nanometers and island-like molecular assemblies during the thin film forming process. Conventionally, modified substrates have been subjected to ex-situ measurements, which require film formation to be temporarily interrupted and the sample to be withdrawn and replaced. In recent years, observing the film growth process in situ has served as a powerful tool for precisely probing elementary processes.

2.2 *Photoelectron Spectroscopy*

As mentioned previously, in an OFET, charges are injected from the electrode to the OSC, followed by charge transport at the OSC/insulator interface. The positions of the energy levels at these interfaces needs to be determined to understand the electronic structure that governs device operation. Photoelectron spectroscopy is one of the most reliable methods for measuring these values [6, 7]. This spectroscopy utilizes the photoelectric effect and electron emission from the surface of a material upon light irradiation (Fig. 3a). A He gas discharge lamp (He I 21.2 eV; He II 40.8 eV) is often used as the photon source in ultraviolet photoelectron spectroscopy (UPS), while an Al K_α X-ray anode is used (1.49 keV) in X-ray photoelectron spectroscopy (XPS). The resultant photoelectrons are analyzed in terms of the emission angle θ and kinetic energy E_{kin} at extremely low pressure. E_{kin} is expressed as $E_{\text{kin}} = h\upsilon - E_{\text{B}} - \Phi$, where $h\upsilon$ is the photon energy, E_{B} the binding energy relative to the Fermi level, and Φ the work function of the sample. Photoelectrons are generated from the energy levels occupied by electrons, as exemplified by the valence bands in an inorganic compound, and the highest occupied molecular orbital (HOMO) and core levels in organic molecules (Fig. 3b). In UPS, secondary electrons with small E_{kin} values are detected owing to inelastic scattering before escaping from the sample surface into the vacuum. The threshold on the low-energy side of a UPS spectrum is referred to as the "secondary electron cutoff," which corresponds to the vacuum level of the sample and is the origin of E_{kin}. The relative positions of the Fermi level of the metal, the HOMO levels of the organic molecule, and the vacuum level can be determined when UPS is applied to an OSC/metal interface [8–10]. Based on these values, the ionization potential and hole injection barrier can also be calculated.

The use of synchrotron radiation in photoelectron spectroscopy is useful for the experiments with a wide range of photon energy, as well as those with a polarized light source. In addition, the angle-resolved method enables the direct observation of energy–momentum relationships. These advanced techniques enable the band

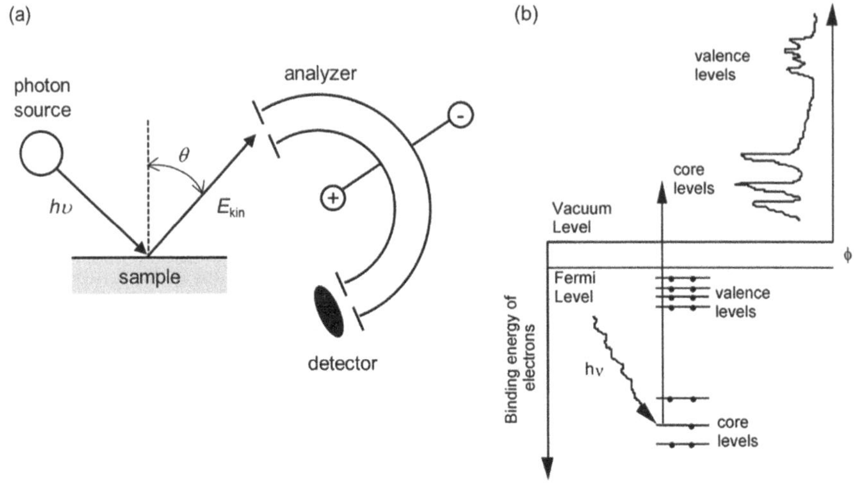

Fig. 3 Schematic representations of **a** a photoelectron spectrometer and **b** photoelectron spectroscopy. Adapted with permission from [7]. Copyright 2010 Elsevier

dispersion in an OSC thin film formed on a substrate to be experimentally determined [11–13].

3 Examples of the Analysis of Small Molecular Organic Semiconductor Films

3.1 DBTDT Derivatives on Substrates

Herein, we introduce a representative example that shows that the crystallinity of a OSC film depends significantly on the underlying substrate. Dibenzo[*d*,*d'*]thieno [3,2-*b*;4,5-*b'*]dithiophene (DBTDT) is a thienoacene-based p-type OSC that exhibits high hole mobility in an OFET [14]. To prepare a bottom-contact OFET, C6-DBTDT, a DBTDT derivative with two hexyl groups (Fig. 4a), was deposited on a SiO_2 substrate with an Au electrode pattern [15], and the SiO_2 region of the substrate was coated with hexamethyldisilazane (HMDS), a common silane coupling reagent. AFM revealed that the thin film surface of C6-DBTDT on HMDS-coated SiO_2 was smooth and showed good continuity (Fig. 4b). In contrast, a morphological discontinuity was observed in film surface on Au. Smaller grains were observed on the Au surface than the SiO_2 surface, which indicated that the film formed on Au is inferior to that formed on SiO_2. This difference was confirmed using microarea X-ray diffractometry (MAXRD). The MAXRD pattern of the C6-DBTDT film on SiO_2 showed a series of peaks assigned to the (00*l*) plane at low angles (Fig. 4c). These correspond to half the *c*-axis length of the unit cell in a previously reported single crystal of C6-DBTDT

[14], and indicated that the C6-DBTDT molecules acquired an edge-on orientation, that is, they are perpendicularly oriented to the substrate (Fig. 4a). Starting peaks at $2\theta = 3.26°$ were also observed for C6-DBTDT films on carbon and Au electrodes. However, the rocking curve of the (φ) peak at $2\theta = 3.26°$ on Au was observed to be substantially broader than those observed for the carbon and SiO_2 substrates (Fig. 4d). These results show that C6-DBTDT molecules are disordered to some degree in the film on Au, consistent with the AFM results.

The insight into the formation of ordered films on carbon presented above has been applied to device fabrication [15]. The solution-based preparation of a carbon electrode is more advantageous than the vacuum thermal evaporation of Au in terms of energy savings. The use of carbon electrodes in bottom-contact OFETs very effectively reduces the contact resistance at the C6-DBTDT/electrode interface, with a level similar to that of a conventional top-contact OFET with Au achieved (Fig. 4e

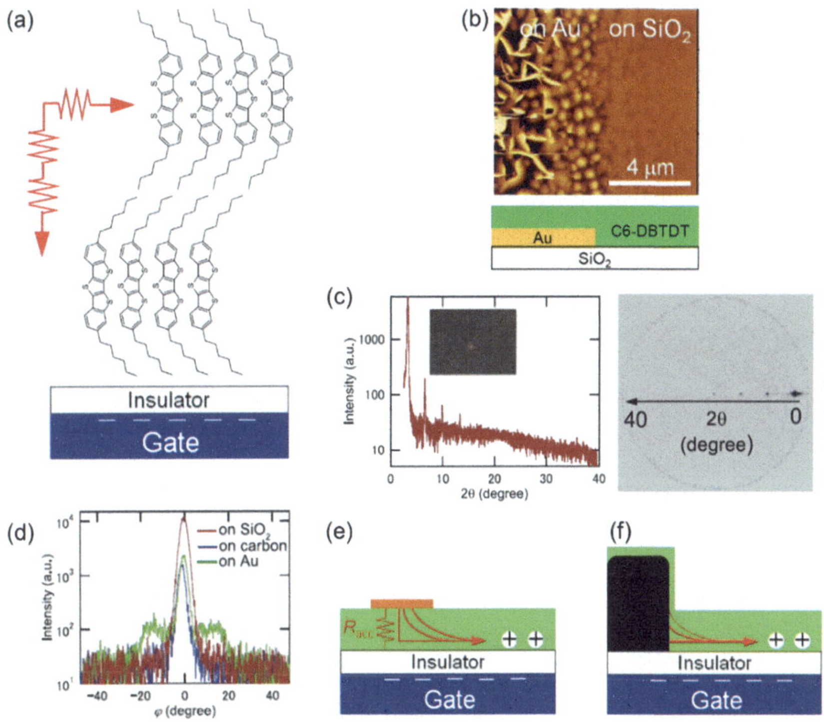

Fig. 4 **a** Schematic depicting a C6-DBTDT film on an insulator and its anisotropic resistivity. **b** AFM image of a C6-DBTDT film in a bottom-contact Au transistor; (inset) schematic side view of the film on an Au/SiO_2 substrate. **c** MAXRD patterns of a C6-DBTDT film on HMDS-treated SiO_2; (inset) optical image of the film in which bright spots correspond to X-ray irradiated regions. **d** Rocking curves (φ) of peaks at $2\theta = 3.26°$ for films on SiO_2, carbon, and Au. Schematic representation of charge injection in **e** a top-contact OFET with Au and **f** a bottom-contact OFET with carbon. Adapted with permission from [15]. Copyright 2015 Elsevier

and f). Access resistance (R_{acc}) was significantly suppressed by the lateral charge injection process at the carbon/OSC interface.

3.2 BTBT Derivatives on Substrates

[1]Benzothieno[3,2-*b*][1]benzothiophene (BTBT) and its derivatives are among the most investigated p-type molecular OSCs owing to their high hole mobilities and long-term stabilities (Fig. 5a) [16–19]. Functional groups, including alkyl and phenyl, are often introduced at the 2- and/or 7-positions of BTBT. BTBT derivatives form herringbone-packed structures in their crystal states. For example, C8-BTBT, a symmetric BTBT derivative with two octyl groups, forms a layered herringbone packing consisting of layers of BTBT cores and alkyl chains (Fig. 5b). Thin films of C8-BTBT were prepared using a solution process or vacuum thermal evaporation, resulting in an edge-on orientation of C8-BTBT with respect to the substrate. This molecular orientation facilitates two-dimensional (2D) charge transport along the BTBT core layer (Fig. 5b), which is a key feature for achieving high mobility during OFET operation.

Although the edge-on orientation of C8-BTBT has often been reported, experimental evidence is based on XRD data acquired for the bulk film. More precisely,

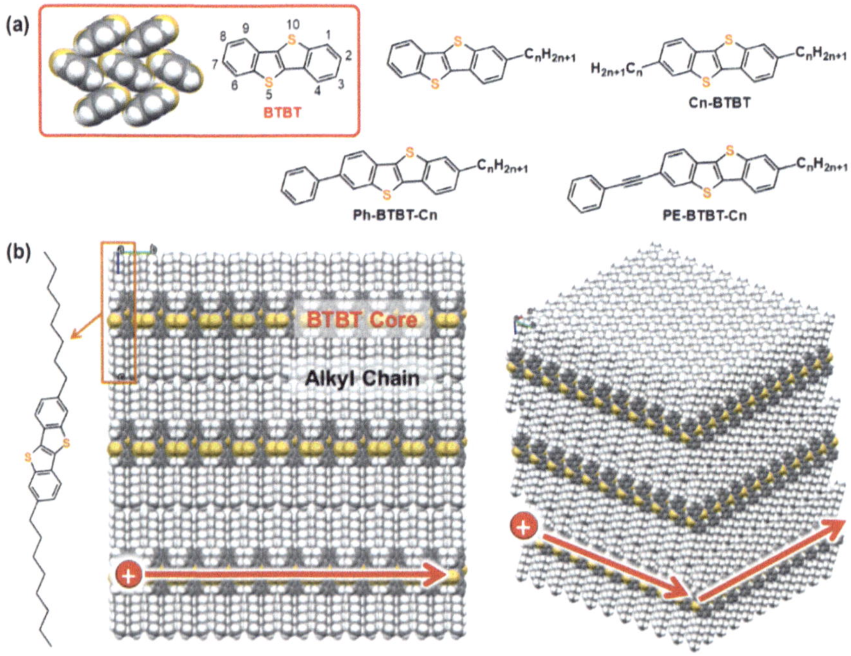

Fig. 5 **a** Structures of BTBT derivatives. **b** Crystal structure of C8-BTBT

the molecular orientation at the OSC/electrode (or OSC/insulator) interface is more critical for OFET performance than the average molecular orientation over the entire film. With an awareness of this point, the interfacial structure and corresponding electronic structure of C8-BTBT/Au were investigated in detail [20]. Thin films of C8-BTBT with various thicknesses were prepared on polycrystalline Au films via vacuum thermal evaporation, after which they were subjected to UPS (Fig. 6). The secondary electron cutoff (E_c) was determined by linear extrapolation, with work function (WF) obtained using the equation: WF = $h\upsilon - E_c$ (Fig. 6a). The WF value decreased by 0.26 eV upon film formation, and further decreased by 0.76 eV as the film thickened from 0.2 to 12.8 nm. In the region of low binding energy, the features of HOMO and HOMO-1 of C8-BTBT became clear upon the increase of the film thickness while the Au 5d peaks became buried (Fig. 6b). However, the corresponding change in the HOMO level was smaller than that observed for the WF (Fig. 6c).

AFM revealed that the first layer of C8-BTBT adopted a lying-down conformation [20] (Fig. 7a). The step height was 1.79 nm in the cross-sectional profile, and the polycrystalline Au film exhibited a root-mean-square roughness of 1.2 nm. The lying-down conformation is reasonable because C8-BTBT is adsorbed by Au through strong S···Au interactions. The step heights of the second and third layers of C8-BTBT were determined to be 2.13 and 2.73 nm, respectively, which approached

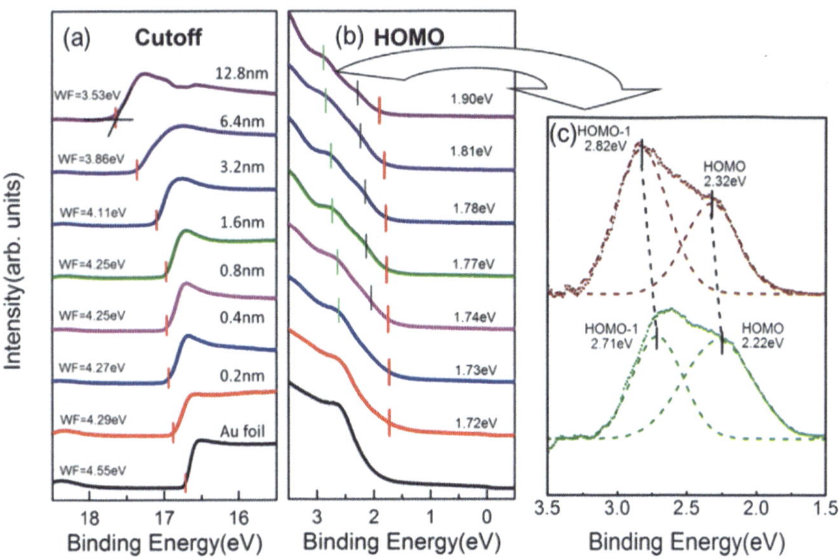

Fig. 6 UPS results for **a** the low-kinetic-energy secondary-electron cutoff region and **b** the HOMO region near the Fermi level as functions of C8-BTBT film thickness on Au. **c** Enlarged views of the HOMO region at thicknesses of (top) 12.8 nm and (bottom) 1.6 nm. Adapted with permission from [20]. Copyright 2017 Elsevier

the *c*-axis length of the unit cell of the previously reported single crystal of C8-BTBT (2.98 nm). These results suggested that C8-BTBT molecules changed their orientation from a lying-down to a standing-up conformation in Stranski–Krastanov growth mode, accompanied by a change from a disordered to an ordered molecular arrangement (Fig. 7c). The energy level alignment at the C8-BTBT/Au interface is shown in Fig. 7b. The vacuum level lowered by 0.26 eV at a film thickness of 0.2 nm, which corresponded to the interfacial dipole generated between Au and C8-BTBT. C8-BTBT/Au had a hole injection barrier of 1.72 eV. The energy levels did not shift much in the 0.2–1.6 nm range. Above 1.6 nm, the vacuum levels and the HOMO levels shifted largely downward, with total decreases of 0.76 and 0.18 eV observed, respectively, as the film thickened from 1.6 to 12.8 nm. This change corresponds to a decrease in the ionization potential of 0.58 eV. An in-depth investigation of how UPS properties depend on C8-BTBT/Au film thickness revealed downward band bending away from the interface.

The interfacial electronic structure of C8-BTBT/SiO_2 has also been investigated [21]. AFM indicated that C8-BTBT exhibited a Vollmer–Weber growth mode on SiO_2. This mode resulted from stronger interactions between C8-BTBT

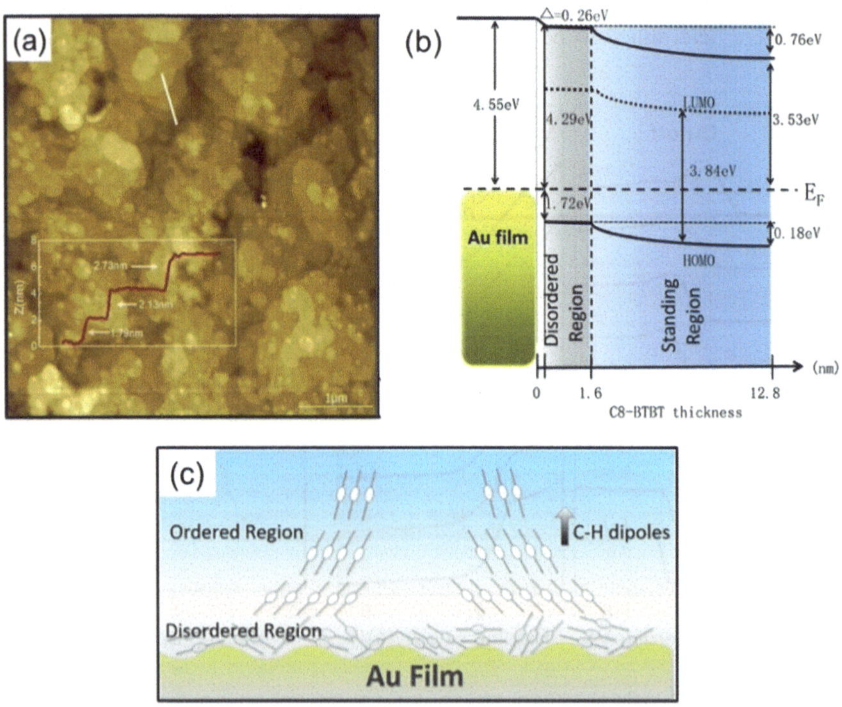

Fig. 7 **a** AFM image of a C8-BTBT film with a thickness of 12.8 nm on Au. **b** Energy level alignment diagrams for C8-BTBT films on Au. **c** Schematic representation of molecular orientations in C8-BTBT films on Au. Adapted with permission from [20]. Copyright 2017 Elsevier

molecules than between C8-BTBT and the SiO_2 substrate. Accordingly, the C8-BTBT molecules maintained a highly ordered standing-up conformation throughout the film. The combined understanding acquired for the C8-BTBT/Au and C8-BTBT/SiO_2 interfaces provides a plausible interpretation of device performance. For top-contact OFETs, the C8-BTBT molecules are highly ordered on SiO_2 due to weak interactions at the interface, whereas ordered structure is maintained upon preparation on top electrodes. However, for bottom-contact OFETs, C8-BTBT molecules are disordered on metal electrodes owing to strong interactions. The superior interfacial structure of SiO_2 compared to those of metals generally results in greater mobility by several orders of magnitude in a top-contact OFET [18]. This interpretation is in good agreement with the interfacial structures of DBTDT on various substrates.

Ideally, a thin OSC film is a single layer of molecular OSC components on a substrate, which has been realized with BTBT derivatives using solution processes [22]. A blade-coating technique with a mixed solution of Ph-BTBT-C6 and Ph-BTBT-C10 was used to afford single-molecule bilayers on a SiO_2/Si wafer. The single-molecule bilayers were characterized by AFM, in-plane film XRD, thin-film X-ray reflectometry, and optical anisotropy measurements, revealing geometrical-frustration features derived from differences in alkyl chain length. This frustration inhibits multilayer crystallization but does not perturb the herringbone packing of the BTBT moieties. These features enable the formation of an ultrathin OSC layer over an area as large as wafer scale while maintaining good 2D charge-transport properties during OFET operation. This solution-based methodology provides important insight into the precise control of molecular assemblies on substrates.

3.3 BTBT Derivatives on Functionalized Self-assembled Monolayers

The preparation of self-assembled monolayers (SAMs) is useful for adjusting surface properties by chemically modifying these surfaces with functionalized constituent molecules. SAMs are prepared on the surfaces of SiO_2 gate insulator films using a silane coupling reagent such as HMDS when preparing OFETs [23]. This covering often enhances highly crystalline OSC film growth, prevents charge-trap formation, and improves OFET performance. In addition, SAMs with electron-withdrawing and electron-donating groups accumulate holes and electrons, respectively, in contact with the conduction channels at the OSC/SiO_2 interface. This carrier density control is associated with the electric dipoles derived from SAM substituents and weak charge transfer between the OSC and SAM [24]. Moreover, various functional SAMs have been inserted at OSC/insulator (or OSC/electrode) interfaces, especially SAMs that are photo-responsive and redox-active [25].

Here, we discuss an investigation into a functional SAM buried under an OSC layer and located at the OSC/SiO_2 interface as an example (Fig. 8a) [26]. Hard X-ray photoelectron spectroscopy (HAXPES) is a useful technique for detecting photoelectrons emitted from core levels that are deeper than those detectable using general soft X-ray radiation [27]. Moreover, HAXPES is useful for obtaining information on deep areas of samples using higher X-ray energies and associated photoelectrons with prolonged inelastic mean free paths. This method was applied to a SAM prepared on the surface of SiO_2/Si with a silane coupling reagent bearing a ferrocene (Fc) moiety. A Fe 1s peak with a binding energy of 7114.8 eV was observed with a shoulder at 7119.8 eV in the HAXPES spectrum at a photon excitation energy of 8 keV (Fig. 8b). When HAXPES was applied to the Fc SAM@SiO_2/Si under the C8-BTBT film prepared by vacuum thermal evaporation, Fe 1s spectra were successfully acquired for films with thicknesses ranging from 50 to 80 nm. The binding energies of the Fe 1s peaks of the pristine and underlying Fc SAM were comparable, which indicated that the junction with the OSC layer had almost no impact on the electronic structure of the Fe in the SAM. Interestingly, an OFET containing the Fc SAM exhibited a noticeable source-drain current hysteresis, which diminished after the second sweep. However, the pronounced hysteresis was recovered when an opposite bias was applied to the gate voltage. The HOMO level of C8-BTBT (− 5.7 eV) is 0.9 eV lower than that of Fc (− 4.8 eV). Hence, controlled hysteresis via gate bias was attributable to the reversible trapping and detrapping of hole carriers, accompanied by a reversible change (between Fe(III) and Fe(II)) in the oxidation state of Fc. In situ HAXPES will provide further information on underlying SAM behavior during device operation. Such an in-depth investigation into interfaces is crucial for adding new features to OFET devices, such as nonvolatile memory, photo-switchability, and external stimuli-responsiveness.

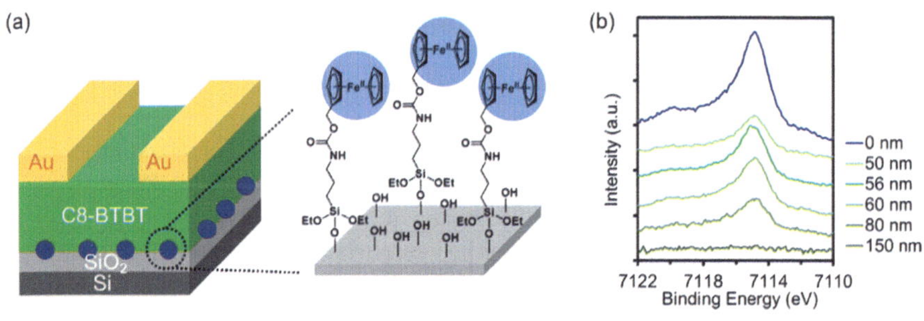

Fig. 8 **a** OFET-containing redox-active Fc SAMs. **b** Fe 1s HAXPES spectra of Fc SAM and C8-BTBT thin films formed on Fc SAM@SiO_2/Si with various thicknesses (50–150 nm). Adapted with permission from [26]. Copyright 2020 American Chemical Society

3.4 BTBT Derivatives with a Focus on Liquid Crystal Properties

Liquid crystalline OSCs have attracted great interest because they combine conventional OSC properties with liquid crystal properties, including controlled molecular orientation/packing and the formation of defect-free large aligned monodomains [28–31]. In this subsection, we focus on the liquid crystal properties of BTBT derivatives and their use in OFETs.

Ph-BTBT-C10 was synthesized by asymmetrically substituting BTBT with phenyl and decyl groups (Fig. 5a) [32]. The C10-BTBT symmetric counterpart bearing two decyl groups exhibited a smectic A liquid–crystal phase. The C10-BTBT crystal film melted into droplets above 100 °C accompanied with a phase transition to the liquid-like smectic A phase. On the other hand, Ph-BTBT-C10 exhibited an additional smectic E phase that resembled a crystal phase in the 142–210 °C temperature range. The emergence of the smectic E phase is advantageous in terms of the thermal durability of the film. Bilayer steps of 5.7 nm were observed in the AFM height profile of the Ph-BTBT-C10 film after thermal annealing at 120 °C (Fig. 9a). In addition, a peak was observed at $2\theta = 3.4°$ for the pre-annealed film, and a different peak was observed at $2\theta = 1.7°$ after annealing (Fig. 9b). Moreover, the differential scanning calorimetry (DSC) curve exhibited a broad exothermic peak in the 70–120 °C temperature range during heating and then an intense peak at 144 °C ascribable to the phase transition from the crystalline to the smectic E phase (Fig. 9c). These results suggest that annealing causes a monolayer-to-bilayer structural transformation that serves as a crystal-to-crystal phase transition. The resulting crystal phase has the same structure as the single crystal structure obtained by recrystallization from an organic solution, with a head-to-head orientation at the interlayer and herringbone-packed Ph-BTBT moieties in the layer observed. Strikingly, bottom-contact OFETs prepared using the Ph-BTBT-C10 polycrystalline film showed high hole mobilities over 10 $cm^2 V^{-1} S^{-1}$ after annealing at 120 °C. Smectic E liquid crystals are promising for overcoming current problems associated with crystalline OSC films during OFET fabrication, including inhomogeneity, defects, and low thermal durability.

Regarding the smectic E phase, an interesting solid crystal phase was revealed by microcrystal electron diffractometry of PE-BTBT-C6, an asymmetric BTBT derivative bearing phenylethynyl (PE) and hexyl groups, using cryo-transmission electron microscopy (cryo-TEM) [33] (Fig. 10). An ultrathin single crystal was prepared on a holey carbon film coated on a TEM grid by casting a toluene solution of PE-BTBT-C6 (Fig. 10a). The maximum periodic length was in good agreement with the molecular length based on the spacing between the spots along the c^* axis in the electron diffraction pattern (Fig. 10b and c). The PE and hexyl groups exhibited positional disorder (Fig. 10d) to afford disordered layered herringbone packing (Fig. 10e). The PE and hexyl groups are of approximately the same length, which is the key factor in this unique packing. Combined with the results for the other unsymmetrical BTBT derivatives, the disordered crystal phase was concluded to serve as the frozen solid state of the smectic E phase. Deep structural investigations of liquid crystal phases are

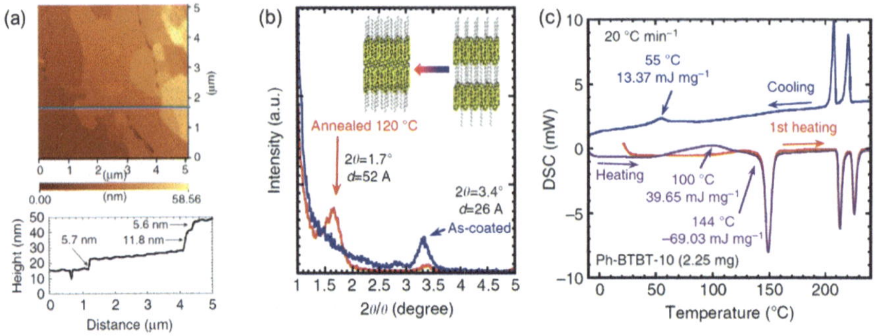

Fig. 9 **a** AFM image of a Ph-BTBT-C10 film after annealing at 120 °C. **b** XRD patterns of the as-coated Ph-BTBT-C10 and after annealing. **c** DSC curves of Ph-BTBT-C10 acquired at 20 °C min^{-1}. Adapted from [32]

expected to facilitate the establishment of molecular designs and OFET applications for liquid crystalline OSCs.

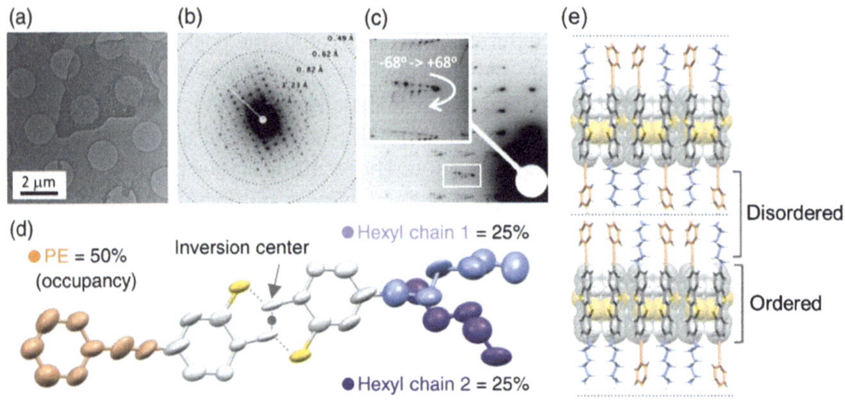

Fig. 10 **a** TEM image and **b** diffractogram of rotated electron diffraction patterns of a single crystal of PE-BTBT-C6. **c** Enlarged diffractogram showing the spot spacing along the *c** axis. **d** Disordered structure of PE-BTBT-C6 determined by microcrystal 3D electron diffractometry. **e** Schematic representation of the disordered layered-herringbone packing of PE-BTBT-C6. Adapted with permission from [33]. Copyright 2022 American Chemical Society

3.5 Estimation of Schottky Barrier at an Organic Semiconductor/Metal Interface by Accumulated Charge Measurement

The junction interface between OSCs and metal electrodes has been studied using mainly spectroscopy, as explained above. Spectroscopy allows the determination of ionization energy and electron affinity, which can be combined with the work function of the metal to estimate the energy barrier at the interface. In 2016, Tajima et al. reported on accumulated charge measurement (ACM) [34, 35]. This is a new method for determining the Schottky barrier value at the OSC-metal junction interface. This electrical measurement method is based on the principle of a capacitor. The main feature is that measurements can be conducted under atmospheric conditions while maintaining the actual device structure, for instance organic transistors.

Figure 11 shows the principle of ACM. The device has a sandwich structure consisting of a bottom electrode M1, insulating layer, OSC, and top electrode M2 (Fig. 11a). This structure is similar to that of a top-contact organic transistor. An oscillating voltage V_a is applied to the device; at $V_a = 0$ V, the potentials between M1 and M2 coincide. In this state, a built-in electric field exists, which originates from the difference in the work functions and from the trapped charges near the insulating layer (Fig. 11b). The methods to solve this problem have already been reported [36], but the analysis presented here assumes that no built-in voltage exists.

When V_a is applied, charge Q_{acc} accumulates on both sides of M2 and M1 (Fig. 11c). The applied voltage V_a is divided into voltage drops V_I at the insulation layer and V_i at the semiconductor. That is, $V_a = V_I + V_i$. According to Gauss's law, this can be expressed as $V_I = V_a - Q_{acc}/C_I$, where C_I is the capacitance of the insulation layer. From this, V_i can be expressed as $V_i = V_a - Q_{acc}/C_I$.

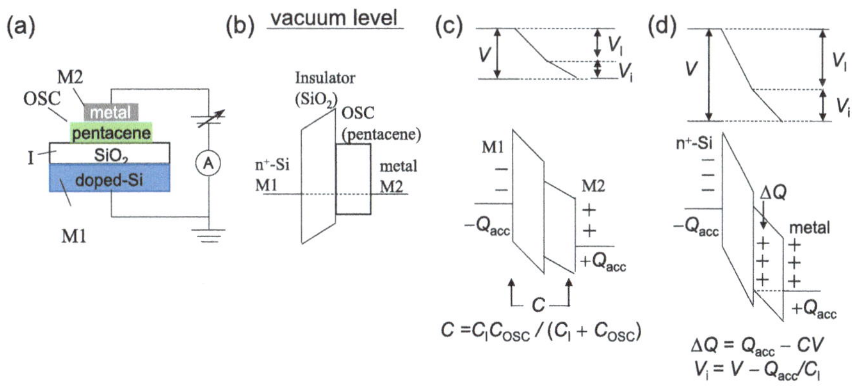

Fig. 11 **a** Schematic diagram of the device structure. Energy-level diagrams for the cases where **b** the applied bias voltage is zero, **c** charges accumulated at the metal/pentacene interface, and **d** charge injection into pentacene occurred. ΔQ indicates the degree of charge injection. Copyright 2017 American Chemical Society

As V_a increases, V_i also increases. Eventually, when V_i exceeds the charge-injection barrier (Schottky barrier) at the OSC/M2 interface, charge carriers are injected into the semiconductor (Fig. 11d). The degree of charge injection ΔQ is expressed as $\Delta Q = Q_{acc} - CV_a$.

Because C is the capacitance of the capacitor connecting the insulating layer I in series with the semiconductor layer OSC, C is expressed as follows:

$$C = C_I C_{OSC}/(C_I + C_{OSC}) \tag{1}$$

Therefore, by plotting V_{OSC} on the horizontal axis and ΔQ on the vertical axis, the threshold V_i at which ΔQ increases rapidly is confirmed. The value of V_i at that point can be determined as the charge-injection barrier at the metal/OSC interface.

To calculate ΔQ in ACM, the degree of charges Q_{acc} that accumulated in the device should be determined. Q_{acc} is obtained from the integral of the observed displacement current. In this experiment, four different voltage oscillation patterns (ZP, ZN, PZ, and NZ) are applied to accurately determine Q_{acc} (Fig. 12a). ZP is a repeated voltage application pattern of 0 V → positive voltage → 0 V → positive voltage → 0 V. In contrast, PZ is an application pattern of positive voltage → 0 V → positive voltage → 0 V → positive voltage. The voltage patterns of ZP and PZ oscillating on the negative side are ZN and NZ, respectively. Figure 12b shows the displacement current for an ideal capacitor, where the waveforms are steplike.

The accumulated charges Q_{acc} in the PZ and NZ modes are the result of applying voltage V_a for a long period of time and injecting sufficient charges into the OSC. Therefore, it is considered that the injected charge is not completely ejected in the

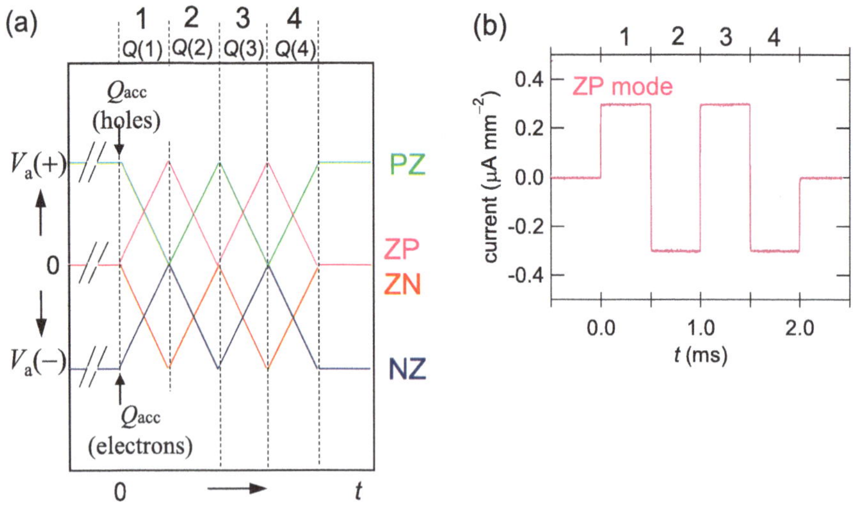

Fig. 12 **a** Bias voltage sweep patterns used in ACM. **b** Time dependence of the displacement current for the metal/SiO_2 capacitor at $|V| = 0.5$ V. The figure corresponds to $C_{SiO2} = 0.30$ nF/mm^2. Copyright 2017 American Chemical Society

subsequent extraction process owing to electron traps and other effects in the organic thin film. This implies that the integral values of the displacement currents of the first waveforms of the PZ and NZ modes cannot be used directly as the accumulated charge quantity Q_{acc}. Therefore, for each Q_{acc} of the PZ and NZ modes, the integral values of the respective displacement currents for the ZP and ZN modes are used for correction. In ACM, voltage oscillations must be repeated several times to establish a state of thermal equilibrium in which the waveforms of the PZ and ZP modes converge to the same shape. In Fig. 12a, $Q_{\mathrm{acc}}(\mathrm{PZ})$ is expressed using the charge balance of the voltage oscillation of ZP as

$$Q_{\mathrm{acc}}(\mathrm{PZ}) = Q(\mathrm{PZ}(1)) + Q(\mathrm{PZ}(2)) + Q(\mathrm{PZ}(3)) - \left[Q(\mathrm{ZP}(1)) + Q(\mathrm{ZP}(2)) + Q(\mathrm{ZP}(3)) + Q(\mathrm{ZP}(4))\right] \quad (2)$$

Q_{acc} in electron injection can also be obtained using a similar analysis of the displacement currents in the NZ and ZN modes. These equations are correct even when carriers are trapped at the interface owing to charge traps.

Figure 13a shows the displacement current waveform of a device with pentacene as the OSC. At $|V| = 0.5$ V, the waveform has the step-type characteristic of a capacitor. This indicates that the pentacene film behaves as an insulator and no charge injection into the pentacene occurs. Figure 13b shows the displacement current waveforms at $|V| = 5$ V. Here, in the PZ and ZP mode voltage oscillation patterns, the waveforms change and the current density increases; in the NZ and ZN mode voltage oscillations, the waveforms remain step-type and unchanged. This indicates that only holes are injected into the pentacene as charge carriers. Figure 13c shows the amount of accumulated (injected) charge plotted against the oscillation voltage V_{a}. The slope of this graph represents the capacitance C. In the region $V_{\mathrm{a}} < 0$ V, the slope is constant at $C = 0.20$ nF/mm^2. This indicates that no electron injection occurs and that the pentacene film behaves as an insulator. In contrast, for $V_{\mathrm{a}} >$ 0 V, the slope of the graph increases, reaching a capacitance of $C = 0.30$ nF/mm^2 for the oxide silicon substrate. This is the value for only the silicon oxide film as the insulating layer in the capacitor; hence, holes are injected into the pentacene thin film and accumulated up to near the surface of the silicon oxide substrate. Figure 13d shows a plot of the internal voltage V_{i} applied to the pentacene versus the degree of accumulated charges ΔQ. This figure is a reanalysis of Fig. 13c based on the theoretical equation, and its behavior is essentially the same. In the region $V_{\mathrm{a}} < 0$ V, ΔQ changes minimally, whereas for $V_{\mathrm{a}} > 0$ V, ΔQ increases sharply around $V_{\mathrm{i}} =$ 0.2 V. Because of the increased hole injection at $V_{\mathrm{i}} = 0.2$ V, the hole-injection barrier ϕ_{hole} at the Au-pentacene interface can be determined as $\phi_{\mathrm{hole}} = 0.2$ eV. Improved methods have been also reported [37–40] and further applications are expected.

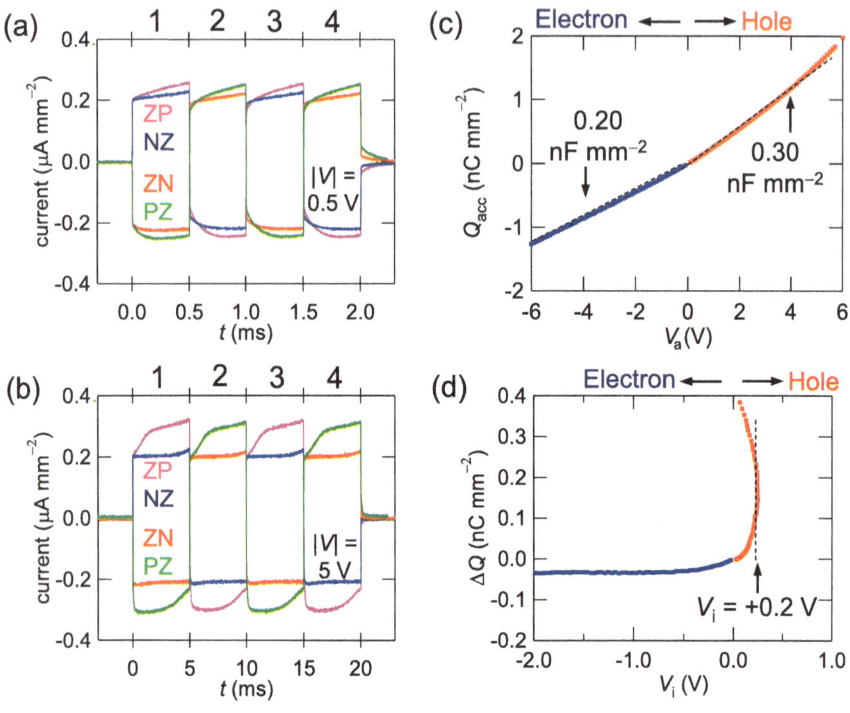

Fig. 13 Time dependence of displacement current measurements for the Au/pentacene interface. **a** $|V| = 0.5$ V and **b** $|V| = 5$ V. **c** Q_{acc} as a function of V_a. The dotted lines correspond to series capacitance densities of 0.20 nF/mm^2 and 0.30 nF/mm^2. **d** ΔQ plotted as a function of V_i. Copyright 2017 American Chemical Society

4 Summary

In this chapter, we discussed analysis methods used to evaluate OSC/electrode and OSC/insulator interfaces, with a focus on OFET applications. XRD is useful for evaluating the crystallinities of OSC thin films formed on substrates as well as molecular orientations in the films. AFM effectively complements this insight by providing a microscopic view that enables various interfacial aspects to be characterized in detail. The obtained interfacial structural insight is closely related to the electronic structure at the interface, which governs OFET device performance. UPS is useful for determining the positions of the energy levels of the components of the interfaces. ACM is a new method for determining the Schottky barrier at the OSC/electrode interface in an OFET device under atmospheric conditions based on the capacitor principle. Recent characterization examples for interfaces containing small-molecule OSCs have also been discussed, along with the relationship between interfacial structure and device performance. New features will be added to OFET devices through the multifunctionalization of OSCs, the use of multilayering techniques with solution processes, and sophisticated interfacial engineering using SAMs. Hence, further advancements in analytical methods are required to elucidate structure–function correlations at a higher level.

Acknowledgements The authors would like to thank Prof. Hiroyuki Tajima for useful ACM discussions.

References

1. S. Ogawa, *Organic Electronics Materials and Devices* (Springer, Tokyo, 2015)
2. G. Zhang, C. Xie, P. You, S. Li, *Introduction to Organic Electronic Devices* (Springer, Singapore, 2022)
3. M. Fahlman, S. Fabiano, V. Gueskine, D. Simon, M. Berggren, X. Crispin, Interfaces in organic electronics. Nat. Rev. Mater. **4**, 627–650 (2019)
4. C. Wang, H. Dong, W. Hu, Y. Liu, D. Zhu, Semiconducting π-conjugated systems in field-effect transistors: a material odyssey of organic electronics. Chem. Rev. **112**, 2208–2267 (2012)
5. H. Sirringhaus, 25th Anniversary Article: Organic field-effect transistors: the path beyond amorphous silicon. Adv. Mater. **26**, 1319–1335 (2014)
6. F. Reinert, S. Hüfner, Photoemission spectroscopy—from early days to recent applications. New J. Phys. **7**, 97 (2005)
7. Y. Gao, Surface analytical studies of interfaces in organic semiconductor devices. Mater. Sci. Eng. R **68**, 39–87 (2010)
8. H. Ishii, K. Sugiyama, E. Ito, K. Seki, Energy level alignment and interfacial electronic structures at organic/metal and organic/organic interfaces. Adv. Mater. **11**, 605–625 (1999)
9. Y. Nakayama, S. Kera, N. Ueno, Photoelectron spectroscopy on single crystals of organic semiconductors: experimental electronic band structure for optoelectronic properties. J. Mater. Chem. C **8**, 9090–9132 (2020)
10. S. Olthof, The impact of UV photoelectron spectroscopy on the field of organic optoelectronics—a retrospective. Adv. Opt. Mater. **9**, 2100227 (2021)
11. S. Kera, H. Yamane, N. Ueno, First-principles measurements of charge mobility in organic semiconductors: valence hole–vibration coupling in organic ultrathin films. Prog. Surf. Sci. **84**, 135–154 (2008)
12. M. Ohtomo, T. Suzuki, T. Shimada, T. Hasegawa, Band dispersion of quasi-single crystal thin film phase pentacene monolayer studied by angle-resolved photoelectron spectroscopy. Appl. Phys. Lett. **95**, 123308 (2009)
13. H. Yamane, S. Kera, N. Ueno, Photoelectron spectroscopy on single crystals of organic semiconductors: experimental electronic band structure for optoelectronic properties. J. Mater. Chem. C **8**, 9090–9132 (2020)
14. Y. Miyata, E. Yoshikawa, T. Minari, K. Tsukagoshi, S. Yamaguchi, High-performance organic field-effect transistors based on dihexyl-substituted dibenzo[*d*,d']thieno[3,2-b;4,5-b']dithiophene. J. Mater. Chem. **22**, 7715–7717 (2012)
15. T. Kadoya, O. Pitayatanakul, T. Mori, Suppression of access resistance using carbon electrodes in organic transistors based on alkyl-substituted thienoacene. Org. Electron. **21**, 106–110 (2015)
16. H. Ebata, T. Izawa, E. Miyazaki, K. Takimiya, M. Ikeda, H. Kuwabara, T. Yui, Highly soluble [1]Benzothieno[3,2-*b*]benzothiophene (BTBT) derivatives for high-performance, solution-processed organic field-effect transistors. J. Am. Chem. Soc. **129**, 15732–15733 (2007)
17. T. Izawa, E. Miyazawa, K. Takimiya, Molecular ordering of high-performance soluble molecular semiconductors and re evaluation of their field-effect transistor characteristics. Adv. Mater. **20**, 3388–3392 (2008)
18. H. Minemawari, Y. Toshikazu, H. Matui, J. Tsutsumi, S. Haas, R. Chiba, R. Kumai, T. Hasegawa, Inkjet printing of single-crystal films. Nature **475**, 364–367 (2011)

19. P. Xie, T. Liu, J. Sun, J. Yang, Structures, properties, and device applications for [1]Benzothieno[3,2-*b*]Benzothiophene derivatives. Adv. Funct. Mater. **32**, 2200843 (2022)
20. S. Wang, D. Niu, L. Lyu, Y. Huang, X. Wei, C. Wang, H. Xiea, Y. Gaoa, Interface electronic structure and morphology of 2,7-dioctyl[1]benzothieno[3,2-*b*]benzothiophene (C8-BTBT) on Au film. Appl. Surf. Sci. **416**, 696–703 (2017)
21. L. Lyu, D. Niu, H. Xie, Y. Zhao, N. Cao, H. Zhang, Y. Zhang, P. Liu, Y. Gao, The correlations of the electronic structure and film growth of 2,7-diocty[1]benzothieno[3,2-*b*]benzothiophene (C8-BTBT) on SiO_2. Phys. Chem. Chem. Phys. **19**, 1669–1676 (2017)
22. S. Arai, S. Inoue, T. Hamai, R. Kumai, T. Hasegawa, Semiconductive single molecular bilayers realized using geometrical frustration. Adv. Mater. **30**, 1707256 (2018)
23. S. Casalini, C.A. Bortolotti, F. Leonardi, F. Biscarini, Self-assembled monolayers in organic electronics. Chem. Soc. Rev. **46**, 40–71 (2017)
24. S. Kobayashi, T. Nishikawa, T. Takenobu, S. Mori, T. Shimoda, T. Mitani, H. Shimotani, N. Yoshimoto, S. Ogawa, Y. Iwasa, Control of carrier density by self-assembled monolayers in organic field-effect transistors. Nat. Mater. **3**, 317–322 (2004)
25. H. Chen, W. Zhang, M. Li, G. He, X. Guo, Interface engineering in organic field-effect transistors: principles, applications, and perspectives. Chem. Rev. **120**, 2879–2949 (2020)
26. T. Ikeda, K. Tahara, T. Kadoya, H. Tajima, N. Toyoda, S. Yasuno, Y. Ozawa, M. Abe, Ferrocene on insulator: silane coupling to a SiO_2 surface and influence on electrical transport at a buried interface with an organic semiconductor layer. Langmuir **36**, 5809–5819 (2020)
27. C. Weiland, A.K. Rumaiz, P. Pianetta, J.C. Woicik, Recent applications of hard X-ray photoelectron spectroscopy. J. Vac. Sci. Technol. A **34**, 030801 (2016)
28. M. O'Neill, S.M. Kelly, Ordered materials for organic electronics and photonics. Adv. Mater. **23**, 566–584 (2011)
29. M. Funahashi, Nanostructured liquid-crystalline semiconductors—a new approach to soft matter electronics. J. Mater. Chem. C **2**, 7451–7459 (2014)
30. T. Kato, M. Yoshio, T. Ichikawa, B. Soberats, H. Ohno, M. Funahashi, Transport of ions and electrons in nanostructured liquid crystals. Nat. Rev. Mater. **2**, 17001 (2017)
31. H. Iino, J. Hanna, Liquid crystalline organic semiconductors for organic transistor applications. Polym. J. **49**, 23–30 (2017)
32. H. Iino, T. Usui, J. Hanna, Liquid crystals for organic thin-film transistors. Nat. Commun. **6**, 6825 (2015)
33. S. Inoue, K. Nikaido, T. Higashino, S. Arai, M. Tanaka, R. Kumai, S. Tsuzuki, S. Horiuchi, H. Sugiyama, Y. Segawa, K. Takaba, S. Maki-Yonekura, K. Yonekura, T. Hasegawa, Emerging disordered layered-herringbone phase in organic semiconductors unveiled by electron crystallography. Chem. Mater. **34**, 72–83 (2022)
34. H. Tajima, F. Miyao, M. Mizukoshi, S. Sato, Determination of charge injection barrier using the displacement current measurement technique. Org. Electron. **34**, 193–199 (2016)
35. T. Kadoya, M. Otsuka, A. Ogino, S. Sato, T. Yokomatsu, K. Maenaka, J. Yamada, H. Tajima, Estimation of charge-injection barriers at the metal/pentacene interface through accumulated charge measurement. J. Phys. Chem. C **121**, 2882–2888 (2017)
36. H. Tajima, K. Yoshida, S. Sato, T. Kadoya, J. Yamada, Estimation of the charge injection barrier at a metal/organic semiconductor interface based on accumulated charge measurement: the effect of offset bias voltages. J. Phys. Chem. C **121**, 14725–14730 (2017)
37. S. Shimomoto, T. Kadoya, T. Tanimura, K. Maenaka, T. Yokomatsu, T. Komino, H. Tajima, Accumulated charge measurement: control of the interfacial depletion layer by offset voltage and estimation of band gap and Electron injection barrier. J. Phys. Chem. C **125**, 1990–1998 (2021)
38. H. Tajima, T. Kadoya, K. Yamaguchi, Y. Omura, T. Oda, A. Ogino, Thermal and non-thermal equilibrium processes of charge extraction in accumulated charge measurement (ACM). J. Appl. Phys. **130**, 195501 (2021)
39. T. Oda, K. Yamaguchi, T. Kadoya, H. Tajima, Measurement of electron injection barriers in OS/Au (OS=phthalocyanine and pentacene) using accumulated charge measurement. Org. Electron. **120**, 106827 (2023)

40. H. Tajima, T. Oda, T. Kadoya, Nonthermal equilibrium process of charge carrier extraction in metal/insulator/organic semiconductor/metal (MIOM) junction. Magnetochemistry **9**, 180 (2023)

Electrochemical Measurement

Kazuya Kubo

1 Fundamental Process of Electrolysis

Cyclic voltammetry is an electrochemical measurement which determines redox potentials of the materials. If intensity of current is measured while electrical potentials are swept between the two electrodes, current–potential curves are obtained. The current–potential curves, which are called voltammograms, give us important information about the redox potentials. The cyclic voltammetry is measured by three-electrode systems in the electrolytic solutions. In this system, electric energy facilitates oxidation–reduction reactions. The electric energy is controlled by electrode potential, and current I represents the velocity of the redox reactions. In order to understand the redox process, the fundamental processes of the electrochemical reactions around vicinities of the electrode are reviewed.

Figure 1 shows a two-electrode system of electrolysis. Two electrodes connecting a direct current circuit are immersed into the electrolytic solution. At first step, weak voltage is applied between the electrodes. Then, ions in the electrolytic solution move momentarily, and charging current flows during approximately 0.01 s. As a result of the movement of the ions, electric double layers are formed within a range of approximately 1×10^{-9} m from the electrode surfaces. The ions with negative charges exist in the electric double layer on the surface of anode. The sum of the charges on the ions in the layer is equal to the charges on the electrode. In the case of the cathode, the ions with positive charges exist in the electric double layer. The total amounts of negative charges in the cathodic boundary are balanced to total amounts of positive charges on the ions.

At the second step, increasing voltage between the electrodes, the electric energies attain sufficient energy to transfer electrons to the anode from the solvent, solute or

K. Kubo (✉)
Graduate School of Science, University of Hyogo, 3-2-1 Kouto, Kamigori-cho, Ako-Gun, Hyogo, Japan 678-1297
e-mail: kubo@sci.u-hyogo.ac.jp

M. Funahashi (ed.), *Functional Soft Matter II*,
https://doi.org/10.1007/978-981-92-0100-6_8

Fig. 1 Fundamental process of electrolytic reaction. Reproduced and adapted from Ref. [1, p. 12], with modifications

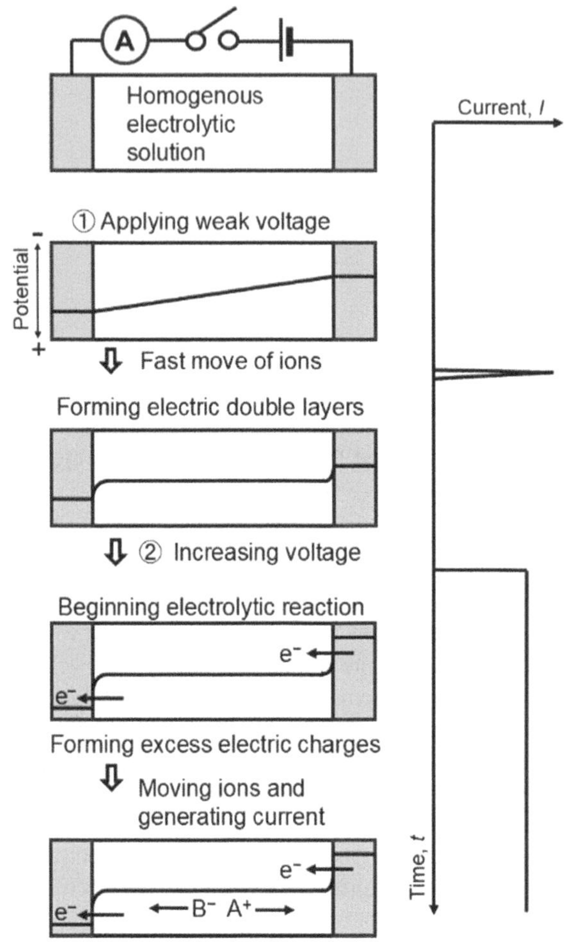

electrode itself. Simultaneously, the cathode receives electrons from other components such as the solvent, solute, and electrode itself, among the solvent, solute, and electrode itself, the substance which is most easily oxidized transfers electrons to the anode, and the substance which is most easily reduced receives electrons from the cathode. In this step, electrolysis begins. Examples of anodic and cathodic reactions are shown below.

Anodic reaction

$$2H_2O \rightarrow O_2 + 4H^+ + 4e^- \tag{1}$$

$$2Cl^- \rightarrow Cl_2 + 2e^- \tag{2}$$

Cathodic reaction

$$2H_2O + 2e^- \rightarrow H_2 + 2OH^- \quad (3)$$

$$Cu^{2+} + 2e^- \rightarrow Cu \quad (4)$$

As reactions progress, positive charges increase, i.e., negative charges decrease, in the electric double layer on the anodic electrode. In contrast, negative charges increase, i.e., positive charges decrease, in the double layers on the cathodic electrode. Accordingly, the electric charges having matching signs crowd in the electric double layers near the anode and cathode in a moment. However, natural phenomena are governed by the principle of electric neutrality. Therefore, the states of electric double layers in which large amounts of electric charges exist are extremely unstable.

At the final step, the positive and negative ions migrate to cathode and anode, respectively, to maintain electric neutrality of the systems. At this point, current begin to flow. Applying large voltage to the electrolytes enough to cause electrolytic reactions, moving the ions to dissipate the excess charges at boundary surfaces, electric current flows.

2 Standard Electrode Potential

In principle, the voltammetry should be able to be measured by cyclic sweeping the potentials between the two electrodes, shown in Fig. 1, and recording the resulting current intensities. However, in order to define the accurate potentials applied to the electrolytic solutions, it is not sufficient to simply define the electric potentials between the two electrodes. To understand the electrical potentials defined in the voltammetry, it is necessary to explain a standard electrode potential.

The electric potential is determined by a potential difference between the two electrodes, but how is the potential difference calculated? The potential difference is explained by using electrolysis of water. The reaction that produces hydrogen H_2 and oxygen O_2 during electrolysis of dilute sulfuric acid is described by the following reaction equation:

$$2H_2O \rightarrow 2H_2 + O_2 \quad (5)$$

The reaction equation indicates an up-hill reaction generating 1 mol of oxygen with + 474.26 kJ of the difference of reaction Gibbs energy $\Delta_r G°$ based on the standard Gibbs free energy of formation $\Delta_f G°$ of H_2O (− 237.13 kJ mol^{-1}). Equation 5 can be divided to following equations. Please refer to the specialized books for detail explanations of thermodynamics.

- Cathodic reaction (reduction)

$$4H^+ + 4e^- \rightarrow 2H_2 \quad (6)$$

- Anodic reaction (oxidation)

$$2H_2O \rightarrow O_2 + 4H^+ + 4e^- \quad (7)$$

In this reaction, generating 1 mol of oxygen, 4 mol of electrons move. The electric energy ΔE can be obtained by following equation containing Faraday constant F (96,485 C mol^{-1}):

$$474260 = \Delta E \times 4 \times 96485$$
$$\Delta E \approx +1.23\,V$$

Difference between the electrical potentials of Eqs. 6 and 7 is + 1.23 V. However, the electric potentials of Eqs. 6 and 7 themselves are uncertain. In order to evaluate abilities of materials donating or accepting electrons, it is convenient to determine the electrical potentials of each reaction. In order to evaluate the magnitudes of the electric potentials, it is necessary to define the origin. When determining the origin, the most appropriate reaction is the equilibrium reaction accepting and donating electrons between proton H^+ and hydrogen H_2 (Eq. 8):

$$2H^+ + 2e^- \leftrightarrows H_2 \quad (8)$$

Values of $\Delta_f G^\circ$ of proton H^+ and hydrogen H_2 are zero, and $\Delta_f G^\circ$ is an equal footing with electric energy. The energies on both sides of the reaction 8 are equal. If the electric potential at the certain position of electron is E°, 2 mol of electrons have the electric energy of $2FE^\circ$. Thus, balance of the energies can represent following equation:

$$2\Delta_f G^\circ(H^+) - 2FE^\circ = \Delta_f G^\circ(H_2) \quad (9)$$

$$E^\circ = 0\,V$$
$$\because\ \Delta_f G^\circ(H^+) = \Delta_f G^\circ(H_2) = 0$$

In brief, the potential of the electron required to establish the equilibrium reaction is appropriate as the origin of the electric potential. The potentials of the equilibrium reactions determined based on such considerations are called the standard electrode potentials of the substances.

Here is one example of calculation of the standard electrode potential for the following equation.

$$Fe^{3+} + e^- \leftrightarrows Fe^{2+} \tag{10}$$

If the potential of the electron e^- for the equation is represented with $E°(Fe^{3+}/Fe^{2+})$, the following equation holds as Eq. 9.

$$\Delta_f G°(Fe^{3+}) - FE°(Fe^{3+}/Fe^{2+}) = \Delta_f G°(Fe^{2+}) \tag{11}$$

Substituting each $\Delta_f G°$ value (Fe^{3+}: – 4.7 kJ mol^{-1}, Fe^{2+}: – 78.90 kJ mol^{-1}), + 0.77 V of the standard electrode potential $E°(Fe^{3+}/Fe^{2+})$ is obtained. The numbers of effective digits depend on the precision of the experiments. A pair like Fe^{3+} and Fe^{2+} is called a redox pair. Substances that accept and donate electrons are called oxidant (Fe^{3+}) and reductant (Fe^{2+}), respectively. In Eqs. 8 and 10, electrons represent same symbol e^-. However, it should be noted that each electron has a different energy.

3 Standard Electrode Potential: Aspect of on the Surface of Electrodes

Electrons in Eqs. 6–8 exist in the electrodes. Assuming the electrodes metals, the equilibrium reactions accepting and donating electrons that occur on the surfaces of the electrodes can be described as follows. In metals, the electrons exist up to the Fermi levels. Fermi energy is represented by symbols ε_F (energy) or E_f (electric potential). There are no electrons above the Fermi levels ε_F. Electrons outside the metals can flow into any level above the Fermi levels ε_F. On the other hand, among the electrons in the metals, the electrons exiting in the vicinity of the Fermi levels can transfer to other substances; the electrons in lower level cannot transfer. The electrons in Eqs. 6 and 8 can transfer between the electrode and another substance at applied potentials E_f. As shown in Fig. 1, when applying the electric potentials, the electric double layers are formed on the surfaces of the electrodes. If the electric potentials are applied between the two electrodes as shown in Fig. 2, how can we investigate the electron transfers between the anode and the electrolyte solution? In order to explain the electron transfer, it is necessary to control the potential between the points and B in Fig. 2, i.e., between the electrode and the electrolyte. The cell potential V is $V_1 + V_2$. However, there are no way to know how the potential V is distributed between V_1 and V_2. Furthermore, the distribution may also change over time. Applying electric potentials between two electrodes does not result in controlling the potentials in the system. Thus, for the actual potential measurements between the electrodes, a reference electrode is required.

If there is a warranty that the potential profile between the points P and Q would not change when applying the cell potential V, the potentials V will vary depending on the potential profile between the points A and B as shown in Fig. 2. In other words, the potential profile between the points A and B can be controlled by magnitude of the cell potential V. Electrodes with the characteristics are called reference electrodes

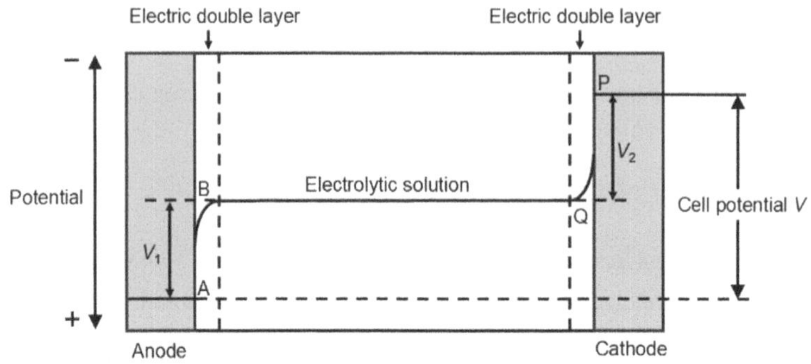

Fig. 2 Profile of the electric potential in the electrochemical cell. Reproduced and adapted from Ref. [1, p. 40], with modifications

which are represented by the symbol R. In the case of Fig. 2, the reference electrodes are constructed by the cathode and electrolytic solutions. Electrodes whose potential vary because of the electron transfers are called working electrodes, i.e., the anode in Fig. 2, which are represented by the symbol W. Varying the potential between the electrodes R and W, essentially, the cell potentials V changes only in the electric double layer on the working electrode W.

4 Reference Electrodes

4.1 Standard Hydrogen Electrode

In previous section, we discuss that the potential of the electron e^-, which establishes the equilibrium reaction shown in Eq. 8, is appropriate as the origin of the electric potentials. If an electrode system based on the equilibrium reaction 8 could be materialized, we would be able to effectively connect the theory and our electrochemical experiments. Fortunately, the ideal electrode can be obtained, which is called a standard hydrogen electrode (SHE). The electrode can be constructed by bubbling hydrogen gas at 1 atom into an aqueous solution with a pH = 0.00 immersing a platinum wire (Fig. 3a). The equilibrium reaction 8 proceeds on the surface of the platinum, but because both the right- and left-hand reactions are fast, the potential difference at the electric double layer dose not fluctuate much. In order to reduce current density, a surface of the platinum wire is coated by a platinum black, increasing the surface area of the platinum wire several thousand times. Salt bridges are glass or vinyl tubes filled with agar containing potassium chloride (KCl), and its function is to allow ionic current to flow while preventing the transfer of substances between the electrolytic solution and the aqueous solution of the reference electrode. SHE is still adopted as the origin for electric potentials, but because actual experiments require the constant availability of hydrogen cylinders, SHE is no longer used

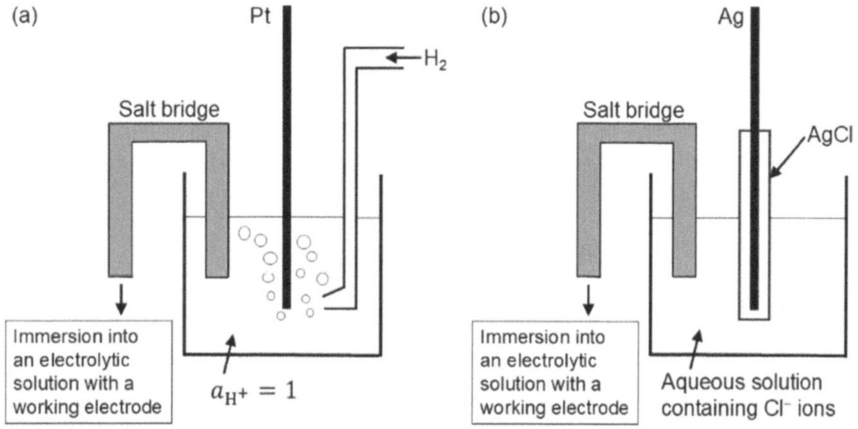

Fig. 3 Schematic drawings of **a** standard hydrogen electrode and **b** Ag/AgCl electrode. a_{H^+} represents the activity of H^+. Reproduced and adapted from Ref. [1, p. 41], with modifications

much. When the electrode potential is at 1.23 V versus, SHE, it is shown as follows:

$$+1.23 \text{ V versus SHE} \tag{12}$$

4.2 *Silver-Silver Chloride (Ag/AgCl) Electrode*

A thin film of AgCl can be formed on the surface of a sliver wire by electrochemical oxidation of the silver wire in a KCl solution. A system of aqueous solution containing Cl^- ions with the Ag wire coated by AgCl is called silver-silver chloride (Ag/AgCl) electrode (Fig. 3b). In the boundary surface between the thin layer of AgCl and the electrolytic solution, a reaction involving three substances, electrons, silver, and silver chloride, proceeds rapidly (Eq. 13).

$$\text{AgCl} + e^- \leftrightarrows \text{Ag} + \text{Cl}^- \tag{13}$$

Originally, the solution containing $[Cl^-] = 1.00$ M is standard. However, it is difficult to keep the concentration of the electrolyte constant, so in typical experiments, the Ag/AgCl wire is immersed in a saturated KCl solution. When the potentials of the Ag/AgCl electrodes are measured against SHE, the potentials are

$$+0.222 \text{ V versus SHE } \left(\left[\text{Cl}^-\right] = 1.00 \text{ M}\right) \tag{14}$$

$$+0.199 \text{ V versus SHE (Saturated KCl solution)} \tag{15}$$

4.3 Calomel Electrode

Calomel electrode is composed of mercury chloride (I) on mercury in contact with an aqueous KCl solution. Calomel is a common name for Hg_2Cl_2. As with the Ag/AgCl electrode, an equilibrium reaction involving three substances, electrons, mercury, and mercury chloride(I), proceeds both the right- and left-hand rapidly (Eq. 16).

$$Hg_2Cl_2 + 2e^- \leftrightarrows 2Hg + 2Cl^- \tag{16}$$

Potentials of calomel electrodes against SHE are

$$+0.268\,\text{V versus SHE}\ \left(\left[Cl^-\right] = 1.00\,\text{M}\right) \tag{17}$$

$$+0.241\,\text{V versus SHE (Saturated aqueous KCl solution)} \tag{18}$$

The correlation between the potentials of the three reference electrodes is shown in Fig. 4. We can convert the potential based on the Ag/AgCl electrode into that based on SHE by using the diagram. The reference electrodes such as SCE and Ag/AgCl are also adapted to electrochemical measurements in organic solvents. However, since solvation energies of the ions differ between the organic solvent and the aqueous solution used in the reference electrodes, the potential differences between the liquids, i.e., liquid junction potentials, should be corrected. For the electrochemical measurements in the organic solutions, the reference electrodes composed of the sliver wire soaking into the organic solution such as acetonitrile containing 0.01–0.1 M of Ag^+ ions are commercially available.

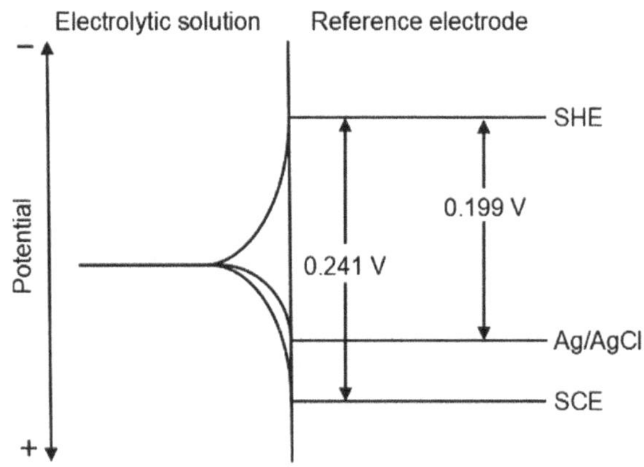

Fig. 4 Schematic diagram of the correlation of the potentials between the three reference electrodes

5 True Nature of E°

Because of the experimental nature of the previous two sections, one might misunderstand that the E° values are actual measured potentials. However, the E° values are calculated values obtained from thermodynamic data, $\Delta_f G^\circ$. For example, the E° value of the following Eq. 19 can be obtained as an ideal value when the activities of all chemical substances in the equation are assumed to be 1.

$$NO_3^- + 4H^+ + 3e^- \leftrightarrows NO(g) + 2H_2O \tag{19}$$

Using concentrations and partial pressure instead of the activities, concentrations of $[NO_3^+]$ and $[H^+]$ are 1 M. The partial pressure p_{NO} is 1 atm. However, these concentrations and partial pressures are not real, but rather virtual conditions in which there is no interaction between the ions and molecules at all. For example, actual acidic solutions, the magnitude of the interactions between the ions and molecules change depending on the type of counter anions, and the potentials also vary. Therefore, the actual measured potentials cannot be used as a standard. Relationships between the standard electrode potentials and measured potentials will be discussed later.

Although the standard electrode potentials are ideal values, they provide important information in electrochemical measurements including cyclic voltammetry. Some important knowledge obtained by the standard electrode potentials will be explained. Following explanations, the activities of all substances involved in electron transfer equilibrium reactions are assumed to be 1.

5.1 Abilities of Donating and Accepting Electrons for Ions and Molecules

In redox pairs where the E° values are negative and have large absolute values, the reductants are more easily oxidized, and the oxidants are less easily reduced. For example, the E° value of the following redox pair is negative and has one of the largest absolute values.

$$Li^+ + e^- \leftrightarrows Li \quad E^\circ\left(Li^+/Li\right) = -3.045\,\mathrm{V\ versus\ SHE} \tag{20}$$

In the case of the redox pair, the reductant and oxidant are Li and Li^+, respectively. The reductant can donate one electron easily, and the oxidant hardly accept one electron. On the other hand, although the E° value of the next redox pair is has one of the largest absolute values, the E° value has positive sign.

$$F_2 + 2H^+ + 2e^- \leftrightarrows 2HF \quad E^\circ(F_2/HF) = +3.053\,\mathrm{V\ versus\ SHE} \tag{21}$$

The reductant and oxidant are HF and F_2, respectively. The reductant can hardly donate one electron easily, and the oxidant easily accepts one electron.

5.2 Direction of Chemical Reaction

The correlation between the two electrochemical equilibria is discussed below.

$$5\left[\mathrm{Fe(CN)_6}\right]^{3-} + 5e^- = 5\left[\mathrm{Fe(CN)_6}\right]^{4-} \quad E^\circ = +0.36\,\mathrm{V}\ \text{versus SHE} \tag{22}$$

$$[\mathrm{MnO_4}]^- + 8\mathrm{H}^+ + 5e^- = \mathrm{Mn}^{2+} + 4\mathrm{H_2O} \quad E^\circ = +1.51\,\mathrm{V}\ \text{versus SHE} \tag{23}$$

The electrons transfer from a lower potential to a higher potential. Therefore, the redox reaction proceeds as $[Fe(CN)_6]^{4-}$ donates electrons and $[MnO_4]^-$ accepts them (see the following scheme).

$$5\left[\mathrm{Fe(CN)_6}\right]^{4-} \rightarrow 5\left[\mathrm{Fe(CN)_6}\right]^{3-} + 5e^- \tag{24}$$

$$[\mathrm{MnO_4}]^- + 8\mathrm{H}^+ + 5e^- \rightarrow \mathrm{Mn}^{2+} + 4\mathrm{H_2O} \tag{25}$$

Combining the two electrochemical equilibria, the overall reaction equation can be written as follows:

$$5\left[\mathrm{Fe(CN)_6}\right]^{4-} + [\mathrm{MnO_4}]^- + 8\mathrm{H}^+ \rightarrow 5\left[\mathrm{Fe(CN)_6}\right]^{3-} + \mathrm{Mn}^{2+} + 4\mathrm{H_2O} \tag{26}$$

In chemical reactions involving multiple electrochemical equilibria, it is easy to determine the direction in which the reaction proceeds by considering which redox pair will donate electrons by using an axis of potential where the top is negative, and the bottom is positive. Figure 5 depicts the spontaneous direction of electron transfer between the ferricyanide and permanganate half-reactions, based on their relative standard reduction potentials.

5.3 Required Voltage for Electrolysis

The following redox reaction 27 can be decomposed into two half-reaction Eqs. (28 and 29).

$$2\mathrm{Ag} + \mathrm{Cd}^{2+} \rightarrow 2\mathrm{Ag}^+ + \mathrm{Cd} \tag{27}$$

$$\mathrm{Cd}^{2+} + 2e^- \rightarrow \mathrm{Cd} \quad E^\circ = -0.403\,\mathrm{V}\ \text{versus SHE} \tag{28}$$

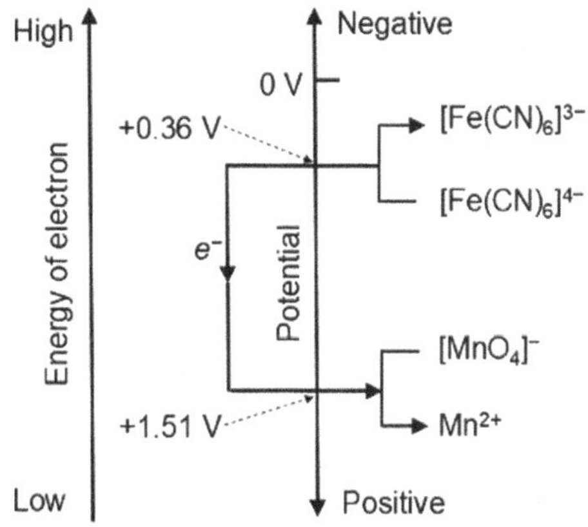

Fig. 5 Spontaneous electron transfer between permanganate and ferricyanide half-reactions illustrated via directional flow diagram

$$2Ag \rightarrow 2Ag^+ + 2e^- \quad E^\circ = +0.799\,V \text{ versus SHE} \tag{29}$$

The half-reactions indicated that the electrons should be transferred from the positive side of the potential to the negative side. In other words, the redox reaction would not proceed unless energy is applied, since the reaction involves the electron transfer processes from the stable to unstable states. In order the reaction to proceed with electrochemical energy, an electric potential greater than the difference in standard electrode potentials ΔE° must be applied between the two electrodes. In the case of reaction 27, the reaction proceeds when the electric potential above 1.2 V is applied between the electrodes. However, if reaction phases with high activation energy are included in middle steps, the potentials required for electrolysis would also be higher because the overvoltage is larger. For example, the ΔE° value in the electrolysis of water is 1.23 V, but an applied potential of 2 V or higher is required for the actual electrolysis of water to proceed sufficiently.

5.4 Determination of E° *for Electron Transfer Equilibrium*

From the two conditions shown in following Eqs. 30 and 31, what calculation would you use to find E for the redox reaction 32?

$$Cu^{2+} + 2e^- \rightarrow Cu \quad E^\circ = +0.337\,V \text{ versus SHE} \tag{30}$$

$$Cu^+ + e^- \rightarrow Cu \quad E^\circ = +0.520\ V \text{ versus SHE} \tag{31}$$

$$Cu^{2+} + e^- \rightarrow Cu^+ \quad E^\circ = ? \tag{32}$$

The redox reaction 32 can be derived by subtracting both sides of the half-reactions 31 from 30, respectively. However, since the energies of the electrons in each half-reaction equation are different, the E° value of reaction 32 cannot be obtained by similar subtraction. The E° value must be obtained by using Gibbs energies of each chemical species.

$$\Delta_f G^\circ(Cu^{2+}) - 2 \times 0.377F = \Delta_f G^\circ(Cu) = 0 \tag{33}$$

$$\Delta_f G^\circ(Cu^{+}) - 0.520F = \Delta_f G^\circ(Cu) = 0 \tag{34}$$

$$\Delta_f G^\circ(Cu^{2+}) - FE^\circ(Cu^{2+}/Cu^{+}) = \Delta_f G^\circ(Cu^{+}) \tag{35}$$

A simple calculation yields $E^\circ(Cu^{2+}/Cu^{+}) = +0.154$ V.

6 Formal Potential $E^{\circ\prime}$

The standard electrode potential E° is not an actual value. Therefore, even if the potentials of certain solutions in their equilibrium states are measured, the obtained potentials are always different from the standard electrode potentials. Therefore, the actual measured potential is denoted by the symbol $E^{\circ\prime}$, and is called the formal potential or conditional potential. Examples of formula potentials $E^{\circ\prime}$ and standard electrode potentials E° are summarized in Table 1. The formal potentials obtained from precise measurements can be used as quasi-standards of potentials and about 200 examples are described in various literatures [2–4].

Table 1 Standard electrode potentials E° and formal potentials $E^{\circ\prime}$

Electrochemical equilibrium	E° (V) versus SHE	$E^{\circ\prime}$ (V) versus SHE
$Ag^+ + e^- = Ag$	+ 0.799	+ 0.792 (in 1 M, $HClO_4$)
		+ 0.770 (in 1 M, H_2SO_4)
$Ce^{4+} + e^- = Ce^{3+}$	+ 1.71	+ 1.70 (in 1 M, $HClO_4$)
		+ 1.60 (in 1 M, HNO_3)
		+ 1.44 (in 1 M H_2SO_4)
$Fe^{3+} + e^- = Fe^{2+}$	+ 0.771	+ 0.710 (in 0.5 M, HCl)
		+ 0.530 (in 10 M, HCl)
		+ 0.680 (in 1M, H_2SO_4)
$Pd^{2+} + 2e^- = Pd$	+ 0.915	+ 0.792 (in 4 M, $HClO_4$)
$Pb^{2+} + 2e^- = Pb$	− 0.126	− 0.320 (in 1 M, CH_3COONa)

Adapted from Ref. [5]

As shown in Table 1, there are no formal potentials that match the standard electrode potentials. The formal potentials themselves also vary from several tens of mV to 300 mV depending on the type of the electrolytes and the concentrations. In some systems, the potentials fluctuate by nearly 1000 mV. Therefore, it is risky to judge the electron donor or acceptor ability of chemical substances present in solutions and/or in the air based solely on the standard electrode potentials. That said, it is rare for the differences between the formal potentials $E^{\circ\prime}$ and the standard electrode potentials E° to exceed 300 mV. Therefore, for chemical substances with a difference of more than 300 mV in standard electrode potential, it is acceptable to use the values of the standard electrode potential to discuss the ranking of oxidizing and reducing strengths of the substances.

7 Nernst Equation

In earlier sections, the activities of all substances constituting the electron transfer equilibrium equations were assumed to be 1. Naturally, the ratio of activities, a_O/a_R between an oxidant, O, and reductant, R, is also 1. If the ratio a_O/a_R is larger than 1, the oxidant can easily accept electrons from the electrode. Therefore, the potential of the electrode E shifts to a positive direction. In this section, the difference between the electrode potential E and the standard electrode potentialE° is discussed quantitatively when the ratio of activity a_O/a_R is not 1.

In a typical chemical equilibrium reaction 36, the Gibbs energy of 1 mol of a substance with an activity, a, also known as the chemical potential, can be expressed by Eq. 37. μ° is the chemical potential when the activity is 1.

$$p\mathrm{P} + q\mathrm{Q} + \cdots \text{(original system)} \leftrightarrows x\mathrm{X} + y\mathrm{Y} + \cdots \text{(product)} \tag{36}$$

$$\mu = \mu^{\circ} + RT \ln a \tag{37}$$

By using the relationship that the Gibbs energy of the original system is equal to the Gibbs energy of the resulting system, the law of mass action can be derived with the equilibrium constant, K (Eq. 38).

$$K = \frac{a_{\mathrm{X}}{}^{x} \cdot a_{\mathrm{Y}}{}^{y} \cdot \cdots}{a_{\mathrm{P}}{}^{p} \cdot a_{\mathrm{Q}}{}^{q} \cdot \cdots} \tag{38}$$

Electrons and ions involved in electrochemical equilibrium are electrically charged. Additionally, in both solutions and solids, an electrical potential E are formed at every point in space. Therefore, charged particles possess, in addition to their intrinsic energy, the electrical energy ε indicated by Eq. 39. Symbol Q denotes electric charges.

$$\varepsilon = EQ + \text{constant} \tag{39}$$

If the valence of particle i is z_i, for example, z_i of Na^+ is + 1 and z_i of $CO_3{}^{2-}$ is – 2, the charge of 1 mol of the particles is z_iF C. Therefore, the electrical energy of 1 mol of particles located at an electric potential E_1, measured from a reference point, can be expressed as z_iFE_1. Therefore, when dealing with electrochemical equilibrium involving charged particles, it is necessary to use the following equation, which adds electrical energy z_iFE_1 to the chemical potential μ_i.

$$\widetilde{\mu}_i \equiv \mu_i + z_iFE_i = \overset{\circ}{\mu_i} + RT\ln a_i + z_iFE_i \tag{40}$$

$\widetilde{\mu}$ is referred to as the electrochemical potential of charged particles. However, when considering the chemical equilibrium of a homogeneous solution, the potential E of the solution remains the same at all locations. Therefore, using either $\widetilde{\mu}$ or μ does not affect the result. By applying the law of mass action 38 and Eq. 40 to the electron transfer equilibrium Eq. 41 at the interface and organizing the results, the following Eq. 42 is obtained.

$$p\mathrm{P} + q\mathrm{Q} + ne^- = x\mathrm{X} + y\mathrm{Y} \tag{41}$$

$$E = E^\circ + \frac{RT}{nF}\ln\frac{a_{\mathrm{P}}{}^p \cdot a_{\mathrm{Q}}{}^q}{a_{\mathrm{X}}{}^x \cdot a_{\mathrm{Y}}{}^y} \tag{42}$$

If we consider the equilibrium simplified with the oxidant, O and the reductant, R, the equation corresponding to Eq. 44 is as follows.

$$\mathrm{O} + ne^- = \mathrm{R} \tag{43}$$

$$E = E^\circ + \frac{RT}{nF}\ln\frac{a_{\mathrm{O}}}{a_{\mathrm{R}}} \tag{44}$$

Equations 42 and 44 are referred to as Nernst equation. The Nernst equation was originally proposed by Walther Nernst in 1889 to describe the electrode potential as a function of ion concentration [6–8].

The Nernst equation provides quantitative information on the following two aspects.

1. The electrode potential changes in response to variations in the activity (or con centration) of a species.
2. The activity (or concentration) of a species changes in response to variations in the electrode potential.

The following section provides an explanation of electrolytic current.

8 Electrolytic Current Determined by Electrode Potential

8.1 Two-Electrode System

As shown in Sect. 4, if one electrode is used as the working electrode and the other as the reference electrode, the potential of the working electrode should theoretically be controllable. However, since the electrolyte solution also has resistance R, when a current I flows through the electrolyte solution, an IR voltage drop occurs within the solution. Therefore, even if a voltage ΔE is applied between the two electrodes, the effective voltage will be $\Delta E - IR$. Additionally, when a large current flows, the potential of the reference electrode itself may also shift. In order to accurately define the potential, it is necessary to minimize the current in the potential control circuit to reduce the IR voltage drop as much as possible. The three-electrode system circuit is essential for this purpose.

8.2 Three-Electrode System

In a three-electrode system, in addition to the working electrode W and the reference electrode R, an auxiliary electrode A is introduced, as shown in Fig. 6. A very small current is passed between the working electrode W and the reference electrode R to define the potential of the working electrode W, while the majority of the current flows between the working electrode W and the auxiliary electrode A. Additionally, the tip of the salt bridge of the reference electrode R is placed near the working electrode W to reduce the IR voltage drop. The device that automatically performs these operations is called a potentiostat, and using this device allows the potential of the working electrode W to be controlled within $\pm$ 1 mV.

8.3 Potential and Current

The electrode potential can be controlled using the potentiostat. Next, let's examine how the electrode potential affects the electrolytic current—in other words, how it influences the rate of the electrode reaction. Let's consider a redox system where the left side represents the reactant system and the right side represents the product system (Eq. 45). Here, we focus on the anodic reaction, treating the reactant system as the reducing agent.

$$\mathrm{R} \leftrightarrows \mathrm{O} + ne^- \tag{45}$$

Initially, this system is in equilibrium at the standard electrode potential E°. At that moment, the activation energy G^* is equal in both directions. When the potential

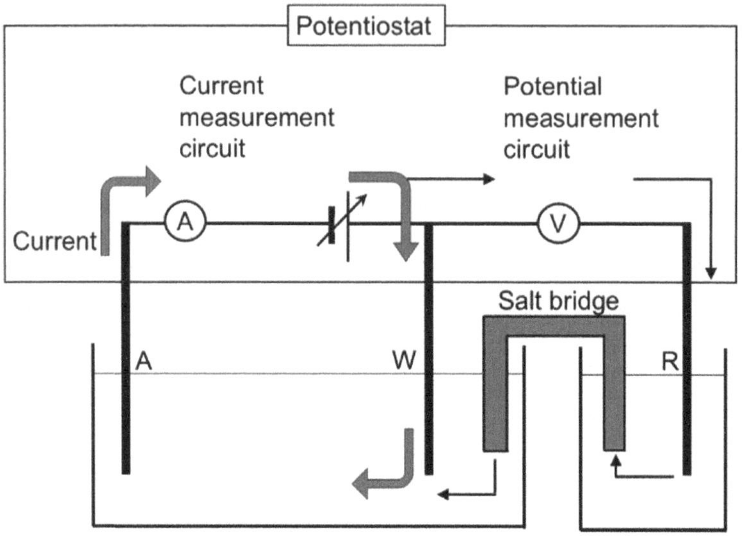

Fig. 6 Three-electrode system for the electrochemical measurement. Reproduced and adapted from Ref. [1, p. 62], with modifications

is shifted by η in the positive direction, i.e., polarization, the energy of the electron e^- in the product system changes from $-nFE$ to $-nF(E+\eta)$, causing the energy curve to sink by $-nF\eta$, as shown in Fig. 7.

The activation energy of this reaction decreases in the forward direction, meaning the anode reaction, and increases in the reverse direction, meaning the cathode reaction. The sum of the decrease and increase is equal to $nF\eta$. Therefore, if a fraction α ($0 < \alpha < 1$) of $nF\eta$ is allocated to the forward reaction and the remaining $1-\alpha$ to the cathode reaction, the activation energy changes as follows. If the polarization η is not too large, α is approximately 0.5.

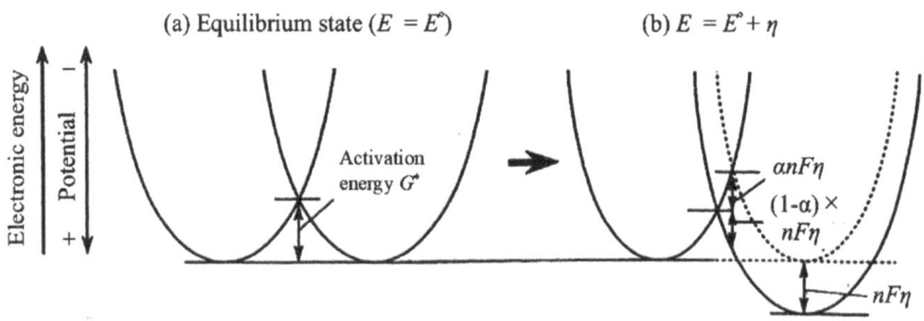

Fig. 7 Schematic illustration of potential shift and activation energy barrier. Reproduced and adapted from Ref. [1, p. 63], with modifications

$$\text{Anodic reaction}: \; G* \rightarrow G*-\alpha nF\eta \tag{46}$$

$$\text{Cathodic reaction}: \; G* \rightarrow G*-(1-\alpha)nF\eta \tag{47}$$

The magnitude of the electrolytic current I (simply expressed as $A = \mathrm{C\,s^{-1}}$) can be represented as the product of the following three quantities:

1. *the amount of substance undergoing electron transfer at the electrode per second* (unit: mol s^{-1}),
2. *number of electrons exchanged per particle, n, and*
3. *the Faraday constant, F* (unit: C mol^{-1}).

Furthermore, quantity 1 is proportional to the electrode area A cm^2 and the concentration c mol cm^{-3} of the reactant present near the electrode surface. The proportionality constant k is referred to as the rate constant of electron transfer. Considering the units of quantity 1, the unit of k is cm s^{-1}.

The total current I is the sum of the anodic (oxidation) current I_a and the cathodic (reduction) current I_c. The anodic current I_a is proportional to the product of the rate constant k_a and the concentration of the reduced species c_R, while the cathodic current I_c is proportional to the product of the rate constant k_c and the concentration of the oxidized species c_O. By convention, taking the anodic current as positive, Eq. 48 is obtained:

$$I = I_a + I_c = nFA(k_a c_R - k_c c_O) \tag{48}$$

Because electrochemical cells and electrodes used in electrochemistry are typically on the centimeter scale, it is common to express dimensions such as height and surface area in centimeter-based units rather than in SI units based on meters.

The value obtained by dividing the current I in Eq. 46 by the area A is called the current density, which is denoted by the lowercase letter i (unit: A cm^{-2}). Current density i is expressed by Eq. 49.

$$i = i_a + i_c = nF(k_a c_R - k_c c_O) \tag{49}$$

When the electrode potential is at its equilibrium value, E°, anodic and cathodic currents of exactly the same magnitude pass through the electrode surface in opposite directions. This current is referred to as the exchange current density. In this chapter, it is assumed that both c_a (reductant concentration) and c_o (oxidant concentration) remain constant at all times. Furthermore, for simplicity, we assume $c_a = c_o = c$. Under this condition, the activation energy is the same for both the forward and reverse reactions, and the rate constant can be expressed as Eq. 50, where "eq" indicates equilibrium

$$k_a = k_c = k_{eq} = k_0 \exp\left(-G^*/RT\right) \tag{50}$$

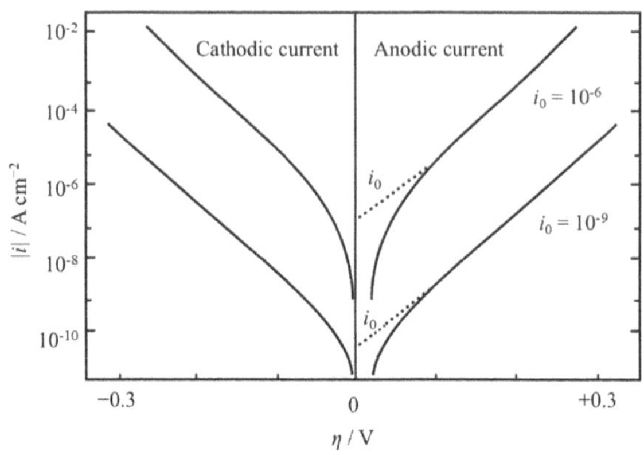

Fig. 8 Dependence of current density i on overpotential η under conditions of $\alpha = 0.5$ and $n = 2$. Reproduced and adapted from Ref. [1, p. 65], with modifications

Therefore, the exchange current density i_o can be expressed as Eq. 51:

$$i_0 = nFck_{eq} = nFck_0 \exp(-G^*/RT) \tag{51}$$

When the electrode potential is shifted by an overpotential η from its equilibrium value $E°$, the activation energies change as shown in Eqs. 46 and 47. Assuming $c_a = c_o = c$, the current density can be expressed by the following equation:

$$i = i_0\left[\exp(\alpha nF\eta/RT) - \exp\{-(1-\alpha)nF\eta/RT\}\right] \tag{52}$$

Equation 52 is known as the Butler–Volmer equation [9–11]. The anodic current i_a and cathodic current i_c, as functions of the overpotential η, are shown in Fig. 8.

The measured current value is the sum of the anodic current i_a and cathodic current i_c, as shown in Eq. 52; in terms of absolute values, it is their difference. However, when the overpotential η becomes sufficiently large—meaning the electrode potential deviates significantly from the equilibrium value $E°$—either i_a or i_c dominates, and the other can be neglected. In this case, the current becomes an exponential function of η, and log i varies linearly with η. This relationship is known as the Tafel relation. For example, when η is sufficiently large and positive, the anodic branch of Eq. 52,

$$\ln i_a = \ln i_0 + \alpha nF\eta/RT \tag{53}$$

represents the Tafel relation [8, 10–13]. At a temperature of $T = 298.15$ K, (ln 10) × RT/F becomes approximately 0.059 V. By converting the natural logarithm into a common (base-10) logarithm, Eq. 53 becomes

$$\log_{10} i_a = \log_{10} i_0 + \alpha n\eta/0.059 \tag{54}$$

By plotting the Tafel relation and extrapolating the line to the equilibrium potential ($\eta = 0$), the magnitude of the exchange current density i_0 can be determined.

Although i_0 cannot be measured directly—since it represents current flowing in both directions—it is a crucial parameter that reflects the performance of the electrode.

The overpotential η corresponding to a specified current density i is referred to as the overvoltage at that current density. The overvoltage serves as an indicator of the "slowness" of the reaction or the electrode itself. When the electrode material is changed, the ease with which processes such as molecular adsorption, desorption, bond cleavage, and surface diffusion occur also changes, resulting in variations in the overvoltage. Table 2 presents examples of overvoltages for hydrogen and oxygen evolution in water electrolysis.

When the electrode potential is altered, the energy difference between the reactant system and the product system involved in electron transfer changes easily. This variation leads to increases or decreases in the activation energy, thereby affecting the magnitude of the current (i.e., the reaction rate). The value of the activation energy varies depending on the specific reaction, and even for the same reaction, it differs depending on the electrode material. A "slow electrode" is essentially an electrode with a high activation energy, and to drive electrolysis on such an electrode, an extra amount of electrical energy (overvoltage) must be applied.

Now, when examining Fig. 8 and Eq. 54, one may notice something peculiar. According to the figure and equation, shifting the electrode potential by only 0.5 V would cause the current to increase by eight orders of magnitude (a hundred million times). However, in reality, such a current does not flow. Therefore, there must be some inconsistency. In fact, Eq. 54 and similar expressions are derived under the assumption that the concentration of reactants at the electrode surface remains constant. In other words, it is assumed that, regardless of how large the current becomes, the chemical species are never depleted. In electrode reactions involving a solvent like water (H_2O), this approximation holds reasonably well, as water molecules are present in essentially infinite quantity near the electrode. However, when solutes participate in electron transfer, the situation changes significantly. In the next section, we will consider how the electrolysis process changes when the quantity of reactants is finite.

Table 2 Overvoltages for hydrogen and oxygen evolution in water electrolysis (approximate values at $i = 1\ \mathrm{mA\ cm^{-2}}$)

Hydrogen evolution (in acidic solution)		Oxygen evolution (in alkaline solution)	
Electrode	Absolute value of overvoltage (V)	Electrode	Absolute value of overvoltage (V)
Pt	0.10	RuO_2	0.20
Fe	0.40	Fe	0.45
Cu	0.60	Ni	0.55
Zn	0.70	Graphite	0.60
Pb	0.80	Ag	0.60
Hg	1.00	Pt	0.75

Adapted from Ref. [1]

9 Mass Transport-Limited Electrolysis Current

The rate of a process that proceeds through multiple steps depends on the speed of the slowest step. Since this step governs the overall rate, it is referred to as the rate-determining step. In the previous section, we focused on the current governed by electron transfer at the electrode surface as the rate-determining process. In this section, we examine the current determined by mass transport, with particular emphasis on the electrode reactions of solutes.

9.1 Diffusion

The transport of substances occurs through a physical process known as diffusion. The amount of substance transported per unit time and unit area by diffusion is called the flux, J (unit: mol cm^{-2} s^{-1}), and it is proportional to the concentration gradient dc/dx (unit: mol cm^{-4}). This relationship is known as Fick's First Law, and the proportionality constant D (unit: cm^2 s^{-1}) is called the diffusion coefficient [8, 14]. Since the concentration gradient can be either positive or negative, and the flux J is directed toward regions of lower concentration, a minus sign is placed on the right-hand side of the equation.

$$J = -D(\mathrm{d}c/\mathrm{d}x) \tag{55}$$

Let us illustrate the rate of diffusion using a model. As shown in Fig. 9, consider a solution contained in a vessel with a cross-sectional area of 1 cm^2, where the concentration of a substance changes linearly from 0 to 0.1 M (10^{-4} mol cm^{-3}) over a distance of 10^{-3} cm (10 μm). This region is referred to as the diffusion layer. The concentration gradient dc/dx is 10^{-1} mol cm^{-4}. Since the diffusion coefficient D of molecules or ions in solution is typically on the order of 10^{-5} cm^2 s^{-1} (see next section), the flux J is -10^{-6} mol cm^{-2} s^{-1}. The quantity -10^{-6} mol corresponds to the amount of substance contained within Region B of Fig. 9, assuming its thickness is 0.01 cm. Thus, the substance is transported leftward at an average velocity of 0.01 cm s^{-1}. Whether in electrode reactions or in general chemical reactions, when the overall rate is governed by diffusion, the process is called a diffusion-limited reaction. Diffusion can occur even in the absence of a concentration gradient, and in typical chemical reactions, this is often the case. Once diffusion begins, boundary ① shifts leftward and boundary ② shifts rightward. As a result, the diffusion layer thickens, the concentration gradient diminishes, and the flux gradually weakens. In the case illustrated in Fig. 9, if a liquid layer ten times thicker than the diffusion layer moves leftward every second, the flow pattern changes almost instantaneously.

Fig. 9 Conceptual model of mass transport by diffusion. Reproduced and adapted from Ref. [1, p. 74], with modifications

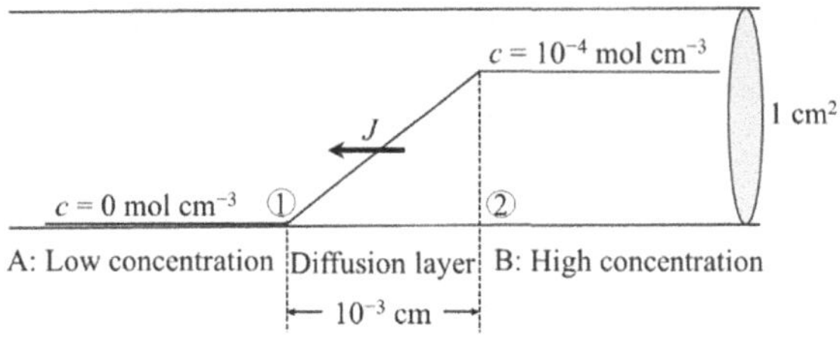

9.2 *Dynamics at the Electrode Interface*

Consider immersing an electrode in a solution containing a reductant R at bulk concentration c_{Rb} (where "b" denotes the bulk of the solution), and setting the electrode potential to E. Let E be a value that allows the following oxidation reaction to proceed smoothly, with O as the oxidized form:

$$\mathrm{R} \rightarrow \mathrm{O} + \mathrm{n}e^{-} \tag{56}$$

In other words, we are considering the case where the overpotential $\eta = E + E^{\circ}$ is sufficiently large. Let us now examine what happens over time.

(1) Charging of the Electric Double Layer

As discussed in Sect. 1, the charging of the electric double layer occurs first. If the electrolyte concentration is sufficiently high, this process is completed in about 0.01 s, preparing the environment for the electrochemical reaction to take place. Such transient currents are referred to as non-Faradaic currents.

(2) Electron Transfer-Limited Electrode Reaction

Reaction 56 begins. At the interface, the reductant R is consumed and the oxidant O increases. The current density i (Faradaic current) [8, 15] at this point is given by the generalized Butler–Volmer equation for the case $c_R \neq c_O$, assuming $\alpha = 0.5$:

$$i = nFk_0\left[c_R\exp(nF\eta/2RT) - c_O\exp\{-nF\eta/2RT\}\right] \tag{57}$$

When $\eta \gg 0$, the first term inside the brackets dominates and can be considered alone for practical purposes. As the reaction proceeds, c_R decreases (Fig. 10). The surface concentration of R (c_R) continues to decrease—effectively approaching zero—until the molar ratio (activity ratio) c_O/c_R reaches the value prescribed by the Nernst equation. In that case, the total current ($\propto c_R$) would also be expected to approach zero—but in practice, this does not occur.

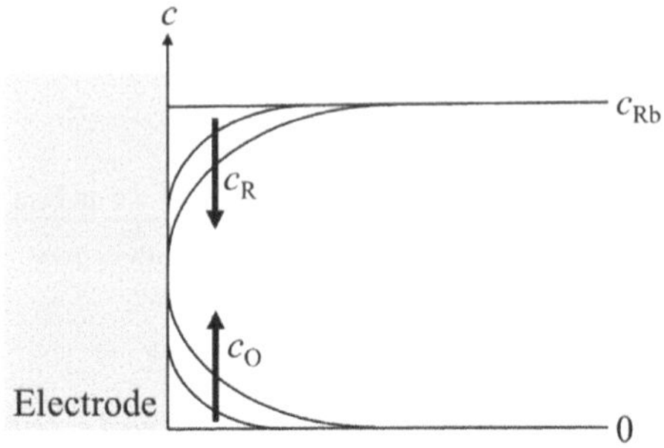

Fig. 10 Concentration changes at the electrode interface during electron transfer

9.3 Diffusion-Limited Electrode Reactions

The surface concentration of the reductant R decreases to nearly zero within approximately one second. However, as soon as a concentration gradient forms near the surface, *R* begins to diffuse toward the electrode, and the resulting current is governed primarily by this diffusion process. In other words, the nature of the current changes significantly at this stage.

To express the diffusion-limited current density *i*, let us consider one-dimensional diffusion with the electrode surface defined as $x = 0$. According to Eq. 55, the flux of *R* (J_R, in mol cm^{-2} s^{-1}) impacting the electrode surface per unit time and per unit area is proportional to the concentration gradient (dc_R/dx) at $x = 0$:

$$J_R = -D_R(dc_R/dx)_{x=0} \tag{58}$$

Assuming that oxidation occurs instantly upon contact with the surface, the anodic current density *i* can be expressed as

$$i = nFJ_R - nFD_R|dc_R/d_x|_{x=0} \tag{59}$$

As diffusion (electrolysis) progresses, the spatial distribution of R's concentration evolves as illustrated in Fig. 11. The diffusion layer widens, the concentration gradient at the surface decreases, and the electrolytic current diminishes accordingly. This phenomenon can be described as follows.

The concentration of the reductant *R*, c_R, is a function of distance *x* from the surface and time *t*, expressed as $c_R(x, t)$, and its behavior obeys Fick's Second Law [8, 14]:

$$\frac{\partial c_R(x,t)}{\partial t} = \frac{D_R \partial^2 c_R(x,t)}{\partial x^2} \tag{60}$$

By solving this partial differential equation under suitable boundary conditions and substituting the solution into Eq. 59, the current density *i* is derived as

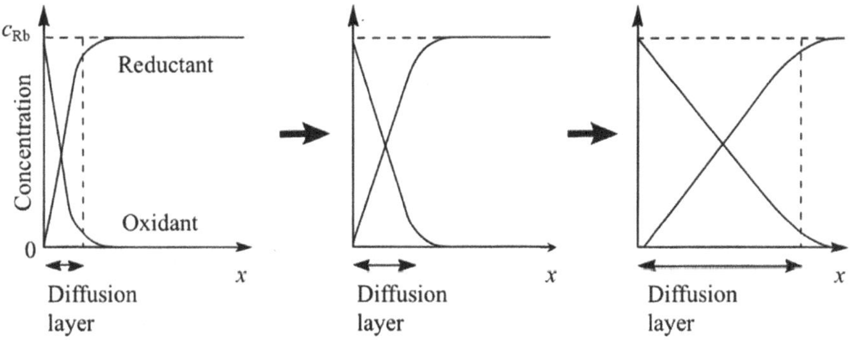

Fig. 11 Time-dependent growth of the diffusion layer during electrode reaction. Reproduced and adapted from Ref. [1, p. 76], with modifications

$$i(t) = nFc_R D_R^{1/2} \pi^{1/2} t^{-1/2} \tag{61}$$

Thus, the current decreases in inverse proportion to the square root of time t. Equation 61 is known as the Cottrell equation [8, 16–18]. If the bulk concentration c_{Rb} of the reductant is known, the value of the diffusion coefficient D_R can be estimated from the slope of the experimental line plotted as shown in Fig. 12.

Once the electrode potential is set to a value that enables rapid electron transfer, three types of current flow in succession (Fig. 13):

1. Non-Faradaic current associated with charging the electrical double layer (occurring on the order of 0.01 s),
2. Faradaic current limited by the rate of electron transfer (lasting for about 1 s), during which the surface concentration of the reactant decreases from its initial bulk concentration c_b to nearly zero, and
3. Faradaic current governed by diffusion processes.

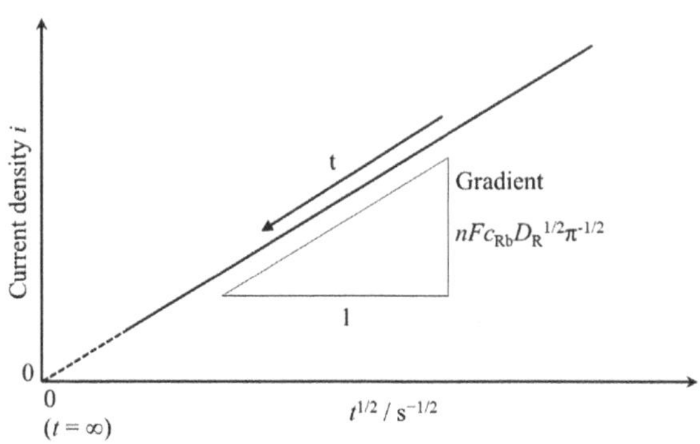

Fig. 12 Diffusion-limited current as a function of time reproduced and adapted from Ref. [1, p. 77], with modifications

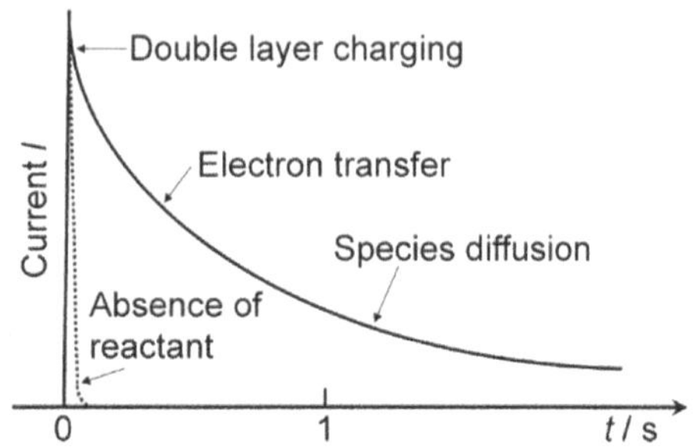

Fig. 13 Model of I–t curve generated by electrode reaction of a solute. Reproduced and adapted from Ref. [1, p. 78], with modifications

9.4 Electrophoretically Determined Current

No matter how fast electron transfer is, and how abundant the reactant is near the electrode surface, electrolysis current does not flow unless there are ions available to neutralize the surplus charge generated during electron transfer. What happens when the ion concentration is extremely low? Electrolysis in ultrapure water illustrates this well. Although H_2O molecules (the reactant) are abundantly present near the electrode and sufficient electric energy is supplied, the lack of counter ions (H^+ or OH^-) makes it difficult for current to flow.

The applied voltage in electrolysis is primarily used to mobilize ions in the solution. Because ion velocity is proportional to the electric field, the relationship between electrolysis current and voltage becomes linear. Most of the applied voltage is dropped across the electrolyte, resulting in relatively low energy efficiency for the electrochemical process. For these reasons, electrolysis in which electrophoresis becomes the rate-determining step is generally not employed.

9.5 Thermal Motion

Thermal motion of particles plays an important role in understanding electrode reactions and chemical reactions. Here, we provide a brief overview of particle thermal motion.

1. **Particle Velocity**

One mole of particles moving at velocity v possesses kinetic energy equal to $(1/2)Mv^2$, where M is the molar weight in kg mol^{-1}. According to statistical mechanics, translational motion of one mole of a substance at absolute temperature T is associated with an energy of $(3/2)RT$. Therefore, the following relationship holds

$$Mv^2 = 3RT \tag{62}$$

This equation allows for an approximate estimation of the average particle velocity. For example, at 25 °C, the average velocities of N_2 and H_2O molecules are approximately 520 m s^{-1} and 640 m s^{-1}, respectively. These values are largely similar whether in liquid or gaseous phases.

2. **Collision Frequency**

Particles moving at extremely high speeds collide with other particles shortly after traveling even small distances. The number of collisions a single particle makes with others per second can be estimated based on the density of the population and the molecular size.

Assuming that both N_2 and H_2O molecules are spherical with a diameter of 0.4 nm, we can calculate that, at room temperature, an N_2 molecule (in gas phase at 1 atm) undergoes approximately 900 million collisions per second, while an H_2O molecule (in the liquid phase) experiences roughly 15 trillion collisions per second. For a solute at a concentration of 1 mol L^{-1}, the collision frequency reaches the order of 10^{11} per second.

3. **Diffusion Distance**

Due to their intense thermal motion, particles constantly change position. Diffusion refers not only to movement caused by concentration gradients, as discussed in the previous section, but also to movement driven purely by thermal motion (see Fig. 14). In three dimensions, the average distance a particle travels in t seconds is given by the following equation:

$$\bar{x} = \sqrt{6Dt} \tag{63}$$

The diffusion coefficient D varies significantly depending on the substance or system [2, 17–22]. Small molecules and ions dissolved in solution typically have diffusion coefficients of $D = (1\text{–}3) \times 10^{-5}$ cm^2 s^{-1}, corresponding to an average displacement of about 0.01 cm in one second. This is comparable to the distance traveled in diffusion driven by concentration gradients; however, in the case of thermal motion, the movement lacks directional flow.

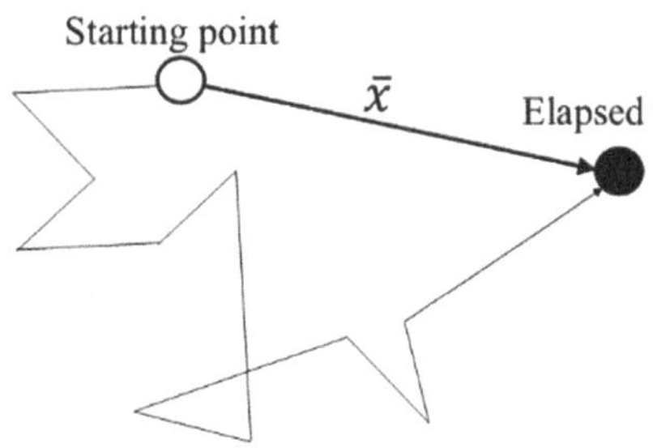

Fig. 14 Diffusion due to thermal motion

9.6 Thermal Motion and Reaction Rate

1. **Ordinary Chemical Reactions**

For the reaction between molecules A and B, expressed as

$$\mathrm{A} + \mathrm{B} \rightarrow \mathrm{P} + \mathrm{Q} + \cdots,$$

the rate is given by the following equation:

$$\frac{\mathrm{d}[\mathrm{A}]}{\mathrm{d}t} = -k[\mathrm{A}][\mathrm{B}] \tag{64}$$

The unit of the rate constant k is $\mathrm{M}^{-1}\ \mathrm{s}^{-1}$, which represents the number of certain events occurring per second per 1 mol/L concentration. These events correspond to collisions between molecules that lead to overcoming an energy barrier. If the probability of overcoming the barrier is unity—meaning the activation energy is zero—then k equals the collision frequency itself. At a concentration of 1 mol/L, the collision frequency is on the order of $10^{11}\ \mathrm{s}^{-1}$. Thus, the maximum reaction rate constant in nature is approximately:

$$k \approx 10^{11}\ \mathrm{M}^{-1}\ \mathrm{s}^{-1} \tag{65}$$

2. **Electron Transfer Reactions in Solution**

In electron transfer reactions, the solvent environments of the reactants and products differ, so activation energy E^* inevitably exists. For small organic molecules, E^* is typically around 20 kJ mol^{-1}. Therefore, the maximum rate constant is estimated as the product of $10^{11}\ \mathrm{M}^{-1}\ \mathrm{s}^{-1}$ and $\mathrm{e}^{-E^*/RT}$, resulting in a value of roughly $10^{8}\ \mathrm{M}^{-1}\ \mathrm{s}^{-1}$.

3. **Electrode Reactions**

The rate constant for electron transfer at an electrode is expressed in cm s^{-1}. The value corresponding to collision velocity is about 10^4 cm s^{-1}. For the same reaction, the activation energy at the electrode is considered to be half that in solution, so the maximum rate constant is approximately 10^2 cm s^{-1}.

At the electrode interface, particles exhibit a variety of dynamic behaviors that influence current flow. Even in the absence of an electrode, a single solute particle undergoes about 100 billion collisions per second with other particles—and under favorable conditions, a reaction may occur. In the next section, we explore electrode reactions where both potential and mass transport interplay.

10 Voltammetry

10.1 Static and Dynamic Potentials

As discussed previously, electrochemical systems drive redox reactions using electrical energy. The electrode potential E serves to control this energy, while the current I represents the rate at which the reaction proceeds. Up to the previous section, we assumed that the electrode potential is held constant and examined how the current varies with time under that condition.

Now, what happens when the electrode potential E is continuously varied? How does the reaction rate (i.e., the current I) change in response, and what insights can be gained from such measurements? In this section, we explore the realm of dynamic potentials.

Gradually varying the potential over time is known as "potential scanning" or "potential sweeping." By recording the current while scanning the potential, a plot of current versus potential—known as a current–potential curve—can be obtained. This measurement technique is called voltammetry, and the resulting current–potential profile is referred to as a voltammogram.

Voltammetry is typically performed under conditions where the solution remains still, known as stationary voltammetry. In special cases, such as with a rotating electrode or under conditions of stirring or forced convection, the technique is classified as convective voltammetry.

In this section, we focus on the most versatile method: stationary voltammetry conducted by repetitively scanning the potential over a defined range. This technique is called cyclic voltammetry, abbreviated as CV. Unless otherwise noted, the term "voltammetry" will refer specifically to cyclic voltammetry throughout the remainder of this text.

10.2 Fundamentals of Voltammetry

Measurement Setup

To perform voltammetry, the potentiostat shown in Fig. 6 must be connected to a function generator that produces various potential waveforms over time, as well as to a recorder that captures current–potential curves or current–time profiles (Fig. 15). The working electrode (W) may be either a reactive electrode (e.g., zinc) that itself undergoes electrochemical reactions, or an inert electrode (such as platinum, gold, or glassy carbon) that remains unreactive over a wide potential range. Inert electrodes are typically used to investigate the behavior of redox-active species in solution.

For the reference electrode (R), an Ag/AgCl electrode or a saturated calomel electrode (SCE) is commonly used (see Sect. 1.2). The auxiliary electrode (A), which

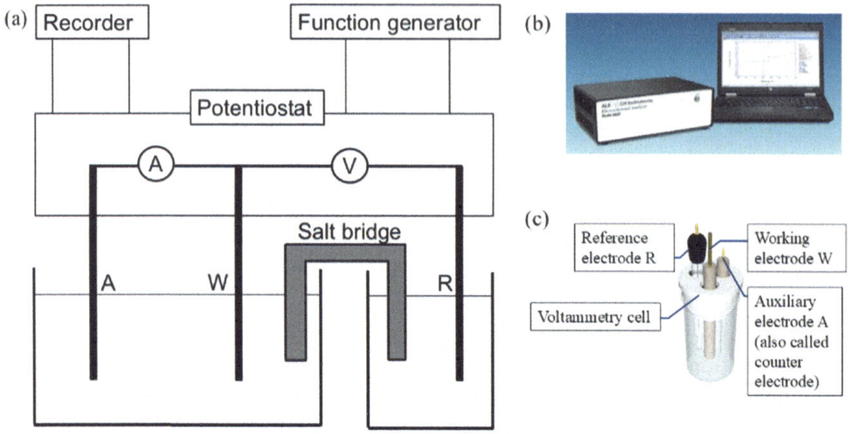

Fig. 15 **a** Schematic diagram of a voltammetric measurement circuit, **b** potentiostat and data acquisition PC, and **c** physical appearance of the measurement cell. Figures (b) and (c) reproduced from the official website of Bioanalytical Systems Inc. (BAS Inc.) [23]

serves as the current sink, should be an inert electrode with as large a surface area as possible in order to minimize resistance and avoid interference with potential control.

Electrolyte Solution

The solvent used may be water or an organic solvent such as acetonitrile. A supporting electrolyte that remains stable within the potential range relevant to electron transfer of the analyte is dissolved at a concentration of 0.1–1 M. A high concentration of supporting electrolyte increases the conductivity of the solution, allowing the electric field to rapidly extend to the electrode surface. Additionally, when the reactant is ionic, its transference number decreases, resulting in near-pure diffusion-based transport to the electrode surface.

To this prepared electrolyte solution, the analyte is added at a concentration of several millimoles. Since dissolved oxygen can interfere with reduction currents, it should be removed by purging the solution with an inert gas such as nitrogen or argon. When volatile solvents are used, care must be taken to prevent changes in analyte concentration over time.

Background Measurement

Prior to dissolving the analyte, a solution containing only the supporting electrolyte is prepared, and the potential is scanned over a range in which no redox reactions of the solvent or electrolyte occur. When a fixed potential is applied, no Faradaic current flows between the electrode and the electrolyte. In this case, the system is equivalent to a circuit consisting of the solution resistance R and the electrical double-layer capacitance C connected in series. The charging current decays with a time constant RC (typically around 0.01 s), as expressed by Eq. 66:

$$I(t) = (V/R)e^{-t/RC} \tag{66}$$

However, if the potential is scanned as shown in Fig. 16a, a continuously varying potential is applied to the electrode, resulting in ongoing charging of the double layer. In this case, the relationship between current and time is given by

$$I = \left[\{(E_i/R) - vC\}\exp(-t/RC)\right] + vC \tag{67}$$

At the onset of the scan, a current of E_i/R flows, and after approximately $3 \times RC$, the system reaches a steady-state current of magnitude vC. If the scan is reversed at potential E_λ, the current also reverses direction. A plot of the forward and reverse scans yields the current–potential curve shown in Fig. 16d. Note that the charging current is proportional to the scan rate v.

In actual measurements, the capacitance C may vary with the applied potential, and additional electrolysis currents from impurities may also be present. The combined contributions from these sources are referred to as residual current.

By conducting such background measurements, one can determine the potential window in which the solvent and supporting electrolyte remain stable, as well as evaluate the degree of purification of these components and the extent to which dissolved oxygen has been removed.

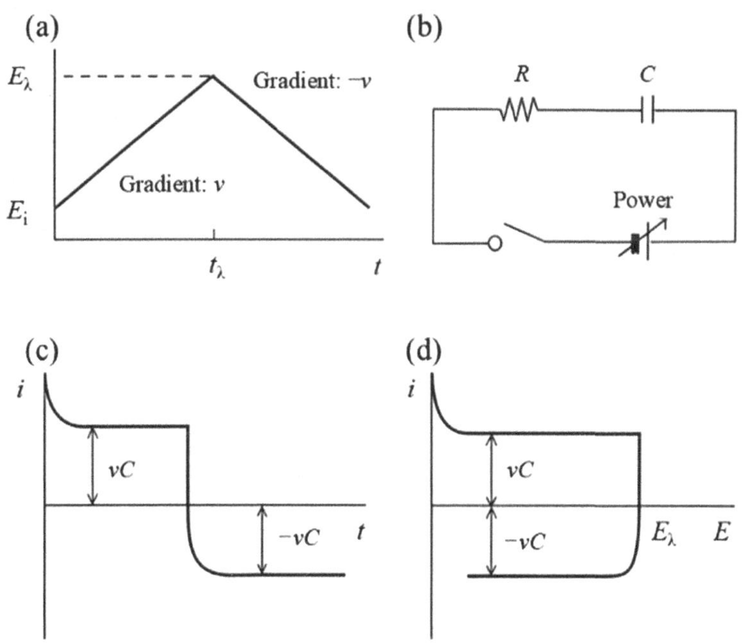

Fig. 16 Cyclic voltammogram of a solution containing only supporting electrolyte. Reproduced and adapted from Ref. [1, p. 88], with modifications

Measurement of the Analyte

The analyte R is dissolved into the previously prepared solution containing only the supporting electrolyte. Let us consider the case where R undergoes the following reversible electron transfer reaction:

$$R \leftrightarrows O + ne^- \tag{68}$$

Assuming that the electron transfer kinetics are sufficiently fast, the concentrations of the reductant (R) and oxidant (O) at the electrode surface are expected to change rapidly in accordance with the Nernst equation (Eq. 69) when the electrode potential E is applied. Here, $E^{\circ\prime}$ denotes the formal potential.

$$E = E^{\circ\prime} + (RT/nF)\ln(c_O/c_R) \tag{69}$$

Scanning the potential back and forth over the range $E_i \sim E^{\circ\prime} \sim E_\lambda$ results in the current response illustrated in Fig. 17c and d. Under these conditions, the equivalent circuit is represented in Fig. 17b, where the Faradaic current flows through the reaction resistance R_f. This resistance is not constant—it is considered to vary inversely with the magnitude of the current. It is desirable for the applied electrode potential E (referenced against the reference electrode) to drop almost entirely across the parallel circuit consisting of R_f and the double-layer capacitance C. To achieve this, the supporting electrolyte concentration is increased and the electrode configuration is optimized to minimize the solution resistance R, thereby reducing the IR drop between the working and reference electrodes.

A detailed analysis of the voltammogram shown in Fig. 17d will be presented in the following section.

10.3 Analysis of the Voltammogram

Processes at the Interface

In an electrolytic solution containing species R, as defined in Eq. 68, suppose the potential of the working electrode (W) is scanned from $E_i = E^{\circ\prime} - 0.2$ V to $E_\lambda = E^{\circ\prime} + 0.2$ V. As a result, the voltammogram shown in Fig. 18 is obtained. Under the assumption that electron transfer is sufficiently fast, the current I is governed primarily by mass transport and can be expressed by the following equation:

$$I = nFAD_P|dc_P/dx|_{x=0} \ (P = R \text{ or } O) \tag{70}$$

Based on this equation, we can trace the changes from point a to point j on the voltammogram.

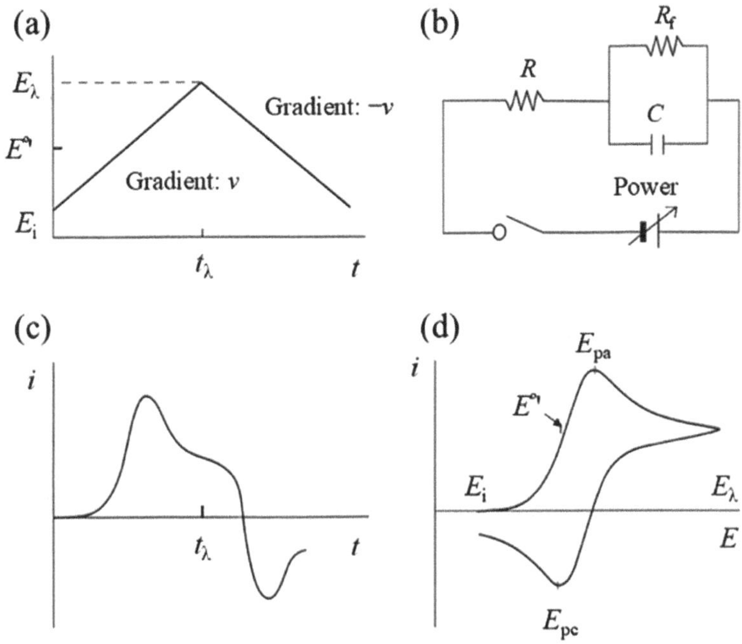

Fig. 17 Cyclic voltammogram of a solution containing electroactive species (cf. Fig. 16). Reproduced and adapted from Ref. [1, p. 89], with modifications

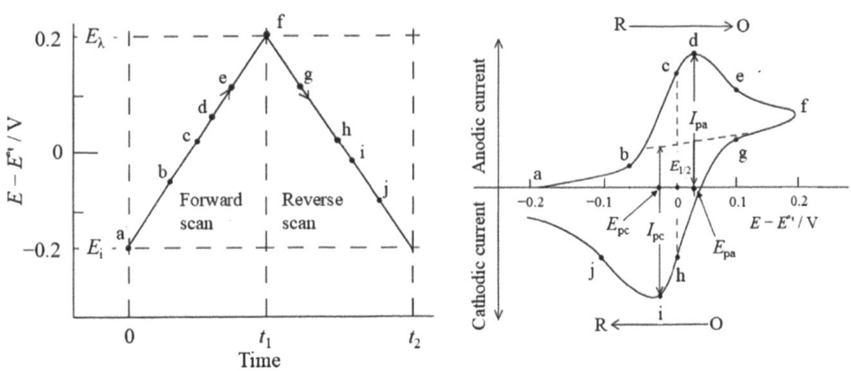

Fig. 18 **a** Potential scan profile, and **b** cyclic voltammogram resulting from the scan shown in (**a**). Reproduced and adapted from Ref. [1, p. 90–91], with modifications

(1) Point a

From Eq. 70 and assuming $n = 1$, the ratio $c_O/c_R \approx 4 \times 10^{-4}$, indicating that only R is present near the electrode surface.

(2) a → d

The surface concentrations of R and O change in accordance with the Nernst equation (Eq. 69). As time progresses, the diffusion layer expands. However, the effect of the increasing concentration gradient of R (the "c_R phenomenon") dominates, causing the current I to increase approximately exponentially. Thus, from point a to point d, the current is primarily determined by the change in potential E.

When the potential is stepped from E_i to b → d, the resulting current can be described by

$$I(t) = nFAD_R c_{Rb}(\pi D_R t)^{-1/2} \times \left[1 + (D_R/D_O)^{1/2}\exp\{(nF/RT)\left(E^{\circ\prime} - E\right)\}\right]^{-1} \quad (71)$$

(3) d → f

At point e, the potential E is approximately 0.1 V above $E^{\circ\prime}$. According to Eq. 69 with $n = 1$, $c_O/c_R \approx 500$, so c_R at the electrode surface can be considered nearly zero. Under these conditions, the current is governed solely by the elapsed time since c_R became zero, independent of both the potential E and the scan rate v.

In Eq. 71, the exponential term approaches zero, so the bracketed expression becomes approximately 1, and the resulting current–time curve after the potential step matches the Cottrell equation (Eq. 61).

This illustrates that the dominant factors shaping the voltammogram shift drastically around the peak current at point d.

(4) f → g

After the potential reversal, the system continues as described in (3), with the diffusion layer continuing to expand over time and the current gradually decreasing.

(5) g → i

The oxidized species O, generated during the forward scan, is reduced back to R in accordance with the Nernst equation (Eq. 69), producing a reduction current. This current, like that observed during the a → d scan, is mainly dictated by the applied potential E.

(6) i → j → initial potential E_i

This segment reflects the reverse of the d → e → f process, with the current decreasing as the diffusion layer continues to grow.

These changes result from the concentration profiles of R and O depicted in Fig. 18. By comparing Figs. 18 and 19, one can clearly understand the physical basis behind the distinctive shape of the voltammogram.

Information Provided by the Voltammogram

For an O/R system exhibiting sufficiently fast electron transfer (see Fig. 13), the voltammogram allows the definition of several parameters: the anodic peak current

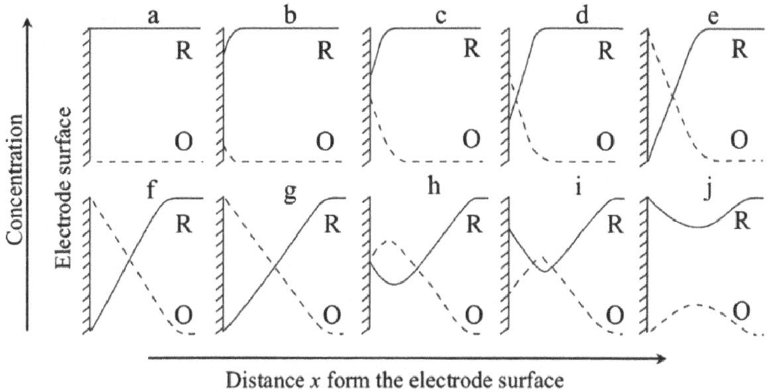

Fig. 19 Concentration profiles of oxidized and reduced species at selected points in the voltammogram shown in Fig. 16. Reproduced and adapted from Ref. [1, p. 82], with modifications

I_{pa}, anodic peak potential E_{pa}, the potential at which the current is half of I_{pa}, $E_{p/2}$, the half-wave potential $E_{1/2}$, cathodic peak current I_{pc}, and cathodic peak potential E_{pc}. These values are related to the number of electrons transferred n, the diffusion coefficient D in cm^2 s^{-1}, the scan rate v in V s^{-1}, and the concentration c in mol cm^{-3} by the following equations:

$$\begin{aligned} I_{pa} &= 0.4463nFAc_{Rb}(nF/RT)^{1/2}v^{1/2}D_R^{1/2} \\ &= 2.69 \times 10^5 n^{3/2}Ac_{Rb}v^{1/2}D_R^{1/2}(25\,°C) \end{aligned} \tag{72}$$

$$E_{pa}(mV) = E_{1/2} + 28.5/n(25\,°C) \tag{73}$$

$$E_{p/2}(mV) = E_{1/2} - 28.5/n\,(25\,°C) \tag{74}$$

The average of E_{pa} and E_{pc} is referred to as the half-wave potential, denoted as $E_{1/2}$. This value is related to the formal potential $E^{o\prime}$ from the Nernst equation (Eq. 69), as follows:

$$E_{1/2} = E^{o\prime} - (RT/nF)\ln(D_O/D_R)^{1/2} \tag{75}$$

If $D_R \approx D_O$, then $E_{1/2} \approx E^{o\prime}$. Thus, the voltammogram provides access to the key quantity $E^{o\prime}$.

In reverse scans, behavior depends on the switching potential E_λ. If $E_\lambda > E^{o\prime} + 0.2$ V, the difference between the anodic and cathodic peak potentials (the peak separation ΔE_p) becomes

$$\Delta E_p(mV) = E_{pa} - E_{pc} = 57/n(25\,°C) \tag{76}$$

ΔE_p reflects the reversibility of the electrode reaction: for a reversible one-electron transfer ($n = 1$), ΔE_p equals 57 mV. As irreversibility increases, this value becomes larger.

Voltammetry of Ferrocene

Ferrocene (Fc) is a well-known compound that undergoes a fast and reversible one-electron transfer ($n = 1$). Its redox equilibrium with ferrocenium (Fc^+) is described by the following reaction:

$$Fc \leftrightarrows Fc^+ + e^- \quad (77)$$

Figure 20 shows voltammograms obtained by dissolving 0.1 M $LiClO_4$ (supporting electrolyte) and 1 mM ferrocene in dehydrated, purified acetonitrile, followed by degassing with argon. Measurements were carried out using a platinum disk as the working electrode, a saturated calomel electrode (SCE) as the reference electrode, and a platinum wire coil as the auxiliary electrode, while scanning the potential from $E_i = +\ 100$ mV to $E_\lambda = +\ 600$ mV at various scan rates.

As the scan rate v increases, the peak current rises while the peak potential remains nearly constant, yielding a peak separation ΔE_p of approximately 60–70 mV. Although this is slightly larger than the theoretical value of 57 mV for a reversible one-electron transfer ($n = 1$), the deviation is attributed to *IR* drop effects in real measurements. Therefore, the response can be considered nearly reversible.

From the values of E_{pa} and E_{pc}, the half-wave potential $E_{1/2}$ ($\approx E^{\circ\prime}$) is determined to be + 310 mV versus SCE.

At higher scan rates, the charging current of the electrical double layer increases, resulting in broader waveforms near the switching potential (+ 600 mV). Furthermore, plotting the anodic peak current I_{pa} versus the square root of the scan rate reveals a linear relationship in accordance with Eq. 72 (Randles–Ševčík equation)

Fig. 20 Cyclic voltammograms of ferrocene in acetonitrile solution for various sweep rates. Reproduced and adapted from Ref. [1, p. 94], with modifications

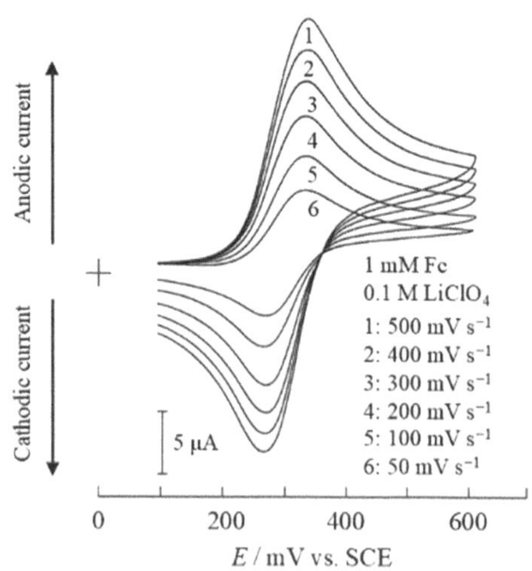

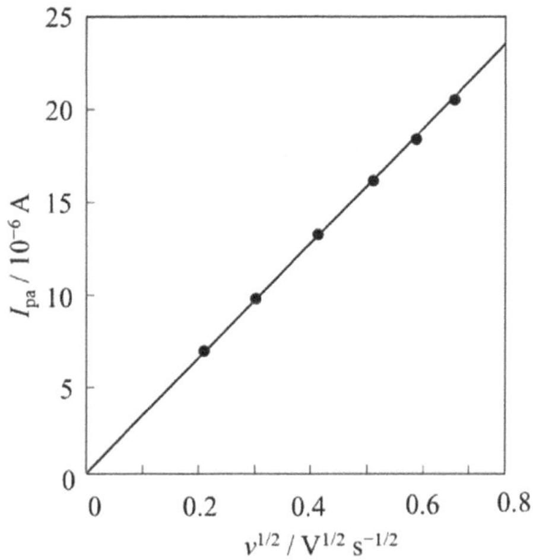

Fig. 21 Linear regression plot of peak current (I_p) versus the square root of sweep rate ($v^{1/2}$) based on the Randles–Ševčík equation. Reproduced and adapted from Ref. [1, p. 94], with modifications

[8, 24–26], allowing the diffusion coefficient D of ferrocene to be determined (see Fig. 21).

Voltammograms of Irreversible Systems

The voltammograms discussed thus far have pertained to systems with sufficiently rapid electron transfer. When the electron transfer becomes slower, the shape of the voltammogram deviates from the form shown in Fig. 18. As the scan rate v increases, the anodic peak potential E_{pa} shifts positively and the cathodic peak potential E_{pc} shifts negatively, causing the peak separation ΔE_p to widen progressively.

At a fixed scan rate, this effect becomes more pronounced as the standard rate constant of electron transfer k_0 decreases (Fig. 22). This is because when the potential is set to a given value E, the surface concentration ratio of the oxidized and reduced species does not immediately reflect the Nernst equation. A more positive potential is required to bring the ratio to its expected value.

A similar deviation occurs when the solution resistance is large: the true electrode potential differs from the applied potential by the amount of IR drop. As the scan

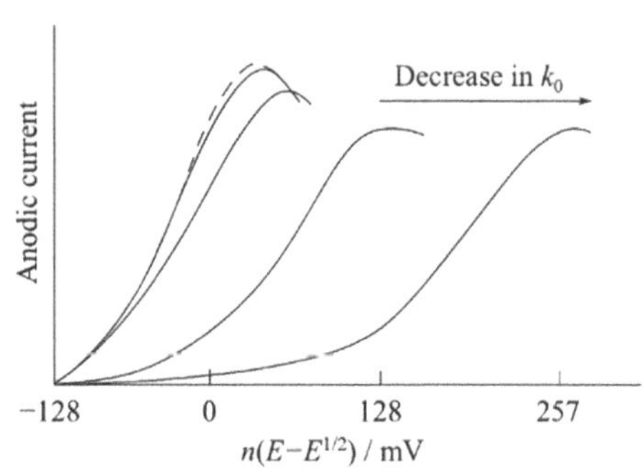

Fig. 22 Effect of standard heterogeneous electron transfer rate constant k_0 on cyclic voltammogram shape (dashed line: reversible response). Reproduced and adapted from Ref. [1, p. 95], with modifications

rate v increases, the current I also becomes larger, resulting in greater distortion of the voltammogram—an effect similar to that observed when k_0 is small.

Reversibility of Electrode Reactions

The closer the voltammogram is to its ideal form, the higher the reversibility of the electron transfer. The degree of reversibility is determined by the standard rate constant of electron transfer k_0, in cm s^{-1}, the scan rate v, and the number of electrons involved in the reaction n, with the following criteria:

Reversible: $k_0 > 0.3\ (nv)^{1/2}$,
Quasi-reversible: $0.3\ (nv)^{1/2} > k_0 > 2 \times 10^{-5}\ (nv)^{1/2}$,
Irreversible: $k_0 < 2 \times 10^{-5}\ (nv)^{1/2}$.

Given typical scan rates ranging from 1 mV s^{-1} to 10 V s^{-1}, systems with $k_0 > 10^{-2}$ cm s^{-1} are generally considered reversible, while those with $k_0 < 10^{-4}$ cm s^{-1} are deemed irreversible. A reversible response is also referred to as a Nernstian response.

10.4 *Effect of Electrode Size*

Voltammetry Using Microelectrodes

The voltammetry techniques discussed thus far have typically involved electrodes ranging from centimeters to millimeters in size. However, when the electrode size is reduced to the micrometer scale, the shape of the resulting voltammogram changes significantly. Figure 23 presents voltammograms for ferrocene dissolved in acetonitrile, measured using two electrodes: one with a radius $r = 10$ µm (a), and another with $r = 0.8$ mm (b). While Fig. 23b displays a conventional reversible voltammogram, Fig. 23a shows no distinct peak currents; instead, the forward and reverse scans trace nearly identical paths. This response is characteristic of micrometer-scale electrodes, commonly referred to as microelectrodes.

For reversible electron transfer reactions, voltammograms recorded with microdisk electrodes typically exhibit the following behavior. At sufficiently high potentials relative to $E^{\circ\prime}$, the current reaches a limiting value I_{lim}, expressed as

$$I_{\text{lim}} = 4nFD_{\text{R}}c_{\text{Rb}}r \tag{78}$$

If the current I equals half of I_{lim}, the corresponding potential is defined as the half-wave potential $E_{1/2}$, and the following equation holds

$$E = E_{1/2} + (RT/nF)\ln[I/(I_{\text{lim}} - I)] \tag{79}$$

By substituting Eq. 78 into this relationship, the current–potential dependence becomes

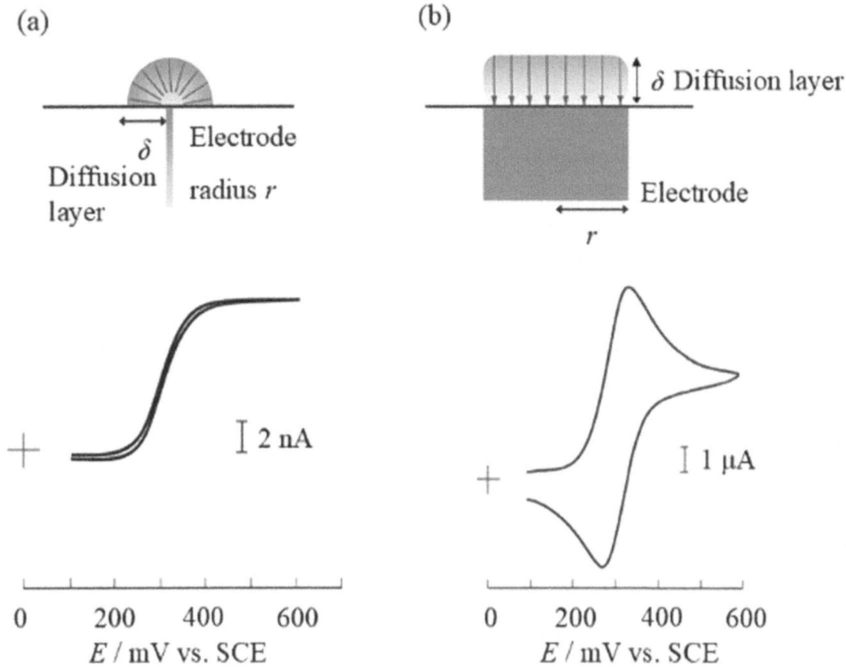

Fig. 23 Comparison of cyclic voltammograms obtained using electrodes of different sizes. Reproduced and adapted from Ref. [1, p. 97], with modifications

$$I = 4nFD_R c_{Rb} r[1 + \exp\{(nF/RT)(E_{1/2} - E)\}]^{-1} \tag{80}$$

It is noteworthy that the current I is independent of the scan rate.

The characteristic features of a reversible voltammogram at a microelectrode are as follows.

1. The limiting current I_{lim} is proportional to the number of electrons n, the diffusion coefficient D, and the electrode radius r.
2. The current I varies within a narrow potential window of approximately 150 mV.
3. A linear relationship exists between the potential E and $\ln[I/(I_{lim} - I)]$, from which n can be determined via the slope.

Measurements using microelectrodes are applied in the following areas.

1. localized analysis such as monitoring concentration changes in photosynthetic products and neurotransmitters,
2. precision measurement of diffusion coefficients, taking advantage of steady-state current behavior,
3. investigation of electrochemical phenomena in high-resistance solutions, cryogenic solvents, or polymeric solid electrolytes, utilizing the small current and minimal IR drop, and

4. lifetime measurement of unstable intermediates, enabled by reduced charging currents and high-speed operation.

Why the Voltammogram Changes

Let r be the radius of the electrode, and δ the thickness of the diffusion layer through which the reactant is transported to the electrode surface. For large electrodes, $\delta \ll r$, meaning that the reactants diffuse almost perpendicularly to the electrode surface. This is referred to as linear diffusion. The current is described by Eq. 70, and since dx ($\approx \delta$) increases proportionally to $(Dt)^{1/2}$, both the oxidative and reductive currents decrease over time after reaching their respective peaks. This results in a voltammogram like that shown in Fig. 23b. Additionally, the current is proportional to the electrode surface area A.

In contrast, for microelectrodes, when the electrolysis time is long (i.e., the scan rate v is low), the condition $\delta > r$ is established. In this case, the shape of the diffusion layer approaches that of a hemisphere, and the reactants diffuse radially from the hemisphere surface toward the microelectrode. This is known as radial diffusion or hemispherical diffusion. While δ still increases proportionally to $(Dt)^{1/2}$—as in the case of linear diffusion—the surface area of the hemisphere also grows with time, thereby increasing the supply of reactants.

Because the rate of reaction (i.e., consumption) at the electrode surface has a physical upper limit, a point is eventually reached where the supply rate and reaction rate (current) become balanced, resulting in a steady-state diffusion layer and current. At this equilibrium, the radius of the diffusion layer (defined as the point where $c_R = 0.9\ c_{Rb}$) is approximately 6.5 times the electrode radius r.

It is also worth noting that the current on a microdisk electrode is not distributed uniformly across its surface, but rather is concentrated near the circumference. Consequently, the limiting current I_{lim} is proportional not to the surface area A, but to the radius r.

Correlation Between Time and Space

In voltammetry, the observed behavior is influenced by the transport of chemical species. Even when investigating the same redox system, the shape of the voltammogram changes depending on the electrode size and scan rate. A similar phenomenon can be observed when examining how the current evolves after a potential step, as a function of electrode size.

Figure 24 shows the current–time curves obtained when the potential of a disk electrode is stepped to a value sufficient to oxidize the reductant R. These measurements are known as chronoamperometry. For microelectrodes, following the initial charging current of the electrical double layer, a steady limiting current appears, described by Eq. 78. In contrast, conventional (macro) electrodes exhibit current behavior governed by the Cottrell equation (Eq. 61).

This difference arises from the relationship between the diffusion distance of the species (proportional to $(Dt)^{1/2}$) and the electrode size r. By introducing a dimensionless quantity, $\tau = 4Dt/r^2$, the current observed at a disk electrode can be expressed

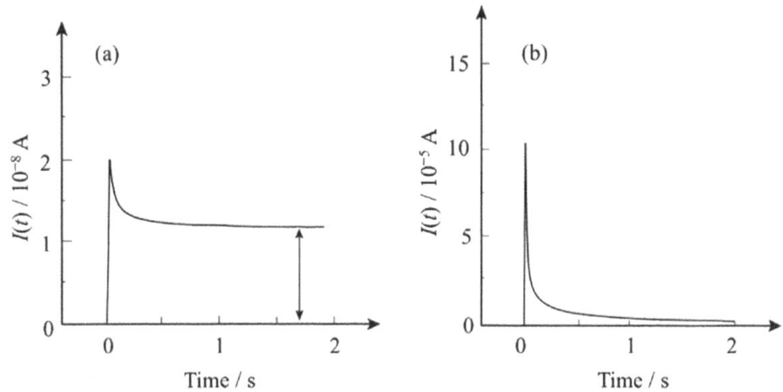

Fig. 24 Chronoamperometric responses illustrating hemispherical vs. planar diffusion at (a) micro- and **b** macroelectrodes. Reproduced and adapted from Ref. [1, p. 99], with modifications

by a unified equation:

$$I/(4nFDcr) = 0.7854 + 0.8862\tau^{-1/2} + 0.2146\exp(-0.7823\tau^{-1/2}) \quad (81)$$

When plotted, Eq. 81 (Aoki's equation) yields the curve shown in Fig. 25. For $\log\tau \ll 0$, the behavior coincides with the Cottrell equation, while for $\log\tau \gg 0$, it reflects the limiting current described by Eq. 78 for microelectrodes [8, 26, 27].

In other words, when varying the diffusion coefficient D, observation time t, or electrode radius r, systems with similar values of $4Dt/r^2$ will exhibit similar current behavior—even across different scales of measurement.

Fig. 25 Log–log plot of normalized chronoamperometric current based on Aoki's equation, illustrating the transition from linear to radial diffusion. Reproduced and adapted from Ref. [1, p. 99], with modifications

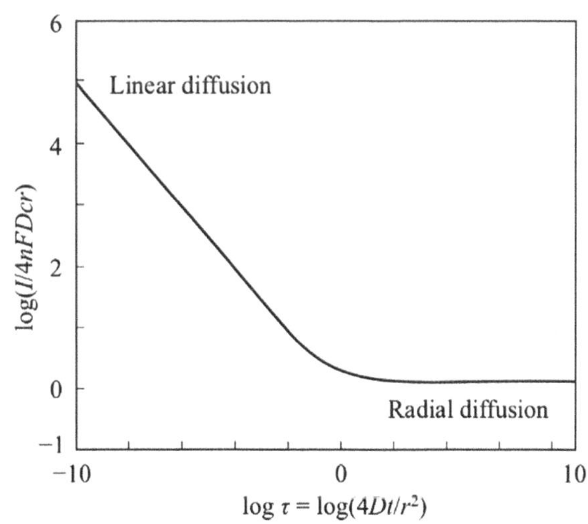

11 Estimating Molecular Orbital Energies from Cyclic Voltammetry

11.1 Understanding HOMO, LUMO Levels, and Energy Gaps

CV and Molecular Orbitals

Cyclic voltammetry (CV) is an electrochemical technique used to study oxidation and reduction behavior of molecules. By sweeping the potential at a constant rate and recording the current response, one can determine at which potentials electron transfer occurs. This chapter explains how oxidation and reduction potentials obtained from CV can be used to approximately determine the energy levels of the HOMO and LUMO orbitals, as well as their energy gap.

Relationship Between Molecular Orbitals and Electrochemical Reactions

Electrons in a molecule are primarily located in the HOMO (Highest Occupied Molecular Orbital). During oxidation, electrons are removed from this orbital. In contrast, during reduction, electrons are added to the LUMO (Lowest Unoccupied Molecular Orbital), which acts as the accepting orbital.

Therefore, the onset potential for oxidation corresponds to the HOMO level, while the onset potential for reduction corresponds to the LUMO level.

Linking Potential and Energy: Why Do Volts Correspond to Electronvolts?

The potential measured in CV (in volts, V) is proportional to the energy (in joules) required to move an electron through an electrical potential. This relationship is described as

$$\text{Energy (J)} = nF \times E(\text{V}) \tag{82}$$

where

n: number of electrons transferred (typically 1),

F: Faraday constant (approximately 96,485 C/mol).

Molecular orbital energies, however, are typically expressed in electronvolts (eV). One electronvolt is the energy gained by a single electron moving across a 1-V potential difference:

$$1\,\text{eV} = 1.602 \times 10^{-19}\,(\text{J}) \tag{83}$$

In other words, the volt scale corresponds directly to electron energy levels, allowing us to use potential values as estimates of orbital energies. Since electrochemical measurements reflect the energy required for electron transfer, the measured potentials can be converted to molecular orbital energies in eV.

Approximating HOMO and LUMO Energy Levels

Oxidation and reduction potentials obtained by CV, when referenced against an internal standard such as ferrocene, can be converted to vacuum energy levels using the following approximations:

$$E_{\mathrm{HOMO}} = -\left(E_{\mathrm{pa}} + 4.8\,\mathrm{eV}\right) \tag{84}$$

$$E_{\mathrm{LUMO}} = -\left(E_{\mathrm{pc}} + 4.8\,\mathrm{eV}\right) \tag{85}$$

Here, the value "4.8 eV" represents the energy difference between the Fc/Fc^{+} redox couple and the vacuum level [28]. This offset depends on the electrochemical setup, so using ferrocene as an internal reference helps ensure consistency and reliability.

It is important to remember that potential (V) and energy (eV) have different physical units, and this is not a simple mathematical addition. Instead, the potential difference reflects a proportional energy change per electron, which forms the theoretical basis for estimating orbital energies.

Variations in Scaling: Alternate Conventions

Different studies use slightly different offset values for vacuum-level conversion. For example, in polymer-based materials, some researchers use

$$E_{\mathrm{HOMO}} = -\left(E_{\mathrm{pa}} + 5.1\,\mathrm{eV}\right) \tag{86}$$

$$E_{\mathrm{LUMO}} = -\left(E_{\mathrm{pc}} + 5.1\,\mathrm{eV}\right) \tag{87}$$

This "5.1 eV" offset reflects an alternative assumption regarding the energy of ferrocene relative to vacuum. Cardona et al. highlight that this conversion is widely used in the context of conjugated polymers for solar cells [29]. They caution that comparing HOMO/LUMO levels based on different conventions may lead to confusion.

To support student understanding, this textbook introduces both the standard " + 4.8 eV" correction and the alternative " + 5.1 eV" approach, along with literature references, to emphasize the importance of recognizing scale differences in orbital energy estimation.

Calculating the HOMO–LUMO Energy Gap

The energy gap ΔE reflects the energy required for electronic excitation and is estimated as

$$\Delta E = E_{\mathrm{LUMO}} - E_{\mathrm{HOMO}} = E_{\mathrm{pc}} - E_{\mathrm{pa}} \tag{88}$$

The difference between the oxidation and reduction potentials measured by CV corresponds to the approximate HOMO–LUMO gap.

11.2 Practical Use and Limitations

This method offers a simple way to estimate electronic structure from CV data, but it is fundamentally an approximation. For precise orbital analysis, additional techniques such as UV–vis spectroscopy, photoelectron spectroscopy (UPS), or theoretical calculations (e.g., DFT) should be used.

Energy correction values depend on the reference electrode, solvent, and electrolyte composition, so comparisons should be made with standardized references. Moreover, CV results may be affected by reaction reversibility and diffusion control, requiring careful interpretation of potential values.

12 Summary

This chapter introduced the fundamental principles and techniques of electrochemical measurements, with a particular focus on cyclic voltammetry. By examining the behavior of molecules and polymers at their frontier orbitals, readers learned how oxidation and reduction processes are governed by electron transfer and electrode potentials. The concept of standard electrode potentials E° and formal potentials $E^{\circ\prime}$ was explored in detail, along with the role of reference electrodes, the formation of electric double layers, and the onset of electrolysis.

Key theoretical frameworks—including the Nernst equation, Butler–Volmer equation, Tafel relation, Cottrell equation, Randles–Sevcik equation, and Aoki's equation—were introduced to describe current–potential relationships, mass transport, and reaction kinetics. The distinctions between Faradaic and non-Faradaic currents and between diffusion-controlled and electron transfer-controlled reactions were emphasized throughout. The use of microelectrodes was also highlighted, revealing how electrode size influences voltammetric behavior and enables steady-state measurements.

Finally, the chapter addressed how oxidation and reduction potentials obtained from CV can be used to estimate HOMO and LUMO energy levels and their gap—a concept that bridges electrochemical measurement with molecular electronic structure. By integrating theory, experimental design, and practical interpretation, this chapter provides a foundation for applying electrochemical methods to soft materials and advanced molecular systems.

To support further understanding, a selection of introductory references is provided at the end of this chapter [30–33].

References

1. T. Watanabe, M. Kanemura, H. Masuda, M. Watanabe, in *Kiso Kagaku Ko–su Denki Kagaku [Electrochemistry: Fundamental Chemistry Course]*. ed. by H. Inoue, T. Kitamori, M. Komiyama, K. Takagi, S. Hirano (Maruzen, Tokyo, 2001)
2. W.M. Haynes (ed.), *CRC Handbook of Chemistry and Physics*, 97th edn. (CRC Press, Boca Raton, 2016)
3. A.J. Bard, R. Parsons, J. Jordan (eds.), *Standard Potentials in Aqueous Solution* (Marcel Dekker, New York, 1985)
4. S.G. Bratsch, Standard electrode potentials and temperature coefficients in water at 298.15 K. J. Phys. Chem. Ref. Data **18**, 1–21 (1989). https://doi.org/10.1063/1.555839
5. The Chemical Society of Japan (ed.), *The Chemical Handbook (Kagaku Binran)*, Basic Edition, 5th edn. (Maruzen, Tokyo, 2004)
6. W. Nernst, Zur Elektromotorischen Wirksamkeit der Ionen. Z. Phys. Chem. **4**, 129–181 (1889)
7. W. Nernst, Über die Löslichkeit von Mischkrystallen. Z. Phys. Chem. **9**, 137–142 (1892)
8. A.J. Bard, L.R. Faulkner, *Electrochemical Methods: Fundamentals and Applications*, 2nd edn. (Wiley, New York, 2001)
9. J.A.V. Butler, *The Kinetics of Chemical Change* (Methuen, London, 1924)
10. M. Volmer, T. Erdey-Gruz, Zur Theorie der Wasserstoffüberspannung. Z. Phys. Chem. **150**, 203–213 (1922)
11. E.J.F. Dickinson, A.J. Wain, The Butler-Volmer equation in electrochemical theory: origins, value, and practical application. J. Electroanal. Chem. **872**, 114145 (2020). https://doi.org/10.1016/j.jelechem.2020.114145
12. J. Tafel, Über die Polarisation bei elektrolytischer Gasentwicklung. Z. Phys. Chem. **50**, 641–712 (1905). https://doi.org/10.1515/zpch-1905-5043
13. G.T. Burstein, A hundred years of Tafel's Equation: 1905–2005. Corros. Sci. **47**, 2858–2870 (2005). https://doi.org/10.1016/j.corsci.2005.07.002
14. A. Fick, Ueber diffusion. Ann. Phys. **94**, 59–86 (1855). https://doi.org/10.1002/andp.18551700105
15. D.C. Grahame, The electrical double layer and the theory of electrocapillarity. Chem. Rev. **41**, 441–501 (1947). https://doi.org/10.1021/cr60130a002
16. F.G. Cottrell, Der Reststrom bei galvanischer Polarisation, betrachtet als ein Diffusionsproblem. Z. Phys. Chem. **42**, 385–431 (1903). https://doi.org/10.1515/zpch-1903-4229
17. G. Inzelt, Chronoamperometry, chronocoulometry, and chronopotentiometry, in *Encyclopedia of Applied Electrochemistry*. ed. by R. Varma, E. Rogers (Springer, New York, 2014), pp.207–214
18. N.P. Bansal, J.A. Plambeck, Corrections to the Cottrell equation for spherical and cylindrical electrodes. Can. J. Chem. **56**, 155–160 (1978). https://doi.org/10.1139/v78-024
19. T.R. Marrero, E.A. Mason, Gaseous diffusion coefficients, in *CRC Handbook of Chemistry and Physics*, 52nd edn., ed. by W.M. Haynes (CRC Press, Boca Raton, 1972)
20. National Institute for Materials Science (NIMS), Diffusion Database: Kakusan. https://mits.nims.go.jp/kakusan/. Accessed July 2025
21. D.J. Fisher (ed.), *ASM Handbook: Volume 22—Metals Process Modeling* (ASM International, Materials Park, OH, 2010)
22. DDBST GmbH, DDB-DIF: Diffusion Coefficient Database. https://ddbst.com/ddbdif.html. Accessed July 2025
23. BAS Inc., Electrochemical Instrumentation Product Page. https://www.bas.co.jp/1626.html and https://www.bas.co.jp/1405.html. Accessed July 2025
24. J.E.B. Randles, Kinetics of rapid electrode reactions. Discuss. Faraday Soc. **1**, 11–19 (1947). https://doi.org/10.1039/DF9470100011
25. A. Ševčík, Oscillographic polarography with a rotating electrode. Collect. Czech. Chem. Commun. **13**, 349–377 (1948). https://doi.org/10.1135/cccc19480349

26. O.A. González-Meza, E.R. Larios-Durán, A. Gutiérrez-Becerra, N. Casillas, J.I. Escalante, M. Bárcena-Soto, Development of a Randles–Ševčík-like equation to predict the peak current of cyclic voltammetry for solid metal hexacyanoferrates. J. Solid State Electrochem. **23**, 3123–3133 (2019). https://doi.org/10.1007/s10008-019-04410-6
27. K. Aoki, K. Tokuda, Theory of chronoamperometry at microdisk electrodes. J. Electroanal. Chem. **225**, 19–32 (1987). https://doi.org/10.1016/0022-0728(87)80002-5
28. C.M. Cardona, W. Li, A.E. Kaifer, D. Stockdale, G.C. Bazan, Electrochemical estimation of the HOMO–LUMO gap in organic semiconductors. Adv. Mater. **23**, 2367–2371 (2011). https://doi.org/10.1002/adma.201004554
29. J. Pommerehne, H. Vestweber, W. Günther, R.E. Muhrt, H. Bässler, M. Porsch, J. Daub, Efficient organic electroluminescent devices using doped polymer layers. Adv. Mater. **7**, 551–554 (1995). https://doi.org/10.1002/adma.19950070608
30. P.H. Rieger, *Electrochemistry*, 2nd edn. (Chapman & Hall, New York, 1994)
31. P.A. Christensen, A. Hamnett, *Techniques and Mechanisms in Electrochemistry* (Blackie Academic & Professional, Glasgow, 1994)
32. K.B. Oldham, J. Myland, A. Bond, *Electrochemical Science and Technology: Fundamentals and Applications* (Wiley, Chichester, 2011)
33. K.B. Oldham, J.C. Myland, *Fundamentals of Electrochemical Science* (Academic, San Diego, 1994)

Topics

Recent Advancements in π-Conjugated Polymers for Semiconductor Device Applications

Satoru Inoue

1 Introduction

Recently, π-conjugated polymers with polymer main chains composed of π-conjugated compounds have garnered significant research interest as promising materials for the next-generation plastic electronics industry. General polymeric materials, such as synthetic rubber, photoresists, and biopolymers, typically act as insulators based on their electric properties. However, several types of polymers, such as polyacetylene and polythiophene, which contain π electrons that are spread throughout the main chain, can achieve excellent conductivity even without the conductive material additives such as metals. This is attributed to the properties of π electrons, which can be freely transported in intra- and inter-chains through the overlap of π-electron orbitals at the proper molecular arrangements in the solid-state. Π electrons that are loosely bound to molecules exhibit a high sensitivity and rapid response to light irradiation and the application of electric fields. Therefore, the application of π-conjugated polymers in new-type optoelectronic devices is expected.

A series of π-conjugated polymers are often referred to as conductive polymers; however, they are classified as intrinsic semiconductors because of their organic nature. Doped systems, such as iodine and acidic polymers, are employed when polymers are used as conductors. Pioneering discoveries regarding conductive polymers by Heeger, MacDiarmid, and Shirakawa were conducted in the 1970s [1–4]. The Nobel Prize in Chemistry was awarded to the aforementioned authors in 2000, and ongoing research efforts have been dedicated to the development of conductive materials [5]. Since the 1990s, the research focus has expanded to the development of semiconductor materials with the emergence of organic semiconductor (OSC) devices, organic field-effect transistors (OFETs), organic light-emitting diodes (OLED), and organic photovoltaics (OPVs).

S. Inoue (✉)
Innovation Center for Organic Electronics (INOEL), Yamagata University, Yamagata, Japan
e-mail: satoru.inoue@yz.yamagata-u.ac.jp

M. Funahashi (ed.), *Functional Soft Matter II*,
https://doi.org/10.1007/978-981-92-0100-6_9

π-Conjugated polymers are beneficial for electronic devices because they can dissolve in solvents. Therefore, devices can be produced using solution-printing techniques, such as spin coating, bar coating, screen printing, and inkjet printing, under low-temperature conditions (below 200 °C). This facilitates the fabrication of flexible devices on plastic substrates, which typically have low thermal resistance but are indispensable for producing next-generation flexible and stretchable devices such as flexible displays and wearable sensors. Plastic substrates are not suitable for inorganic semiconductors because their fabrication process requires high temperatures (exceeding 350 °C). Recently, this solution-printing technology, known as printed electronics, has been shown to reduce the environmental impact during device manufacturing and effective for producing large-area devices using plastic substrates. However, π-conjugated polymers generally form complicated molecular arrangements in the solid state. In electronic devices with a semiconducting layer composed of π-conjugated polymers, the rate of electron-flow (carrier conductivity) is primarily determined by the arrangements of the π-conjugated components. Therefore, the improvement of electronic device performance using π-conjugated polymers critically depends on controlling the chemical and/or conformational structure of the polymer chains and the resulting molecular alignments (crystallinity) within the thin film.

This chapter reviews poly(3-alkylthiophene)s (P3ATs), which are the most fundamental materials in π-conjugated polymers, to develop an understanding of the relationship between complicated molecular arrangements and OSC characteristics. In addition, this chapter introduces the recent trends in polymer design strategies for improving device performance.

2 Fundamentals and Microstructures in P3AT

The chemical structure of P3AT consists of a repeating unit of a 3-alkylthiophene monomer (Fig. 1). Since the application in OFETs was first reported by Tsumura, Koezuka, and Ando in 1986 [6], extensive research has been conducted on thin-film properties and device physics. In P3AT, the molecular weight and regioregularity are known to critically affect the formation of the thin-film morphology. As discussed later, the degree of crystallinity in the solid state varies significantly owing to differences in the symmetry of the molecular structure [7, 8]. Regiocontrolled P3AT was successfully synthesized by a coupling reaction using the corresponding halogenated alkylthiophene and appropriate transition metal catalysts (Ni, Pd) [9, 10]. Currently, highly regiocontrolled (regioregularity: RR > 98%) poly(3-hexylthiophene) (P3HT) is commercially available.

P3AT possesses a distinctive rod-like structure that is not found in other conventional organic polymers such as polystyrene and polymethyl methacrylate. This is because the repeating thiophene rings are connected by sp^2 carbon atoms. When each thiophene ring forms a coplanar arrangement, the rigidity of the polymer is strengthened by the effect of the arrangement on the stabilization of the π-conjugation along

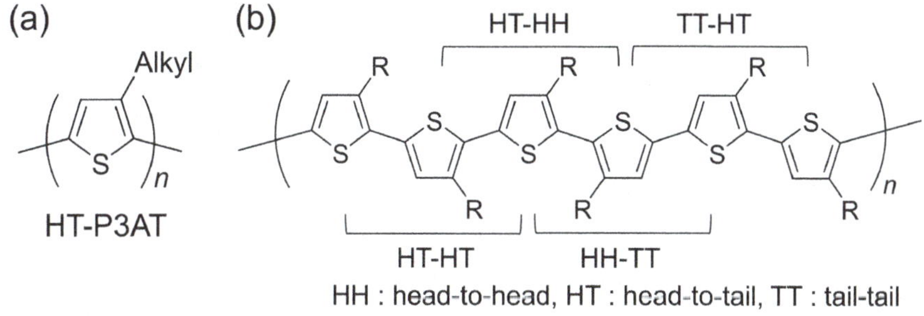

Fig. 1 **a** Chemical structure of head-to-tail (HT) P3AT. **b** Possible combination of regioisomers

the main chain elongation direction. However, the solid-state conformation of P3AT does not exclusively form a coplanar structure because torsional freedom of the main chains is present.

P3AT forms a complicated microstructure (often referred to as a semicrystalline structure) that combines crystalline and amorphous domains (Fig. 2). The crystalline domain is organized to form a lamellar structure, in which π-stacked polythiophene backbones alternate with layers of more or less ordered alkyl side chains [11, 12]. The XRD patterns of the P3AT films exhibited evenly spaced reflections corresponding to the first-, second-, and higher-order reflections owing to the lamellar structure. The observed π-stacking period was 0.38–0.39 nm. The layer-stacking period increases linearly with the substituted alkyl chain length n and depends on the molecular weight [13, 14]. For P3HT, the most well-known polythiophene material, the period was 15–17 nm. This study details cases of π-conjugated polymers that form "semicrystalline" molecular arrangements in the solid-state.

Considering the degree of carrier transport in each region, the crystalline region can be assumed to enhance carrier mobility compared with the amorphous region. However, both crystalline and amorphous regions coexist throughout the carrier transport region of the device known as the channel. Moreover, external factors, such as the solvent effect [15] and interfacial interactions [16], that adopt the solution process also affect the uniformity of semicrystalline thin films. Therefore, considerations

Fig. 2 Schematics of **a** the lamellar structure and **b** microstructure of the semicrystalline thin film of the π-conjugated polymer

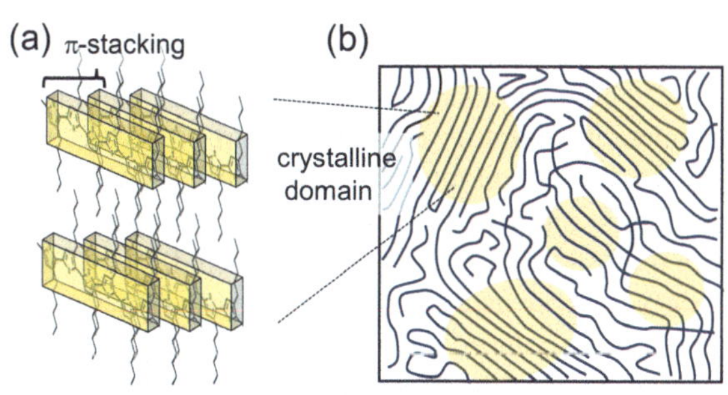

of the molecular design and film formation process are required to improve OSC devices.

3 Relationship Between Carrier Transport Properties and Thin-Film Microstructure in OFET and OPV Devices

OFETs are the most fundamental components of OSC devices. A schematic of the device structure is shown in Fig. 3. By applying a voltage to the gate, the carrier is injected from the source electrode and transported to the semiconductor/insulator interface. Therefore, OFETs are primarily used as switching or amplifying devices. Sirringhaus et al. achieved a significant increase in the mobility of OFET devices using P3HT [8, 17]. The key to improve performance is the control of two distinct orientations in the crystalline domains on the substrate, referred to as "edge-on" and "face-on" orientations (Fig. 3b). The OFET device requires high in-plane charge transport between the source and drain electrodes. The edge-on arrangement, where the π-stacking direction is parallel to the substrate, is more suitable for charge transport than that of the face-on arrangement. The edge-on or face-on orientation could be selected by controlling the molecular weight and regioregularity of P3HT. When polymers with a low molecular weight and high regioregularity (HT; 96%) formed an edge-on arrangement, the OFET devices with these films exhibited mobilities of up to 0.1 cm^2/Vs. In contrast, polymers with high molecular weight and low regioregularity (HT; 81%) formed face-on arrangements, and their mobilities were lower (approximately 2×10^{-4} cm^2/Vs) [8]. Comparing polymers containing similar regioregularities, those with high-molecular-weight P3HT formed lower crystallinity devices than those formed from lower-molecular-weight P3HT; however, those formed from lower-molecular-weight P3HT exhibited higher mobilities. This mechanism can be explained by the fact that higher-molecular-weight P3HT has well-connected crystalline grains, whereas low-molecular-weight P3HT has more defined grain boundaries, limiting effective carrier transport [18, 19].

Additionally, Noriega et al. proposed the requirement for high-mobility films by considering the perspective of the paracrystallinity model described by the structural disorder in an imperfect crystal [20]. The study investigated the relationship between polymer chain length, crystallinity, and mobility for various π-conjugated polymers including P3HT. This report proposes that the unifying requirement for semicrystalline thin films with high carrier mobility is not required for macroscopic crystallinity; however, it is required for the formation of interconnected aggregates, even if those that are small and disordered.

OPV devices using π-conjugated polymers were assessed as energy conversion devices with reduced environmental impact; this was accomplished by increasing the proportion of their renewable energy utilization. The device utilizes a photoactive OSC layer, which is composed of a mixture of p-type (donor-type) and n-type

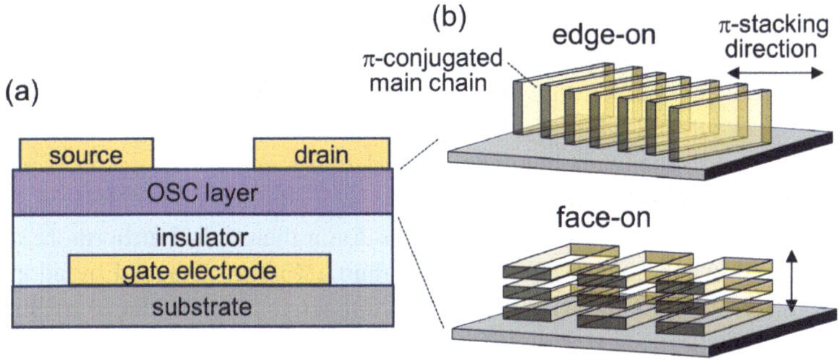

Fig. 3 Schematics of **a** the typical configuration of OFET devices (bottom-gate top-contact type) and **b** edge-on (top) and face-on (bottom) polymer arrangements within OSC layers

(acceptor-type) semiconductors and is positioned between the cathode and anode (Fig. 4). Anodes are typically transparent electrodes, such as indium tin oxide (ITO), that are used to capture external light from this side and initiate the photoelectric conversion process. When light irradiates an OPV device, an electron transition within acceptor molecules from the highest occupied molecular orbital (HOMO) to the lowest unoccupied molecular orbital (LUMO) occurs, resulting in exciton generation. The photogenerated excitons immediately move to the donor/acceptor interface, and the excited electrons in the donor are transferred to the LUMO of the acceptor, where a charge-separated state is formed. At this point, the donor and acceptor form radical cations and anions, respectively. Through the conduction of holes and electrons to their respective electrodes, the external circuit gains a photocurrent, facilitating the function of the OPV device.

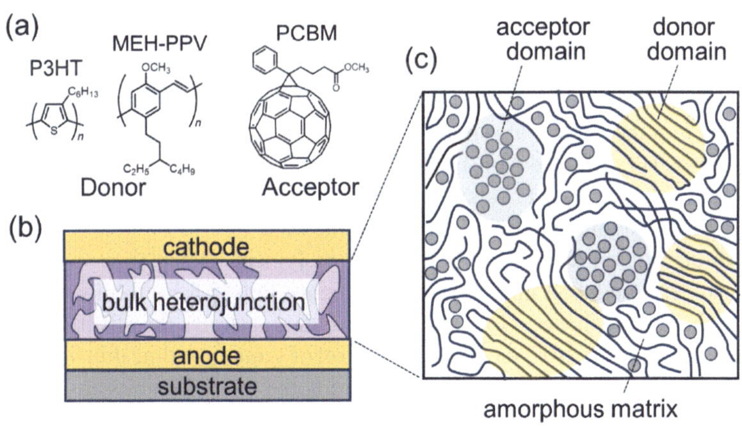

Fig. 4 **a** Chemical structures of P3HT, MEH-PPV, and PCBM. **b** Schematic of the typical configuration of a bulk heterojunction OPV device. **c** Schematic of a bulk heterojunction microstructure

A prototypical sample of this device was developed in 1986 by Tang and featured a two-layer p-n junction structure using low-molecular-weight materials [21]. Thereafter, the effectiveness of π-conjugated polymers for OPV devices was demonstrated by Sariciftci et al. and Morita et al. via the utilization of poly[2-methoxy-5-(2-ethylhexyloxy)-1,4-phenylenebinylene] (MEH-PPV) and P3AT, respectively [22, 23]. Extended π-conjugated compounds can intrinsically function as electron donors; thus, π-conjugated polymers are ideal electron-donor materials. Furthermore, a high-power conversion efficiency (PCE) (greater than 1%) was achieved by adopting a bulk heterojunction structure (Fig. 4) as the photoactive layer produced by blending soluble fullerene derivatives (PCBM) as electron acceptors [24]. The bulk heterojunction structure facilitates the formation of larger charge donor/acceptor interfaces for charge separation than those of the two-layer heterojunction structure. Moreover, the structure promotes efficient charge separation of photogenerated excitons by the formation of phase-separated donor/acceptor domains at the tens-of-nanometers scale.

The modulation of the morphology of P3HT has significantly contributed to the advancement of the PCE [25]. The crucial factor is the control of semicrystallinity using regiocontrolled and molecular-weight-controlled polymers to promote efficient charge separation and carrier transport. Consequently, the conversion efficiency increased from approximately 1% to approximately 5% [26–28]. The bulk heterojunction structure in the P3HT-PCBM system is a standard configuration underlying recent advances in the development of high-performance OPV materials.

4 Recent Advances in Polymer Design for High-Performance OSCs

Currently, a wide variety of π-conjugated polymers have been developed that enhance the mobility of OFETs to 10 cm^2/Vs [29, 30] and PCEs of OPV devices to above 18% [31, 32]. These advancements are critically supported by the polymer design strategy that focuses on the control of semicrystallinity, as elucidated through research findings using P3HT. This section briefly introduces three concepts as examples of these advanced strategies: (1) alkyl side-chain engineering, (2) donor-acceptor-type polymers, and (3) intrachain conformational locking.

Substituted alkyl chains in conjugated polymers affect their solubility in organic solvents and their molecular arrangements. Poly(2,5-bis(3-alkylthiophen-2-yl)thieno[3,2-b]thiophenes) (PBTTT) (Fig. 5a) demonstrates the effective coupling of long alkyl chains with the morphology control of semicrystalline thin films for efficient carrier transport in OFETs [33, 34]. Herein, the repeating units are centrosymmetric, which suppresses any disorder due to regioregularity. Moreover, the introduction of unsubstituted thieno[3,2-b]thiophene units enabled the production of a free volume for alkyl chain interdigitation between neighboring polymers. The interdigitation of alkyl side chains promotes the formation of a mesophase above the

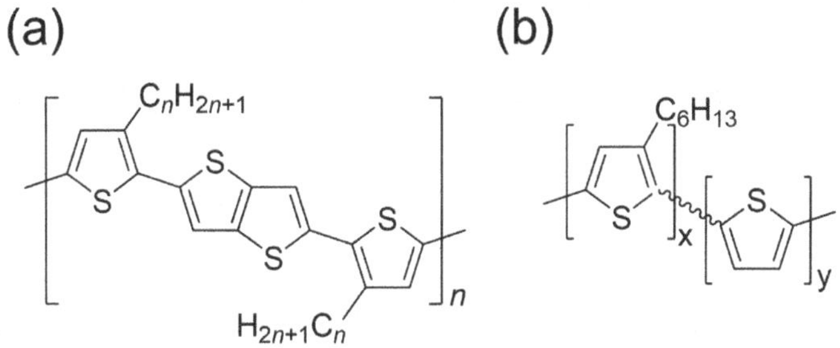

Fig. 5 Chemical structure of **a** PBTTT (n = 10, 12, 14) [33] and **b** a random copolymer of polythiophene [35]

melting temperature of alkyl chains. This contrasts with P3HT, in which the alkyl side chains are significantly disordered and do not interdigitate. The regrowth of the more ordered alignments of polymers within the semicrystalline thin film is made possible by the post-annealing (processing above this phase transition and below the backbone melt transition) of the film, resulting in the more ordered film structures and higher OFET carrier mobilities (up to 0.7 cm^2/Vs) than P3HT. Post-annealing is a commonly used technique to improve the crystallinity of films.

Decreasing the alkyl chain density in films of π-conjugated polymers is effective for improving carrier transport properties. This is demonstrated in random copolymers composed of alkyl-substituted and non-substituted thiophenes as repeating units [35] (Fig. 5b) where the alkyl chain density is controlled by varying the copolymerization ratio. Although the copolymer with a high ratio of non-substituted thiophene units exhibited low crystallinity, OFET carrier mobility, which reached 1.37 cm^2/Vs, was enhanced by increasing the ratio of non-substituted thiophene units. This enhancement was attributed to the increased main chain planarity due to the suppression of the steric hindrance of the alkyl chains and the existence of localized aggregates in the amorphous regions, which produced enhanced intra- and interchain charge transport, respectively.

As demonstrated by PBTTT, modulating the π-conjugated main chains has become the mainstream approach in recent developments of π-conjugated polymers. In this context, the most widely adopted strategy is the donor-acceptor (D-A)-type conjugated polymer design (Fig. 6a). The chemical structure of D-A-type conjugated polymers consists of an alternating array of electron-donating (D) and electron-accepting (A) units [36–39]. Due to the difference in electron density, D-A type conjugated polymers usually enhance π-π interactions, including the charge transfer interactions that are beneficial to charge transport, irrespective of intra- or intermolecular arrangements. Although enhancing the intermolecular interactions between the polymer chains often leads to a reduction in solubility, the introduction of branched alkyl groups (Fig. 6b) can be used in solution processes [29, 30]. For example, by appropriately controlling the branching positions of the alkyl chains, dense π-π stacking of the polymer (3.67 Å) is achieved and the solution-processed solubility was maintained; this resulted in the excellent carrier mobility of OFETs (up

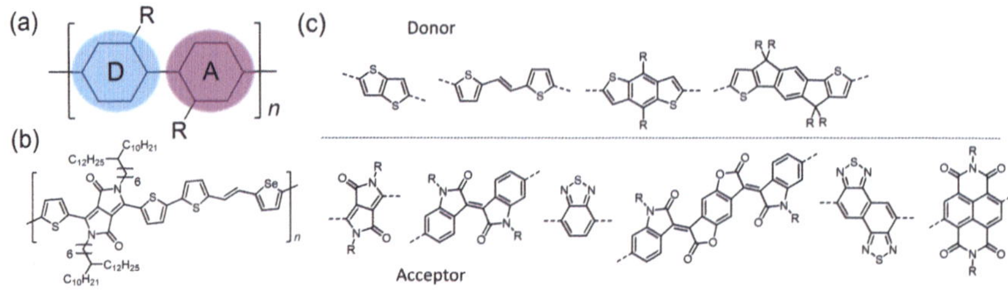

Fig. 6 **a** Schematic of D-A polymers. **b** Chemical structure of D-A polymer with high carrier mobility [29]. **c** Chemical structures of representative donor/acceptor units

to 12 cm^2/Vs). One of the significant advantages of adopting a D-A polymer structure is the modulation of the energy levels of the molecular orbitals by combining the donor and acceptor units. By deepening the LUMO level, efficient electron injection from the electrodes to the OSCs in the OFETs can be accomplished. This results in the fabrication of excellent ambipolar or n-type OFET devices. Furthermore, adjusting the optical gap enables a wide range of photoabsorption to utilize photoconversion in OPV devices [31, 32]. Tuning the absorption wavelength of OSC layers has been a critical challenge in P3HT-PCBM systems, which are limited to approximately 650 nm.

Utilizing a supramolecular interaction has also been proposed to enhance the planarity of the π-conjugation polymer, which contributes to the formation of the close π-π stacking of the polymer main chains. This demonstrates that the introduction of heteroatoms, such as oxygen, nitrogen, and halogens, into the monomer units induces noncovalent interactions between adjacent monomer units [40]. This attractive interaction is effective in locking the main chain because it restricts its rotational freedom. This effect is often referred to as conformational locking. Representative polymers with high planarity are shown in Fig. 7 [41–43]. Solubility can be maintained by the entropic gain that arises from the rotational motion of the single bonds that link the π-cores. This is because supramolecular interactions are typically weaker attractions than those of covalent bonds. In contrast, the π-conjugated polymers easily form a planar conformation that is thermodynamically more stable in the thin-film state. These features are suitable for printed devices requiring excellent solution processability.

5 Summary and Perspective

In this chapter, initially, the characteristics of P3AT were discussed to guide the fundamental understanding of π-conjugated polymers and their device applications. Additionally, recent developments in π-conjugation polymers for applications in OFET and OPV devices were introduced. Although not covered in this chapter, the application of π-conjugated polymers in various electronic devices, such as OLED,

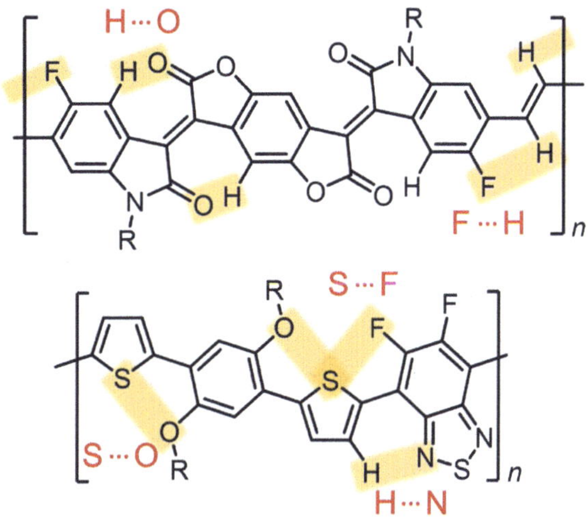

Fig. 7 Representative π-conjugated polymers with interchain conformational locking [41, 43]

photodetectors, conductors, and stretchable devices, is also being assessed. However, π-conjugated polymers are primarily incorporated in these devices to control solid-state morphology. This control is introduced as semicrystallinity based on the molecular design of π-conjugated polymers. The synergy of flexibility and rigidity in organic electronics provided by π-conjugated polymers is referred to as soft matter electronics in a wide range of research fields. Future advancements are expected to include high-quality computational science and technology.

References

1. T. Ito, H. Shirakawa, S. Ikeda, Simultaneous polymerization and formation of Polyacetylene film on the surface of concentrated soluble Ziegler-type catalyst solution. J. Polym. Sci. **12**, 11–20 (1974)
2. H. Shirakawa, E.J. Louis, A.G. MacDiamid, C.K. Chiang, A.J. Heeger, Synthesis of electrically conducting organic polymers: halogen derivatives of polyacetylene, $(CH)_x$. Chem. Commun. 578–580 (1977)
3. C.K. Chiang, C.R. Fincher Jr., Y.W. Park, A.J. Heeger, Electrical conductivity in doped Polyacetylene. Phys. Rev. Lett. **39**, 1098–1101 (1977)
4. A.G. MacDiamid, A.J. Heeger, Organic metals and semiconductors: the chemistry of Polyacetylene, $(CH)_x$, and its derivatives. Syn. Metals **1**, 101–118 (1980)
5. G.K. Sharma, S.L. Joseph, N.R. James, Recent progress in poly (3,4-ethylene Dioxythiophene): polystyrene sulfonate based composite materials for electromagnetic interference shielding. Adv. Mater. Technol. **9**, 231203 (2024)
6. A. Tsumura, H. Koezuka, T. Ando, Macromolecular electronic device: Field-effect transistor with a polythiophene thin film. Appl. Phys. Lett. **49**, 1210–1212 (1986)
7. Z. Bao, A. Dodabalapur, A.J. Lovinger, Soluble and processable regioregular poly(3-hexylthiophene) for thin film field-effect transistor applications with high mobility. Appl. Phys. Lett. **69**, 4108–4110 (1996)

8. H. Sirringhaus, P.J. Brown, R.H. Friend, M.M. Nielsen, K. Bechgaard, B.M.W. Langeveld-Voss, A.J.H. Spiering, R.A.J. Janssen, E.W. Meijer, P. Herwig, D.M. de Leeuw, Two-dimensional charge transport in self-organized, high-mobility conjugated polymers. Nature **401**, 685–688 (1999)
9. T. Chen, X. Wu, R.D. Rieke, Regiocontrolled synthesis of poly(3-alkylthiophenes) mediated by Rieke zinc: Their characterization and solid-state properties. J. Am. Chem. Soc. **117**, 233–244 (1995)
10. Q. Wang, R. Takita, Y. Kikuzaki, F. Ozawa, Palladium-catalyzed dehydrohalogenative polycondensation of 2-bromo-3-hexylthiophene: an efficient approach to head-to-tail poly(3-hexylthiophene). J. Am. Chem. Soc. **132**, 11420–11421 (2010)
11. T.J. Prosa, M.J. Winokur, J. Moulton, P. Smith, A.J. Heeger, X-ray structural studies of poly(3-alkylthiophenes): an example of an inverse comb. Macromolecules **25**, 4364–4372 (1992)
12. T.J. Prosa, M.J. Winokur, R.D. McCullough, Evidence of a novel side chain structure in Regioregular poly(3-alkylthiophenes). Macromolecules **29**, 3654–3656 (1996)
13. S.V. Meille, V. Romita, T. Caronna, A.J. Lovinger, M. Catellani, L. Belobrzeckaja, Influence of molecular weight and Regioregularity on the polymorphic behavior of poly(3-decylthiophenes). Macromolecules **30**, 7898–7905 (1997)
14. A. Zen, M. Saphiannikova, D. Neher, J. Grenzer, S. Grigorian, U. Pietsch, U. Asawapirom, S. Janietz, U. Scherf, I. Lieberwirth, G. Wegner, Effect of molecular weight on the structure and crystallinity of poly(3-hexylthiophene). Macromolecules **39**, 2162–2171 (2006)
15. H. Yang, S.W. LeFevre, C.Y. Ryu, Z. Bao, Solubility-driven thin film structures of regioregular poly(3-hexyl thiophene) using volatile solvents. Appl. Phys. Lett. **90**, 172116 (2007)
16. R.J. Kline, M.D. McGehee, M.F. Toney, Highly oriented crystals at the buried interface in polythiophene thin-film transistors. Nat. Mater. **5**, 222–228 (2006)
17. H. Sirringhaus, N. Tessler, R.H. Friend, Integrated optoelectronic devices based on conjugated polymers. Science **280**, 1741–1744 (1998)
18. R.J. Kline, M.D. McGehee, E.N. Kadnikova, J. Liu, J.M.J. Fréchet, Controlling the field-effect mobility of Regioregular Polythiophene by changing the molecular weight. Adv. Mater. **15**, 1519–1522 (2003)
19. R.J. Kline, M.D. McGehee, E.N. Kadnikova, J. Liu, J.M.J. Fréchet, M.F. Toney, Dependence of Regioregular poly(3-hexylthiophene) film morphology and field-effect mobility on molecular weight. Macromolecules **38**, 3312–3319 (2005)
20. R. Noriega, J. Rivnay, K. Vandewal, F.P.V. Koch, N. Stingelin, P. Smith, M.F. Toney, A. Salleo, A general relationship between disorder, aggregation and charge transport in conjugated polymers. Nat. Mater. **12**, 1038–1044 (2013)
21. C.W. Tang, Two-layer organic photovoltaic cell. Appl. Phys. Lett. **48**, 183–185 (1986)
22. N.S. Sariciftci, L. Smilowitz, A.J. Heeger, F. Wudl, Photoinduced electron transfer from a conducting polymer to buckminsterfullerene. Science **258**, 1474–1476 (1992)
23. S. Morita, A.Z. Zakhidov, K. Yoshino, Doping effect of Buckminster fullerene in conducting polymer: change of absorption spectrum and quenching of luminescence. Solid State Commun. **82**, 249–252 (1992)
24. G. Yu, J. Gao, J.C. Hummelen, F. Wudl, A.J. Heeger, Polymer photovoltaic cells: Enhanced efficiencies via a network of internal donor-acceptor heterojunctions. Science **270**, 1789–1791 (1995)
25. P. Schilinsky, C. Waldauf, C.J. Brabec, Recombination and loss analysis in polythiophene based bulk hetrojunction photodetectors. Appl. Phys. Lett. **81**, 3885–3887 (2002)
26. W. Ma, C. Yang, X. Gong, K. Lee, A.J. Heeger, Thermally stable, efficient polymer solar cells with nanoscale control of the interpenetrating network morphology. Adv. Funct. Mater. **15**, 1617–1622 (2005)
27. G. Li, V. Shrotriya, J. Huang, Y. Yao, T. Moriarty, K. Emery, Y. Yang, High-effiency solution processable polymer photovoltaic cells by self-organization of polymer blends. Nat. Mater. **4**, 864–868 (2005)
28. Y. Kim, S. Cook, S.M. Tuladhar, S.A. Choulis, J. Nelson, J.R. Durrant, D.C. Bradley, M. Giles, I. McCulloch, C.-S. Ha, M. Ree, A strong regioregularity effect in self-organizing conjugated

polymer films and high-efficiency polythiophene: Fullerene solar cells. Nat. Mater. **5**, 197–203 (2006)
29. I. Kang, H.-J. Yun, D.S. Chung, S.-K. Kwon, Y.-H. Kim, Record high hole mobility in polymer semiconductors via side-chain engineering. J. Am. Chem. Soc. **135**, 14896–14899 (2013)
30. A.-R. Han, G.K. Dutta, J. Lee, H.R. Lee, S.M. Lee, H. Ahn, T.J. Shin, J.H. Oh, C. Yang, E-Branched flexible side chain substituted Diketopyrrolopyrrole-containing polymers designed for high hole and electron Mobilities. Adv. Funct. Mater. **25**, 247–254 (2015)
31. L. Zhu, M. Zhang, J. Xu, C. Li, J. Yan, G. Zhou, W. Zhong, T. Hao, J. Song, X. Xue, Z. Zhou, R. Zeng, H. Zhu, C.-C. Chen, R.C.I. MacKenzie, Y. Zou, J. Nelson, Y. Zhang, Y. Sun, F. Liu, Single-junction organic solar cells with over 19% efficiency enabled by a refined double-fibril network morphology. Nat. Mater. **21**, 656–663 (2022)
32. M. Zhang, L. Zhu, G. Zhou, T. Hao, C. Qui, Z. Zhao, Q. Hu, B.W. Larson, H. Zhu, Z. Ma, Z. Tang, W. Feng, Y. Zhang, T.P. Russell, F. Liu, Single-layered organic photovoltaics with double cascading charge transport pathways: 18% efficiencies. Nat. Commun. **12**, 309 (2021)
33. I. McCulloch, M. Heeney, C. Bailey, K. Genevicius, I. Macdonald, M. Shkunov, D. Sparrowe, S. Tierney, R. Wagner, W. Zhang, M.L. Chabinyc, R.J. Kline, M.D. Mcgehee, M.F. Toney, Liquid-crystalline semiconducting polymers with high charge-carrier mobility. Nat. Mater. **5**, 328–333 (2006)
34. R.J. Kline, D.M. DeLongchamp, D.A. Fischer, E.K. Lin, L.J. Richter, M.L. Chabinyc, M.F. Toney, M. Heeney, I. McCulloch, Critical role of side-chain attachment density on the order and device performance of Polythiophenes. Macromolecules **40**, 7960–7965 (2007)
35. S.Y. Son, Y. Kim, J. Lee, G.-Y. Lee, W.-T. Park, Y.-Y. Noh, C.E. Park, T. Park, High-field-effect mobility of low-crystallinity conjugated polymers with localized aggregates. J. Am. Chem. Soc. **138**, 8096–8103 (2016)
36. H.J. Cheon, T.K. An, Y.-H. Kim, Diketopyrrolopyrrole (DPP)-based polymers and their organic field-effect transistor applications: a review. Macromol. Res. **30**, 71–84 (2022)
37. A. Wadsworth, H. Chen, K.J. Thorley, C. Cendra, M. Nikolka, H. Bristow, M. Moser, A. Salleo, T.D. Anthophoulos, H. Sirringhaus, I. McCulloch, Modification of Indacenodithiophene-based polymers and its impact on charge carrier mobility in organic thin-film transistors. J. Am. Chem. Soc. **142**, 652–664 (2020)
38. X. Cao, Y. Han, Control over the aggregated structure of donor-acceptor conjugated polymer films for high-mobility organic field-effect transistors. Aggregate **2**, e501 (2024)
39. I. Osaka, M. Shimawaki, H. Mori, I. Doi, E. Miyazaki, T. Koganezawa, K. Takimiya, Synthesis, characterization, and transistor and solar cell applications of a Naphthobisthiadiazole-based semiconducting polymer. J. Am. Chem. Soc. **134**, 3498–3507 (2012)
40. N.E. Jackson, B.M. Savoie, K.L. Kohlstedt, M.O. Cruz, G.C. Schatz, L.X. Chen, M.A. Ratner, Controlling conformations of conjugated polymers and small molecules: The role of nonbonding interactions. J. Am. Chem. Soc. **135**, 10475–10483 (2013)
41. T. Lei, X. Xia, J.-Y. Wang, C.-J. Liu, J. Pei, "Conformation locked" strong electron-deficient poly(*p*-PhenyleneVinylene) derivatives for ambient-stable n-type field-effect transistors: synthesis, properties, and effects of fluorine substitution position. J. Am. Chem. Soc. **136**, 2135–2141 (2014)
42. S. Yum, T.K. An, X. Wang, W. Lee, M.A. Uddin, Y.J. Kim, T.L. Nguyen, S. Xu, S. Hwang, C.E. Park, H.Y. Woo, Benzotriazole-containing planar conjugated polymers with noncovalent conformational locks for thermally stable and efficient polymer field-effect transistors. Chem. Mater. **26**, 2147–2154 (2014)
43. M.A. Uddin, T.H. Lee, S. Xu, S.Y. Park, T. Kim, S. Song, T.L. Nguyen, S.-J. Ko, S. Hwang, J.Y. Kim, H.Y. Woo, Interplay of intramolecular noncovalent coulomb interactions for semicrystalline photovoltaic polymers. Chem. Mater. **27**, 5997–6007 (2015)

Photoalignment Polymers

Mizuho Kondo and Nobuhiro Kawatsuki

1 Introduction

The method of transforming a random molecular orientation into an oriented state via the influence of light is denoted photo-orientation. This is a technique that enables positional and angular selective orientation via the irradiation of light under various conditions, and it recently became a critical technique in the fabrication of liquid crystal panels. In addition to light, molecules can also be oriented via stretching, rubbing, friction transfer, magnetic field orientation, nanoimprinting, etc., but the advantages of using light over these methods include the following [1]:

1. Elimination of electrostatic charges and impurities as well as mechanical damage of the surface.
2. A controllable pretilt angle and anchoring energy of the LC cell, as well as its high thermal and UV stability and ionic purity.
3. The possibility to produce structures with required LC director alignment within the selected areas of the cell, thus allowing pixel dividing to enable new special LC device configurations for transflective, multidomain, 3D, and other new display types.
4. A potential increase if manufacturing yield, especially in LCD with active matrix addressing, where the pixels of a high-resolution LCD screen are driven by thin film transistors on a silicon substrate.
5. New advanced applications of LCs in fiber communications, optical data processing, holography, and other fields, where the traditional rubbing LC alignment is not possible due to the sophisticated geometry of the LC cell and/or high spatial resolution of the processing system.
6. The ability for efficient LC alignment on curved and flexible substrates.

M. Kondo (✉) · N. Kawatsuki
Department of Applied Chemistry, Graduate School of Engineering, University of Hyogo, Himeji, Hyogo, Japan
e-mail: mizuho-k@eng.u-hyogo.ac.jp

M. Funahashi (ed.), *Functional Soft Matter II*,
https://doi.org/10.1007/978-981-92-0100-6_10

7. Manufacturing of new optical elements for LC technology, such as patterned polarizes and phase retarders, tunable optical filters, polarization non-sensitive optical lenses, with voltage-controllable focal distance, etc.

These advantages have led to extensive research and 40 years of development in the field of photoalignment, which is now widely used instead of conventional rubbing in LCD alignment during the fabrication of large TVs, smartphones, and other LCDs.

2 Principles of Photoalignment

Photoalignment can be broadly classified into two types: those based on photochemical reactions and those based on photophysical processes. To date, most studies have been concerned with producing molecular orientation via photochemical reactions, and few examples of the control of orientation without structural changes in the molecules are reported. Most of this chapter therefore focuses on photochemical reaction-based photoalignment.

2.1 Axis-Selective Photoreaction and -Reorientation

When inducing anisotropy in polymer thin films, where molecular motion is somewhat restricted, the anisotropic optical absorption efficiencies of the molecules are exploited, i.e., when a polymer film containing photoresponsive moieties is irradiated with linearly polarized (LP) light, the molecules with light absorption axes parallel to the polarized electric field preferentially photoreact (Fig. 1a). This axis-selective photoreaction endows the film with optical and mechanical anisotropies, which can be functionalized via photo-orientation. The induction of anisotropy via photoreaction began around 1920, when Weigert discovered a phenomenon (Weigert effect) where a polymer film wherein dye molecules are dispersed exhibits anisotropy in the direction perpendicular to the polarized electric field when irradiated with LP light [2]. This is because molecules that align with the transition moment of the dye molecules preferentially undergo photoreactions, and anisotropy can be induced using photoisomerization, -lysis, and -crosslinking reactions.

Another crucial means of inducing anisotropy is photoreorientation: Todorov et al. reported that irradiating polyvinyl alcohol films, wherein the azo dye methyl orange was dispersed, with LP light at 488 nm caused contrasting in−/decreases in polarization transmission perpendicular and parallel to the polarization electric field, respectively. The azo dyes in−/decreased polarization transmittance in the directions perpendicular and parallel to the polarization field, respectively. Azo dyes have transition moments in the directions of their long molecular axes. The irradiation of LP light of a specific wavelength (e.g., the emission line of a mercury vapor lamp at

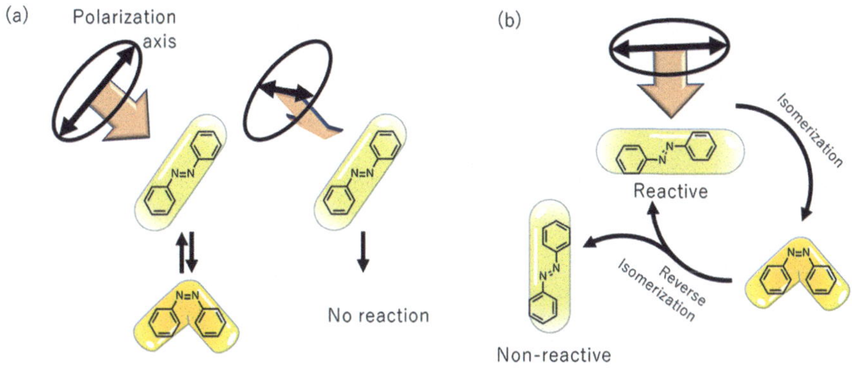

Fig. 1 Schematics of **a** axis-selective photoreaction and **b** photoreorientation

365 nm) with an electric field vector parallel to these long molecular axes causes photoisomerization from the *trans* form to the *cis* form The *cis*-isomer spontaneously undergoes reverse isomerization to the *trans*-isomer because it is thermally unstable, but molecules aligned perpendicular to the polarizing electric field during reverse isomerization do not re-isomerize. As a result, as shown in Fig. 1b, the photoisomerization of methyl orange induces molecular rotational motion and the dye reorients perpendicular to the polarizing electric field. In the material reported by Todorov et al., the dichroism (D) was considerably reduced when light irradiation ceased [3]. However, Natanshon et al. reported that the reduction in D could be alleviated via the bonding of azobenzene to the polymer main chain, leading to more efficient orientation control [4].

The anisotropy produced by an axis-selective photoreaction can be observed in the polarized light absorption spectrum and expressed by the D, as follows:

$$D = (A_{\perp} - A_{\parallel})/(A_{\perp} + A_{\parallel})$$

where $A_{\perp}$ and $A_{\parallel}$ are the absorbances perpendicular and parallel to the electric field vector of LP light to induce photoisomerization, respectively. Table 1 compares the D values obtained by irradiating polymethacrylate thin films bearing various photoreactive groups with LP light to induce axis-selective photoreactions. Each group is bound to polymethacrylate at the wavy line position, and the D values of the materials are not very high at approximately 0.02–0.05.

2.2 *Photoalignment in a Liquid Crystal*

For light-induced molecular orientation to display practical significance, the light-generated anisotropy must be increased, and liquid crystallinity is critical. Liquid crystals are typical groups of molecules that exhibit cooperative phenomena with

Table 1 *D* values of polymethacrylate thin films containing photoreactive groups

Photoreactive group	Photoreaction		*D*
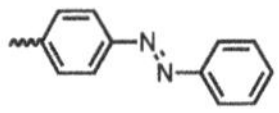	Photoisomerization		0.03–0.05
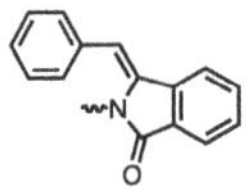 [5]	Photoisomerization	Photodimerization	0.04–0.04
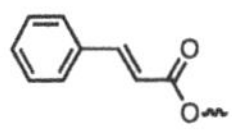	Photoisomerization	Photodimerization	0.02–0.04
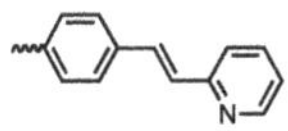 [6]	Photoisomerization	Photodimerization	0.02–0.04
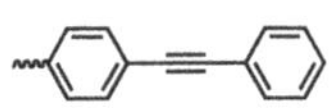 [7]		Photodimerization	0.025–0.03
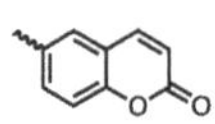 [8]		Photodimerization	0.02–0.04

respect to their positions and orientations. The cooperative phenomenon is as follows: when the states of several molecules or atoms in a molecular (atomic) population are altered by an external field, the event is transmitted, similar to dominoes, from one molecule to another, changing the state of the entire molecular (atomic) population. Via this cooperative phenomenon, changing the states of several molecules and atoms with weak stimuli is possible, thereby ultimately inducing a change in the state of the entire population, i.e., the entire system can be converted to an oriented state using the cooperative phenomenon, which is initiated by the molecular orientation generated via light irradiation, and high anisotropy can be realized (Fig. 2) [9]. Depending on the involvement of the liquid crystal in the secondary amplification of light-induced molecular orientation, the system can be split into interface and integrated alignment types.

3 Alignment Films Produced Using Photochemical Reactions

Depending on the manner in which the liquid crystal is involved in the secondary amplification of light-induced molecular alignment, two types of alignment systems have been reported: interfacial and integrated alignment. In interfacial alignment, the

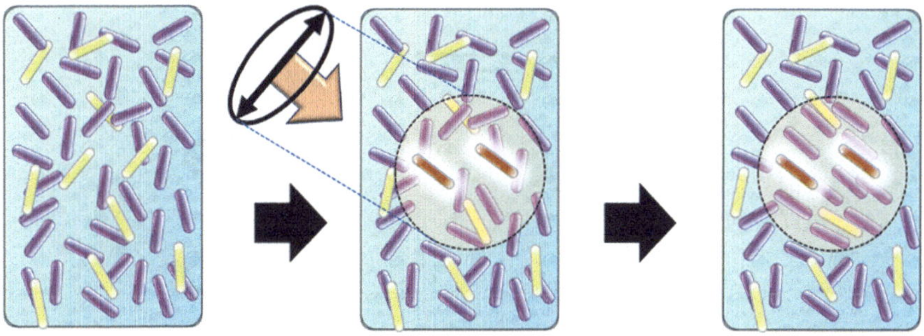

Fig. 2 Schematic of liquid crystal photoalignment based on the cooperative phenomenon

anisotropy generated in the thin film with a photoreactive material is brought into contact with the target liquid crystal, and the change in the orientation of the liquid crystal in the vicinity of the interface is transferred to the entire system. Integrated alignment describes a system wherein molecules that can be aligned via light irradiation are chemically bonded or physically mixed with the liquid crystal material and integrated, and the changes in the orientations of the light-responsive molecules in the system are reflected in the entire system. Various polymeric materials have been investigated using both methods.

3.1 Interfacial Alignment Based on Photoisomerization

In interfacial orientation, the anisotropy generated within the thin film with a photoactive material is brought into contact with the target liquid crystal, and the change in the orientation of the liquid crystal in the vicinity of the interface is transferred to the entire system. The first report of interfacial photoalignment was reported by Ichimura et al. in 1988 [10]. Based on this report, when a nematic liquid crystal is sandwiched between modified glass substrates wherein azobenzene has been introduced via silylation, homeotropic orientation is induced when the azobenzene on the substrate surfaces is *trans*, and planar orientation is induced when photoisomerization is caused by light irradiation (Fig. 3a). The authors denote such photoreactive surface layers "command surfaces."

Subsequently, the planar orientations of interface-oriented films with azobenzene-doped polyimide films as their liquid crystal alignment layers were controlled via linear polarization by Gibbons et al. (Fig. 3b) [11]. Photoalignment technologies with command surfaces have been expanded from switching between homeotropic and planar orientations to arbitrary horizontal orientation control (Fig. 3c). Several of these photoalignment technologies are used as alignment films in the initial orientation of small-molecule liquid crystals and the orientation of polymerizable liquid crystals in the manufacturing of LCDs and other applications.

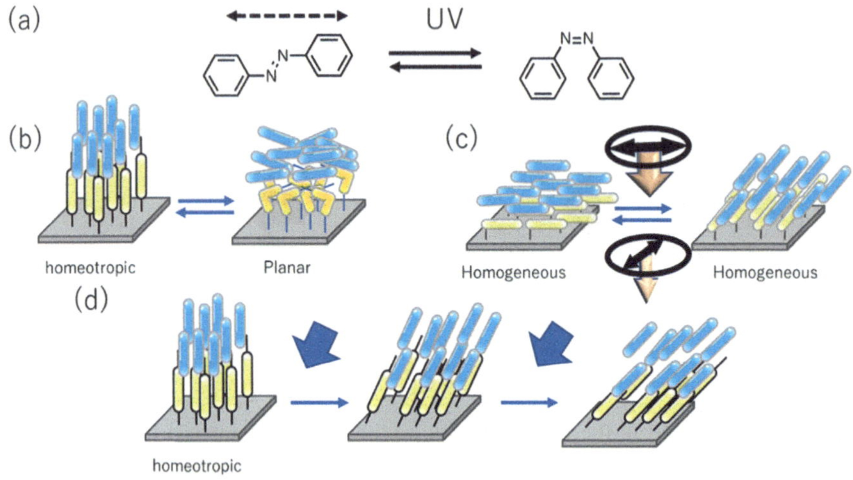

Fig. 3 **a** Axis-selective isomerization of azobenzene. The changes in the alignment direction of the liquid crystal using an azobenzene command surface: **b** out-of-plane-to-in-plane, **c** in-plane-to-in-plane, and **d** out-of-plane-to-tilt. The black-dotted double-headed arrow shown in (a) indicates the polarization axis of LP light

Tilt orientation via photoisomerization has also been reported in alignment films [12]. Azo dyes are not easily photoisomerized by LP light with an electric field vector that is not parallel to their long molecular axes, i.e., isomerization efficiency decreases not only when the short molecular axis is parallel to the electric field vector, but also when reoriented in the direction of light propagation. Therefore, the continuous irradiation of unpolarized light from an oblique direction will eventually shift azobenzene parallel to the direction of irradiation, inducing a 3D orientation (Fig. 3d). Tilt orientation is a crucial technique in generating tilt angles in LCDs.

3.2 Interfacial Alignment Based on Photodimerization

The first example of the use of photodimerization reactions in photoalignment was the use of poly(vinyl cinnamate) (PVCi) derivatives by Schadt et al. [13]. PVCi undergoes [2 + 2] photocrosslinking when the direction of the polarization of the irradiating light is parallel to the conjugated double bond of the side-chain styryl group (Fig. 4a). This photoreaction results in the axially selective photodimerization of the cinnamic acid moiety and optical anisotropy on the thin-film surface. Other polymer materials utilizing axial selective photodimerization have been reported, such as coumarin (Fig. 4b), chalcone (Fig. 4c), styrylpyridine (Fig. 4d), diphenylacetylene (Fig. 4e), and anthracene (Fig. 4f). The liquid crystals are then oriented via anisotropic interactions with the liquid crystals on the thin-film surface.

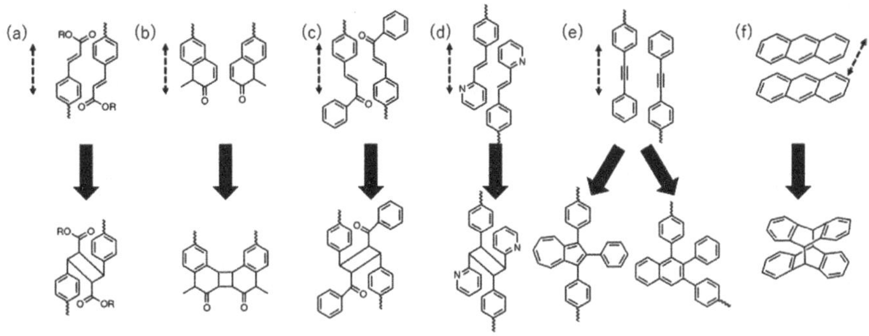

Fig. 4 Axis-selective photodimerization. The black-dotted double-headed arrows indicate the polarization axes of LP light

When photoalignment is performed using polymers that photodimerize, the direction of orientation may vary depending on the polymer structure, exposure dose, and dominant photoreaction. For the polymers shown in Table 2a–c, the liquid crystals are oriented perpendicular to the polarization axis in (a), perpendicular and parallel at low and high exposure doses, respectively, in (b), and parallel in (c) [14]. This is interpreted by the following mechanism: In (a), there are no spacers between the polymer backbone and the photoreactive double bond is on the polymer backbone side. Therefore, the photoreaction products have an elongated structure perpendicular to the polarization axis and are vertically oriented. Furthermore, reorientation by photoisomerization of cinnamate functions effectively. As a consequence, the vertical orientation is predominant. Conversely, in (c), the alkyl chains and benzene rings are arranged between the photoreactive double bond and the polymer backbone, resulting in a parallel orientation because the photoreactant is elongated in a direction parallel to the polarization axis. The structure of (b) is interpreted as intermediate between (a) and (c), with a large contribution from the reorientation component in the early stages of the reaction, but dimer orientation becomes dominant as the exposure increases.

In the photoreaction of cinnamic acid, the reactivity is enhanced by the addition of Michler's ketone (MK) and other triplet sensitizers [15]. MK is excited by light at a wavelength of 365 nm and it can therefore induce photodimerization, even for cinnamic acid derivatives that do not absorb at 365 nm. MK exhibits a high absorption efficiency in the direction of its long molecular axis and it can be used for linear polarized light at 365 nm. MK displays a high absorption efficiency in the longitudinal direction of the molecule and axially selective photoabsorption when 365 nm LP light is used. The sensitizer excited via this process rapidly transfers energy to the cinnamate moiety before orientation relaxation occurs in the alignment film, retaining its orientation data, and photoisomerization or dimerization proceeds. As a result, liquid crystal orientation in the direction perpendicular to polarization is induced (Fig. 5).

Table 2 Effect of the connection position on the photoalignment behavior of the polymer with cinnamate side chains

Polymer structure		Photoreaction product	Alignment	
			Low	High
(a)	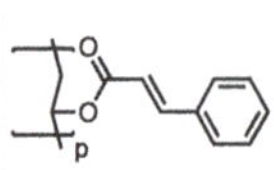	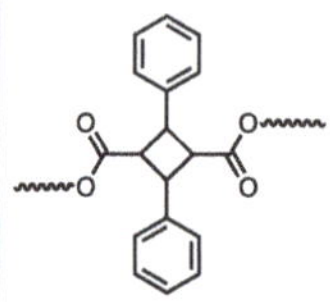	Perpendicular	Perpendicular
(b)	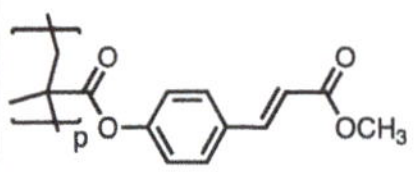	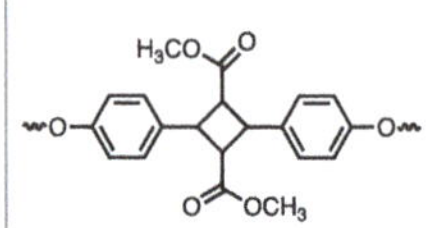	Perpendicular	Parallel
(c)	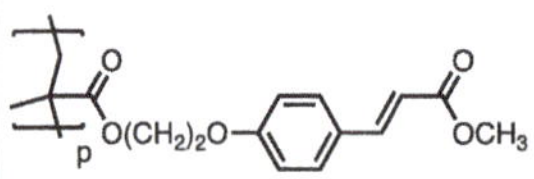	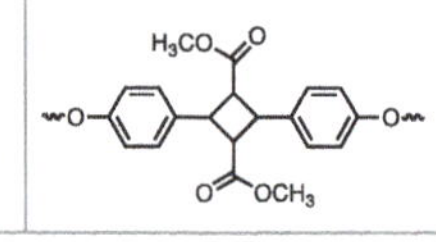	Parallel	Parallel

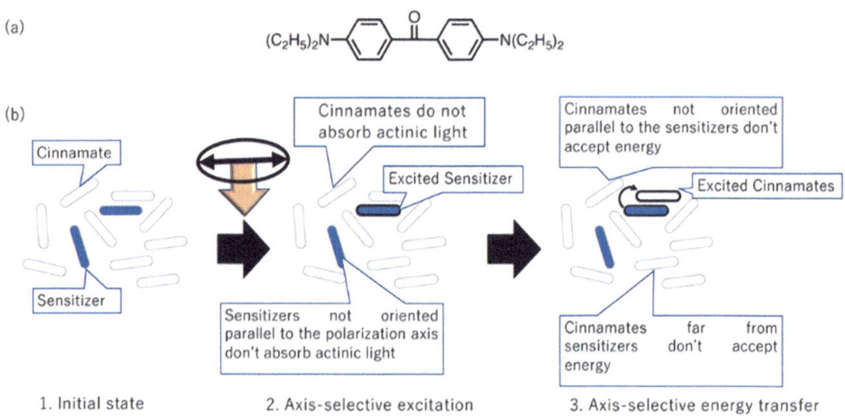

Fig. 5 Molecular structure of the triplet sensitizer Michler's ketone (**a**). Schematic of axis-selective energy transfer (**b**)

3.3 *Interfacial Alignment Based on Photocleavage*

Polyimide displays a high heat resistance and it is commonly used as an alignment film in liquid crystal panels. Controlling the orientation of this material utilizing light is considered useful, and numerous studies have been conducted to impart photoresponsiveness by introducing functional groups with axis-selective photoresponses. Hasegawa et al. were the first to report the orientation control of polyimide films using LP UV light [16]. This involved the axis-selective cleavage of cycloalkanes (Fig. 6)

Fig. 6 Axis-selective photocleavage of polyimides. The black-dotted double-headed arrow indicates the polarization axis of LP light

[17] within the polyimide backbones [18] that were parallel to the polarization axis within the film, and the unreacted polyimide induced an orientation perpendicular to the polarization axis. Other proposed reaction mechanisms include photooxidative degradation [19] and methods originating from additional chloromethyl side chains [20]. Polyimide with photoreactive groups has been used practically as a liquid crystal alignment film, as it also displays the properties of polyimide as a rubbing-oriented alignment film for use in LCDs.

3.4 Interfacial Orientation Based on Photorearrangement

Photo-Fries rearrangement reactions occur in phenyl ester-containing compounds upon UV light irradiation. The Fries rearrangement is generally a rearrangement reaction from a phenyl ester to a hydroxyketone using a catalyst, such as a Lewis acid, but it also proceeds under light. A light absorption axis is observed in the orthogonal direction relative to the phenyl ester, and the reaction is axially selective. Yaroshchuk et al. reported that LP light irradiation of poly(1-naphthyl methacrylate) induced an axis-selective photoreorientation reaction, leading to liquid crystal photoorientation (Fig. 7) [21]. The Fries reaction does not require a reactive double bond to increase π-conjugation, thus shifting the absorption of the molecule into the deep UV region. As poly(1-naphthyl methacrylate) exhibits a poor thermal stability in aligned films, they also attempted to improve its thermal stability by introducing tetrahydrophthalic acid or maleimide and crosslinking after orientation.

Fig. 7 Axis-selective Fries rearrangement of poly(1-naphthyl methacrylate)

3.5 Integrated Orientation Based on Photoisomerization

In integrated orientation, the molecules in the film are reoriented using LP light irradiation, enabling the application of the film as a stand-alone retardation film. In the amplification behavior of the integrally oriented type, coordinated molecular orientation amplification of the mesogens is induced in copolymerized films with non-photosensitive mesogens. As the introduction of non-photosensitive mesogens facilitates the adjustment of the light absorption efficiency and thermal properties, in addition to the cooperative imparting and orientation of various properties derived from the non-photosensitive mesogens, various combinations of materials are under investigation.

3.5.1 Azobenzene

Liquid crystalline polymers bonded with azobenzene are examples of integrally oriented photoalignment materials, and the holographic material reported by Wendorff et al. is considered the first of this type [22]. In this material, azobenzene initially generates anisotropy and the non-photoactive mesogens simultaneously undergo reorientation via cooperative motion with the azobenzene moieties at temperatures above the glass transition temperature.

In integrally oriented films, heating may also promote self-assembly. Stumpe et al. reported that the use of side-chain liquid crystalline polymers considerably amplifies the orientation effect and enables the maintenance of the oriented structure in the liquid crystalline state (Fig. 8) [23].

Thermal orientation amplification via self-assembly or coordinated orientation amplification among copolymers is possible within these materials after photoreorientation to produce slight optical anisotropy. Research regarding photoalignment

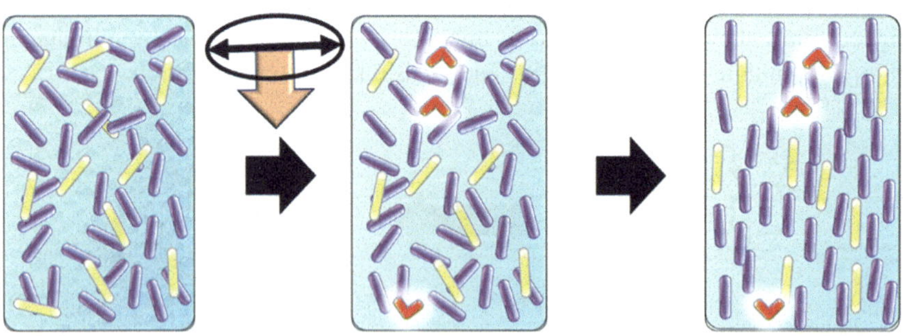

Fig. 8 Axis-selective isomerization of azobenzene liquid crystals and the subsequent thermal amplification of integral alignment

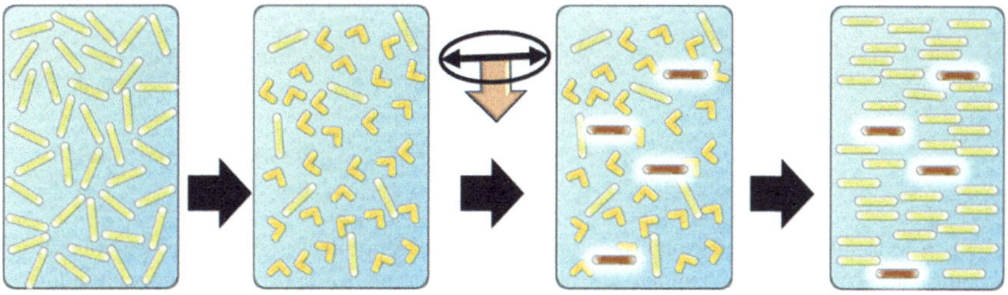

Fig. 9 Schematic of orientation amplification via the axis-selective photoisomerization of the *cis* form of azobenzene

using molecules with azobenzene groups that react via axis-selective photoisomerization is particularly abundant, and numerous applications in optical recording and holography that exploit reversibility based on polarized light reactions are reported.

The amplification of anisotropy using the photoreactions of *cis* substrates has also been reported. In- and out-of-plane reorientation behaviors have been studied upon irradiation with LP light of various wavelengths, with the photoreaction activities of the *trans* and *cis* isomers depending on the wavelength of the light. The *cis* form of azobenzene displays slight absorption at a wavelength of approximately 633 nm. Therefore, when the azobenzene in the system is photoisomerized to the *cis* form via UV light irradiation until a photosteady state is reached and then irradiated with LP light (wavelength = 633 nm), axis-selective reverse isomerization induces anisotropy by producing azobenzene parallel to the polarization axis. This effect is amplified via thermal treatment (Fig. 9) [24].

3.5.2 *N*-Benzylideneaniline (NBA)

NBA, as shown in Fig. 10a, exhibits photoisomerization similar to that of azobenzene. The *cis* form of NBA, which is generated via UV light irradiation, is thermally unstable and reverts to the *trans* form as soon as light irradiation ceases. Figure 10b shows a liquid crystal polymer with NBA in the side chain that undergoes orientation amplification similar to that of polymers containing azobenzene groups when irradiated with LP light and heated. The effects of polarized light irradiation and thermal alignment amplification are almost unchanged when the C=N bond to the polymer main chain is reversed, as shown in Fig. 10c [25].

The imino groups, which comprise the photoresponsive backbone, are easily synthesized via the condensation reactions of phenylaldehyde and aromatic amines. Therefore, when a polymer with phenylamine or benzaldehyde in the side chain is subjected to a counterpart small molecule, NBA is formed in situ, e.g., as shown in Fig. 11; when a small-molecule solution is pattern-coated using an inkjet printer, NBA is formed only in the coated area, and photoreactivity can be generated at any position [26].

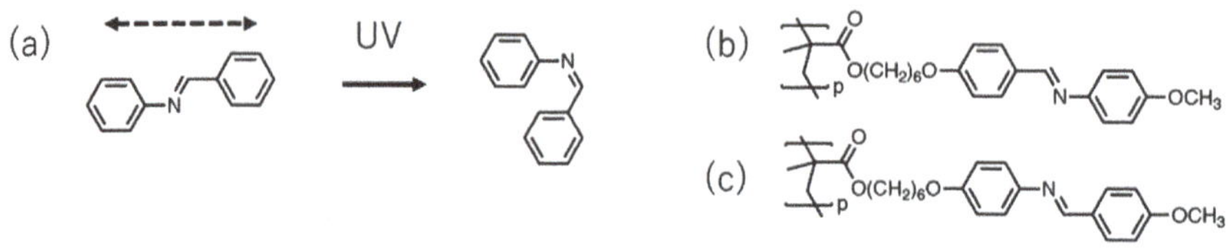

Fig. 10 **a** Axis-selective photoisomerization of NBA. **b** NBA liquid crystal polymer with integrated orientation and **c** its structural isomer. The black-dotted double-headed arrow shown in **a** indicates the polarization axis of LP light

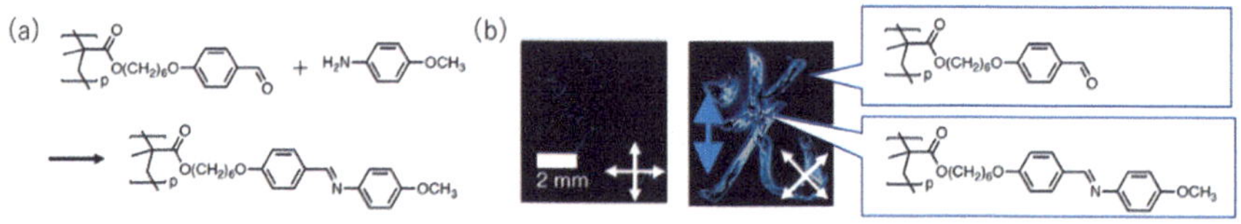

Fig. 11 Polymeric reaction to form NBA moieties using a phenylaldehyde polymer (**a**) and a polarized-light micrograph of a regioselectively formed photoaligned film (**b**). The white arrows indicate the directions of the polarizer of the polarization microscope. The gray arrows indicate the orientation directions of NBA [26]

3.6 Integrated Orientation Based on Photodimerization

The film of the liquid crystalline polymer shown in Fig. 12a was irradiated with LP UV light and then heated at the liquid crystal temperature. As shown in Fig. 12b, the spectral changes in the polarization absorption reveal a slightly faster decrease in absorption in the parallel direction after the photoreaction. Additionally, a large increase in absorption in the parallel direction and a decrease in the perpendicular direction are observed after heating at the liquid crystal temperature. The average of the parallel and perpendicular absorption after orientation amplification increases compared to the absorption before orientation amplification, indicating that the mesogens that are oriented in the out-of-plane direction before the reaction are also oriented in the in-plane direction during self-assembly. A plausible mechanism is illustrated in Fig. 12c. The orientation amplification in the direction parallel to the polarization electric field is due to (1) the axis-selective photocrosslinking of the mesogens that become anchors when heated at the liquid crystal temperature after the photoreaction and (2) the thermal self-assembly of the unreacted mesogens in the direction of the photocrosslinked mesogens (parallel to the polarization electric field).

In such a self-assembly, the positional relationship between the light-reactive component and mesogen is critical, e.g., the compound shown in Fig. 12a, which is a polymer wherein the biphenyl moiety and cinnamic acid are joined by a two-carbon alkylene spacer that does not exhibit orientation amplification upon heating, even if an axis-selective photoreaction is induced. This is likely because the photo-oriented components are bound slightly bent with respect to the biphenyl and do

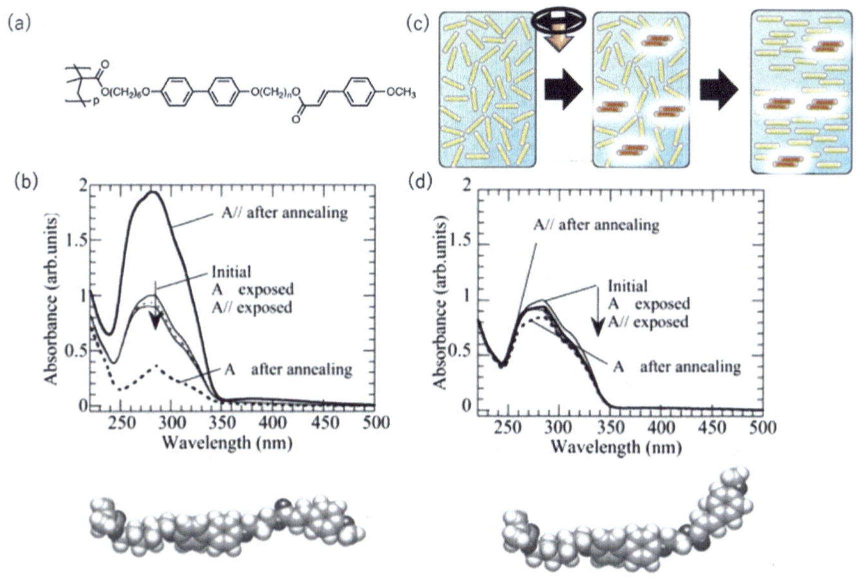

Fig. 12 Molecular structure of a liquid crystal polymer that undergoes integrated orientation via photodimerization (**a**). Changes in the polarization absorption spectrum of a film formed using **a** after irradiation with LP light and subsequent heating (**b**). Schematic of the mechanism of integrated orientation (**c**). Changes in the polarization absorption spectrum of a polymer with an extended alkylene spacer adjacent to the cinnamic acid unit shown in **a** and **d**. The figures at the bases of **b** and **d** show the simulated structure of each monomer [27]

not act as anchors to promote self-assembly, even after axis-selective crosslinking (Fig. 12d) [27]. The material shown in Fig. 12a, which contains a biphenyl mesogen directly bound to cinnamate, can also exhibit an in-plane orientation parallel to the polarizing electric field. This material can also display 3D molecular reorientation with mesogens tilted in the direction of the light source when irradiated with p-polarized UV light from an oblique direction [28].

For the material shown in Fig. 13a, orientation is amplified in the direction perpendicular to the polarizing electric field when the degree of photoreaction is small. This may be because the photoreacted mesogens reduce the purity of the liquid crystal in the parallel direction rather than acting as anchors when the degree of photoreaction is small and favor thermal orientation amplification in the vertical direction. This results in an enhanced liquid crystallinity (easier self-assembly) when promoting self-assembly (Fig. 13b) [29]. As a result, molecular orientation can be amplified in the perpendicular and parallel directions depending on the level of exposure in this compound; however, the anisotropy is not very large due to the aggregation of mesogens during heating. Increased levels of vertical amplification have been reported in materials with shortened spacers between their mesogens and polymer backbones and

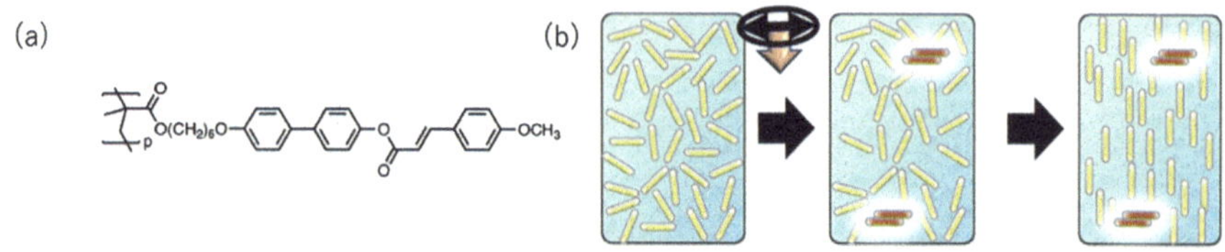

Fig. 13 Liquid crystalline polymer with integrated orientation perpendicular to the direction of polarization upon photodimerization (**a**) and its mechanism (**b**)

those copolymerized with non-liquid crystalline monomers. The orientation amplification is also strongly affected by the lengths of the terminal substituents and spacers connecting the polymer main chain and mesogens.

Unlike the reversal in the orientation direction observed in the interface-oriented type, the reversal in the integration-oriented type depends strongly on the molecular shape of the photoproduct and interface.

On the film, small-molecule liquid crystals are oriented in the same direction as the molecular orientation. Molecularly oriented liquid crystal alignment films display higher orientation-regulating forces than that of PVCi used in interface orientation [30]. The orientation direction is virtually parallel to the direction of light-induced molecular orientation, and the tilt angles on obliquely oriented films can be controlled. As the direction of molecular orientation can be controlled by the level of exposure, pattern exposure using a photomask enables the fabrication of alignment films with controlled orientation directions without changing the direction of the polarizing electric field employed, and pattern-oriented liquid crystal cells can easily be fabricated [31].

For the compounds shown in Figs. 12 and 13, an overlap in light absorption between the biphenyl and photoresponsive moieties is observed, resulting in decreased photoreaction efficiencies. Triplet photosensitizers can be used to suppress this effect. Photosensitization promotes the photoreactions of cinnamic acid derivatives (Fig. 14a). Particularly when a photosensitizer with high absorption anisotropy is used, high optical anisotropy can be realized at a low energy due to the formation of orientation-selective exciplexes comprising the axially photoexcited sensitizers and photoresponsive liquid crystal mesogens [32]. In addition, materials copolymerized with photosensitizing biphenyl monomers exhibit improved axial selectivities compared to those of general triplet sensitizers (Fig. 14b) [33].

Using polymers with cinnamic acid units at the ends of their side chains, as shown in Fig. 14c, to reduce the level of light exposure by rendering the light absorption more efficient is also possible. The carboxylic acid moieties at the end of the side chains form hydrogen-bonded dimers, which easily exhibit liquid crystalline properties, despite the low aspect ratio of the molecular structure. In this film, which is photoreactivated by approximately 5% at ~ 10 mJ cm^{-2}, the optical anisotropy can be increased by heating. The low energy required for reorientation is due to the absence of biphenyl mesogens and the improved photoreactivity [34]. When the exposure dose on this film is increased, the molecular orientation becomes parallel

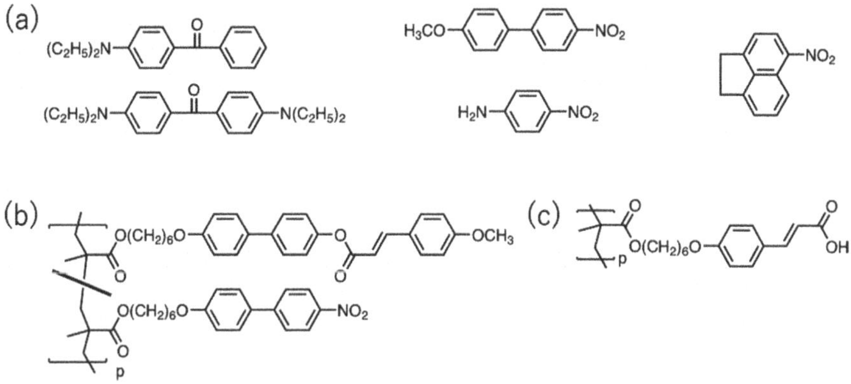

Fig. 14 Structures of small-molecule triplet photosensitizers (**a**). Molecular structure of a copolymer containing mesogens with triplet sensitizing effects (**b**). Hydrogen-bonded photoresponsive liquid crystal polymer (**c**)

to the polarization axis, as in the biphenyl-containing compound, but the anisotropy decreases. Although the small-molecule liquid crystals on this film also orient in the same direction as the molecular orientation, the orientation behaviors of the liquid crystals upon approaching the exposure dose that inverts the molecular orientation and at high levels of exposure are not simple. This may be due to the interactions of the reoriented mesogens with the photoreacted mesogens.

In addition to cinnamic acid esters, photoresponsive liquid crystalline polymers with coumarins [35] or anthracenes [36] at the ends of their side chains can also be used due to their photo-orientation and amplification behaviors, as shown in Fig. 15. In the case of anthracene, the orientation behavior in response to a polarizing electric field is reversed depending on the substitution position.

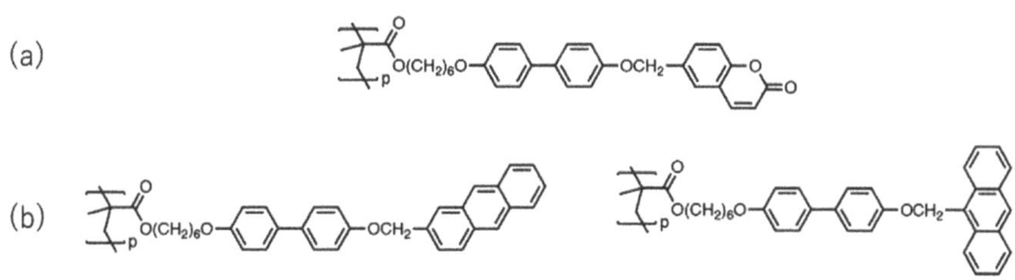

Fig. 15 Liquid crystalline coumarin polymers with integrated orientation via photodimerization (**a**). Molecular structures of liquid crystalline anthracene-bearing polymers

3.7 Integrated Orientation Based on Photorearrangement

When the benzoate ester shown in Fig. 16a is irradiated with LP light at a wavelength of 248 nm, an axis-selective photoreorientation reaction occurs. The products of the Fries rearrangement also function as impurities in the same manner as the previously introduced materials, and the optical anisotropy of the liquid crystal side-chain polymer shown in Fig. 16b is amplified by heating when irradiated with LP light at a wavelength of 248 nm. This polarization irradiation and thermal alignment amplification are almost the same when the ester bond to the polymer main chain is reversed [37]. Conversely, in the structures shown in Fig. 16c–e, where biphenyl is introduced into the mesogen, the amplification behavior differs depending on the position of biphenyl introduction [38]. The polymer with biphenyl introduced at the end of the side chain (Fig. 16c) reorients parallel to the modification axis upon heating at the liquid crystal temperature after the polarized light reaction. In contrast, molecular orientation hardly occurs in materials with biphenyl groups on their main chain sides or those with different ester orientations. This is because the rotation of the biphenyl moiety is inhibited in the product of the photo-Fries rearrangement shown in Fig. 16c, resulting in a linear relationship between the benzene ring and biphenyl. The structure of the photoproduct of Fig. 16c promotes reorientation parallel to polarization. On the other hand, as shown in Fig. 16d and e, the rotation is unlimited in the rearranged products, and the geometric relation of biphenyl and benzene rings is nonlinear. As a result, the structures do not promote orientation in either direction, and self-assembly via heating does not occur.

4 Imparting Photodurability

In light-induced molecular orientation, photofunctional moieties remain in the film after photoalignment. Therefore, in principle, the levels of photodurability of oriented films under UV light are low. To impart photodurability to the photo-oriented film, deactivating or removing the photosensitive moieties is useful.

Polyamic acid-azo copolymer coordination has been reported as a method of deactivating the photoresponsive moieties, where the coordination order realized via the *trans-cis* transformation of the azo fragment is fixed by the simultaneous imidization of the polyamic acid (Fig. 17a) [39]. Similarly, in the copolymerization of azo and cinnamic acid monomers, a technique has been proposed to induce photo-orientation via *trans-cis* isomerization, followed by the fixing of the orientation via the photocrosslinking of cinnamic acid (Fig. 17b) [40].

Another method is to deactivate the photoresponsive component by decomposing it after alignment. NBA is easily hydrolyzed in solution by an acid catalyst. When an alignment film of NBA copolymerized with benzoic acid, which is a liquid crystal with a carboxylic acid, is heated at the liquid crystal temperature under humidity, the benzoic acid moiety acts as an acid catalyst to hydrolyze NBA. As the benzoic

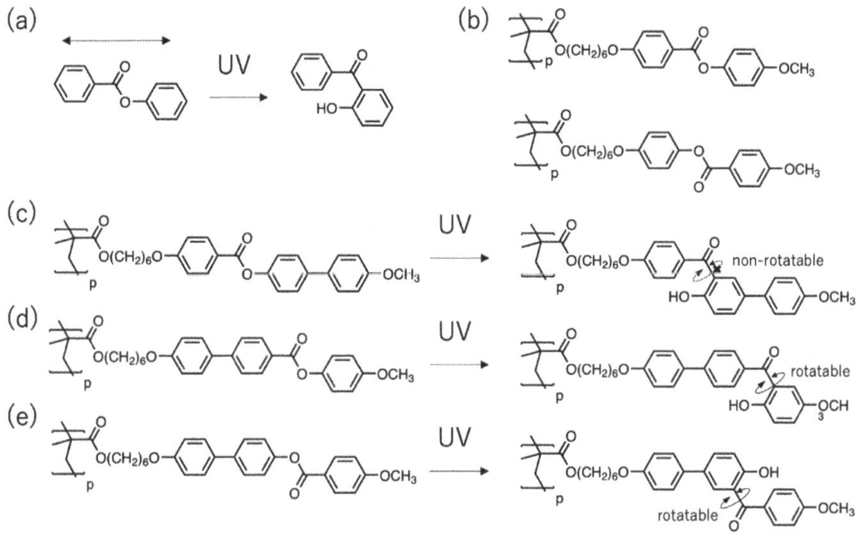

Fig. 16 **a** Axis-selective photo-Fries rearrangement reactions of phenyl esters. **b** Molecular structures of liquid crystalline polymers with integral orientation based on photo-Fries rearrangement reactions. **c** Photo-Fries rearrangement reactions of liquid crystalline polymers with biphenol groups, **d**, **e** and their structural isomers

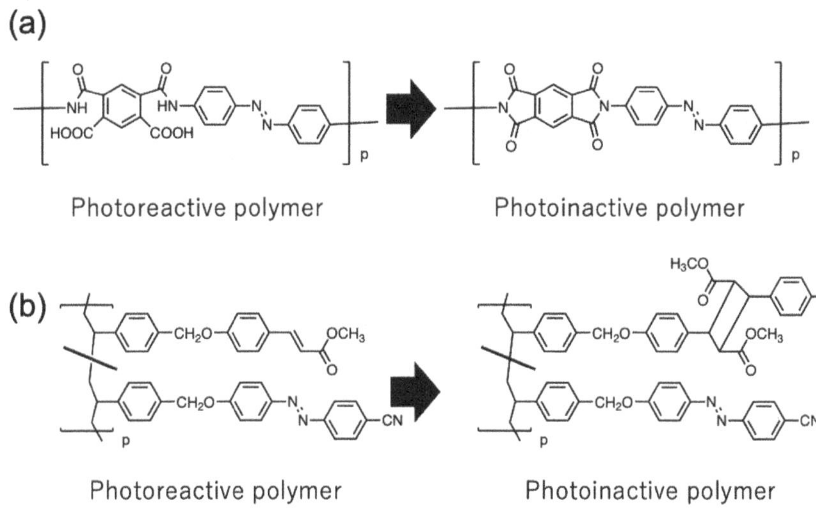

Fig. 17 Molecular structures, which contain azobenzene groups, with photodurability. **a** Material introduced into the polyimide backbone. **b** Copolymer with cinnamic acid

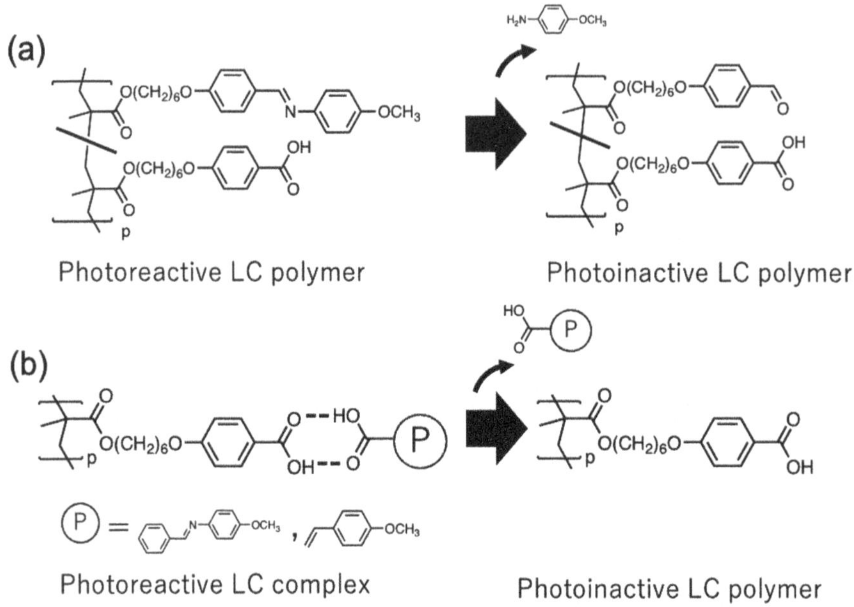

Fig. 18 Schematics of the mechanisms of photodurability **a** based on NBA decomposition and **b** after orientation based on hydrogen-bonding complexes

acid moiety functions as a mesogen; the photoreactivity can be deactivated while maintaining the orientation (Fig. 18a) [41].

Methods are also reported in which the photoreactant is temporarily complexed within the polymer via intermolecular interactions and removed after photoalignment. Benzoic acid derivatives exhibit liquid crystalline properties due to the hydrogen bonding of their carboxylic acids at the molecular ends. As this hydrogen bond can also be complexed with small-molecule carboxylic acids, small molecules of NBA [42] or cinnamic acid [43] that contain carboxylic acids at their ends can be coated onto a benzoic acid polymer and heated to form photoresponsive complexes due to hydrogen bonding. In such a photoresponsive polymer composite, the photoresponsive compound is sublimated and removed during the thermally induced orientation amplification that follows photo-orientation, rendering them durable to UV light (Fig. 18b).

5 Orientation Based on the Free Interface

We have thus far mainly described the method of photoalignment based on the substrate interface, but a method of photoalignment based on the air interface was also recently reported. The orientations of liquid crystal molecules are considerably

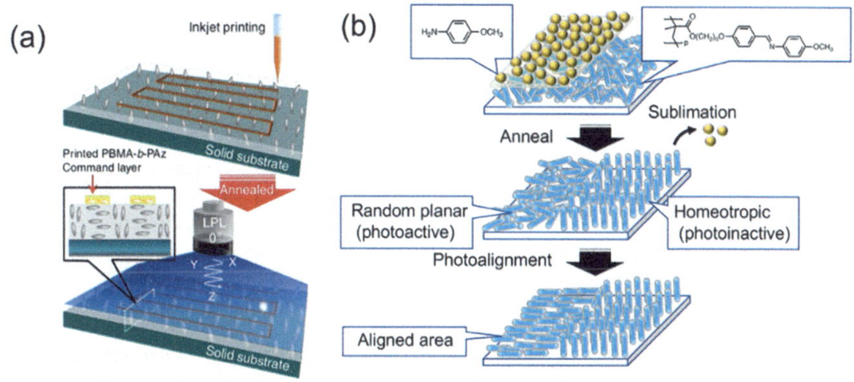

Fig. 19 Schematic of photo-orientation based on the free interface. **a** Position-selective application of a photoresponsive polymer [44]. **b** Selective suppression of the spontaneous orientation of the photoresponsive material

affected by the free interface. Their orientations are generally parallel to the substrate at the substrate interface and perpendicular at the air interface due to the excluded volume effect, and the same trend is observed for side-chain polymer liquid crystals. Controlling photoalignment regioselectively from the free interface side using an azobenzene polymer coated on the surface of a photoinert liquid crystal polymer is possible (Fig. 19a) [44]. For NBA-based materials, we reported a method of providing position-selective photo-orientation via the suppression of spontaneous homeotropic orientation by coating a low-molecular-weight aromatic compound on a material that is prone to such spontaneous orientation due to interaction with the air interface (Fig. 19b) [45].

These methods enable photo-orientation using inkjet or other methods, enabling the control of the fine molecular orientation without using a photomask.

6 Conclusion

Photo-alignable polymers are outlined. Based on axis-selective photoreactions using LP light, liquid crystals can be oriented in various materials to form anisotropic films. The further modulation of the photoreactivity and optical properties is also possible using hydrogen bonding and macromolecular reactions. The production of optical devices with novel functions using these molecularly oriented photoalignment materials is expected.

References

1. V. Chigrinov, V. Kozenkov, H.S. Kwok, *Photoalignment of Liquid Crystalline Materials: Physics and Applications* (Wiley, Hoboken, 2008)
2. F. Weigert, Über einen neuen Effekt der Strahlung. Z. Phys. **85**, 410–427 (1921)
3. T. Todorov, L. Nikolova, N. Tomova, Polarization holography. 1: a new high-efficiency organic material with reversible photoinduced birefringence. Appl. Opt. **23**, 4309–4312 (1984)
4. P. Rochon, J. Gossselin, A. Natansohn, et al., Optically induced and erased birefringence and dichroism in azoaromatic polymers. Appl. Phys. Lett. **60**, 4–5 (1992)
5. D.H. Suh, K. Ichimura, K. Kudo, Polymethacrylates with benzylidenephthalimidine side chains, 2. Photocontrol of alignment of a nematic liquid crystal. Macromol. Chem. Phys. **199**, 375–383 (1998)
6. S. Yamaki, M. Nakagawa, S. Morino, et al., Photochemistry of polyrnethacrylates with styrylpyridine side chains and their photocontrollability of liquid crystal alignment. J. Photopolym. Sci. **12**, 279–282 (1999)
7. M. Obi, S. Morino, K. Ichimura, Photocontrol of liquid crystal alignment by polymethacrylates with diphenylacetylene side chains. Chem. Mater. **11**, 1293–1301 (1999)
8. M. Obi, S. Morino, K. Ichimura, The reversion of photoalignment direction of a liquid crystal induced by a polymethacrylate with coumarin side chains. Macromol. Rapid Commun. **19**, 643–646 (1998)
9. T. Ikeda, Photomodulation of liquid crystal orientations for photonic applications. J. Mater. Chem. **13**, 2037–2057 (2003)
10. K. Ichimura, Y. Suzuki, T. Seki, et al., Reversible change in alignment mode of nematic liquid crystals regulated photochemically by command surfaces modified with an azobenzene monolayer. Langmuir **4**, 1214–1216 (1988)
11. W.M. Gibbons, P.J. Shannon, S.-T. Sun, et al., Surface-mediated alignment of nematic liquid crystals with polarized laser light. Nature **351**, 49–50 (1991)
12. Y. Kawanishi, T. Tamaki, M. Sakuragi, et al., Photochemical induction and modulation of nematic homogeneous alignment by the polarization photochromism of surface azobenzenes. Langmuir **8**, 2601–2604 (1992)
13. M. Schadt, K. Schmitt, V. Kozenkov, et al., surface-induced parallel alignment of liquid crystals by linearly polymerized photopolymers. Jpn. J. Appl. Phys. **31**, 2155–2164 (1992)
14. M. Obi, S. Morino, K. Ichimura, Reversion of photoalignment direction of liquid crystals induced by cinnamate polymer films. Jpn. J. Appl. Phys. **38**, L145–L147 (1999)
15. S. Furumi, K. Ichimura, Highly sensitive photoalignment of liquid crystals through polarization preserved triplet energy transfer. Appl. Phys. Lett. **85**, 224–226 (2004)
16. M. Hasegawa, Y. Taira, Nematic homogeneous photo alignment by polyimide exposure to linearly polarized UV. J. Photopolym. Sci. Technol. **8**, 241–248 (1995)
17. J.A. Moore, A.N. Dasheff, An intrinsically photosensitive polyimide. Chem. Mater. **1**, 163–166 (1989)
18. C.E. Hoyke, E.T. Anzures, Photodegradation of polyimides. III. The effect of chemical composition, radiation source, atmosphere, and processing. J. Polym. Sci. Polym. Chem. **30**, 1233–1245 (1992)
19. Y. Wang, C. Xu, A. Kanazawa, Alignment of a nematic liquid crystal induced by anisotropic photo-oxidation of photosensitive polyimide films. J. Appl. Phys. **84**, 4573–4578 (1998)
20. Z.X. Zhong, X. Li, S.H. Lee, et al., *A New Class of Photoalignable Polyimides with High Photosensitivity. Polyimides and Other High Temperature Polymers*, vol. 3 (2005), pp. 401–423
21. L. Vretik, O.V. Yaroshchuk, V. Zagnii, et al., New Strategy in Development of Liquid Crystal Photoaligning Materials with Reactive C=C Bonds, in *Liquid Crystalline Organic Compounds and Polymers as Materials of the XXI Century: From Synthesis to Applications*, (Transworld Research Network, 2011), pp. 153–189
22. M. Eich, J.H. Wendorff, B. Reck, et al., Reversible digital and holographic optical storage in polymeric liquid crystals. Makromol. Chem. Rapid Commun. **8**, 59–63 (1987)

23. J. Stumpe, L. Muller, D. Kreysig, et al., Photoreaction in mesogenic media, 5†. Photoinduced optical anisotropy of liquid-crystalline sidechain polymers with azochromophores by linearly polarized light of low intensity. Makromol. Chem. Rapid Commun. **12**, 81–87 (1991)
24. E. Uchida, N. Kawatsuki, H. Ono, et al., Extremely high degree photoinduced in-plane reorientation in azobenzene-containing polymer liquid crystal film using 633 nm light. Jpn. J. Appl. Phys. **44**, 570–574 (2005)
25. N. Kawatsuki, H. Matsushita, M. Kondo, et al., Photoinduced reorientation and polarization holography in a new photopolymer with 4-methoxy-*N*-benzylideneaniline side groups. APL Mater. **1**, 022103 (2013)
26. N. Kawatsuki, K. Miyake, M. Kondo, Facile fabrication, photoinduced orientation, and birefringent pattern control of photoalignable films comprised of *N*-benzylideneaniline side groups. ACS Macro Lett. **4**, 764–768 (2015)
27. N. Kawatsuki, R. Tsutsumi, A. Hiraiwa, et al., Thermally enhanced photoinduced in-plane reorientation in photo-cross-linkable polymer liquid crystalline films and its application to linear polarizer. J. Polym. Sci. A Polym. Chem. **46**, 4712–4718 (2008)
28. N. Kawatsuki, et al, Liquid Crystals V. In: Khoo I-C (ed) Proceedings of SPIE series, San Diego, 2001
29. N. Kawatsuki, K. Goto, T. Kawakami, et al., Reversion of alignment direction in the thermally enhanced photoorientation of photo-cross-linkable polymer liquid crystal films. Macromolecules **35**, 706–713 (2002)
30. N. Kawatsuki, K. Hamano, H. Ono, et al., Molecular-oriented photoalignment layer for liquid crystals. Jpn. J. Appl. Phys. **46**, 339–341 (2007)
31. N. Kawatsuki, T. Kawakami, T. Yamamoto, A photoinduced birefringent film with a high orientational order obtained from a novel polymer liquid crystal. Adv. Mater. **13**, 1337–1339 (2001)
32. N. Kawatsuki, T. Tachibana, M.X. An, et al., Photoinduced optical anisotropy based on axis-selective triplet energy transfer and thermally enhanced reorientation in a photo-cross-linkable liquid crystalline polymer film. Macromolecules **38**, 3903–3908 (2005)
33. N. Kawatsuki, K. Kato, T. Shiraku, et al., Photoinduced reorientation and multiple optical data storage in photo-cross-linkable liquid crystalline copolymer films using 405 nm light. Macromolecules **39**, 3245–3251 (2006)
34. E. Uchida, N. Kawatsuki, Photoinduced orientation in photoreactive hydrogen-bonding liquid crystalline polymers and liquid crystal alignment on the resultant films. Macromolecules **39**, 9357–9364 (2006)
35. N. Kawatsuki, K. Goto, T. Yamamoto, Photoinduced anisotropy and photoalignment of nematic liquid crystals by a novel polymer liquid crystal with a coumarin-containing side group. Liq. Cryst. **28**, 1171–1176 (2001)
36. N. Kawatsuki, T. Arita, Y. Kawakami, et al., Thermally stable photoalignment layer of a novel photo-crosslinkable polymethacrylate for liquid crystal display. Jpn. J. Appl. Phys. **40**, L209–L211 (2001)
37. N. Kawatsuki, H. Matsushita, T. Washio, et al., Photoinduced reorientation of a liquid crystalline polymer with phenyl benzoate mesogenic side groups on the basis of an axis-selective photo-fries rearrangement. Macromolecules **45**, 8547–8554 (2011)
38. N. Kawatsuki, H. Matsushita, N. Shin, et al., molecularly oriented liquid crystalline polymer films based on axis-selective photo-fries rearrangement. Mol. Cryst. Liq. Cryst. **583**, 1–9 (2013)
39. K. Usami, K. Sakamoto, Y. Uehara, et al., Stability of azobenzene-containing polyimide film to UV light. Jpn. J. Appl. Phys. **44**, 6703–6705 (2005)
40. H. Kang, D. Kang, J.-C. Lee, Liquid crystal alignment property of polystyrene derivatives containing dual photoreactive side groups. Polymer **50**, 2104–2112 (2009)
41. N. Kawatsuki, S. Inada, R. Fujii, et al., Photoinduced birefringent pattern and photoinactivation of liquid-crystalline copolymer films with benzoic acid and phenylaldehyde side groups. Langmuir **34**, 2089–2095 (2018)
42. R. Fujii, M. Kondo, N. Kawatsuki, Photoinduced orientation of a photoinactive liquid crystalline polymer doped with *n*-benzylideneaniline derivatives. Chem. Lett. **45**, 673–675 (2016)

43. S. Minami, M. Kondo, N. Kawatsuki, Fabrication of UV-inactive photoaligned films by photoinduced orientation of H-bonded composites of non-photoreactive polymer and cinnamate derivative. Polym. J. **48**, 267–271 (2016)
44. K. Fukuhara, S. Nagano, M. Hara, et al., Free-surface molecular command systems for photoalignment of liquid crystalline materials. Nat. Commun. **5**, 3320 (2014)
45. K. Miyake, H. Ikoma, M. Okada, et al., Orientation direction control in liquid crystalline photoalignable polymeric films by adjusting the free-surface condition. ACS Macro Lett. **5**, 761–765 (2016)

Photoresists for Miniaturization of Semiconductor Devices

Masashi Yamamoto

1 Lithography Technology

Semiconductor devices have become absolutely necessary for today's information society. For example, a Central Processing Unit (CPU) that supports the core of a computer's arithmetic functions is constructed with hundreds of millions of transistors per square millimeter or more in the 2020s. One extremely important technology for realizing such high integration of semiconductor devices is miniaturization technology using photolithography. Lithography technology has its origins in engraving and printmaking, which began around the fifteenth century. Moreover, lithography technology has been cultivated through photolithography, which was developed in the nineteenth century. This lithography technology has been applied to microfabrication of semiconductor substrates (Si wafers), leading to the construction of integrated circuits. With the advancement of technology development for miniaturization and mass production, lithography technology has come to be used widely as photolithography technology, which uses polymers (resists) that react to light.

Figure 1 presents a schematic diagram of a typical semiconductor device manufacturing process. (1) Insulating and protective films such as silicon oxide and silicon nitride films, as well as metal films for wiring, are formed on the Si wafer. (2) A resist (photosensitive polymer) is coated onto the Si wafer. (3) The resist is exposed through a mask on which a circuit pattern was drawn. By baking after exposure (Post-Exposure Bake, PEB), diffusion of acids and protons, which are produced from the Photoactive Compounds (PACs) and the Photoacid Generators (PAGs), is promoted in the resist. Generally, exposure and PEB are a set of processes. (4) The exposed Si wafer is developed. This example is a positive-tone resist for which the exposed areas are soluble in a developer. (5) The film formed in step (1) is etched using the

M. Yamamoto (✉)
Department of Electrical and Computer Engineering, National Institute of Technology, Kagawa College, Takamatsu, Japan
e-mail: m-yamamoto@t.kagawa-nct.ac.jp

M. Funahashi (ed.), *Functional Soft Matter II*,
https://doi.org/10.1007/978-981-92-0100-6_11

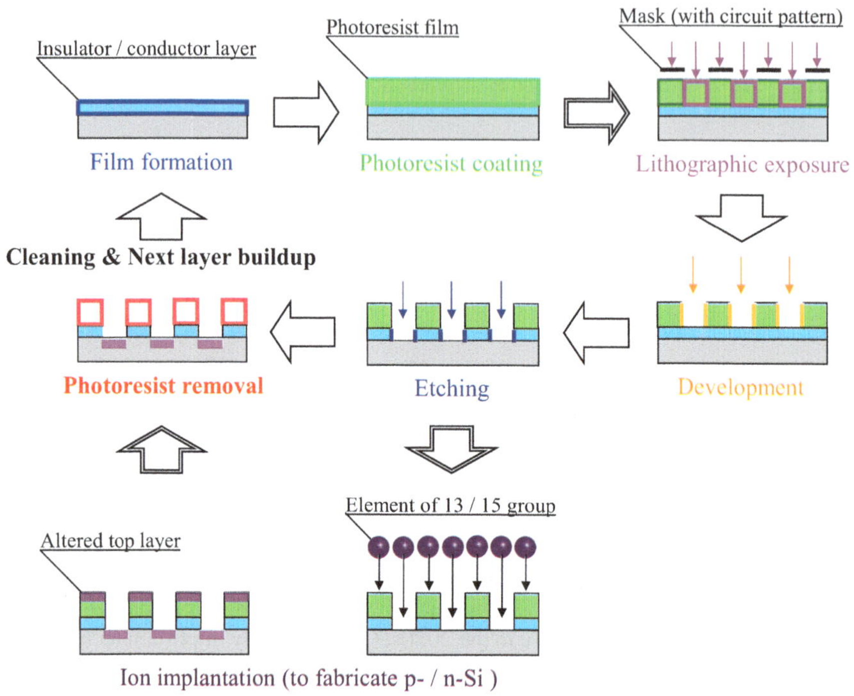

Fig. 1 General semiconductor device manufacturing process

resist as a mask. Then, the circuit pattern is transferred onto the Si wafer. (6) After the resist is removed, the Si wafer is cleaned. Semiconductor devices are manufactured by repeating these steps at least several dozen times. It is noteworthy that the process of forming p-type and n-type semiconductors is called an ion implantation process. Actually, this step is generally performed between steps (5) and (6). To form a p-type semiconductor, B ions are implanted into a Si substrate. To form a n-type semiconductor, P ions and As ions are implanted. At this time, the resist serves as a mask for the implanted ions. The ion-implanted resist changes in quality and becomes difficult to remove. Such a resist is removed using a combination of ashing treatment by oxygen plasma and chemical treatment by a mixture of sulfuric acid and hydrogen peroxide.

2 Resist (Photosensitive Polymer) [1]

Photolithography technology using a resist is an important technology supporting the manufacturing of semiconductor devices. Resists are classified into two types based on their patterning properties: positive-tone photoresists and negative-tone photoresists. In a positive-tone photoresist, the exposed areas dissolve in the developer. The

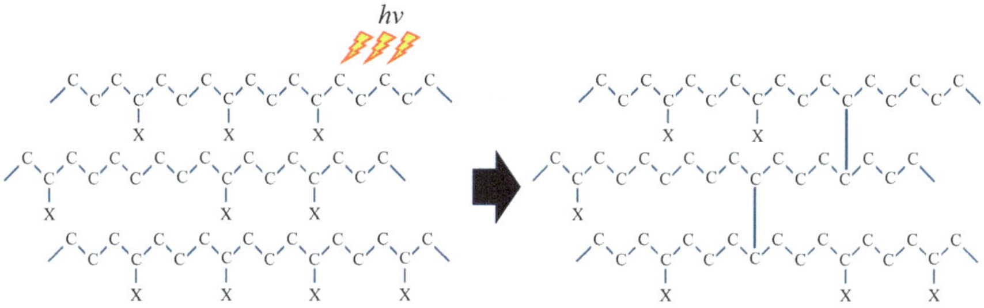

Fig. 2 Illustration of a photochemical process in a photocrosslinkable resist

unexposed areas remain as a resist pattern. A negative-tone photoresist is the opposite of a positive-tone photoresist: a film remains on the exposed area. Their basic principle is the photochemical reaction of the photoresist.

When the photoresist is irradiated with light, a photochemical reaction occurs only in the irradiated area. This reaction changes the photoresist properties (mainly solubility). This change promotes pattern formation by the developer. Typical linear polymers dissolve in appropriate solvents. Solubility is controlled by adding various light-reactive functional groups to the polymers used in photoresists. The following photoresists of four types are introduced as representative examples of photochemical reactions caused by functional groups.

2.1 Photocrosslinkable Resist

Figure 2 shows a schematic for a photoinduced reaction of photocrosslinkable resist. This photoresist is crosslinked by light irradiation. Functional groups that generate new bonds by reacting with light are introduced into polymer side chains. Because crosslinked structures are formed by crosslinking the side chains, the apparent molecular weight becomes enormous. Actually, it becomes insoluble in solvents. A negative pattern is formed because this type of resist is insoluble in the exposed areas.

2.2 Photodegradable (Depolymerized) Resist

As shown in Fig. 3, photodegradable (depolymerized) resist decomposes (depolymerizes) when exposed to light. A functional group that generates active species upon irradiation with light is introduced into the polymer side chain. Generally, in polymeric materials, the degrees of freedom become extremely limited as the molecular weight increases, resulting in an entropically disadvantageous state. Particularly if the entropy loss is much greater than the enthalpy gain associated with polymerization, the polymer can depolymerize near room temperature. This resist is used as a positive

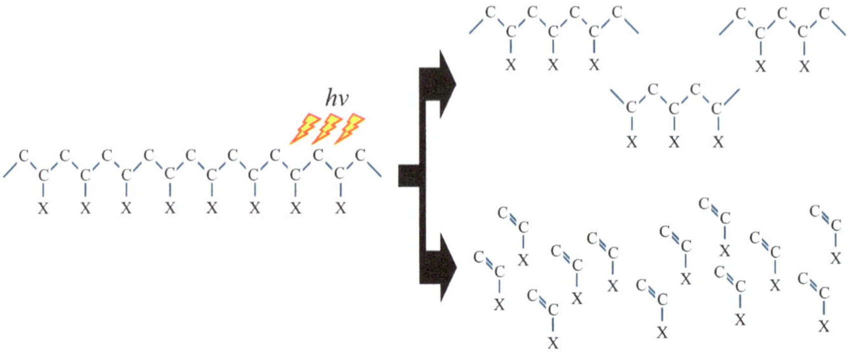

Fig. 3 Illustration of the photochemical process in a photodegradable (depolymerized) resist

resist because it causes depolymerization when exposed to light. If the difference in dissolution rate between the exposed and unexposed areas is sufficiently large, then a pattern can be formed even if complete insolubility such as photocrosslinking is not achieved.

2.3 *Photodenaturing Resist*

Figure 4 shows reaction processes of photodenaturing resists. These photoresists are classifiable into two types. One is that the photoresist itself is denatured by light irradiation, resulting in its solubility changes (Fig. 4a). The other is to change the development characteristics of the photoresist film itself by changing the solubility of a compound mixed in the photoresist using light irradiation (Fig. 4b). The latter is used widely in current processes as a positive-tone novolac photoresist containing a diazonaphthoquinone derivative for the i/g-line.

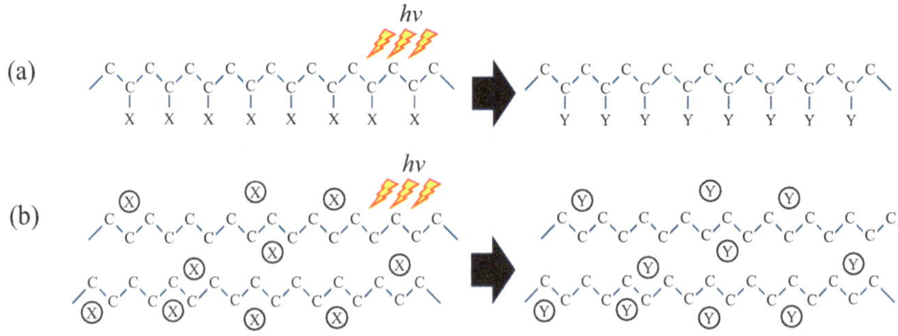

Fig. 4 **a** Photodenaturing resist which itself is denatured by light. **b** Photodenatured resist containing photoreactive components

2.4 *Photopolymerized Resist*

In contrast to photodegradation (depolymerization) type resists, a photopolymerized resist exhibits a photochemical reaction of polymerization. As shown in Fig. 5, the low-molecular-weight monomers are impregnated with matrix polymer. The monomers are polymerized by light irradiation to form polymers. Solubility is altered by changing the molecular weight through polymerization. Generally, it functions as a negative-tone photoresist. Unfortunately, although this type of resist is possible, no practical examples have ever been reported.

3 Physical Properties Required for Resist [1, 2]

To achieve multifunctionality and high performance in semiconductor devices, their manufacturing processes are becoming increasingly diverse and complex. The role of a general photoresist is to serve as a mask. During etching and ion implantation, the substrate must be protected by photoresist. However, after these processes are completed, the photoresist must be removed for subsequent processes. Permanent photoresists, which are never removed, are sometimes used in Microelectromechanical System (MEMS) processes. The properties required of photoresists vary depending on the device. Given considerations of handling, equipment, and the environment, it is important to be able to do processing with a water-soluble developer. Next, the three most important characteristics for general photolithography applications will be described.

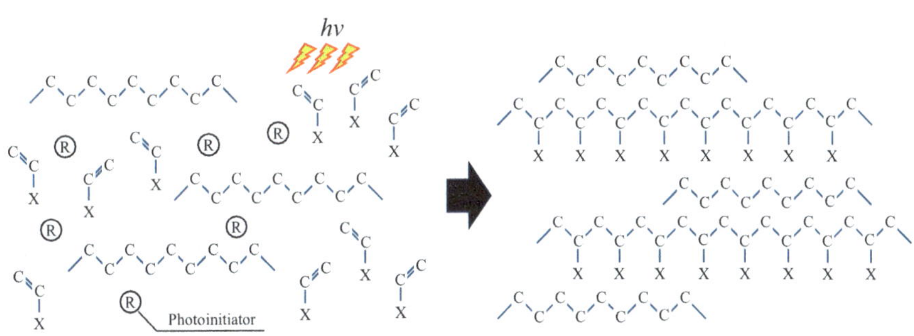

Fig. 5 Illustration for a photochemical process in a photopolymerized resist

3.1 Sensitivity

As described in Sect. 3.1, to manufacture semiconductor devices, the lithography process must be repeated at least several dozen times. A photoresist with high sensitivity can form a pattern with high dimensional accuracy with a small amount of exposure. Therefore, the photoresist sensitivity is a characteristic that affects the time required for the process directly. Figure 6 depicts the sensitivity curves of typical positive-tone and negative-tone photoresists. Generally, a sensitivity curve shows the relation between exposure dose and the normalized residual film thickness. The horizontal axis shows the exposure dose (logarithmic scale). The vertical axis shows the normalized residual film thickness (film thickness before exposure is taken as 100%). Based on the sensitivity curve, positive-tone photoresists and negative-tone photoresists are classifiable. In positive-tone photoresists, the remaining film thickness decreases as the exposure amount increases. Conversely, for a negative-tone photoresist, the residual film thickness increases. Generally, in the case of a positive-tone photoresist, the exposure dose at which the normalized residual film thickness becomes zero is an index of sensitivity. This exposure dose is expressed as Dg^{100}. The contrast of positive-tone photoresist is defined by the gradient ($\gamma = \tan\theta$) near Dg^{100} and the exposure dose difference ($\gamma = 0.5 \log |Dg^{0.5}/Dg^{100}|^{-1}$). In the photoresist development simulator, E_{th} is the exposure dose at which Dg^{100} is achieved with development time of 60 s. This E_{th} is used for various calculations. For negative-tone photoresists, the exposure dose at which the normalized residual film becomes 50% is expressed as Dg^{50}. In general, the contrast of a negative-tone photoresist is defined not only by the gradient around Dg^{50} but also by the exposure dose difference ($\gamma = 0.5 \log |Dg^{0.5}/Dg^{i}|^{-1}$ ($i = 30, 50, 70\%$, etc.)). High-sensitivity photoresists can satisfy Dg^{100} or Dg^{50} with a small exposure dose.

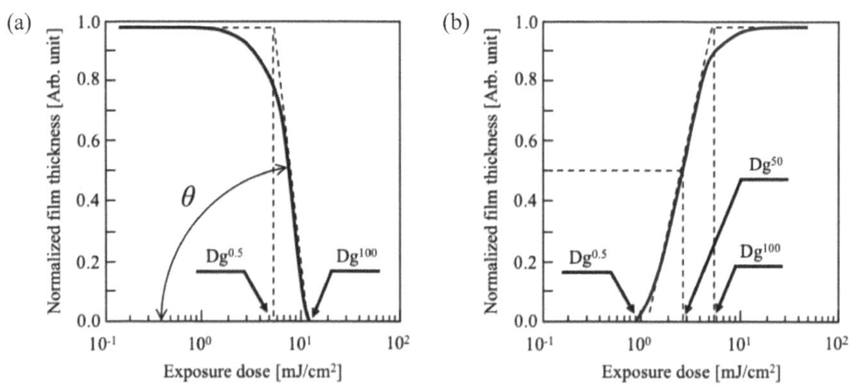

Fig. 6 Sensitivity curves of typical **a** positive-tone and **b** negative-tone photoresist

3.2 Resolution

Resolution is defined as the minimum dimension of a pattern which can be reproduced with a photoresist. However, it is noteworthy that the photoresist resolution varies depending on conditions such as film thickness, exposure, and development. In general, high-resolution photoresists have two main requirements: the photoresist must be developed without swelling, and its contrast must be high. Swelling is a difficulty associated with negative-tone photoresists. Processes using positive photoresists have become the mainstream for increasing resolution.

Resists comprise two basic materials: a polymer material, which is the main component; and a photosensitive material for controlling photosensitivity and solubility. For example, novolac positive-tone photoresists for i/g-line light, which are widely used in current processes, comprise novolac resin and a PAC. As shown in Fig. 7, the unexposed PAC exhibits a dissolution-inhibiting effect, whereas the exposed PAC exhibits a dissolution-promoting effect [3]. For this reason, resolution can also be evaluated as the difference in dissolution rate between the unexposed area and the exposed area [4]. The dissolution rate is expressed by a discrimination curve, as shown in Fig. 8. The contrast can be defined by the slope of the discrimination curve. The total dissolution rate ratio, which is the ratio of the dissolution rate for a completely exposed resist (R_{max}) to that for unexposed resist (R_{min}), might also be chosen as the contrast.

Evaluation of resolution using sensitivity curves and dissolution rate curves is an indirect method. A direct evaluation method is observation of the cross-section of the pattern using an electron beam microscope. The ideal resist pattern is rectangular, but it often becomes trapezoidal because of the penetration of light. The main reason is that the light which passes through the resist film is attenuated by absorption before reaching the bottom. The net exposure amount has a distribution within the film. Standing wave effects depending on the exposure wavelength can also be observed

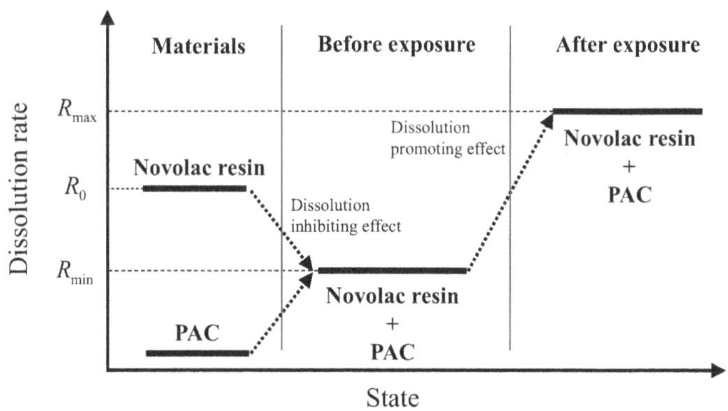

Fig. 7 Dissolution-inhibiting effect when not exposed to light and dissolution-promoting effect when exposed to light

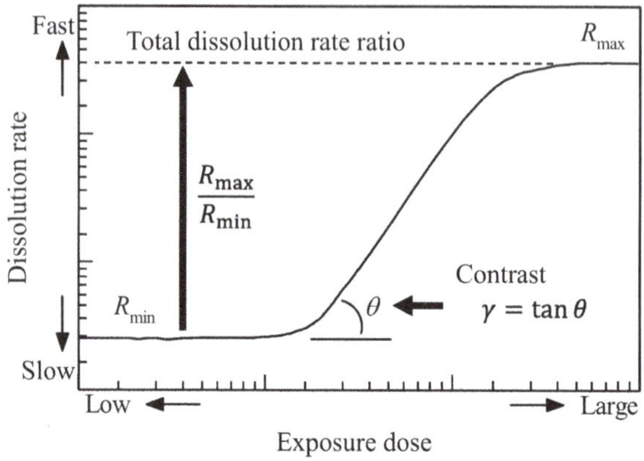

Fig. 8 Discrimination curve [7]

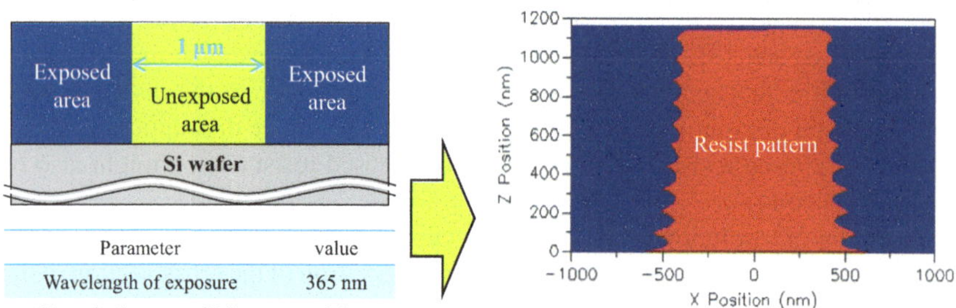

Parameter	value
Wavelength of exposure	365 nm
Numerical aperture (NA)	0.5
Coherence factor	0.5
Resist pattern	Line 1 μm
Resist initial film thickness	1150 nm
Development time	60 s

Parameter	value
Total dissolution rate ratio	369
PAC concentration threshold	0.9
Exposure dose	10 mJ/cm^2

Fig. 9 Simulation evaluation of resist patterns using PROLITH [5–7]

on the pattern sidewalls. A difficulty associated with evaluating resolution through cross-sectional observation is that it is very time-consuming. Therefore, methods using computers (PROLITH) [5, 6] are being used today, as shown in Fig. 9 [7].

3.3 *Etching Resistance*

Photolithography technology transfers a circuit pattern drawn on a mask onto a substrate. The circuit pattern is transferred to the substrate by etching an oxide film or metal film formed on the surface of the substrate. Then, the photoresist becomes a protective film that protects the substrate from etching. There are two types of etching methods. One is wet etching, which uses chemicals such as acids and alkalis.

The other is dry etching, which uses gases such as CF_4 and Cl_2. When classified based on the way the etching progresses, the former is isotropic etching. The latter is anisotropic etching. Along with the progress of miniaturization, the required specifications for processing accuracy have become higher. Therefore, anisotropic etching has become the mainstream.

Reactive ion etching is an example of anisotropic etching that involves plasma. In plasma, active species such as highly reactive ions and radicals are generated. Many organic polymers are degraded by these active species. The circuit pattern cannot be transferred correctly to the substrate if the photoresist pattern is decomposed. A material with high plasma resistance should be selected as the polymer material used for the photoresist. Generally, a material with high carbon content and low oxygen or nitrogen content is preferred. A benzene ring is one example of a structure with high carbon content. Novolac resin, which is the base resin for novolac photoresists, has a benzene ring in its main chain. The etching resistance of resins with benzene rings is about 2–4 times higher than that of resins without them.

4 Typical Resists [1, 2]

4.1 i/g-Line Photoresist

High-pressure mercury lamps exhibit strong bright-line emission spectra such as the g-line (436 nm), the h-line (405 nm), and the i-line (365 nm) [8]. An i/g-line photoresist is sensitive to the bright-line emission wavelength (365–436 nm) of high-pressure mercury lamps.

Kodak Photo Resist (KPR), the actual origin of the word photoresist, is an early and revolutionary photoresist material for i/g-line lights [9]. As shown in Fig. 10, this photoresist has the main component of polyvinyl cinnamate. Cinnamic acid is known to absorb light of approximately 300 nm. It undergoes cyclic dimerization. Therefore, KPR, which has a cinnamic acid group added to the polymer side chain, undergoes a crosslinking reaction when exposed to light. In fact, KPR is a negative-tone resist. KPR itself absorbs light with a wavelength of 350 nm or less [10]. However, the high-pressure mercury lamp used as the exposure light source has no strong bright line at around 350 nm. Therefore, sensitizers are generally used. For example, Michler's ketone, which absorbs light up to 380 nm efficiently, is added to increase the i-line absorption efficiency.

Kodak Thin Film Resist (KTFR), which is composed mainly of cyclized rubber and bisazide compounds, is also known as a negative-tone photoresist material for i/g-lines. When exposed to light, azide produces nitrene, which is extremely active and causes various chemical reactions. It combines cyclized rubber, which is created by cyclizing natural rubber or synthetic rubber with tin, and a bisazide compound that has two azide groups. Thereafter, a crosslinking reaction occurs between the nitrene

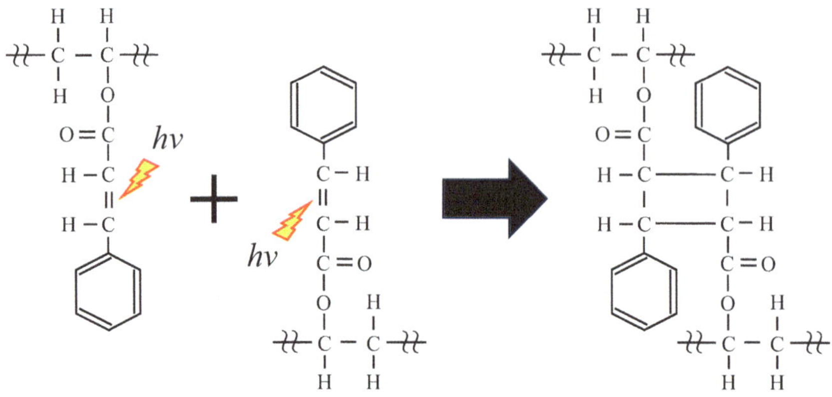

Fig. 10 Kodak Photo Resist (KPR)

generated by light irradiation and the unsaturated groups of the cyclized rubber, consequently producing a negative-tone photoresist.

A difficulty with negative-tone photoresists is that the photoresist pattern swells during wet processing such as development. For this reason, positive-tone photoresists are widely used as photoresists for i/g-line lights. A cresol novolac resin is used as the base resin. Also, a compound having a Diazonaphthoquinone (DNQ) structure is used as the PAC. In fact, DNQ is a diazo derivative of naphthoquinone. This photoresist is classified as a photochromic photoresist. An example of the basic structure of a PAC is 2,3,4,4′-tetrahydroxybenzophenone, as shown in Fig. 11. The DNQ structure which exhibits photosensitivity is attached to the terminal end of tetrahydroxybenzophenone through an ester bond. When this DNQ structure absorbs light, it changes, as shown in Fig. 12. Carbene undergoes a Wolff rearrangement reaction to become ketene [11]. The ketene subsequently reacts with moisture present in the photoresist and/or ambient air to form indene carboxylic acid, a strong acid. When indene carboxylic acid is immersed in an alkaline developer, it dissolves because of an acid–base neutralization reaction. As shown in Fig. 12, in the unexposed area, the DNQ structure exhibits hydrophobicity, which has the effect of suppressing dissolution in alkaline developers. The dissolution rate of the unexposed area is slower than that of the base resin. In the exposed area, indene carboxylic acid has the effect of promoting dissolution in the alkaline developer. The dissolution rate of the exposed area is higher than that of the base resin. An important benefit of a novolac positive-tone photoresist is that it can increase the difference in dissolution rate between exposed and unexposed areas. A photoresist pattern with high contrast is obtainable.

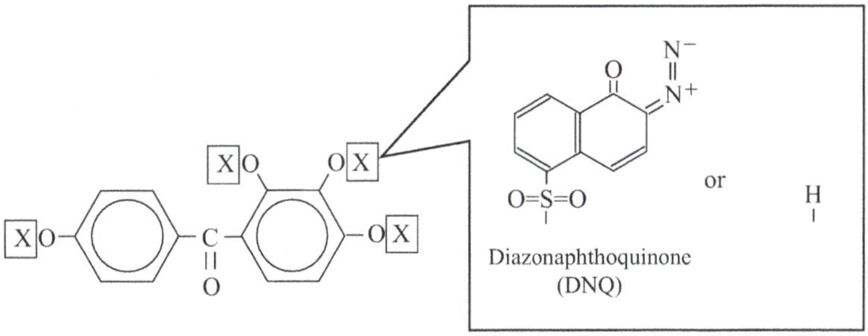

Fig. 11 PAC with naphthoquinone diazide structure. (Example, 2,3,4,4′-tetrahydroxybenzophenone)

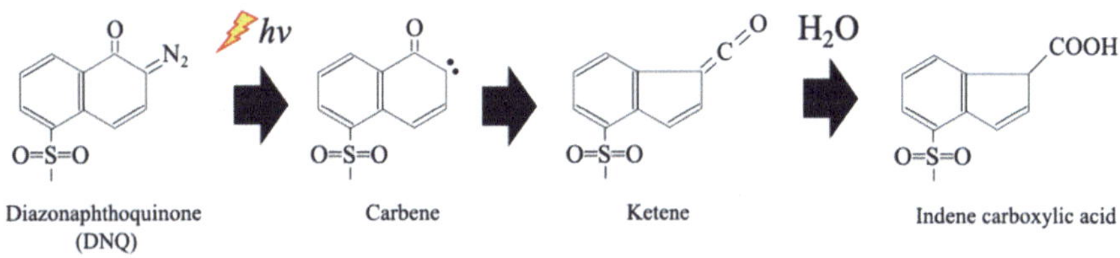

Fig. 12 Photochemical reaction process of DQN structure

4.2 KrF/ArF Photoresist

Rayleigh and Abbe limits are, respectively, indicators of the processing limit of microfabrication [12]. The spatial resolution R is expressed by Eq. (1), where λ represents the wavelength of the exposure light source. Generally called the k factor, k is a constant that represents the degree of influence related to the process.

$$\text{Resolution} : R = k\frac{\lambda}{\text{NA}} \tag{1}$$

As the k factor increases, ensuring optical resolution becomes difficult. The upper limit of the k factor is said to be 0.61. It is possible to reduce the value of k using optical proximity effect correction, phase shift masks, double patterning, anti-reflection coatings, and other means. The numerical aperture (NA) of exposure optical systems in the g-line (435 nm) era was around 0.3. The spatial resolution of the wavelength of the g-line is 0.88 μm. In fact, the various techniques described below have made it possible to achieve miniaturization that exceeds the diffraction limit because of the wavelength of the exposure light source. Herein, the material design of resists will be described, as exposure wavelengths become shorter.

Rare gas excimer lasers have attracted attention because they have a wavelength shorter than the i/g-line. This laser oscillates by exciting a gas such as KrF or ArF [13, 14]. The KrF laser wavelength is 248 nm. The ArF laser wavelength is 193 nm.

It has benefits such as coherence and high output, which are characteristics of lasers. Assuming application of the same exposure optical system as that of the g-line described earlier, it is possible to form a pattern of approximately 0.39 μm with ArF resist. Actually, ArF lasers were the most advanced microfabrication light source until 2019 [15]. Patterns with processing accuracy of the 10 nm generation were fabricated using the techniques described later.

Because KrF/ArF lasers emit light in the vacuum ultraviolet wavelength range, the absorption edge of the material presents a difficulty. Although the amount of exposure can be secured, the photon cost is high. Therefore, the KrF/ArF resist is a chemically amplified resist, as described later. To use KrF/ArF lasers, it is necessary to review not only the material of transmissive optical elements such as lenses, but also the material design of the resist itself. To improve the etching resistance of the resist, a benzene ring structure, which is a resonance-stable structure, is preferable. The i/g-line resist has strong absorption near 240 nm. This absorption is attributable to the benzene ring structure contained in the novolac resin. Because the laser light cannot reach the bottom of the resist film, it is not possible to form a pattern. In a KrF resist, the KrF laser wavelength is slightly longer than the absorption edge of the benzene ring structure. For that reason, a base resin with a benzene ring structure is useful. For example, as shown in Fig. 13, the poly{*p*-(*t*-butoxy)carbonyloxystyrene} (*t*-BOC PHS), which is known to exist, is produced by partly replacing the OH group at the end of the side chain of poly(*p*-hydroxy styrene) (PHS) with a *t*-butoxycarbonyl (*t*-BOC) group [16, 17]. By contrast, in an ArF resist, the ArF laser wavelength is shorter than the absorption edge of the benzene ring structure. For that reason, a base resin with a benzene ring structure cannot be used. A well-known plastic glass, poly(methyl methacrylate) (PMMA), is an example of a material that is transparent to wavelengths of 200 nm or less. However, its low etching resistance presents difficulties [18]. Introduction of an aliphatic material with high carbon content and a condensed ring structure was regarded as a material with high etching resistance and without a benzene ring structure [19]. Etching resistance is often discussed using the Onishi parameter because the Onishi parameter and the etching rate have a roughly linear relation [20]. The Onishi parameter of PMMA is 5.0, but the Onishi parameter of poly(adamantyl methacrylate) (PADMA), which has adamantane as an ester residue, as shown in Fig. 14, is 2.8. The etching resistance of PADMA is comparable to that of novolac resin.

4.3 Electron Beam Resist

Electron beams strongly affect the electronic state of compounds. When high-energy electrons enter a substance, the electrons in the compound are knocked out, thereby producing ions and radicals. The diverse active species thus generated cause a chemical reaction. In electron beam resists, patterns are formed using such reactions to cause crosslinking or decomposition [21]. Electron beam lithography typically uses an electron beam accelerated to 5–30 keV. The wavelengths of the electron waves of

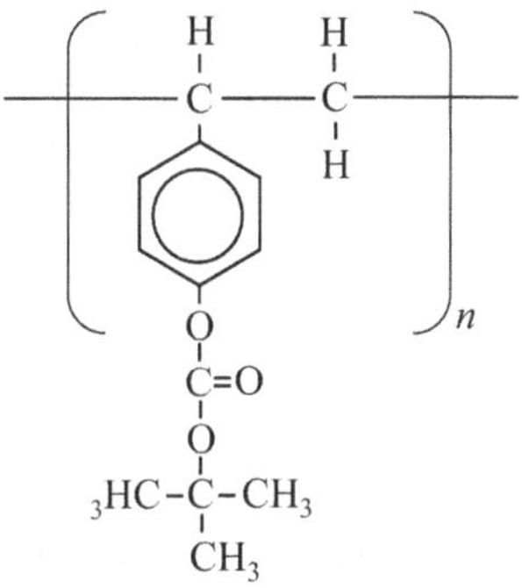

Fig. 13 Molecular structure of a typical base resin for KrF photoresist

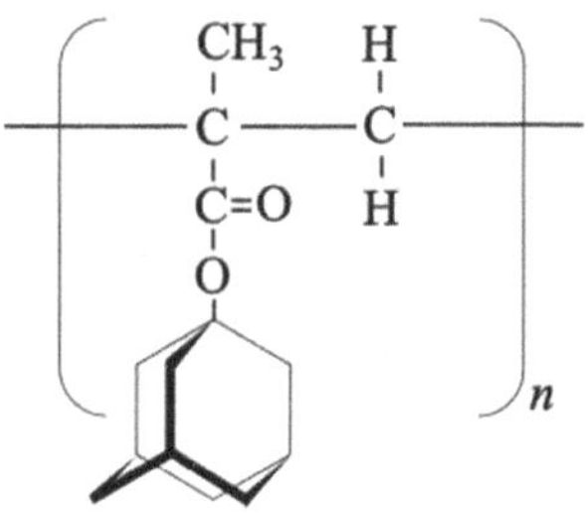

Fig. 14 Molecular structure of a typical base resin for ArF photoresist

this accelerated electron correspond to a few tens of pm. The electron beam can then be focused to a spot diameter of approximately 5 nm [22]. High resolution of less than 1/100 is achieved compared to the resolution of i/g-line photolithography. The typical electron beam resist thickness is approximately 0.5 μm. Although an effect of electron scattering occurs within the resist, forming a pattern of about 5 nm is possible.

Polymers with quaternary carbons in their main chains are known to be decomposed by electron beams. Main chain bonds are broken and decomposed into organic low-molecular-weight compounds. A solvent is used in the developer. A positive pattern is thus obtained. In fact, PMMA has been used as the base resin for such positive resists for electron beams. However, its low sensitivity has presented difficulties. Sensitivity (Dg^{100}) is ~ 500 $\mu C/cm^2$ [23]. To improve sensitivity, polymeric materials with various substituents around the quaternary carbon were synthesized. Poly(hexafluorobutyl methacrylate) and poly(trifluoroethyl chloroacrylate), which have fluoroalkyl groups in their side chains, as shown in Fig. 15, exhibit high sensitivity of 1 $\mu C/cm^2$ (Dg^{100}) or less [24, 25]. Their sensitivity to electron beams is not high. However, the change in solubility associated with the decrease in molecular weight is greater than that of the base resin, PMMA. Bulky functional groups tend to have higher sensitivity to electron beams. Polysulfone polymer is also well known to be decomposed by electron beams [26]. The sensitivity of poly(butene sulfone) to electron beams is higher than that of PMMA. Moreover, it exhibits high sensitivity (< 1 $\mu C/cm^2$, Dg^{100}) similar to those of poly(hexafluorobutyl methacrylate)

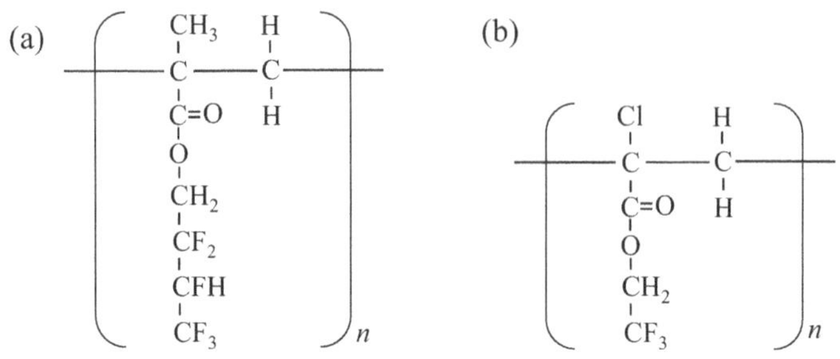

Fig. 15 Base resin of positive resist for electron beams with fluoroalkyl group in the side chain. Molecular structures of **a** poly(hexafluorobutyl methacrylate) and **b** poly(trifluoroethyl chloroacrylate)

and poly(trifluoroethyl chloroacrylate) [27]. Positive resists for electron beam use have reached adequate levels of sensitivity and resolution, but they have low etching resistance. Sensitivity can be increased by decreasing the thickness, but the etching resistance is compromised. Therefore, sensitivity and etching resistance must be improved [28, 29].

As shown in Fig. 16, poly(glycidyl methacrylate) and poly(chloromethyl styrene), which have epoxy rings, are negative resists for electron beams that are crosslinked by irradiation with electron beams. In poly(glycidyl methacrylate), the epoxy ring at the end of the side chain is opened by an electron beam. Then the ring-opened site of the adjacent polymer polymerizes, resulting in crosslinking. A negative pattern with high sensitivity ($< 1\ \mu C/cm^2$, Dg^{50}) is obtainable. Even in poly(chloromethyl styrene), which has a benzene ring, the highly polarizable C-Cl bond in the terminal group CH_2Cl is broken by the electron beam, resulting in CH_2*. In other words, the side chain becomes a benzyl radical. This part crosslinks with the CH_2* site of the adjacent polymer. The sensitivity is of approximately several microcoulombs per square centimeter (Dg^{50}). Thereby, a highly sensitive negative pattern is obtainable. Because it has a benzene ring, it also has high etching resistance.

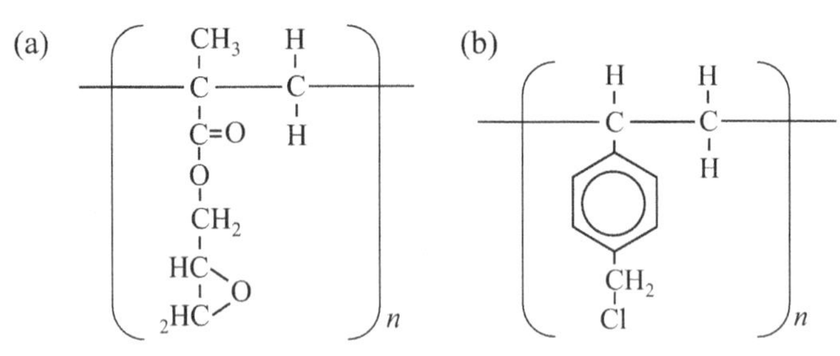

Fig. 16 Base resin for negative resist for electron beams. Molecular structures of **a** poly(glycidyl methacrylate) and **b** poly(chloromethyl styrene)

For electron beam lithography, electron drawing is performed on a resist by controlling a focused electron beam according to a drawing pattern. Electron beam resists can form finer patterns than photoresists such as i/g beams and KrF/ArF. However, electron beam lithography has low throughput because it cannot perform batch exposure as optical lithography can. Electron beam lithography is used exclusively for producing one-of-a-kind "items" such as master masks for optical lithography. Difficulties with developer and etching resistance also arise.

4.4 Chemically Amplified Resist

A chemically amplified resist uses the principle of high sensitivity proposed by Ito et al. (IBM) in the 1980s [16, 17]. As shown in Fig. 17, high sensitivity is achieved by amplifying quantum efficiency. Acid (protons) is generated by light irradiation. The acid acts as a catalyst, causing denaturation and depolymerization, and thereby altering the rate of dissolution into the developer. One proton acts as a catalyst for multiple reactions. This number of catalyst turnovers improves the sensitivity considerably, which is limited by quantum efficiency: a factor that represents the number of photochemical reactions that can be triggered by one incident photon. The photon cost of Extreme Ultraviolet (EUV) light is extremely high. In the cutting-edge ultra-fine process using EUV light, chemically amplified resists are necessary for sensitization using only a small amount of exposure.

Figure 18 portrays a chemically amplified resist consisting of an onium salt and a *t*-BOC PHS-co-PHS copolymer. A proton (H^+) is generated from the onium salt by light irradiation. This H^+ subsequently decomposes the BOC group. The base resin, *t*-BOC PHS, changes to PHS, which is soluble in alkaline developer. The H^+ generated by decomposition of a *t*-BOC group decomposes another *t*-BOC group.

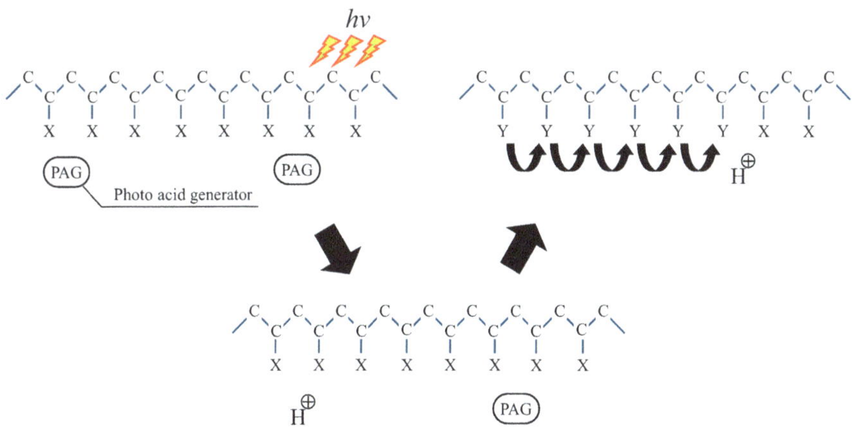

Fig. 17 Principle of high-sensitivity resist

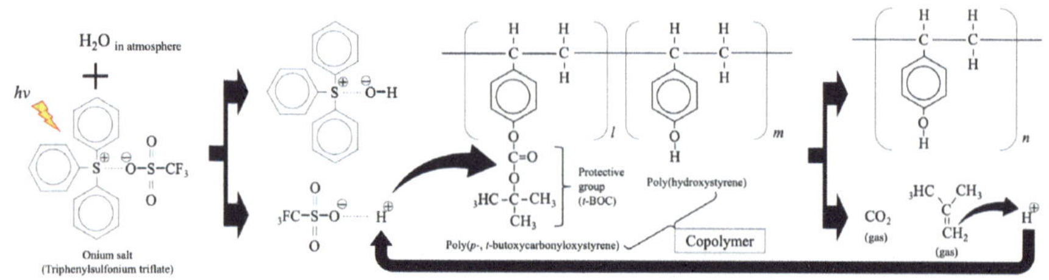

Fig. 18 Chemically amplified positive-tone resist consisting of an onium salt and a *t*-BOC PHS-co-PHS copolymer

Then, a new H^+ is generated again. The decomposition reaction of *t*-BOC groups by H^+ progresses sequentially. In this case, the quantum yield is determined by the absorption efficiency of photons into the onium salt. The apparent quantum yield is improved considerably, and high sensitivity is obtainable. In photolithography using KrF/ArF and EUV, chemically amplified resists are used because the photon cost of the light source is high.

The solubility of a chemically amplified resist in a developer can be changed by altering the modification method using an acid catalyst. For a positive resist, an acetal group or the like is used in addition to the *t*-BOC group described above. For negative resists, crosslinking reactions using melamine compounds and various polarity changing reactions using acids have been devised, as shown in Fig. 19.

Novolac resists for i/g-line lights are sometimes baked after exposure to reduce standing wave effects by dispersing the carboxylic acid generated during exposure. This is designated as PEB. In the case of a novolac resist, patterns can be formed without PEB. However, PEB is obligatory for chemically amplified resists to diffuse the acid generated during exposure and to proceed with the acid-catalyzed reaction sufficiently.

5 Technological Challenges to the Limits of Miniaturization

Miniaturization has been achieved mainly by shortening the projection exposure wavelength and by increasing the NA of the projection lens. The NA of projection lenses started from NA = 0.3 in the g-line era. In an optical system with 193 nm wavelength, the NA reaches 0.9 [30, 31]. The value of NA is expressed as $n \sin \theta$. n is the refractive index of the medium. θ is the incident angle to the photoresist. As shown in Fig. 20, when the exposure atmosphere is air, NA cannot exceed the refractive index of air ($n = 1$). In reality, because θ is less than 90°, NA is less than 1. The technology which overcomes this limitation and which makes NA > 1 possible is immersion exposure technology [31–33]. As shown in Fig. 21, the space between the semiconductor exposure equipment lens and the photoresist is filled with pure water. The refractive index of pure water is 1.33, which is higher than that of air.

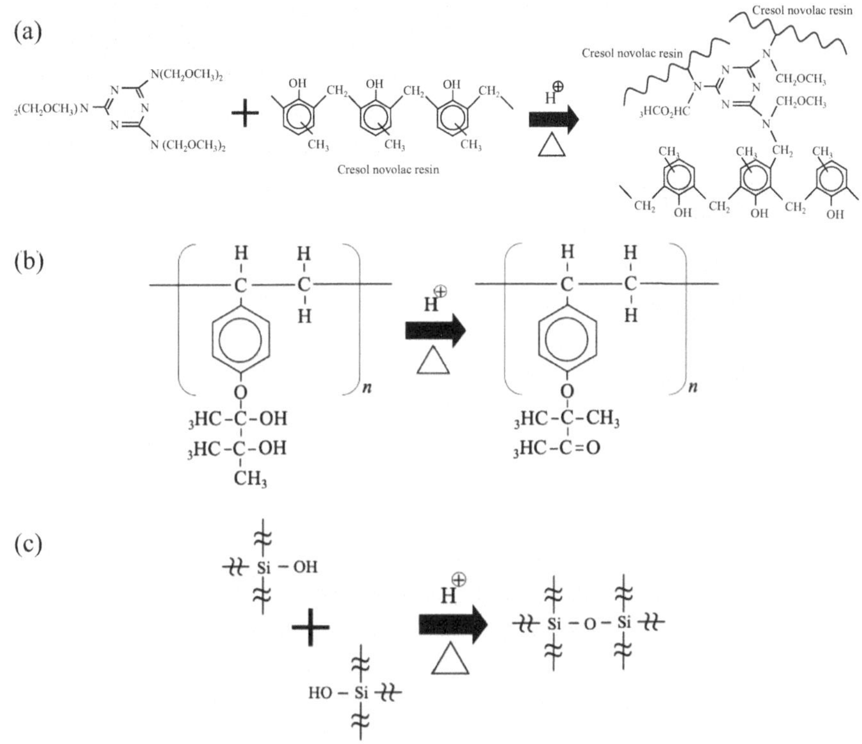

Fig. 19 Chemically amplified negative-tone photoresist. **a** Crosslinking, **b** pinacol rearrangement, and c silanol condensation types

Actually, higher resolution can be achieved using pure water itself, as a lens. Immersion exposure technology has made it possible to realize process nodes below 32 nm using ArF lasers. High NA is not the only benefit of immersion exposure technology. Various techniques cultivated for conventional 193 nm lithography process are also applicable to immersion exposure technology.

Miniaturization of the generation beyond 32 nm cannot be achieved using immersion exposure technology alone. It combines various technologies. Herein, a multiple patterning technology will be introduced. Figure 22 shows Litho-etch Litho-etch (LELE), which is one of the multiple patterning techniques [34–36]. In LELE, the pattern is divided into multiple patterns and overlapped. It is possible to form fine patterns that cannot be achieved with a single exposure. An important difficulty posed by this method is that, as miniaturization progresses, even slight distortions and positional deviations in processing dimensions cannot be tolerated. The accumulation of mutual positional deviations during multiple exposures reduces the processing accuracy considerably. Therefore, there is a limit to miniaturization when using LELE alone. Widely diverse methods are being combined, such as Self-Aligned

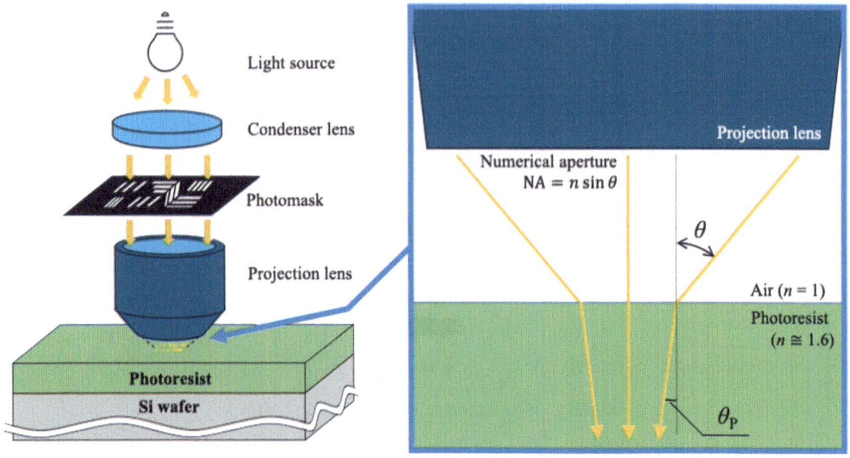

Fig. 20 Exposure using a general projection lens (Case: air atmosphere)

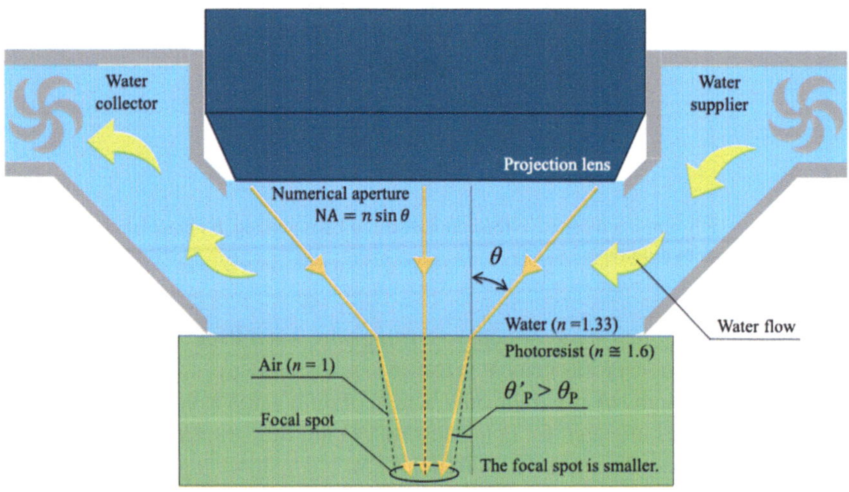

Fig. 21 Immersion exposure technology (Case: pure water atmosphere)

Double Patterning (SADP) [37], Self-Aligned Quadruple Patterning (SAQP) [38], and Self-Aligned Block (SAB) [39].

When it comes to process nodes of the 22 nm generation and beyond, even a combination of these technologies would make the goal difficult to achieve. Therefore, efforts are underway to put EUV lithography technology into practical use. EUV is light with a 13.5 nm wavelength. Shortening the light source wavelength directly affects pattern miniaturization. However, in addition to the need for higher output and longer life of the light source, a difficulty exists that the optical system is limited to a reflective optical system. EUV resists are fundamentally chemically amplified resists

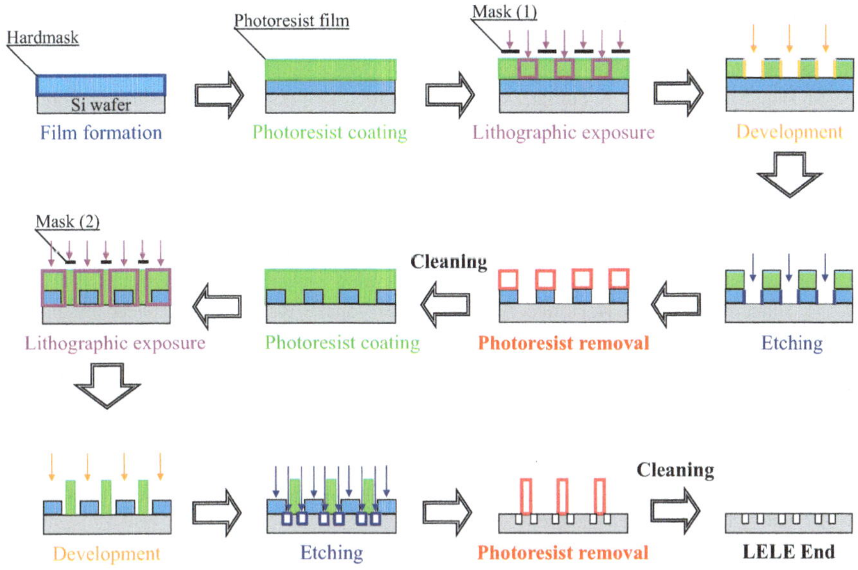

Fig. 22 Litho-etch Litho-etch (LELE) process

[40]. EUV resists are necessary to form high-resolution patterns of 20 nm or less. For this reason, resists made of low-molecular compounds (molecular resists) are attracting attention [41]. With EUV resists based on this concept, negative patterns are obtainable. When using UV (i/g-line) and VUV (KrF/ArF), absorption occurs because of functional groups. By contrast, with EUV, absorption occurs because of elements. Because carbon, oxygen, and hydrogen are absorbed less than other elements, common organic compounds are useful as resist materials. Conventional polymer-based resists include those based on PHS, poly(methacrylic acid-co-lacton), and their copolymers. Chemically amplified resists are constructed by mixing various PAGs with these base resins [40, 42]. In fact, until 2019, patterns were formed using ArF lasers with processing accuracy of the 10 nm generation without using EUV [15]. In addition to major changes in transistor structure, future miniaturization efforts will advance technology to the level of several atoms. Processing accuracy will thereby enter the angstrom generation. Lithography processes play the most important role in semiconductor microfabrication. Although it has been used sparingly until now, EUV is expected to be introduced in earnest for next-generation lithography [40–43].

References

1. Y. Yoshida, T. Hagiwara, T. Takeichi, Y. Tezuka, N. Yonezawa, Y. Nagasaki, S. Ishii, *Polymer Materials Chemistry (Kobunshi Zairyo Kagaku)* (Sankyo Publishing, 2001), pp. 122-134 (in Japanese)

2. H. Hideo, H. Tanaka, M. Hanabata, M. Yamamoto, A. Sekiguchi, S. Toratani, N. Mochiduki, N. Man, *Novolac resist optimization of material and process (Zairyo to process no Saitekika)* (Sankyo Publishing, 2020), pp. 7-90 (in Japanese)
3. M. Hanabata, Y. Uetani, A. Furuta, Design concept for a high-performance positive photoresist. J. Vac. Sci. Technol. B **7**, 640–650 (1989)
4. C.A. Mack, Lithographic optimization using photoresist contrast. Microelectron. Manuf. Technol. **14**, 36–42 (1991)
5. C.A. Mack, Optical microlithography IV. Proc. SPIE **538**, 207–220 (1985)
6. I.Y. Kuzmin, C.A. Mack, Comprehensive simulation of e-beam lithography processes using PROLITH/3D and temptation software tools, 20th annual BACUS symposium on photomask technology. Proc. SPIE **4186**, 503–507 (2000)
7. M. Yamamoto, R. Kitai, H. Horibe, A. Sekiguchi, H. Tanaka, Relationship between dissolution property and molecular weight of positive-tone Novolak resist. J. Photopolym. Sci. & Technol. **22**, 357–362 (2009)
8. M. Ohkahara, What is super high pressure mercury lamp? J. Illum. Engng. Inst. Jpn. **85**, 867–868 (2001)
9. Eastman Kodak Company, *Industrial Uses of Kodak Photo Resist* (Kodak Pamphle Q-24, 1959), pp. 1–20
10. K. Nakamura, S. Kikuchi, The photosensitivity of polyvinyl cinnamate. Nippon Kagaku Zasshi **87**, 930–934 (1966) [in Japanese]
11. L. Wolff, R. Krüche, Über diazoanhydride (1,2,3-oxydiazole oder diazoxyde) und diazoketone. Justus Liebigs Ann. Chem. **394**, 23–59 (1912)
12. A. Yen, Rayleigh or Abbe? Origin and naming of the resolution formula of microlithography. J. Micro Nanolithogr. MEMS MOEMS **19**, 040501 (2020)
13. A.M. Hawryluk, J.A. Mangano, J.H. Jacob, Gain and absorption measurements in a KrF* laser. Appl. Phys. Lett. **31**, 164–166 (1977)
14. Y.A. Kudryavtsev, N.P. Kuzmina, Excimer gas-discharge tunable ArF laser. Appl. Phys. **13**, 107–108 (1977)
15. M. Shirakawa, T. Fujimori, N. Tango, K. Marumo, K. Yamamoto, H. Takahashi, R. Nishio, A. Goto, M. Fujita, Challenges and progress in low defectivity for advanced ArF lithography processes using surface localized material technology. J. Photopolym. Sci. Technol. **30**, 367–372 (2017)
16. H. Ito, C.G. Willson, J.H.J. Frechet, in *1982 Symposium on VLSI Technology. Digest of Technical Papers, Oiso, Japan.* New UV Resists with Negative or Positive Tone (1982), pp. 86–87
17. H. Ito, C.G. Willson, Chemical amplification in the design of dry developing resist materials. Polym. Eng. Sci. **23**, 1012–1018 (1983)
18. M. Ueda, H. Ito, Chemical amplification in resist design. J. Synth. Org. Chem. Jpn. **49**, 437–450 (1991) (in Japanese)
19. T. Ogata, S. Matsumaru, H. Shimizu, N. Kubota, H. Hada, M. Shirai, Effects of protecting group on resist characteristics of acryl polymers for 193 nm lithography. J. Photopolym. Sci. Technol. **17**, 483–488 (2004)
20. H. Gokan, S. Esho, Y. Ohnishi, Dry etch resistance of organic materials. J. Electrochem. Soc. **130**, 143–146 (1983)
21. A.S. Gangnaik, Y.M. Georgiev, J.D. Holmes, New generation electron beam resists: a review. Chem. Mater. **29**, 1898–1917 (2017)
22. K. Yamazaki, H. Namatsu, 5-nm-order electron-beam lithography for nanodevice fabrication. Jpn. J. Appl. Phys. **43**, 3767 (2004)
23. W.H. Wong, E.Y.B. Pun, Exposure characteristics and three-dimensional profiling of SU8C resist using electron beam lithography. J. Vac. Sci. Technol. B **19**, 732–735 (2001)
24. L.F. Thompson, E.M. Doerries, Negative electron resists for direct device lithography: I. Initial material survey. J. Electrochem. Soc. **126**, 1699–1702 (1979)
25. A. Tokunaga, M. Kataoka, Y. Nakayama, F. Soeda, Decomposition products of EBR-9 by X-ray and EB exposure. J. Photopolym. Sci. Technol. **5**, 589–592 (1992)

26. M.J. Bowden, L.F. Thompson, Electron irradiation of poly(olefin sulfones). Application to electron beam resists. J. Appl. Polym. Sci. **17**, 3211–3221 (1973)
27. M. Kakuchi, S. Sugawara, K. Murase, K. Matsuyama, Poly (fluoro methacrylate) as highly sensitive, high contrast positive resist. J. Electrochem. Soc. **124**, 1648–1650 (1977)
28. K. Harada, Additives that improve positive resist durability for plasma etching. J. Electrochem. Soc. **127**, 491–497 (1980)
29. L.F. Thompson, E.D. Feit, R.D. Heidenreich, Lithography and radiation chemistry of epoxy containing negative electron resists. Polym. Eng. Sci. **14**, 529–533 (1974)
30. S. Owa, H. Nagasaka, Y. Ishii, K. Shiraishi, S. Hirukawa, Full-field exposure tools for immersion lithography. Proc. SPIE **5754**, 655–668 (2005)
31. S. Owa, H. Nagasaka, Immersion lithography: Its history, current status and future prospects. Proc. SPIE **7140**, 714015 (2008)
32. M. Switkes, M. Rothschild, Immersion lithography at 157 nm. J. Vac. Sci. Technol. B **19**, 2353–2356 (2001)
33. M. Switkes, M. Rothschild, Resolution enhancement of 157-nm lithography by liquid immersion. Proc. SPIE **4691**, 459–465 (2002)
34. K. Yuan, J.S. Yang, D.Z. Pan, Double patterning layout decomposition for simultaneous conflict and stitch minimization. IEEE Trans. **29**, 185–196 (2010)
35. A.B. Kahng, C.H. Park, X. Xu, H. Yao, in *2008 IEEE/ACM International Conference on Computer-Aided Design, San Jose, California, USA*. Layout Decomposition for Double Patterning Lithography (2008), pp. 465–472
36. C.H. Hsu, Y.W. Chang, S.R. Nassif, in *2009 IEEE/ACM International Conference on Computer-Aided Design—Digest of Technical Papers, San Jose, California, USA*. Simultaneous Layout Migration and Decomposition for Double Patterning Technology (2009), pp. 595–600
37. C. Bencher, Y. Chen, H. Dai, W. Montgomery, L. Huli, 22nm half-pitch patterning by CVD spacer self alignment double patterning (SADP). Proc. SPIE **6924**, 69244E (2008)
38. P. Xu, Y. Chen, Y. Chen, L. Miao, S. Sun, S.W. Kim, A. Berger, D. Mao, C. Bencher, R. Hung, C. Ngai, Sidewall spacer quadruple patterning for 15nm half-pitch. Proc. SPIE **7973**, 79731Q (2011)
39. F. Lazzarino, N. Mohanty, Y. Feurprier, L. Huli, V. Luong, M. Demand, S. Decoster, V.V. Gonzalez, J. Ryckaert, R. Ryoung, H. Kim, A. Mallik, P. Leray, C. Wilson, J. Boemmels, K. Kumar, K. Nafus, A. de Villiers, J. Smith, C. Fonseca, J. Bannister, S. Scheer, Z. Tokei, D. Piumi, K. Barla, Self-aligned block technology: a step toward further scaling. Proc. SPIE **10149**, 1014908 (2017)
40. T. Kozawa, S. Tagawa, Radiation chemistry in chemically amplified resists. Jpn. J. Appl. Phys. **49**, 030001 (2010)
41. D. Yang, S.W. Chang, C.K. Ober, Molecular glass photoresists for advanced lithography. J. Mater. Chem. **16**, 1693–1696 (2006)
42. K. Nishino, K. Maruyama, T. Kimura, T. Kai, K. Goto, S. Sharma, Development of EUV resist for 22nm half pitch and beyond. Proc. SPIE **7969**, 79692I (2011)
43. B.J. Rice, in *Nanolithography*, ed. by M. Feldman. 2—Extreme Ultraviolet (EUV) Lithography (Woodhead Publishing, 2014), pp. 42–79

Liquid Crystals with Electronic Functions

Masahiro Funahashi

1 Electronic Conduction in Liquid Crystal Phases

Electronic charge carrier transport in crystals of aromatic compounds has been studied extensively since the 1960s [1]. In hydrocarbon liquids, electronic charge carrier transport was also reported under strong electric fields [2]. However, it had been difficult to confirm electronic charge carrier transport in fluidic materials under UV light illumination. Conventional liquid crystal phases of low-molecular-weight materials are usually fluidic, and ionic conduction originated from impurities is dominant over electronic processes in the liquid phases. For display applications, suppression of electrical conduction was preferential to improve their energy performance and durability [3].

Figure 1a shows molecular structures of typical liquid crystals studied to aim at electronic charge carrier transport. In the 1970s, Kusabayashi and Yoshino measured carrier mobilities in a nematic phase of azomethine-based liquid crystals **1** and **2** [4, 5]. However, they observed ionic currents derived from impurities. These compounds included a Schiff base which was easily hydrolyzed by a small amount of water to produce ionic impurities. Moreover, these compounds exhibited a nematic phase and smectic phases with low molecular orders (Fig. 1b). Relatively small π-conjugated units could not increase the transfer integral for intermolecular migration of electronic charges.

In the 1990s, electronic conduction of columnar phases of alkylated phthalocyanines **3** was studied by time-resolved microwave conductivity [6]. As shown in Fig. 1b, disk-like π-conjugated units form π-stacks in which electronic charge

M. Funahashi (✉)
Department of Chemical Science and Engineering, Graduate School of Engineering, Kobe University, Kobe, Takamatsu, Japan
e-mail: funahashi.masahiro@phoenix.kobe-u.ac.jp

Health and Medical Research Institute, National Institute of Advanced Industrial Science and Technology, Kagawa, Japan

M. Funahashi (ed.), *Functional Soft Matter II*,
https://doi.org/10.1007/978-981-92-0100-6_12

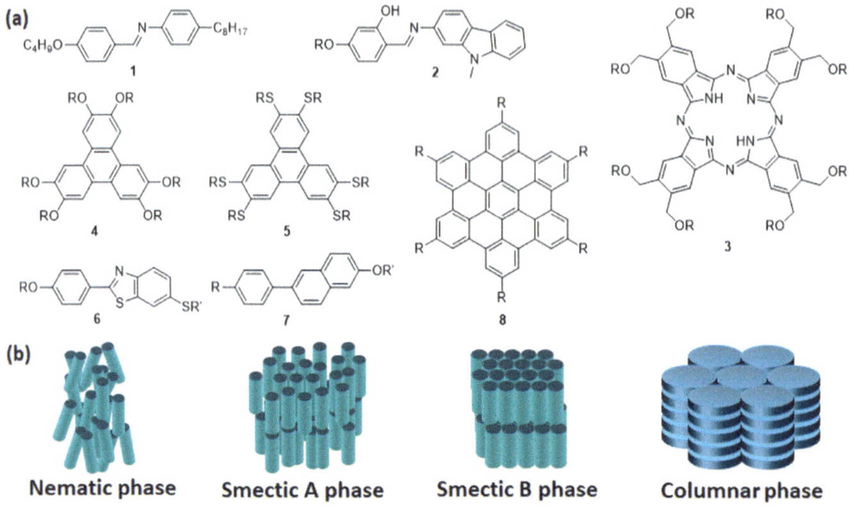

Fig. 1 **a** Molecular structures of typical liquid crystals studied to aim at electronic charge carrier transport. **b** Schematic illustrations for supramolecular aggregation structures of representative liquid crystal phases

carriers are transported in columnar phases. AC conductivity measurements of doped hexaalkoxytriphenylene derivatives **4** revealed a potential of organic semiconductors [7]. By a time-of-flight method, hole mobility up to 0.1 cm^2V^{-1} s^{-1} was observed in columnar phases of hexaalkylthiotriphenylene derivatives **5**, inspiring researchers working in the field of organic electronics [8]. These columnar liquid crystals had extended π-conjugated units which are self-organized in a one-dimensional order to increase transfer integrals remarkably while motion of ions was suppressed in the viscous media.

At the same time, efficient electronic charge carrier transport was confirmed in smectic phases of phenylbenzothiazole and phenylnaphthalene derivatives **6** and **7** which have relatively small π-conjugated units [9, 10]. In the smectic phases, the π-conjugated chromophores form two-dimensional carrier-transporting sheets, as shown in Fig. 1b. Hexabenzocoronene derivatives **8** bearing various functional groups have been designed and self-organization of the electroactive units and functional side chains has been studied extensively [11]. These liquid crystals which are characterized by extended π-conjugated units and electronic charge carrier transport have been called liquid crystalline (LC) semiconductors.

2 Carrier Transport Mechanism in Liquid Crystal Phases

The carrier transport mechanism in the ordered smectic and columnar phases has been studied by a time-resolved microwave conductivity (TRMC) method [12], providing a band mobility as well as a time-of-flight (TOF) technique revealing a drift mobility [13]. The TOF measurements can provide dispersion of density of states (DOS) and charge traps in the liquid crystal phases. As shown in Fig. 2a, electronic carrier transport processes in LC phases are regarded as a hopping transport in disordered systems. The Gaussian and correlated disorder models [14, 15] can reveal the width of the DOS, σ, distribution of transfer integral, Σ, and the preexponential factor μ_0, related to the transfer integral from the temperature- and field-dependence of the carrier mobilities. These disorder models are reviewed in Chap.5 of Part I.

For the analysis based on these models, it is required that the samples exhibit the LC phases over wide temperature ranges, as shown in Fig. 2b. The number of such studies has been limited because most of liquid crystals with extended π-conjugated units crystallize above room temperature.

Haarer reported the hole transport characteristics in the ordered columnar phase of a triphenylene dimer **9** which did not crystallize because of the conformational restriction [16]. The analysis based on the one-dimensional Gaussian disorder model revealed a higher preexponential factor of 0.25 cm^2V^{-1} s^{-1} and a smaller energetic disorder (σ) of 48 meV than those (typically 100 meV) in amorphous organic

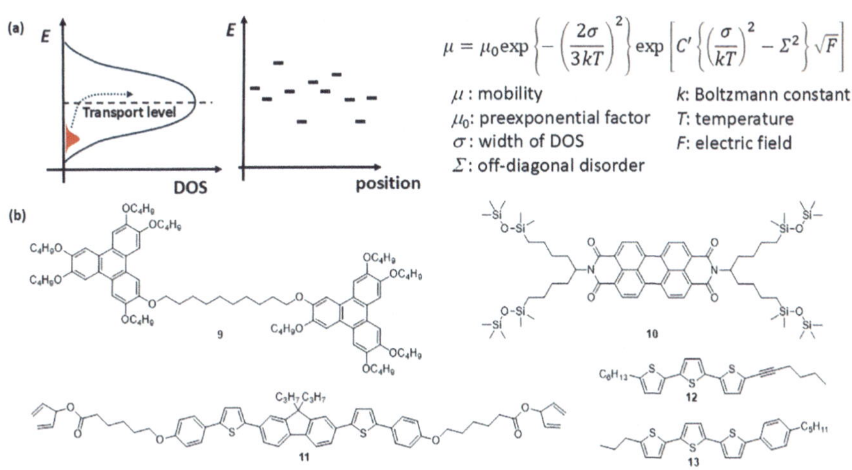

Fig. 2 **a** Schematic illustration for disorder model. Two graphs describe that molecules located in different circumstances induce a dstribution of energy levels, density of states (DOS). Charge carriers are thermallized within the DOS (red area) and excited to the transport level located the center of the DOS assisted by the heat and electric field. Parameters σ and Σ denote the width of the DOS and distribution of transfer integrals caused by fluctuation of intermolecular distances. The formula indicates dependece of a carrier mobility on the electric field and the temperature. **b** Molecular structures of liquid crystalline semiconductors which exhibit mesophases in wide temperature ranges

semiconductors. The π-π stacking structure in the columnar aggregates remarkably increased the transfer integral and decreased energetic disorder, resulting in the high hole mobility on the order of 10^{-2} cm^2V^{-1} s^{-1} at room temperature [17].

Funahashi reported the electron transport characteristics in the ordered columnar phases of perylene bisimide derivative **10** bearing oligosiloxane chains [18]. This perylene bisimide did not crystallize and retain the ordered columnar phase in which one-dimensional π-π stacking aggregates were surrounded by a liquid-like mantle consisting of oligosiloxane moieties. This compound exhibited a temperature- and field-independent electron mobility of 0.1 cm^2V^{-1} s^{-1} above room temperature. Below room temperature, the electron mobilities indicated positive field- and temperature-dependence and the characteristics could be fitted to the one-dimensional Gaussian disorder model, revealing a preexponential factor of 1.2 cm^2V^{-1} s^{-1} and a smaller diagonal disorder of 38 meV.

O'Neill and Kelly reported the analysis of hole transport in the nematic phase of fluorenone derivative **11** which bore lateral alkyl chains on the core unit [19]. The temperature- and field-dependence of the hole mobility provided a better fit to the correlation disorder model, indicating relatively large diagonal disorder of 89 meV.

In the smectic phases, electronic charge carrier transport processes are two-dimensional. Funahashi studied the hole and electron transport processes in the smectic E phase of a terthiophene derivative **12** bearing an alkyl chain as well as an alkynyl group [20]. This asymmetrically substituted terthiophene derivative did not crystallize even when it was cooled to − 100 °C while dialkylterthiophene crystallized above room temperature. This compound also exhibited the temperature- and field-independent mobility for holes and of 2×10^{-2} cm^2V^{-1} s^{-1} for electrons above room temperature. The two-dimensional Gaussian disorder model provided the best fit for the hole and electron transport characteristics. The preexponential parameter was 0.57 cm^2V^{-1} s^{-1} for holes and 0.57 cm^2V^{-1} s^{-1} for electrons. The energetic disorder was 45 meV for both carriers.

A band-like mechanism is proposed for waxy smectic phase of phenylterthiophene derivative **13**. This compound exhibited temperature-independent hole mobility exceeding 0.1 cm^2V^{-1} s^{-1} over 200 K including room temperature [21]. This behavior can be explained by the Hoesterey–Letson polaron theory. In this model, a band conduction affected by phonon scattering and hole trapping on localized states results in the temperature-dependent hole mobility.

3 Electronic Devices Using Liquid Crystalline Semiconductors

3.1 Polarized Light-Emitting Diodes

One of the characteristic features of LC materials is controllability of macroscopic molecular orientation. Uniaxially aligned sandwich-type cells have been produced by a classical rubbing method or photo-alignment technique [22]. When a fluorescent dye is dissolved in the uniaxially aligned medium, linearly polarized fluorescence is emitted from the LC medium.

Figure 3 shows molecular structures of the LC semiconductors applied to LEDs. The first light-emitting diode (LED) to radiate linearly polarized light was based on smectic liquid crystals. Tsutsui synthesized an oxadiazole-based smectic liquid crystal **14** which was injected into a liquid crystal cell with electron and hole injection layers [23]. They observed linearly polarized electroluminescence with a dichroic ratio of 3. Hanna and Funahashi obtained high-quality linearly polarized electroluminescence with a dichroic ratio exceeding 10 from a smectic liquid crystal **15** doped with coumarin dye **16** [24]. In these studies, several μm-thick-cells were used, and high voltages were needed for light emission.

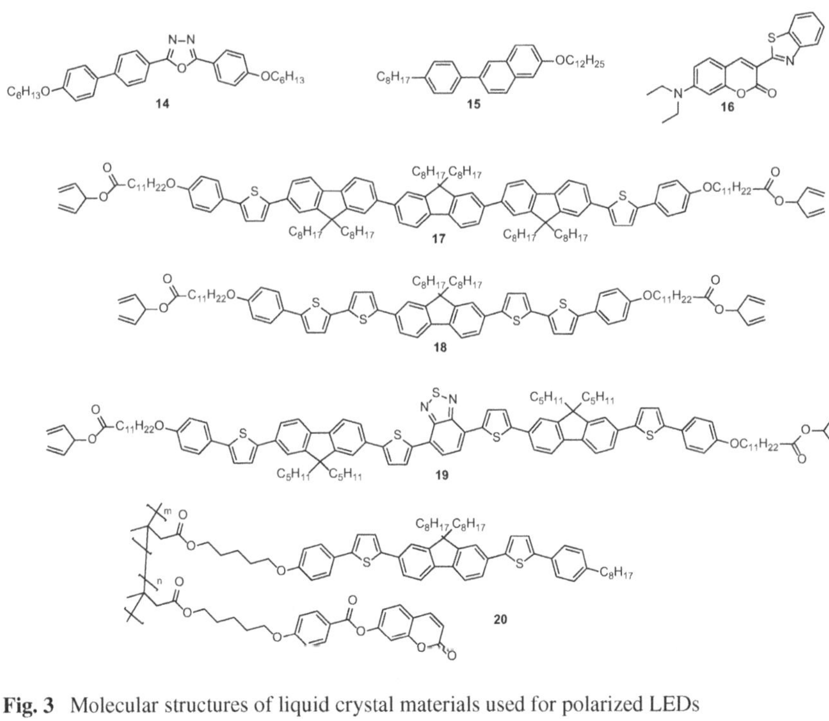

Fig. 3 Molecular structures of liquid crystal materials used for polarized LEDs

Thin films with the thickness around several 10 nm are required for fabrication of practical linearly polarized LEDs. It has been difficult to produce homogeneous thin films of low-molecular-weight liquid crystals by solution processes. Insolubilization of the thin film is also necessary for construction of multilayer devices. In the initial trials, vacuum-deposited columnar liquid crystal films were used for LEDs with the conventional multilayer structure based on vacuum-deposited amorphous films. In the devices, polarized light could not be obtained [25, 26].

O'Neill and Kelly produced linearly polarized LEDs based on LC semiconductors which can be crosslinked by UV light illumination. They produced full color polarized LED devices using red-, green-, and blue-fluorescent LC chromophores **17–19.** Instead of insulative polyimide alignment layers, they synthesized a photo-alignment polymer **20** containing extended π-conjugated units. This photo-alignment polymer also functions as a hole injection layer. The multilayer LEDs emitted linearly polarized light with a dichroic ratio exceeding 10 [27, 28].

3.2 Field Effect Transistors

For device applications of LC semiconductors, control of the molecular orientation is requisite for enhancement of the device performance because LC semiconductors are characterized by an anisotropic carrier transport. In field effect transistors (FETs), charge carriers are transported between source and drain electrodes parallel to the substrate. For columnar liquid crystals, the columnar axis should be aligned parallel to the channel direction [29].

Müllen and coworkers produced uniaxially aligned thin films of the alkylated hexabenzocoronene derivatives **21–23** by a friction transfer technique and a zone cast method, as shown in Fig. 4a [30, 31]. In the friction transfer technique, a heated substrate was rubbed by a Teflon block and an LC solution was spun on the treated substrate. In the zone cast method, an LC solution was coated on a substrate using a nozzle moving unidirectionally. The FETs in which holes could drift through the columnar aggregates indicated the mobilities on the order of 10^{-2} cm^2V^{-1} s^{-1} and the on/off ratio around 10^5.

In contrast to the columnar liquid crystals, classical rubbed polyimide alignment layers are effective for production of uniaxially aligned smectic LC films. Van Breemen and coworkers fabricated field-effect transistors based on alkylated oligothiophene derivatives **24** [32], as shown in Fig. 4b. The LC molecules were oriented parallel to the rubbing direction which was set perpendicular to the channel direction. The FETs indicated the hole mobility of 1×10^{-2} cm^2V^{-1} s^{-1}.

The LC molecules aligned perpendicular to the substrates when a solution of LC semiconductor was spin-coated on the substrates treated by silane coupling reagents. In this alignment, electronic charge carriers were preferentially transported through the smectic layers, as shown in Fig. 4c. Funahashi and coworkers fabricated FETs produced on silicon substrates treated with silane coupling reagents, using phenylterthiophene derivative **25**, indicating the hole mobilities of up to 0.1 cm^2V^{-1} s^{-1}

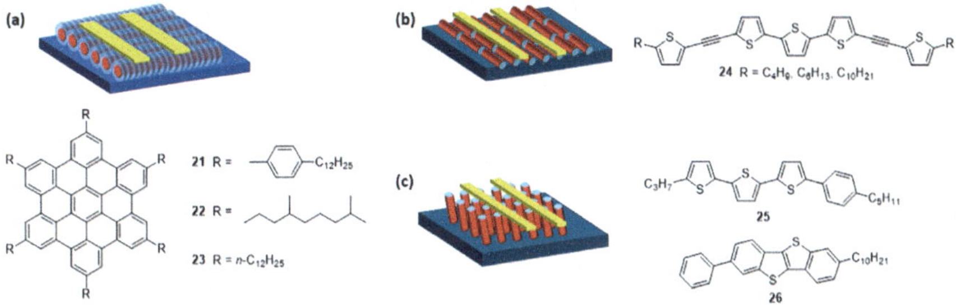

Fig. 4 Molecular structures of liquid crystalline semiconductors applied to field effect transistors and schematics for arrangement of electrodes and supramolecular assemblies of liquid crystal molecules. The first arrangement (a) shows a device using a columnr liquid crystal. Columnar aggregates are aligned uniaxially along the substrate and source and drain electrodes are arranged perpendicular to the columnar axis. The second (b) depicts a device using a smectic liquid crystal. In this case, source and drain electrodes are arranged prallel to the director of the molecules. The third (c) also shows devices using a homeotropically aligned smectic liquid crystal. Reproduced from Ref. [22] by permission of the Institute of Physics

with the on/off ratio around 10^7 [33, 34]. The devices produced on polymer could retain the performance when a strain of 30% was applied to the devices [35]. It should be noted that the FETs worked in the ordered smectic phase but not in the crystal phase.

Iino and Hanna regarded the LC phase as a transient state to produce crystalline thin films of high quality. They synthesized thienothiophene derivatives **26** (Fig. 4c) exhibiting an ordered smectic phase at an elevated temperature and crystallizing at room temperature. They produced LC thin films by the spin-coating method at elevated temperature and cooled the thin film to room temperature. The thin films formed crystalline thin films with low defect density at room temperature. The molecules aligned perpendicular to the substrates in the crystal phase. The devices indicated high hole mobilities exceeding 10 cm^2V^{-1} s^{-1} [36, 37].

3.3 Solar Cells

As conventional organic thin film solar cells, p-n junction is indispensable for enhancement of charge carrier generation in LC solar cells [38, 39]. High carrier mobility and high solubility of LC semiconductors is advantage for solution-processable solar cells. Control of segregation of n-type crystalline semiconductors from type LC semiconductors is significant factor for the improvement of the soler cell performances.

Figure 5 shows molecular structures of the LC semiconductors applied to heterojunction-type solar cells. First example of a thin solar cell using liquid crystalline semiconductors was reported by Müllen and coworkers [40]. They produced a bilayer film consisting of an LC hexabenzocoronene derivative **27** and crystalline

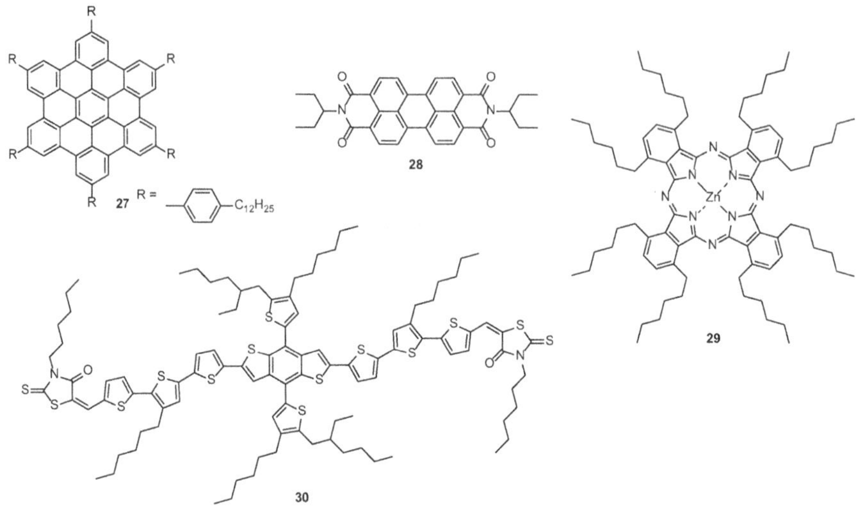

Fig. 5 Molecular structures of liquid crystalline semiconductors applied for bulk heterojunction-type solar cells

perylene bisimide **28**. In a spin-coated film, n-type perylene bisimide crystallized from a mixture of the *p*-type and n-type components to form a vertically segregated structure. Because of the crystallization of the n-type component, the effective photoactive area was increased to enhance the photovoltaic efficiency.

Standard bulk heterojunction solar cells using *p*-type and n-type LC semiconductors were reported by Ozaki and Shimizu [41]. They produced nanosegregated mixtures of *p*-type LC phthalocyanine **29** and n-type crystalline fullerene derivatives. In the columnar phase, the phthalocyanine derivatives which had eight alkyl chains at non-peripheral positions exhibited a high hole mobility exceeding 1 cm^2V^{-1} s^{-1}. The power conversion efficiency was around 3.1% [42].

Sun and coworkers synthesized a nematic LC semiconductor **30** which had a core moiety comprised of benzodithiophene, terthiophene, and rhodamine units [43]. The bulk heterojunction structures were formed from a mixture of the nematic LC semiconductor and a fullerene derivative. The resulting thin-film photovoltaic cell exhibited the PCE of 9.3% with the fill factor of 74%. The nematic liquid crystal exhibited a hole mobility on the order of 10^{-3} cm^2 V^{-1} s^{-1} determined by a space-charge-limited current measurement. This characteristic is suitable for a roll-to-roll production process of organic solar cells.

4 Transfunctional π-Conjugated Liquid Crystals

Higher level functions can be created by coupling of the electronic charge carrier transport of LC semiconductors with other physical properties, such as ionic conduction and ferroelectricity.

Efficient ionic transport as well as long-range charge transfer between electrochemical active sites are required for electrochemical functional materials. Kato and coworkers have successfully developed soft nanosegregated LC functional materials consisting of mutually insoluble components [44, 45]. Ion-conductive and electroactive domains can be constructed in nanosegregated LC phases. Polarization of ionic species in the ion-conductive domains promoted charge carrier injection from the electrodes or electrochemical reaction in the electroactive domains.

Figure 6a shows a schematic illustration and examples of the LC mixed conductor. Funahashi and Kato synthesized phenylterthiophene derivatives **31** bearing an imidazolium moiety [46, 47]. These compounds exhibited a smectic A phase in which ion-transporting and hole-transporting layers were formed alternately. This liquid crystal exhibited electrochromism under the application of a direct current (DC) bias without electrolyte solutions. Beneducci and coworkers reported LC thienoviologens **32** which exhibited electrochemical fluorescence enhancement in the smectic and columnar phases [48].

Funahashi reported electrochromism in the spin-coated thin films of perylene bisimide derivatives **33** and **34** which formed nanosegregated columnar phases consisting of electron-transporting π-stacks and ion-conductive sublayers, as shown

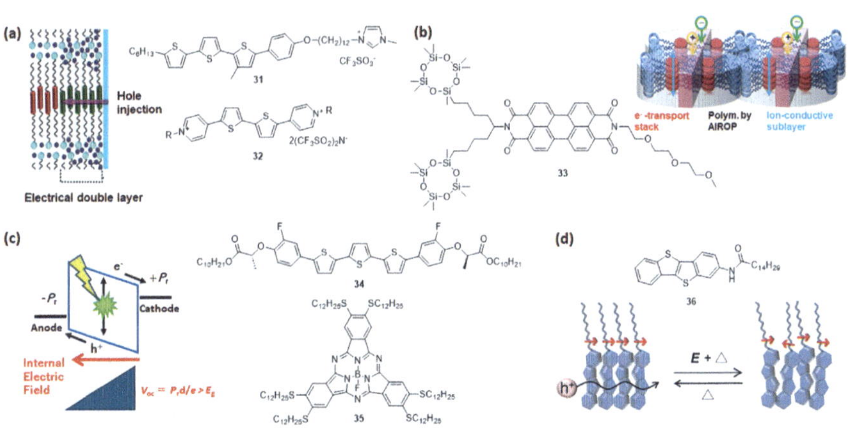

Fig. 6 Schematic illustrations for multifunctional materials and molecular structures of multifunctional liquid crystalline semiconductors. **a** LC mixed conductors. Reproduced from Ref. [46] by permission of the American Chemical Society. **b** Redoxactive columnar liquid crystals. **c** Bulk photovoltaic effect. **d** Switchable LC semiconductor. Reproduced from Ref. [57] by permission of The American Chemical Society

in Fig. 6b [49, 50]. The liquid crystalline perylene bisimide derivatives had polymerizable moieties promoting nanosegregation. Insolubilized thin films which could work in organic electrolyte solutions were obtained by an acid-vapor-induced polymerization method.

An electrical polarization can also be induced by an orientation of dipoles. The internal electric field promotes dissociation of photogenerated excitons, transport of charge carriers, and injection of charge carriers from electrodes.

Figure 6c shows a schematic illustration for bulk photovoltaic effect and the molecular structures of ferroelectric LC semiconductors. In a bulk photovoltaic effect, the spontaneous polarization of the ferroelectrics promotes dissociation of excitons and transport of photogenerated charge carriers. Therefore, higher open-circuit voltage than the bandgap can be generated, and inorganic ferroelectrics such as bismuth ferrite have been studied so far [51, 52]. For organic ferroelectrics, there have been limited studies such as dye-doped poly (vinylidene fluoride) and TTF-chloranil crystals [53].

Funahashi and coworkers synthesized ferroelectric liquid crystals **35** consisting of extended π-conjugated units and chiral alkyl chains. The fullerene-doped ferroelectric liquid crystals exhibit high open-circuit voltages up to 1.2 V [54]. Miyajima and coworkers synthesized ferroelectric columnar liquid crystals comprising a subphthalocyanine core [55]. This compound exhibited the bulk photovoltaic effect for visible-light illumination. Lehmann and Eremin also reported the bulk photovoltaic effect in an LC subphthalocyanine derivative [56].

Akutagawa and coworkers reported an enhancement of electronic charge carrier transport in crystalline films of ferroelectric BTBT **36**, as shown in Fig. 12d [57]. Field-induced orientation of electrical dipoles reduced structural disorder of the BTBT crystals to increase the hole mobility. They suggested a potential application of the BTBT derivatives to memory devices.

5 Conclusion

In the classical concept, liquid crystal materials were dielectrics with optical functions. However, liquid crystals based on extended π-conjugated units can work as semiconductors with flexibility. The liquid crystalline semiconductors are promising materials for the application to lightweight, flexible, and solution-processable electronic devices such as organic light-emitting diodes, field effect transistors, and solar cells. In nanosegregated and polar nanostructures, the electronic charge carrier transport through one- or two-dimensional π-stacks can couple with ionic conduction or ferroelectricity. The multifunctional π-conjugated liquid crystals can provide new phenomena such as the bulk photovoltaic effect and switchable electronic transport.

References

1. R.G. Kepler, Charge carrier production and mobility in Anthracene crystals. Phys. Rev. **119**, 1226–1229 (1960)
2. R.M. Minday, L.D. Schmidt, H.T. Davis, Excess electrons in liquid hydrocarbons. J. Chem. Phys. **54**, 3112–3125 (1971)
3. M. Schadt, Liquid crystal materials and liquid crystal displays. Ann. Rev. Mater. Sci. **27**, 305–379 (1997)
4. K. Yoshino, N. Tanaka, Y. Inuishi, Anomalous carrier mobility in Smectic liquid crystal. Jpn. J. Appl. Phys. **15**, 735–736 (1976)
5. Y. Shimizu, K. Shigeta, S. Kusabayashi, Mesomorphic and Photoconducting behavior of 2-(2-Hydroxy-4-alkoxybenzylideneamino)-9-methylcarbazoles. Mol. Cryst. Liq. Cryst. **140**, 105–117 (1984)
6. P.G. Schouten, J.M. Warman, M.P. de Haas, C.F. van Nostrum, G.H. Gelinck, R.J.M. Nolte, M.J. Copyn, J.W. Zwikker, M.K. Engel, M. Hanack, Y.H. Chang, W.T. Ford, The effect of structural modifications on charge migration in Mesomorphic Phthalocyanines. J. Am. Chem. Soc. **116**, 6880–6894 (1994)
7. N. Boden, R.J. Bushby, J. Clements, Mechanism of quasi-one-dimensional liquid crystals. J. Chem. Phys. **98**, 5920–5931 (1993)
8. D. Adam, P. Schumacher, J. Simmerer, L. Häussling, K. Siemensmeyer, K.H. Etzbach, H. Ringsdorf, D. Haarer, Fast photoconduction in the highly ordered columnar phase of a discotic liquid crystal. Nature **371**, 141–143 (1994)
9. M. Funahashi, J. Hanna, Fast hole transport in a new calamitic liquid crystal of 2-(4′-heptyloxyphenyl)-6-dodecylthiobenzothiazole. Phys. Rev. Lett. **78**, 2184–2187 (1997)
10. M. Funahashi, J. Hanna, Fast ambipolar carrier transport in smectic phases of phenylnaphthalene liquid crystal. Appl. Phys. Lett. **71**, 602–604 (1997)
11. W. Pisula, M. Zorn, J.Y. Chang, K. Müllen, R. Zentel, Liquid crystalline ordering and charge transport in semiconducting materials. Macromol. Rap. Commun. **30**, 1179–1202 (2009)
12. A. Saeki, in *Microwave and Terahertz Spectroscopy. Organic Semiconductors for Optoelectronics*, ed. by H. Naito, (Wiley interscience, 2021)
13. M. Funahashi, in *Time-of-Flight Method for Determining the Drift Mobility in Organic Semiconductors. Organic Semiconductors for Optoelectronics*, ed. by H. Naito, (Wiley interscience, 2021)
14. H. Bässler, Charge transport in disordered organic photoconductors—A Monte Carlo simulation study. Phys. Stat. Solid. B **175**, 15–56 (1993)
15. S.V. Novikov, D.H. Dunlap, V.M. Kenkre, P.E. Parris, A.V. Vannikov, Essential role of correlations in governing charge transport in disordered organic materials. Phys. Rev. Lett. **81**, 4472–4475 (1998)
16. D. Adam, P. Schuhmacher, J. Simmerer, L. Häußling, W. Paulus, K. Siemensmeyer, K.-H. Etzbach, H. Ringsdorf, D. Haarer, Photoconductivity in the columnar phases of a glassy discotic twin. Adv. Mater. **7**, 276–280 (1995)
17. I. Bleyl, C. Erdelen, H.-W. Schmidt, D. Haarer, One-dimensional hopping transport in a columnar discotic liquid-crystalline glass. Phil. Mag. B **79**, 463–475 (1999)
18. M. Funahashi, A. Sonoda, Electron transport characteristics in nano-segregated columnar phases of perylene tetracarboxylic bisimide derivatives bearing oligosiloxane chains. Phys. Chem. Chem. Phys. **16**, 7754–7763 (2014)
19. S.R. Farrar, A.E.A. Contoret, M. O'Neill, J.E. Nicholls, G.J. Richards, S.M. Kelly, Nondispersive hole transport of liquid crystalline glasses and a cross-linked network for organic electroluminescence. Phys. Rev. B **66**, 125107 (2002)
20. M. Funahashi, F. Zhang, N. Tamaoki, J. Hanna, Ambipolar transport in the highly ordered smectic phase of 2-alkyl-5″-alkynyl- terthiophene derivative in a wide temperature range. Chem Phys Chem **9**, 1465–1473 (2008)

21. M. Funahashi, T. Ishii, A. Sonoda, Temperature-independent hole mobility of a smectic liquid-crystalline semiconductor based on band-like conduction. Chem Phys Chem **14**, 2750–2758 (2013)
22. M. Funahashi, Solution-processable electronic and redox-active liquid crystals based on the design of side chains. Flex. Print. Electron. **5**, 043001 (2020)
23. H. Tokuhisa, M. Era, T. Tsutsui, Polarized electroluminescence from smectic mesophase. Appl. Phys. Lett. **72**, 2639–2641 (1998)
24. K. Kogo, T. Goda, M. Funahashi, J. Hanna, Polarized light emission from a calamitic liquid crystalline semiconductor doped with dyes. Appl. Phys. Lett. **73**, 1595–1597 (1998)
25. A. Bacher, I. Bleyl, C.H. Erdelen, D. Haarer, W. Paulus, H.-W. Schmidt, Low molecular weight and polymeric Triphenylenes as hole transport materials in organic two-layer LEDs. Adv. Mater. **9**, 1031–1035 (1997)
26. T. Hassheider, S.A. Benning, H.-S. Kitzerow, M.-F. Achard, H. Bock, Color-tuned electroluminescence from columnar liquid crystalline alkyl Arenecarboxylates. Angew. Chem. Int. Ed. **40**, 2060–2063 (2001)
27. A.E.A. Contoret, S.R. Farrar, P.O. Jackson, S.M. Khan, L. May, M. O'Neill, J.E. Nicholls, S.M. Kelly, G.J. Richards, Polarized electroluminescence from an anisotropic nematic network on a non-contact photoalignment layer. Adv. Mater. **12**, 971–974 (2000)
28. M.P. Aldred, A.E.A. Contoret, S.R. Farrar, S.M. Kelly, D. Mathieson, M. O'Neill, W.C. Tsoi, P. Vlachos, A full-color electroluminescent device and patterned Photoalignment using light-emitting liquid crystals. Adv. Mater. **17**, 1368–1372 (2005)
29. M. Funahashi, Development of liquid-crystalline semiconductors with high carrier Mobilities and their application to thin-film transistors. Polym. J. **41**, 459–469 (2009)
30. A.M. van de Craats, N. Stutzmann, O. Bunk, M.M. Nielsen, M. Watson, K. Müllen, H.D. Chanzy, H. Sirringhaus, R.H. Friend, Meso-epitaxial solution-growth of self-organizing Discotic liquid-crystalline semiconductors. Adv. Mater. **15**, 495–499 (2003)
31. W. Pisula, A. Menon, M. Stepputat, I. Lieberwirth, U. Kolb, A. Tracz, H. Sirringhaus, T. Pakula, K. Müllen, A zone-casting technique for device fabrication of field-effect transistors based on Discotic Hexa-peri-hexabenzocoronene. Adv. Mater. **17**, 684–689 (2005)
32. A.J.J.M. van Breemen, P.T. Herwig, C.H.T. Chlon, J. Sweelssen, H.F.M. Schoo, S. Setayesh, W.M. Hardeman, C.A. Martin, D.M. de Leeuw, J.J.P. Valeton, C.W.M. Bastiaansen, D.J. Broer, A.R. Popa-Merticaru, S.C.J. Meskers, Large area liquid crystal Monodomain field-effect transistors. J. Am. Chem. Soc. **128**, 2336–2345 (2006)
33. M. Funahashi, F. Zhang, N. Tamaoki, High Ambipolar mobility in a highly ordered Smectic phase of a Dialkylphenylterthiophene derivative that can be applied to solution-processed organic field-effect transistors. Adv. Mater. **19**, 353–358 (2007)
34. F. Zhang, M. Funahashi, N. Tamaoki, High-performance thin film transistors from semi-conducting liquid crystalline phases by solution processes. Appl. Phys. Lett. **91**, 063515 (2007)
35. F. Zhang, M. Funahashi, N. Tamaoki, Flexible field-effect transistors from a liquid crystalline semiconductor by solution processes. Org. Electron. **11**, 363–368 (2010)
36. H. Iino, J. Hanna, Availability of liquid crystallinity in solution processing for polycrystalline thin films. Adv. Mater. **23**, 1748–1751 (2011)
37. H. Iino, T. Usui, J. Hanna, Liquid crystals for organic thin-film transistors. Nat. Commun. **6**, 6828 (2015)
38. C. Tang, Two-layer organic photovoltaic cell. Appl. Phys. Lett. **48**, 183–185 (1986)
39. N.S. Sariciftci, L. Smilowitz, A.J. Heeger, F. Wudl, Photoinduced electron transfer from a conducting polymer to buckminsterfullerene. Science **258**, 1474–1476 (1992)
40. L. Schmidt-Mende, A. Fechtenkötter, K. Müllen, E. Moons, R.H. Friend, I.J.D.M. Kenzie, Self-organized Discotic liquid crystals for high-efficiency organic Photovoltaics. Science **293**, 1119–1122 (2001)
41. T. Hori, Y. Miyake, N. Yamasaki, H. Yoshida, A. Fujii, Y. Shimizu, M. Ozaki, Solution Processable organic solar cell based on bulk Heterojunction utilizing Phthalocyanine derivative. Appl. Phys. Exp. **3**, 101602 (2010)

42. Y. Miyake, Y. Shiraiwa, K. Okada, H. Monobe, T. Hori, N. Yamasaki, H. Yoshida, M.J. Cook, A. Fujii, M. Ozaki, Y. Shimizu, High carrier mobility up to 1.4 $cm^2V^{-1}s^{-1}$ in non-peripheral Octahexyl Phthalocyanine. Appl. Phys. Exp. **4**, 021604 (2011)
43. K. Sun, Z. Xiao, S. Lu, W. Zajaczkowski, W. Pisula, E. Hanssen, J.M. White, R.M. Williamson, J. Subbiah, J. Ouyang, A.B. Holmes, W.W.H. Wong, D.J. Jones, A molecular nematic liquid crystalline material for high-performance organic photovoltaics. Nat. Commun. **6**, 6013 (2015)
44. T. Kato, J. Uchida, T. Ichikawa, T. Sakamoto, Functional liquid crystals towards the next generation of materials. Angew. Chem. Int. Ed. **57**, 4355–4371 (2018)
45. T. Kato, M. Gupta, D. Yamaguchi, K.P. Gan, M. Nakayama, Supramolecular association and nanostructure formation of liquid crystals and polymers for new functional materials. Bull. Chem. Soc. Jpn. **94**, 357–376 (2021)
46. S. Yazaki, M. Funahashi, T. Kato, An Electrochromic nanostructured liquid crystal consisting of π-conjugated and ionic moieties. J. Am. Chem. Soc. **130**, 13206–13207 (2008)
47. S. Yazaki, M. Funahashi, J. Kagimoto, H. Ohno, T. Kato, Nanostructured liquid crystals combining ionic and electronic functions. J. Am. Chem. Soc. **132**, 7702–7708 (2010)
48. A. Beneduci, S. Cospito, M. La Deda, L. Veltri, G. Chidichimo, Electrofluorochromism in π-conjugated ionic liquid crystals. Nat. Commun. **5**, 3105 (2014)
49. M. Funahashi, Anisotropic electrical conductivity of n-doped thin films of polymerizable liquid-crystalline perylene bisimide bearing a triethylene oxide chain and cyclotetrasiloxane rings. Mater. Chem. Front. **1**, 1137–1146 (2017)
50. T. Taga, S. Takaoka, S. Uemura, M. Funahashi, Liquid-crystalline perylene bisimide derivatives bearing an azacrown ether ring complexing with alkaline metal ions. Mater. Chem. Front. **6**, 880–890 (2022)
51. T. Choi, S. Lee, Y.J. Choi, V. Kiryukhin, S.-W. Cheong, Switchable ferroelectric diode and photovoltaic effect in $BiFeO_3$. Science **324**, 63–66 (2009)
52. Y.R. Chang, R. Nanae, S. Kitamura, T. Nishimura, H. Wang, Y. Xiang, K. Shinokita, K. Matsuda, T. Taniguchi, K. Watanabe, K. Nagashio, Shift-current Photovoltaics based on a non-Centrosymmetric phase in in-plane ferroelectric SnS. Adv. Mater. **35**, 2301172 (2023)
53. M. Nakamura, S. Horiuchi, F. Kagawa, N. Ogawa, T. Kurumaji, Y. Tokura, M. Kawasaki, Shift current photovoltaic effect in a ferroelectric charge-transfer complex. Nat. Commun. **8**, 281 (2017)
54. M. Funahasahi, High open-circuit voltage under the bulk photovoltaic effect for the chiral smectic crystal phase of a double chiral ferroelectric liquid crystal doped with a fullerene derivative. Mater. Chem. Front. **5**, 8265–8274 (2021)
55. C. Zhang, K. Nakano, M. Nakamura, F. Araoka, K. Tajima, D. Miyajima, Noncentrosymmetric columnar liquid crystals with the bulk photovoltaic effect for organic Photodetectors. J. Am. Chem. Soc. **142**, 3326–3330 (2020)
56. A. Murad, E. Baron, M. Feneberg, M. Baumann, M. Lehmann, A. Eremin, Polarity in liquid crystals formed by self-assembled umbrella-shaped Subphthalocyanine Mesogens. ACS Appl. Mater. Interfaces **16**, 25025–25032 (2024)
57. K. Sambe, T. Takeda, N. Hoshino, W. Matsuda, K. Shimada, K. Tsujita, S. Maruyama, S. Yamamoto, S. Seki, Y. Matsumoto, T. Akutagawa, Carrier transport switching of ferroelectric BTBT derivative. J. Am. Chem. Soc. **146**, 8557–8566 (2024)

Liquid Crystals with Ionic Functions

Takahiro Ichikawa

1 Introduction

Ionic liquid crystals [1–3] are a class of Liquid-Crystalline (LC) materials that have various ionic structures. From the viewpoint of their state, they can be categorized as ionic liquids with LC properties. One of the most attractive characteristics of ionic liquid crystals is their potential to integrate the functions of both a liquid crystal and an ionic liquid. Ionic liquid crystals have extremely diverse molecular structures, and to date, many ionic liquid crystals have been designed and developed. Further, certain reviews have comprehensively described the characteristics of ionic liquid crystals [4–10].

In early studies on ionic liquid crystals, the thermotropic LC behaviors of a series of pyridinium and imidazolium salts were investigated. Fig. 1 presents the molecular structures of representative ionic liquid crystals. For instance, Engberts demonstrated that 4-alkylpyridinium, compound **1**, exhibits mesomorphic properties and forms smectic (Sm) phases [11]. Guillon et al. revealed that a pyridinium derivative, compound **2**, composed of a long alkyl chain at the *N*-position of the pyridinium ring, also exhibits thermotropic LC properties [12]. On the other hand, Seddon and Bruce succeeded in designing compound **3**, an imidazolium-based thermotropic liquid crystal [13]. The formation of the self-organized structures of these molecules involves two important driving forces: (i) the segregation of ionic and nonionic parts at the nanometer scale and (ii) electrostatic interactions between the cationic and anionic parts. Therefore, the length of the alkyl chain and the selection of cation and anion components are important considerations for the development of ionic liquid crystals with desired characteristics. Following the aforementioned initial studies on ionic liquid crystals, various functional ionic liquid crystals have

T. Ichikawa (✉)
Department of Biotechnology and Life Science, Graduate School of Engineering, Tokyo University of Agriculture and Technology, Tokyo, Japan
e-mail: t-ichi@cc.tuat.ac.jp

M. Funahashi (ed.), *Functional Soft Matter II*,
https://doi.org/10.1007/978-981-92-0100-6_13

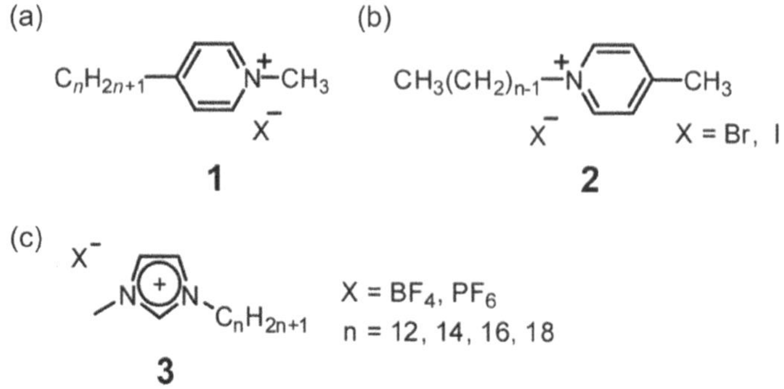

Fig. 1 Molecular structures of a series of ionic liquid crystals, compounds **1–3**, designed at an early stage [11–13]

been designed and developed. In the ensuing section, certain ionic functions of ionic liquid crystals are described.

2 Anisotropic Ion-Conducting Liquid Crystals

By controlling the volume balance between the ionic and nonionic components of ionic liquid crystals, not only layered Sm LC assemblies but also other nanosegregated LC assemblies with curved interfaces can be obtained. For example, wedge-shaped imidazolium compounds $\mathbf{4}_n$, shown in Fig. 2a, are representative molecules that form columnar phases [14, 15]. These salts consist of two incompatible parts: an ionophobic trialkoxy phenyl moiety and an ionophilic imidazolium moiety. These molecules self-organize through nanosegregation between the two incompatible parts and electrostatic interactions between the ionophilic parts, resulting in the formation of hexagonal columnar (Col_h) structures with ion nanochannels in the inner regions of the columns Fig. 2b. The thermotropic LC behaviors of these imidazolium salts are significantly dependent on their anionic species and alkyl chain lengths. For example, $\mathbf{4}_8$ with trioctyl alkyl chains exhibits a Col_h phase between – 29 and 133 °C, whereas an analogous compound with (bis(trifluromethylsulfonyl)imide) (TFSI) anion shows no mesomorphic properties [15]. These results indicate that optimal alkyl chain lengths and an appropriate choice of anionic species are important for designing ionic LC materials with desired properties.

Columnar LC materials can be expected to be used as anisotropic functional materials because of their one-dimensional (1D) structures. It is important to orient the columns homogeneously on the macroscopic scale to observe 1D characteristics. Macroscopic alignment of $\mathbf{4}_n$ was successfully achieved via the mechanical shearing [14], resulting in the alignment of the column axes parallel to the shearing direction. Using this technique, the anisotropic ion-conduction behavior of $\mathbf{4}_n$ was successfully examined. Ionic conductivity was evaluated via the AC impedance method using

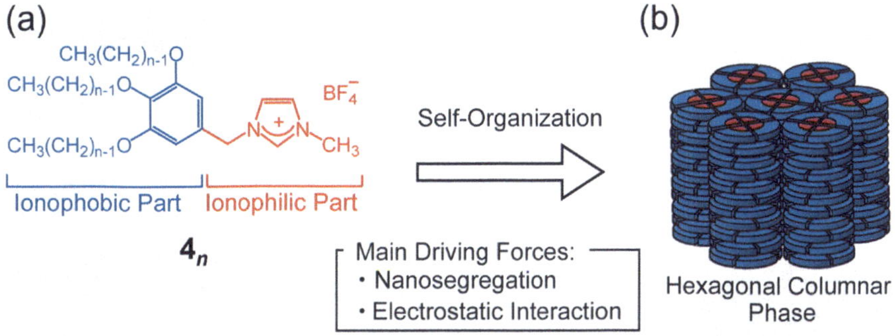

Fig. 2 **a** Molecular structures of a series of wedge-shaped imidazolium compound, $\mathbf{4}_n$, consisting of ionophilic and ionophobic parts [14], and **b** their self-assembled columnar structures. Two main driving forces, viz., nanosegregation between the two immiscible parts and electrostatic interactions between the ionic parts, induce the self-organization of such molecules

cells with comb-like gold electrodes. For measurements, the Col_h LC samples were oriented in two different directions with respect to the electrodes: perpendicular and parallel. Each result corresponds to the ionic conductivity of the sample in parallel and perpendicular directions to the column axis. Fig. 3 presents the anisotropic ionic conductivities of $\mathbf{4}_{12}$ as functions of the measurement temperature. For comparison, the ionic conductivities of $\mathbf{4}_{12}$ in the polydomain Col_h LC state are also plotted. For $\mathbf{4}_{12}$, the ionic conductivities in the parallel direction to the column axis ($\sigma_{\parallel}$) are higher than those in the perpendicular ($\sigma_{\perp}$) one in the Col_h phase, and no anisotropy is observed in the isotropic liquid state. The anisotropy ($\sigma_{\parallel}/\sigma_{\perp}$) of conductivities for the Col_h phase of $\mathbf{4}_{12}$ was calculated to be a constant value of ~ 40. The ionic conductivities in the polydomain Col_h state lie between the $\sigma_{\parallel}$ and $\sigma_{\perp}$ values. This is the first report on the development of 1D ion conductors using organic LC molecules.

3 Anisotropic Ion-Conductive Polymer Membranes

Fixing the columnar LC structures is a prerequisite for the application of ion-conductive columnar LC materials to solid polymer electrolytes. The design of LC monomers and the subsequent in situ photopolymerization in their LC state is a versatile approach to obtain polymeric materials with the LC nanostructures locked in their positions. In 2006, Kato et al. designed and synthesized a wedge-shaped imidazolium salt, **5** with polymerizable acrylate groups at the periphery of its alkyl chains (Fig. 4) [16]. This monomer forms a Col_h phase between 20 and 50 °C upon heating. The self-assembly of the molecules, uniaxial orientation of the assembled structures, and the subsequent in situ photopolymerization of the columnar LC imidazolium salt provided two types of macroscopically oriented nanostructured films. This approach for preparing 1D ion-conducting polymeric films is schematically illustrated in Fig. 4. In one sample, 1D ion nanochannels were oriented perpendicular to the film surface

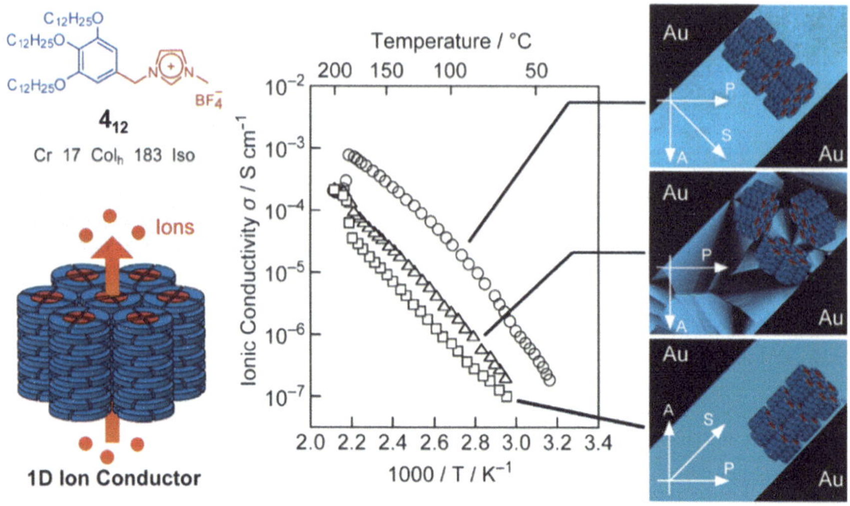

Fig. 3 Temperature-dependent ionic conductivities of wedge-shaped imidazolium salt $\mathbf{4_{12}}$ [14]

(**Film-V**) while the other contained parallelly oriented nanochannels (**Film-P**). The orientation of the columns was achieved by different methods. Whereas the parallel orientation of the columns was achieved by mechanical shearing, their perpendicular orientation to the glass surface was achieved using a glass substrate modified with 3-(aminopropyl)triethoxysilane.

Ion-conduction measurements on the ionic LC films revealed that their ion conductivity in the parallel direction to the column axis is higher than that in the perpendicular one. For example, the ionic conductivities of **Film-P** in the parallel and perpendicular directions to the column axis are $\sigma_{\parallel} = 4.1 \times 10^{-4}$ S cm^{-1} and $\sigma_{\perp} = 2.8 \times 10^{-7}$ S cm^{-1} at 100 °C. The magnitude of anisotropy ($\sigma_{\parallel} / \sigma_{\perp}$) is almost constant at 1.4×10^3. These values are 10–100 times higher than those obtained for the unpolymerized sample of the columnar LC materials. These results indicate that fixing the aligned columnar structure of the LC monomer by in situ photopolymerization is an effective strategy for enhancing the anisotropic physicochemical properties of ionic LC materials. The polymerized films of ionic LCs with oriented nanochannels of well-defined sizes can be applied not only as solid ion conductors but also as other functional solid materials, such as size-selective separation membranes.

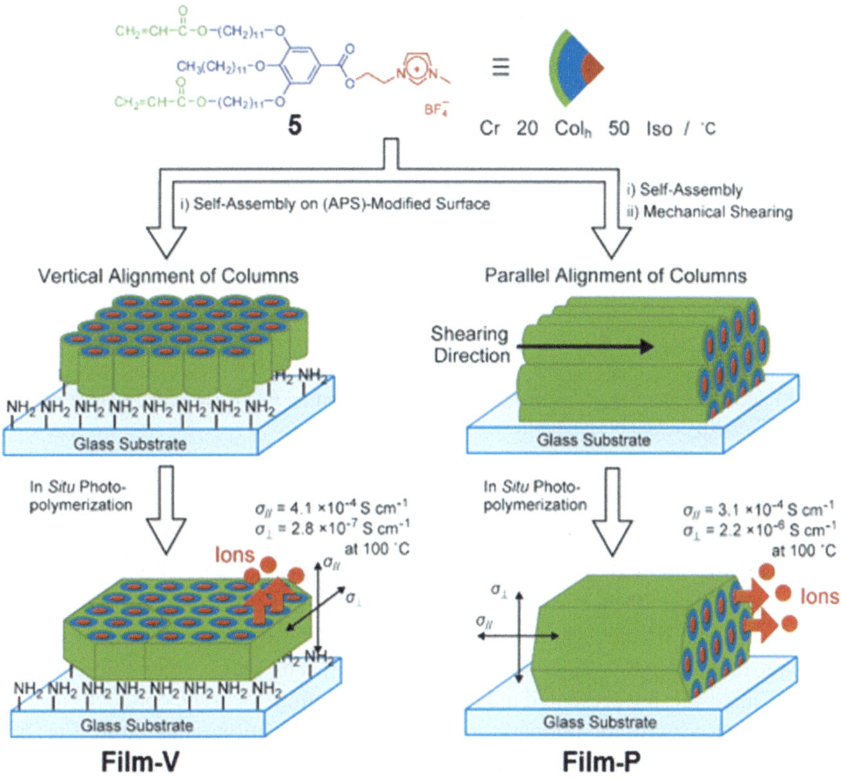

Fig. 4 Schematic illustration of the preparation of 1D ion-conducting polymer films [16]. Self-assembly and the subsequent photopolymerization of Col$_h$ LC imidazolium salt **5**. Vertical orientation of the columns was achieved using a 3-(aminopropyl)triethoxysilane-modified glass substrate. Parallel orientation of the columns was induced by mechanical shearing. In situ polymerization of the Col$_h$ LC materials in the oriented states resulted in two types of 1D ion-conductive polymer films

4 Three-Dimensional (3D) Ion-Conducting

Bicontinuous cubic (Cub$_{bi}$) LCs are a class of nanosegregated LCs with cubic periodicity and 3D continuity. Owing to their unique nanostructures, Cub$_{bi}$ LCs have attracted increasing attention as transporting materials, such as ion-conducting materials and size-selective separation membranes. In 2004, Wiesner et al. designed a series of wedge-shaped dendrimer-type compounds, **6**, with a polar oligoethylene moiety Fig. 5a [17]. Compound **6** was found to form Cub$_{bi}$ LC assemblies in the presence of lithium salts. In the Cub$_{bi}$ assemblies, the oligoethylene moiety of **6** formed 3D ion nanochannels that served as effective ion-transport paths.

The potential utility of Cub$_{bi}$ LCs as ion-transporting materials has also been demonstrated using a class of wedge-shaped ionic liquid crystals that form Cub$_{bi}$ phases [18]. Wedge-shaped ammonium compounds, **7$_n$** and phosphonium

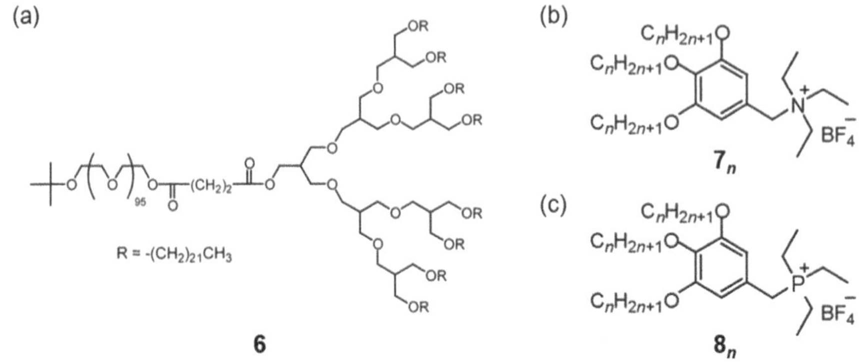

Fig. 5 Molecular structures of: **a** extended amphiphilic dendron **6** [17]; **b** wedge-shaped ammonium salts $\mathbf{7}_n$ [18]; and **c** phosphonium salts $\mathbf{8}_n$ [19]

compounds, $\mathbf{8}_n$ Fig. 5b, c were found to form Cub_{bi} LC assemblies depending on their alkyl chain lengths. For example, $\mathbf{7}_{10}$ forms a Cub_{bi} phase between 49 and 82 °C while $\mathbf{7}_{12}$ forms Cub_{bi} and Col phases. The formation of Cub_{bi} structures was confirmed by reconstructing a 3D electron density map based on synchrotron X-ray diffraction patterns, and the 3D electron density map revealed the formation of 3D-branched ionic nanochannels and a sheath domain Fig. 6a [19]. The ionic conductivities of the Cub_{bi} LC materials are presented in Fig. 6b. Upon heating, the ionic conductivities of $\mathbf{7}_{10}$ (○) increased with the temperature up to its isotropization temperature Fig. 6b. The highest ionic conductivity achieved for $\mathbf{7}_{10}$ is 7.7×10^{-5} S cm^{-1} at 80 °C. Subsequently, the conductivity decreased abruptly during the phase transition of $\mathbf{7}_{10}$ to its isotropic state. This behavior is attributed to the breakup of the 3D nanochannel networks. These results demonstrate the efficiency of the 3D bicontinuous channel structure in facilitating ion transportation in the material. On the other hand, the ionic conductivities of $\mathbf{7}_{12}$ (△) increased at the Cr–Cub_{bi} transition at 32 °C but decreased at the Cub_{bi}–Col_h transition at 49 °C Fig. 6b. For example, the conductivity of $\mathbf{7}_{10}$ at 48 °C in the Cub_{bi} phase is 1.0×10^{-5} S cm^{-1}, while its conductivity at 52 °C in the Col_h phase is 2.2×10^{-6} S cm^{-1}. It means that the ionic conductivities of $\mathbf{7}_{12}$ in the Cub_{bi} phase are higher than those in the polydomain Col_h phase. The higher ionic conductivity of the Cub_{bi} phase is attributed to its 3D nanochannel network structure. In the Cub_{bi} phase, ions can migrate efficiently through the ion channels, even in the presence of structural defects, because the 3D-branched channels in the Cub_{bi} phase form alternative pathways that bypass structural defects. However, in the Col_h phase, the transportation of ions may be blocked by structural defects, such as channel ends at the domain boundaries, because the ions cannot bypass the defects owing to the straightness of the 1D channels.

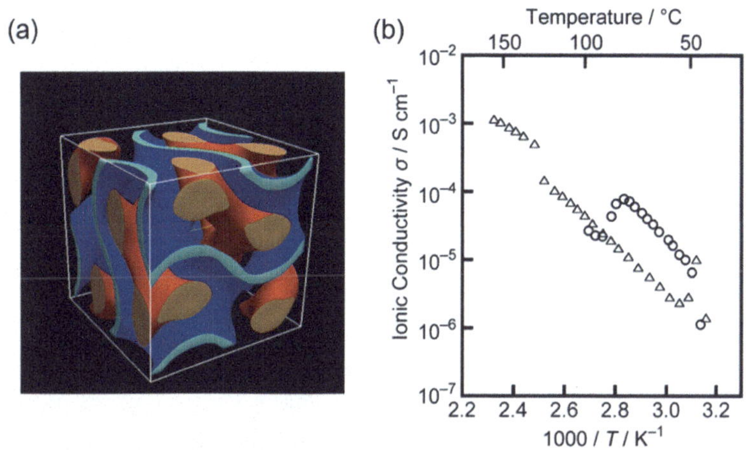

Fig. 6 **a** Electron density map ρ obtained from small-angle X-ray diffraction intensities of a powder pattern of $\mathbf{7}_{10}$ in the $Ia\bar{3}d$ Cub_{bi} phase at 20 °C [19]. Copyright 2012 American Chemical Society. **b** Arrhenius plots of the ionic conductivities of $\mathbf{7}_{10}$ (circles) and $\mathbf{7}_{12}$ (triangles) [18]

4.1 D Nanochannel Membranes

The interest in reverse osmosis and nanofiltration membranes has been increasing because of the rapid increase in the demand for purified water on a global scale. As one of the promising materials for constructing such membranes, nanostructured liquid crystals have garnered significant attention. Figure 7 shows molecular structures of nanostructured liquid crystals exhibiting lyotropic or thermotropic Cub_{bi} phases. Gin et al. designed a series of polymerizable gemini-type ionic molecules **9–10** that form lyotropic Cub_{bi} phases in the presence of a suitable amount of water [20–22]. Because the polymer films obtained from **9** could not withstand high pressures, supported crosslinked Cub_{bi} membranes were prepared using a microporous, hydrophilic, and polyethylene-fiber support. The supported crosslinked Cub_{bi} membranes have uniform pores of 0.75 nm diameter and can be used as effective water nanofiltration membranes. Kato et al. developed nanostructured membranes using polymerizable ionic liquid crystal **11**, which forms a thermotropic Cub_{bi} LC phase [23]. The obtained membranes exhibited salt-rejection properties and unique ion selectivity [24]. As these Cub_{bi} LCs provide 3D interconnected nanochannels, the need to control the LC domains is eliminated. The mechanisms of the salt-rejection properties and unique ion selectivity of these materials have been examined using state-of-the-art technologies [25] and computer simulations [26]. Recently, these types of nanostructured LC polymer membranes have been applied in various domains, such as in the removal of viruses [27, 28] and as nanostructured actuators [29].

These nanostructured membranes can be developed using a combination of anisotropic liquid crystals and specific alignment control techniques. For example,

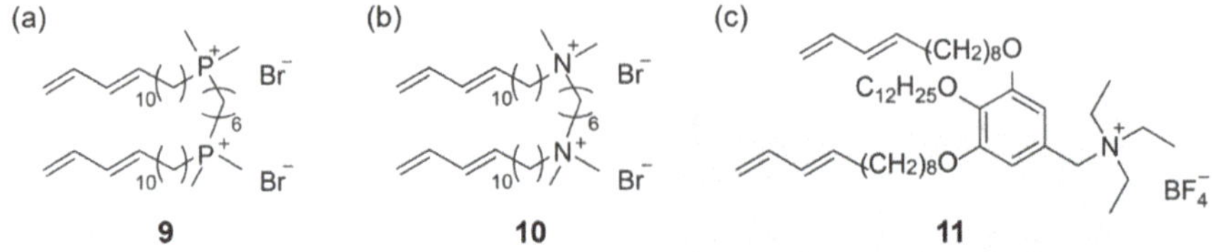

Fig. 7 Molecular structures of ionic liquid crystals that form $\mathrm{Cub_{bi}}$ phases: **a** gemini-type phosphonium salt **9** that forms lyotropic $\mathrm{Cub_{bi}}$ phases [20]; **b** gemini-type ammonium salt **10** that forms lyotropic $\mathrm{Cub_{bi}}$ phases [21, 22]; and **c** wedge-shaped ammonium salt **11** that forms thermotropic $\mathrm{Cub_{bi}}$ phases [23]

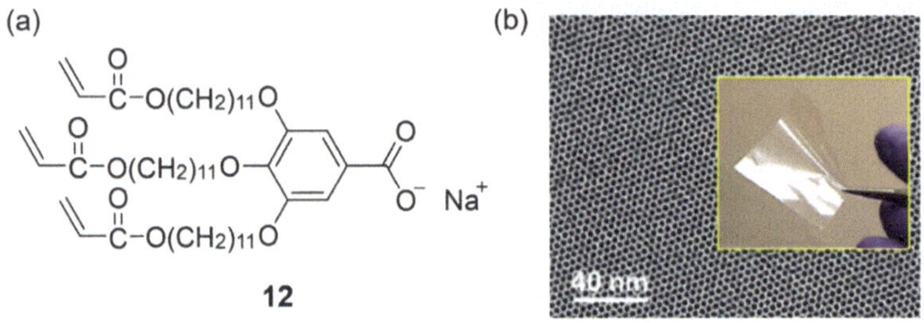

Fig. 8 **a** Molecular structure of wedge-shaped LC monomer **12** consisting of a small hydrophilic head and a large hydrophobic moiety with polymerizable acrylate groups [30]. **b** Photograph and a TEM image of a free-standing crosslinked polymer membrane of **12** with vertically aligned pores [30]. Copyright 2014 American Chemical Society

Osuji et al. [30, 31] created free-standing crosslinked polymer membranes with vertically aligned 1D nanochannels through magnetic field alignment and photo-crosslinking of thermotropic Col_h and water-based lyotropic Col_h phases formed by a wedge-shaped LC monomer, **12** Fig. 8a. Such polymer membranes have a high areal pore density of 2.5×10^{12} cm^2, as directly determined by Transmission Electron Microscopy (TEM) studies Fig. 8b. Nanostructured membranes with continuous-oriented nanochannels are highly attractive as size-selective separation membranes for various applications, including wastewater treatment, water softening, and in the separation of small molecular contaminants.

5 Summary

The molecular design of representative functional ionic liquid crystals is described. Depending on their molecular structures, functional ionic liquid crystals exhibit a variety of nanosegregated liquid-crystalline phases, including smectic, bicontinuous cubic, columnar, and micellar cubic phases. The design of the ionic components of such materials is an important factor in controlling their mesophase pattern and phase-transition temperatures, as well as their specific functions. Ionic liquid crystals exhibit certain anisotropic ionic functions, such as ion conductivity, and by taking advantage of their nanosized channel structures, they can be utilized as separation membranes with specific chemical- and/or size-selectivity. The prospective functions and applications of these ionic liquid crystals are still under investigation, and continued studies are expected to result in innovative ionic liquid crystal candidates with advanced functions and properties.

References

1. K. Binnemans, Ionic liquid crystals. Chem. Rev. **105**, 4148–4204 (2005). https://doi.org/10.1021/cr0400919
2. C.M. Gordon, J.D. Holbrey, A.R. Kennedy, K.R. Seddon, Ionic liquid crystals: Hexafluorophosphate salts. J. Mater. Chem. **8**, 2627–2636 (1998). https://doi.org/10.1039/A806169F
3. J.D. Holbrey, K.R. Seddon, The phase behaviour of 1-alkyl-3-methylimidazolium tetrafluoroborates; ionic liquids and ionic liquid crystals. J. Chem. Soc. Dalton Trans., 2133–2140 (1999). https://doi.org/10.1039/A902818H
4. K. Goossens, K. Lava, C.W. Bielawski, K. Binnemans, Ionic liquid crystals: Versatile materials. Chem. Rev. **116**, 4643–4807 (2016). https://doi.org/10.1021/cr400334b
5. K.V. Axenov, S. Laschat, Thermotropic ionic liquid crystals. Materials **4**, 206–259 (2011). https://doi.org/10.3390/ma4010206
6. L. Douce, J.-M. Suisse, D. Guillon, A. Taubert, Imidazolium-based liquid crystals: A modular platform for versatile new materials with finely tuneable properties and behaviour. Liq. Cryst. **38**, 1653–1661 (2011). https://doi.org/10.1080/02678292.2011.610474
7. S. Chen, S.H. Eichhorn, Ionic discotic liquid crystals. Isr. J. Chem. **52**, 830–843 (2012). https://doi.org/10.1002/ijch.201200046
8. T. Kato, M. Yoshio, T. Ichikawa, B. Soberats, H. Ohno, M. Funahashi, Transport of ions and electrons in nanostructured liquid crystals. Nat. Rev. Mater. **2**, 17001 (2017). https://doi.org/10.1038/natrevmats.2017.1
9. T. Ichikawa, T. Kato, H. Ohno, Dimension control of ionic liquids. Chem. Commun. **55**, 8205–8214 (2019). https://doi.org/10.1039/c9cc04280f
10. N. Kapernaum, A. Lange, M. Ebert, M.A. Grunwald, C. Haege, S. Marino, A. Zens, A. Taubert, F. Giesselmann, S. Laschat, Current topics in ionic liquid crystals. Chem Plus Chem **87**, e202100397 (2022). https://doi.org/10.1002/cplu.202100397
11. E.J.R. Sudhölter, J.B.F.N. Engberts, W.J. de Jeu, Thermotropic liquid-crystalline behavior of some single- and double-chained pyridinium amphiphiles. Phys. Chem. **86**, 1908–1913 (1982). https://doi.org/10.1021/j100207a035
12. D.W. Bruce, S. Estdale, D. Guillon, B. Heinrich, Mesomorphic N-alkylpyridinium dodecylsulphates. Liq. Cryst. **19**, 301–305 (1995). https://doi.org/10.1080/02678299508031983
13. C.J. Bowlas, D.W. Bruce, K.R. Seddon, Liquid-crystalline ionic liquids. Chem. Commun., 1625–1626 (1996). https://doi.org/10.1039/CC9960001625
14. M. Yoshio, T. Mukai, H. Ohno, T. Kato, One-dimensional ion transport in self-organized columnar ionic liquids. J. Am. Chem. Soc. **126**, 994–995 (2004). https://doi.org/10.1021/ja0382516
15. M. Yoshio, T. Ichikawa, H. Shimura, T. Kagata, A. Hamasaki, T. Mukai, H. Ohno, T. Kato, Columnar liquid-crystalline imidazolium salts. Effects of anions and cations on mesomorphic properties and ionic conductivities. Bull. Chem. Soc. Jpn. **80**, 1836–1841 (2007). https://doi.org/10.1246/bcsj.80.1836
16. M. Yoshio, T. Kagata, K. Hoshino, T. Mukai, H. Ohno, T. Kato, One-dimensional ion-conductive polymer films: Alignment and fixation of ionic channels formed by self-organization of polymerizable columnar liquid crystals. J. Am. Chem. Soc. **128**, 5570–5577 (2006). https://doi.org/10.1021/ja0606935

17. B.-K. Cho, A. Jain, S.M. Gruner, U. Wiesner, Mesophase structure–mechanical and ionic transport correlations in extended amphiphilic dendrons. Science **305**, 1598–1601 (2004). https://doi.org/10.1126/science.1100872
18. T. Ichikawa, M. Yoshio, A. Hamasaki, T. Mukai, H. Ohno, T. Kato, Self-organization of room-temperature ionic liquids exhibiting liquid-crystalline bicontinuous cubic phases: Formation of nano-ion channel networks. J. Am. Chem. Soc. **129**, 10662–10663 (2007). https://doi.org/10.1021/ja0740418
19. T. Ichikawa, M. Yoshio, A. Hamasaki, S. Taguchi, F. Liu, X. Zeng, G. Ungar, H. Ohno, T. Kato, Induction of thermotropic bicontinuous cubic phases in liquid-crystalline ammonium and phosphonium salts. J. Am. Chem. Soc. **134**, 2634–2643 (2012). https://doi.org/10.1021/ja209010m
20. M. Zhou, P.R. Nemade, X. Lu, X. Zeng, E.S. Hatakeyama, R.D. Noble, D.L. Gin, New type of membrane material for water desalination based on a cross-linked bicontinuous cubic lyotropic liquid crystal assembly. J. Am. Chem. Soc. **129**, 9574–9575 (2007). https://doi.org/10.1021/ja073067w
21. E.S. Hatakeyama, B.R. Wiesenauer, C.J. Gabriel, R.D. Noble, D.L. Gin, Nanoporous, bicontinuous cubic lyotropic liquid crystal networks via polymerizable Gemini ammonium surfactants. Chem. Mater. **22**, 4525–4527 (2010). https://doi.org/10.1021/cm1013027
22. B.R. Wiesenauer, D.L. Gin, Nanoporous polymer materials based on self-organized, bicontinuous cubic lyotropic liquid crystal assemblies and their applications. Polym. J. **44**, 461–468 (2012). https://doi.org/10.1038/pj.2012.15
23. T. Ichikawa, M. Yoshio, A. Hamasaki, J. Kagimoto, H. Ohno, T. Kato, 3D interconnected ionic nano-channels formed in polymer films: Self-organization and polymerization of thermotropic bicontinuous cubic liquid crystals. J. Am. Chem. Soc. **133**, 2163–2169 (2011). https://doi.org/10.1021/ja106707z
24. M. Henmi, K. Nakatsuji, T. Ichikawa, H. Tomioka, T. Sakamoto, M. Yoshio, T. Kato, Self-organized liquid-crystalline nanostructured membranes for water treatment: Selective permeation of ions. Adv. Mater. **24**, 2238–2241 (2012). https://doi.org/10.1002/adma.201200108
25. R. Watanabe, T. Sakamoto, K. Yamazoe, J. Miyawaki, T. Kato, Y. Harada, Ion selectivity of water molecules in subnanoporous liquid-crystalline water-treatment membranes: A structural study of hydrogen bonding. Angew. Chem. Int. Ed. **59**, 23461–23465 (2020). https://doi.org/10.1002/anie.202008148
26. Y. Ishii, N. Matubayasi, G. Watanabe, T. Kato, H. Washizu, Molecular insights on confined water in the nanochannels of self-assembled ionic liquid crystal. Sci. Adv. **7**, eabf 0669 (2021). https://doi.org/10.1126/sciadv.abf0669
27. T. Sakamoto, K. Asakura, N. Kang, R. Kato, M. Liu, T. Hayashi, H. Katayama, T. Kato, Development of liquid-crystalline smectic nanoporous membranes for the removal of SARS-CoV-2 and waterborne viruses. J. Mater. Chem. A **11**, 22178–22186 (2023). https://doi.org/10.1039/D3TA02705H
28. D. Kuo, T. Sakamoto, S. Torii, M. Liu, H. Katayama, T. Kato, Removal of viruses from their cocktail solution by liquid-crystalline water-treatment membranes. Polym. J. **54**, 821–825 (2022). https://doi.org/10.1038/s41428-022-00631-1
29. S. Cao, C. Liu, M. Yoshio, Ionic electroactive PEDOT: PSS/liquid-crystalline polymer electrolyte actuators: Photopolymerization of zwitterionic columnar liquid crystals complexed with a protic ionic liquid. Mater. Chem. Front. **7**, 2828–2838 (2023). https://doi.org/10.1039/D3QM00192J
30. X. Feng, M.E. Tousley, M.G. Cowan, B.R. Wiesenauer, S. Nejati, Y. Choo, R.D. Noble, M. Elimelech, D.L. Gin, C.O. Osuji, Scalable fabrication of polymer membranes with vertically aligned 1 nm pores by magnetic field directed self-assembly. ACS Nano **8**, 11977–11986 (2014). https://doi.org/10.1021/nn505037b
31. X. Feng, S. Nejati, M.G. Cowan, M.E. Tousley, B.R. Wiesenauer, R.D. Noble, M. Elimelech, D.L. Gin, C.O. Osuji, Thin polymer films with continuous vertically aligned 1 nm pores fabricated by soft confinement. ACS Nano **10**, 150–158 (2016). https://doi.org/10.1021/acsnano.5b06130

Functional Liquids Based on Organic Materials

Kyosuke Isoda

1 Functional Soft Materials Based on Organic Molecules

1.1 π-Conjugated Molecules and their Solid-State Materials

Since the early twenty-first century, significant advancements in synthetic organic chemistry, analytical chemistry, and methodologies for fabricating electronic devices have developed compounds and materials tailored for organic electronics. The core skeleton of organic electronics is functional soft material, including polymers, liquid crystals, surfactants, and gels, primarily comprising π-conjugated molecules. In the field of organic electronics, these compounds can provide applications in bulk materials such as Organic Light-Emitting Diodes (OLEDs), Organic Field-Effect Tansistors (OFETs), and organic photovoltaic cells. Notably, recent years have seen the commercialization of products applied in displays and cell phones based on OLED technology, signifying the maturation and widespread adoption of organic electronics.

The key compounds in organic electronics are π-conjugated molecules, composed of sp or sp^2 hybrid orbitals and p orbitals occupied by π electrons, whose features are closely associated with the Highest Occupied Molecular Orbital (HOMO) and the Lowest Occupied Molecular Orbital (LUMO) levels around frontier orbitals. The band gap energy between HOMO and LUMO levels corresponds to the visible-light region, with the nature of the π electrons influencing luminescence, charge transport properties, magnetism, and more.

K. Isoda
Department of Applied Chemistry, Faculty of Science and Engineering, Toyo University, Kawagoe, Saitama, Japan

K. Isoda (✉)
Organic Materials Chemistry Group, Sagami Chemical Research Institute (SCRI), Ayase, Kanagawa, Japan
e-mail: k-isoda@sagami.or.jp; isoda032@toyo.jp

M. Funahashi (ed.), *Functional Soft Matter II*,
https://doi.org/10.1007/978-981-92-0100-6_14

There are two main strategies for controlling the physical properties and characteristics of π-conjugated molecules: synthetic organic chemistry and self-assembly. In the former, carbon-atom-based π-conjugated molecules can adjust HOMO and LUMO levels through structural modifications, polymerization to extend conjugation, and introduction of hetero elements (B, N, O, S, P, etc.). This method can directly tune a molecule's properties. Conversely, self-assembly stacks π-conjugated molecules to create overlap between p orbitals, which is crucial for device properties like OFETs. High performance necessitates precise molecular alignment at the nanolevel to enhance the transfer integral between p orbitals of one another. In OLEDs, maintaining molecular dispersion and preventing aggregation is vital for improving color purity of emitted colors.

As mentioned, molecular design for bulk materials in the solid state must consider both the molecular character and the expected aggregate structure. Additionally, achieving high reproducibility is crucial for demonstrating homogeneous properties during thin film fabrication. However, completely suppressing the formation of crystal polymorphisms, defects, and grain boundaries remains challenging. Consequently, functional liquid materials have emerged in recent years as alternatives to solid materials, offering the advantage of maintaining physical properties in the bulk state without significant changes [1–6].

1.2 Liquid Materials

Functional liquid materials refer to fluids that exhibit liquid or supercooled liquid properties at room temperature (around 25 °C). Among these, ionic liquids are the most typical examples, composed of cationic and anionic species in Fig. 1 [7]. Due to their low vapor pressure and specific polarity, they find utility not only as reaction solvents and separation materials in organic synthesis, but also as electrolytes with ionic conductivity, thus serving as battery materials. In contrast, π-conjugated liquid materials have emerged in the twenty-first century of which, unlike ionic liquids, are non-ionic liquids in Fig. 1 [1–6]. Traditionally, liquefaction of π-conjugated molecules posed challenges due to their strong intermolecular interactions. However, through appropriate chemical modifications, it is now feasible to maintain π-conjugated molecules in a liquid state at room temperature.

π-Conjugated liquid materials offer distinct advantages over solid-state bulk materials, including their versatile, reproducible, and defect-free usability, anytime, anywhere, and by anyone. Moreover, their utilization without organic solvents aligns with environmentally Sustainable Development Goals (SDGs), garnering significant attention [8–10]. This section predominantly delves into liquid materials comprising π-conjugated frameworks, elucidating molecular design principles, analytical methodologies for fluidic properties, and their diverse applications in organic electronics. Topics covered include dimensionality, luminescent properties, stimulus-responsive behaviors, and the integration of liquid materials with computational chemistry.

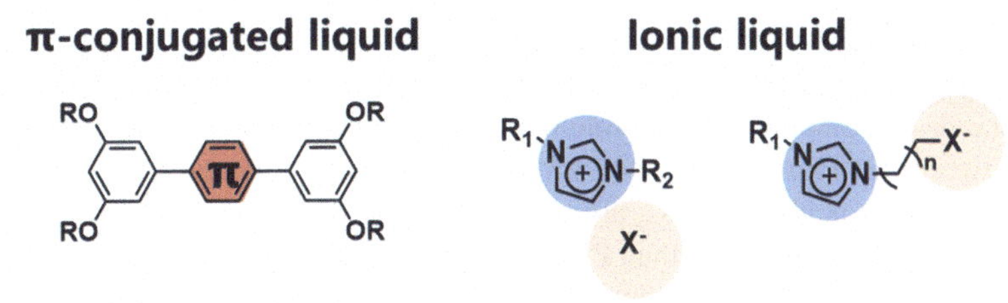

Fig. 1 Typical molecular structures of *p*-conjugated liquids and ionic liquids

1.3 Molecular Design Guidelines toward Functional Liquid Based on π-Conjugated Framework at Room Temperature

To advance the development of functional liquid materials comprising π-conjugated frameworks, it is crucial to suppress intermolecular interactions among the π-conjugated molecules. This method requires the incorporation of sterically bulky and fluidic properties such as branched alkyl chains, linear alkyl chains, and oxyethylene chains into the π-conjugated framework. Practical examples illustrating this concept are depicted in Fig. 2.

The following are some examples for organic liquid materials. Pyromellitic diimide (**PMDI**) is a well-known π-conjugated molecule recognized for its high chemical and thermal stability, planarity, and cohesive strength. These properties have led to the successful development of polyimides for engineering plastics in both academic research and industry. This study explores the impact of introducing linear and branched alkyl chains to **PMDI** on its thermal properties in Fig. 3a. Linear alkyl chain-appended **PMDI** (**PMDI-C8**) exhibits a significant decrease in melting point (m.p.) to 180 °C, compared to polyimide with a glass transition temperature (T_g) of 400 °C. Also, **PMDI-C6C4**, **PMDI-C8C6**, and **PMDI-C10C8** with branched alkyl chains show m.p. in the range of 40–70 °C, lower than **PMDI-C8** with linear

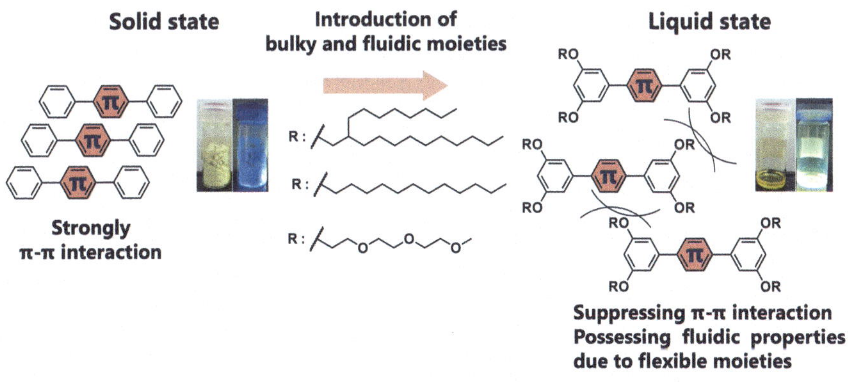

Fig. 2 Practical examples to design and develop π-conjugated liquid materials

alkyl chains in Fig. 3a [11]. These findings indicate that both types of alkyl chains effectively suppress intermolecular interactions among π-conjugated molecules, with molecular design playing a crucial role in lowering the m.p. For example, the difference between the m.p. and b.p. for **PMDI-C8** is approximately 6 °C, while for **PMDI-C10C8**, it is about 34 °C, allowing it to adopt the supercooled liquid state over a wider temperature range than **PMDI-C8** in Fig. 3b. Moreover, mixing multiple materials proves to be an effective method for lowering the m.p. **PMDI-MIX1**, a 1:1 (mol) mixture of **PMDI-C8C6** and **PMDI-C10C8**, exhibits the m.p. of 26 °C and the f.p. of 11.2 °C. This is attributed to the inclusion of an "impurity" with a different alkyl chain geometry, which inhibits crystallization and lowers both temperatures.

(a)

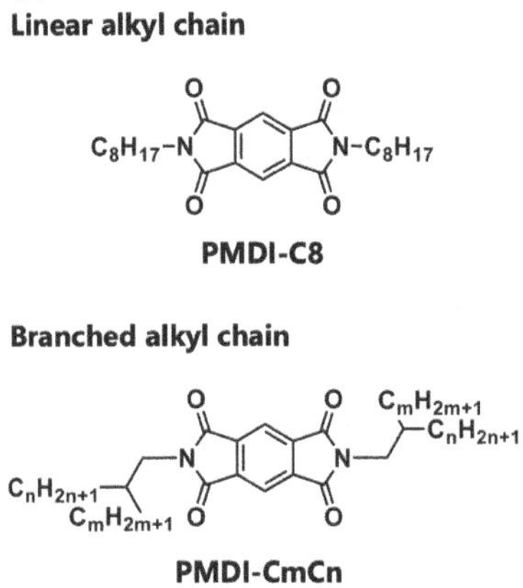

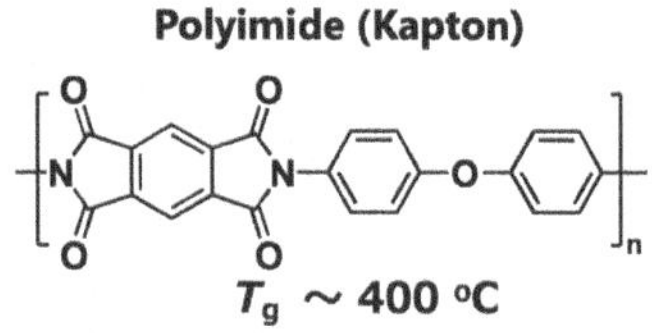

Compounds		m.p. / °C	f.p. / °C
PMDI-C6C4	m = 6, n = 4	66.8	47.3
PMDI-C8C6	m = 8, n = 6	38.9	23.9
PMDI-C10C8	m = 10, n = 8	50.8	6.4
PMDI-MIX1 (C8C6/C10C8 = 1:1 (mol))	-	26.0	11.2
PMDI-C8	-	179.3	172.8

(b)

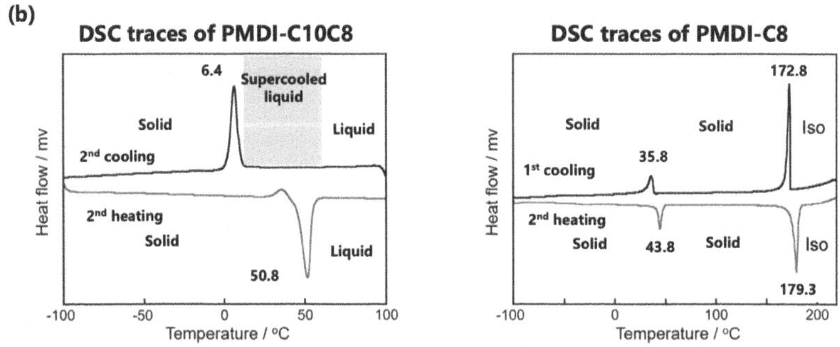

Fig. 3 **a** Molecular structures of **PMDI** derivatives. DSC traces of **b PMDI-C10C8** and **c PMDI-C8** at 10 K/min

1.4 Analytical Methods

The fluidic properties of liquid materials can be assessed through various analytical techniques including thermal analysis, microscopic observation, structural analysis, and rheological measurements. Thermal analysis, particularly using Differential Scanning Calorimetry (DSC), allows for the measurement of m.p., f.p., and T_g, although caution is needed due to the influence of sweep rate variability. Optical Microscopic (OM) and Polarized OM (POM) observations, coupled with a hot stage, enable the detection of birefringence from aggregates and facilitate in situ observation of state changes with temperature variation. Structural analysis, often conducted by *X*-Ray Diffraction (XRD), can investigate the sample's state based on the presence or absence of diffraction peaks originating from aggregate structures. Rheological measurements assess the storage and loss moduli to characterize solid and liquid properties across various frequencies and temperatures. Additionally, spectroscopic measurements such as UV-*vis* absorption and luminescent spectra can be performed on samples coated on quartz or KF plates, like solid materials.

2 Functional Liquid Materials

Recently, various molecular designs have been proposed for functional liquid materials. As discussed above, liquid materials incorporated with π-conjugated molecules achieve fluidity by suppressing π-π interactions through the incorporation of bulky and fluidic substituents. This essence can suppress intermolecular interaction between π-conjugated cores, making molecules fluidic. While this molecular design poses a drawback for their use as carrier-transporting materials, it offers the advantage of enabling the creation of homogeneous coating films with excellent reproducibility, mitigating defects and grain boundaries in solid materials. Additionally, lower-dimensional liquid materials with independent core structures have been developed, featuring cores with cage structures that possess discrete spaces in a liquid state. These studies illustrate the diverse dimensionality of liquid materials, ranging from discrete materials devoid of intermolecular interactions between cores to materials that exhibit functionality while maintaining some level of interaction.

Moreover, stimuli-responsive liquid materials are emerging as a counterpoint to the stimuli-responsive solid materials that have attractive attention in recent years. Stimuli-responsive solid materials can undergo significant changes in properties such as luminescence color before and in response to external stimuli like pressure, shear, heat, light, vapor, humidity, and others. However, they face an inherent problem in that it is difficult to maintain homogeneous luminescence properties due to crystal polymorphism formation during thin-film fabrication. Conversely, stimuli-responsive liquid materials provide reproducible, defect-free usability, accessible anytime and anywhere. Some of these materials not only change color or luminescence color in response to external stimuli but also induce and amplify Circularly

Polarized Luminescence (CPL). Furthermore, the development of stimuli-responsive liquid materials through combining computational and experimental science has been underway in recent years and will be discussed herein.

2.1 Organic Electronics

In 1999, Hendrickx et al. reported on the ellipsometric determination of the electric-field-induced birefringence of the photorefractive dye 2-dihexylamino-7-(dicyanomethylidene)-3,4,4a,5,6-pentahydronaphthalene (**DHADC-MPN**) in a liquid carbazole derivative, **Cz** in Fig. 4 [12]. Although **DHADC-MPN** has been dispersed in polymer materials such as polyvinylcarbazole (**PVK**) for measurement, it has also been successfully measured using liquid **Cz** media with a lower T_g. Subsequently, Wada et al. reported on the hole mobility in liquid **Cz**, of which hole mobility is estimated to be 4×10^6 cm^2/Vs measured by a time-of-flight experiment upon the application of an electric field of 2.5×10^5 V/cm, of which mobility was more than that of **PVK** (6×10^7 cm^2/Vs). The charge-transporting mobility in liquid **Cz** is enhanced due to both a larger transfer integral between carbazole cores and changes in the distribution of the excimer trapping sites [13].

Furthermore, liquid **Cz** derivatives have been developed for OLED applications. Adachi et al. firstly reported on OLEDs utilizing liquid **Cz** as a light-emitting layer,

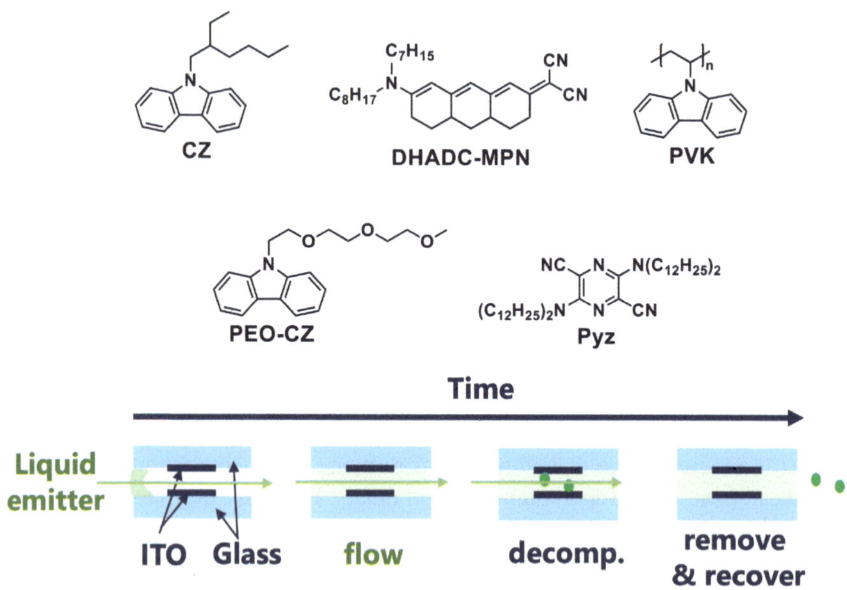

Fig. 4 Molecular structures of liquid materials applied in organic electronics (top) and concept for liquid OLEDs (bottom)

which are expected to possess their flexible and degradation-free characteristics [14, 15]. Liquid OLEDs inherently prevent detachment or peeling of the liquid-emitting layer from the electrodes. Liquid OLEDs can restore emission properties by replacing a liquid emitter with a fresh one without disassembling the device structure in Fig. 4. Additionally, incorporating a liquid carbazole tethered to a poly (ethylene oxide) chain (**PEO-Cz**) can enhance carrier injection compared to devices utilizing **Cz** alone. This improvement is attributed to the high polarity of **PEO-Cz**, allowing it to be doped with tetrabutylammonium chloride as an electrolyte. Consequently, the external electroluminescence quantum efficiency in the liquid **PEO-Cz** system is 25 times higher than that in the liquid **Cz** one without an electrolyte [16]. In addition, Matsumoto et al. in 2022 reported a liquid pyrazine (**Pyz**) with a fluorescence quantum yield as high as 0.59. Regarding OLED applications, efficient yellow electroluminescence was observed at 593 nm. The device exhibited a 10 V turn-on voltage and reached a maximum brightness of 336 cd/m^2 at 14 V [17].

Professors Robert Carl (Rice University), Richard Smoley (Rice University), and Harold Kroat (Sussex University) were awarded the Nobel Prize in Chemistry in 1996 for their discovery of fullerene. The application of fullerene, which has been a hot topic in the early twentieth century, particularly, to liquid materials have been reported by Nakanishi et al. in Fig. 5. Rheological measurements have confirmed that liquid fullerene **C60,** based on fulleropyrrolidine substituted with a 2,4,6-tris (alkyloxy) phenyl, exhibits liquid behavior, indicated by a loss modulus exceeding the storage modulus at 25 °C. The frequency dependence of liquid **C60** also suggests the presence of multiple components with long relaxation times in each sample, potentially indicating cluster formation and intricate dynamics. CV measurements have also shown that it exhibits redox properties even in the liquid state. Furthermore, **C60** possesses the high hole mobility, reaching 0.03 cm^2 V^{-1} s^{-1} at 20 °C by using time-of-flight measurement. This value is comparable to those of methanofullerene derivatives in polymer blends [18] as well as smectic liquid crystalline phases of organic-conjugated oligomers [19].

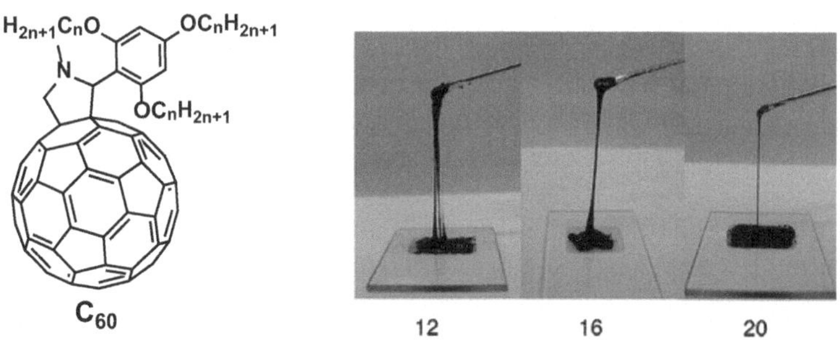

Fig. 5 Molecular structure of C60 and photographs of C60 derivatives having various alkyl chains at room temperature. Reproduced with permission from ref. [20]. Copyright 2006 American Chemical Society

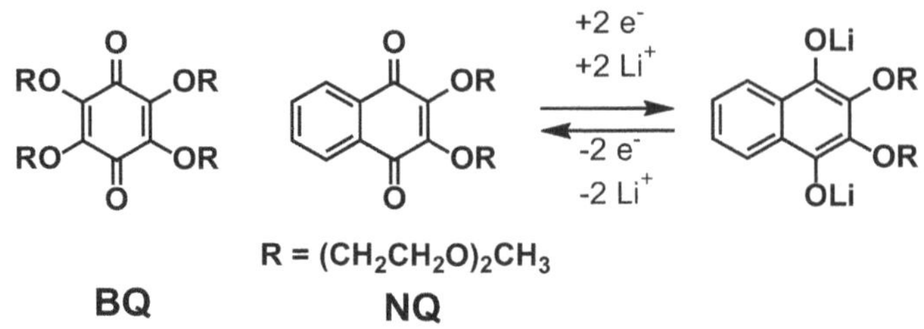

Fig. 6 Molecular structures of liquid quinone and their charge/discharge processes

Yoshida et al. reported redox-active liquid quinones (**BQ**, **NQ**) for application in redox flow batteries in 2017 [21] in Fig. 6. Liquid **BQ** and **NQ** tethering diethylene glycol monomethyl ether groups can dissolve supporting electrolytes. Consequently, Li-ion batteries fabricated using **NQ** with $LiBF_4$/propylene carbonate show favorable performance in terms of voltage, capacity, and energy. This study suggests the potential of solvent-free organic redox flow batteries utilizing liquid redox-active materials.

Porphyrins are compounds found in the pigment components of important biological molecules such as hemoglobin (the pigment in blood), cytochromes (respiratory pigments), chlorophyll (the pigment in plants), and other compounds involved in crucial redox reactions within the body. These characteristics have promoted to develop porphyrin derivatives, which are widely applied in functional electronic materials. In 2010, Maruyama et al. reported on liquid porphyrin (**POR1**) [22]. Spectroscopic measurements indicate the presence of donor-acceptor or π–π interactions between the liquid **POR1** and fullerene, implying the potential utility of this liquid as a dispersant medium for fullerene C_{60} or carbon nanotube. Thereafter, Nakanishi et al. have developed liquid **POR2** [10]. The utilization of a shielded π-conjugated core within liquid **POR2** with trapped charge can provide an unprecedented liquid-electret in Fig. 7a, b. This liquid-electret was primarily devised to show its piezoelectric and electroacoustic functionalities. Shielding the π-core with hydrophobic and insulating bulky alkyl chains facilitated the storage of electric charges within liquid **POR2**. Morisue et al. have developed their own liquid porphyrins (**POR3**, **4**), such as morphological control of fullerene C_{60} [23] and how coordinated metal of core on porphyrin influences antifreezing effect of supercooled liquid and glass in Fig. 7a, c [24].

Furthermore, Shinohara and Nakanishi et al. have achieved the development of liquid materials comprising π-conjugated polymers. The modulation of viscoelastic properties and luminescent colors can be achieved through alterations in the alkyl chain and by selecting π-conjugated units [25]. Also, they have recently reported on the application of polymeric liquids to electrets [26].

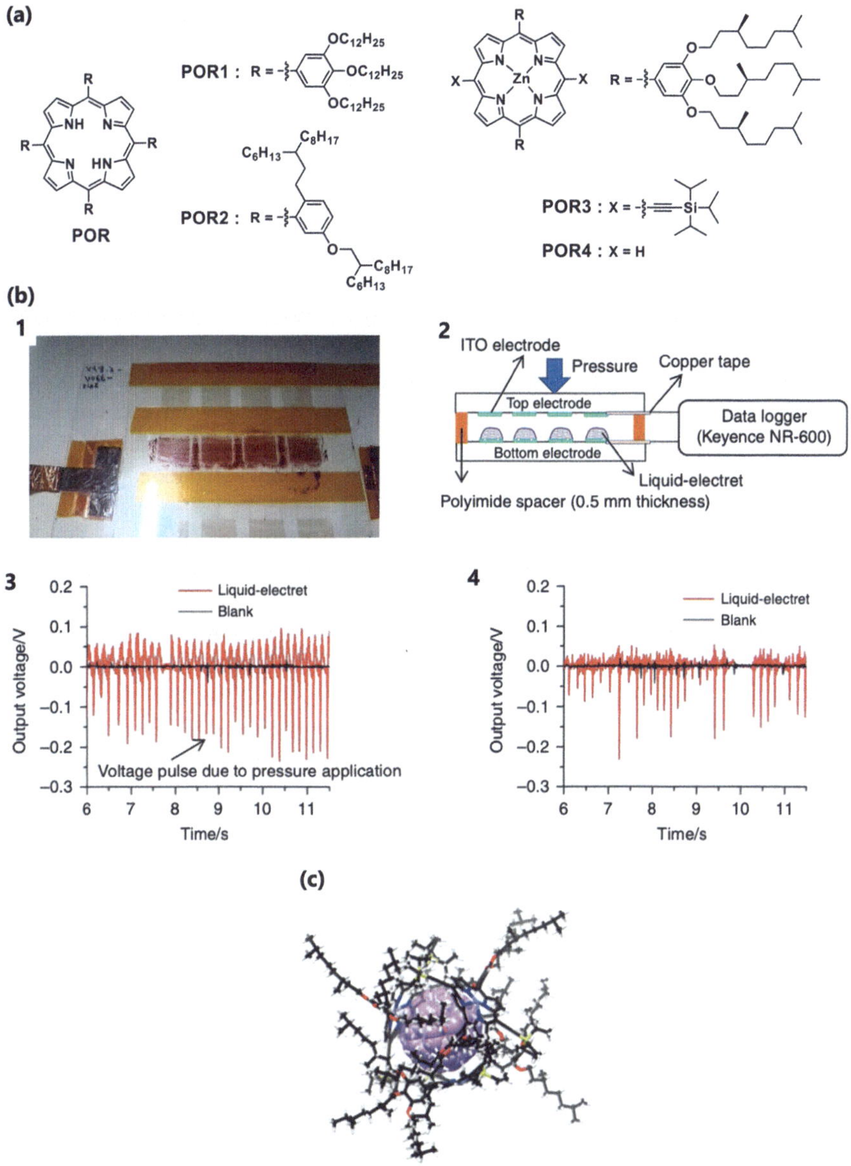

Fig. 7 **a** Molecular structures of liquid porphyrins (PORs). **b** Liquid electret characteristics exhibited by POR2; photograph of the prototype liquid-electret device based on POR2 (b1); schematic diagram of the cross-sectional view of the liquid-electret device (b2). Output voltage characteristics of the liquid-electret device based on POR2 in its initial state (b3) and after 7 days (b4). Reproduced with permission from ref. [10]. Copyright 2019 Springer Nature Publishing AG. **c** Schematic models of POR3/C_{60} produced using the MM+ force field. Reproduced with permission from ref. [23]. Copyright 2020 American Chemical Society

2.2 Molecules Having Cavity

Porous solids such as zeolites [27], Metal–Organic Frameworks (MOFs), [28, 29] and Covalent-Organic Frameworks (COFs) [30, 31] are useful for molecular separation, adsorption, and catalysis, etc., in the fields of inorganic chemistry, coordination chemistry, organic chemistry, and materials science. However, a porous solid can be limited by its intrinsic nature of the solid. The value of these materials stands to be significantly augmented through the broadening of their utilization scope and application environments. Therefore, materials' designed to liquefy porous solids have attractive attention. In 2007, James et al. in Queen's University Belfast reported on the concept of porous liquids, [32] which can be classified into liquids with permanent porosity, different from porous solids composed of continuously connected spaces in Fig. 8. However, the above porous liquid based on zeolite does not have a completely independent space. Thus, liquid materials with independently individual spaces have been developed in recent years, incorporating cyclodextrin, calixarene, pillar arene, and cage as core frameworks.

Ogoshi et al. have engineered a liquid material at room temperature through the chemical modification of pillar [5] arene and cyclodextrin in Fig. 9. These liquids can behave as host molecules capable of encapsulating various substrates [33, 34]. For instance, pillar [5] arene type liquids tethered with PEG can form rotaxane structures, where alkyl chains penetrate the cavity within [33]. In addition, XRD measurements for alkyl-penetrated **Pillar [5] arene** indicated a structural liquid with a periodic structure. Furthermore, it was found that pillar [6] arene transforms from a structural liquid to a solid with a lamellar structure upon exposure to cyclohexane vapor, and these behaviors are reversible [34].

James et al. have developed the rigid organic iminospherand cage (**Cage 1**) as a liquid material with independent space [35]. This report shows a significant decrease in the melting point by the introduction of alkyl chains. **Cage 1** exhibits the m.p. at ca 50 °C upon heating and adopts a supercooled liquid upon cooling, whereas the cage without alkyl chains has the melting point > 300 °C. Similar behavior is observed in rheological measurements, where the shear modulus (Pa) for the loss modulus surpasses that of the storage one above 50 °C. Additionally, at temperatures above 80 °C, the material behaves as a Newtonian liquid [35]. Furthermore,

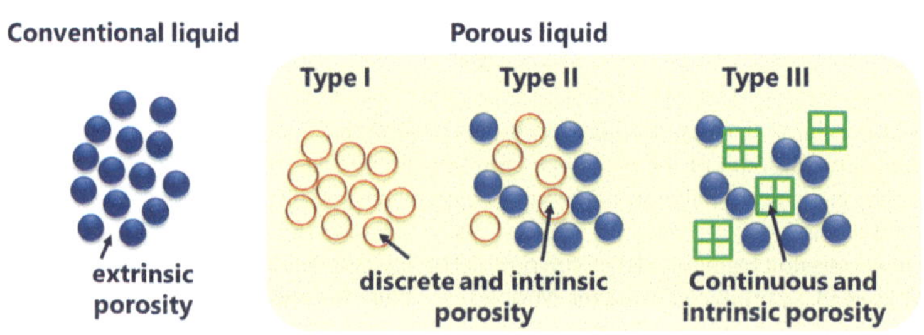

Fig. 8 Concept of porous liquids

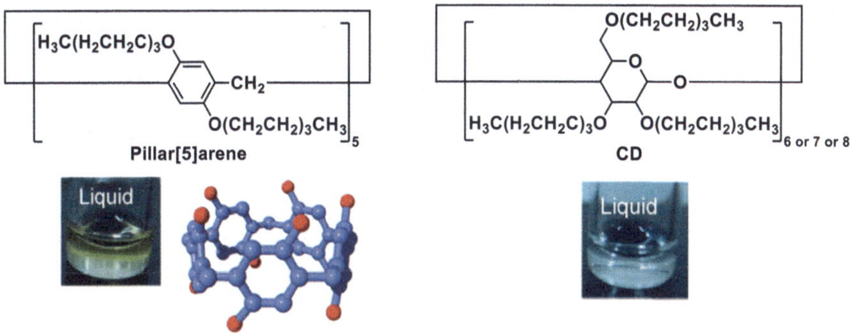

Fig. 9 Liquid materials based on Pillar [5] arene (left) and cyclodextrin (CD) (right). Reproduced with permission from ref. [33]. Copyright 2012 American Chemical Society

James et al. developed a **Cage 2** substituted with a crown ether instead of an alkyl chain for **Cage 2** in 2015 [36]. **Cage 1** can adopt a liquid state at room temperature when mixed with crown ether, and Molecular Dynamics (MD) simulations have verified that the molecule possesses independent space in solution, which includes gas molecules such as N_2, CH_4, and CO_2. Moreover, the resulting gas-loaded **Cage 2** can selectively include compounds via size-dependent recognition. As elucidated earlier, liquid materials based on solid porous materials, by enhancing their fluidic properties, hold promise for diverse applications across various fields in the future. In addition, the development of MOF [37–39] and cage [40] molecules has been reported in recent years, and further development is expected in the future.

To date, liquid materials have predominantly involved compounds developed through organic synthesis. However, there have also been reports on the liquefaction of biomolecules such as proteins. Mann et al. have prepared a liquid material based on Ferritin, an iron-binding protein complex comprising 24 proteins, where there is a cage-like structure in a globular protein complex in Fig. 10. The process involves Ferritin being cationized through reaction with *N*, *N*-dimethyl-1,3-propanediamine (**DMPA**) to afford cationized Ferritin (**C-Fn**). Subsequently, **C-Fn/S** is prepared through complexing cationized Ferritin with an anionic surfactant (**S**). DSC measurements have validated that the phase transition of **C-Fn/S** from a solid state to a liquid one occurs at approximately 30 °C.

2.3 *Photofunction*

With the advancements in optoelectronics, luminescent dyes have emerged as crucial materials. Until now, they exist in a solid state due to their composition of π-conjugated molecules; however, recent years have seen the emergence of room-temperature liquid materials in this field. These liquid materials not only possess inherent luminescent properties but also serve as dispersible media for organic and

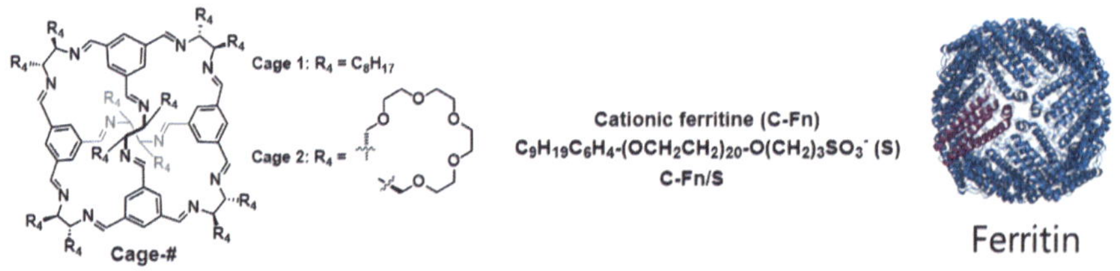

Fig. 10 Molecular structures of Cage-1, Cage-2, and C-Fn/S, and structure of mouse ferritin complex 1 lb3. Reproduced with permission from ref. [41]. Copyright 2003 Springer

inorganic substrates. The preparation of a homogeneous state in mixing solid materials presents significant challenges, whereas homogeneous composites based on liquids as media can show unique properties distinct from both liquids and substrates, thus paving the way for novel composite material development.

In 2012, Nakanishi et al. reported liquid oligophenylenevinylene (**OPV**), which uses a partial framework of poly (phenylenevinylene) as a well-known luminescent polymer in Fig. 11a [42]. The inherent blue emission for **OPV** can be turned into a white emission by incorporating rubrene or Aluminum tris (8-hydroxyquinolinate) (**Alq**$_3$), offering potential applications as a security ink. Subsequently, Nakanishi et al. devised liquid anthracene (**AN1**), enabling control of emission color through adjusting the quantity of **EP-AN** added in Fig. 11b. Moreover, the introduction of **Eu** into the mixture can engineer a thermo-responsive liquid material [43]. At lower temperatures, the energy transfer from the mixture to **Eu** dominantly occurs to show red emission derived from **Eu**, of which efficiency can be controlled upon heating. Also, liquid anthracene (**AN2**), reported by Yanai et al., can disperse **POR-Pt**, which, under excitation by the green laser at 532 nm, can be converted to 433 nm as blue emission, meaning photon upconversion in Fig. 11c [44]. This is because that two triplet excited states derived from **POR-Pt** as donor transfer or migrate within the singlet-state acceptor **AN2**, both of which collide with each other to generate a single excited state of the acceptor. The engagement of the energy migration process in Triplet-Triplet Annihilation (TTA) offers a novel perspective for designing advanced photon upconversion systems in academic and industrial research.

In recent years, various room-temperature phosphorescent materials have been reported with studies extending to solid-state materials in Fig. 12. Babu et al. presented findings on a bromonaphthalimide derivative (**NDI1**) to show room-temperature phosphorescence, which can be tuned by the addition of aldehyde derivatives [45]. Au et al. have reported on a liquid material (**P4OH**) as a super-cooled liquid, achieved by incorporating a hydroxy group at the terminus of the alkyl chain. These liquids exhibit green phosphorescence [46]. Tani et al. have reported that an asymmetrical 1,2-diketone with both brominated thiophene and furan forms an organic RTP liquid (**Diketone**). The crystallization of **Diketone** can be restricted by pricking or shearing stress to attain a supercooled liquid state for several months [47]. Other liquid mixtures have been developed to facilitate Charge Transfer (CT) and exciplex formation. Babu et al. reported on solvent-free organic

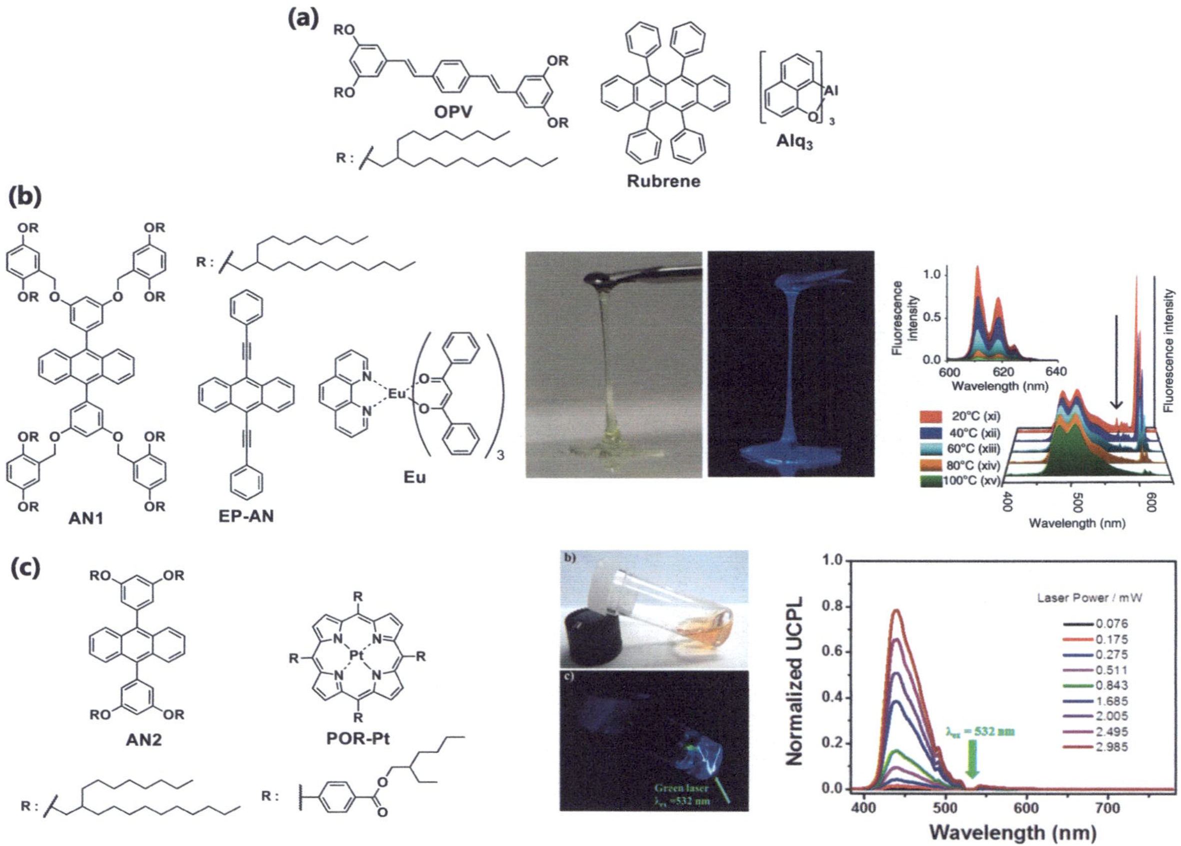

Fig. 11 **a** Molecular structure of OPV liquid. **b** Molecular structure of AN1, its photographs under room and black light, and fluorescent spectra. Reproduced with permission from ref. [43]. Copyright 2003 Springer Nature Publishing AG. **c** Molecular structure of AN2, its photographs under room and black light, and fluorescent spectra. Reproduced with permission from ref. [44]. Copyright 2013 American Chemical Society

liquid hybrids (**NDI2** and **NA**) demonstrating stable donor-acceptor interaction. The complex formation by using two liquid molecules was elucidated through ^{1}H NMR measurements, as well as NOESY and ROESY analyses. Additionally, CT liquid exhibited stable behavior even in column chromatography [48]. Furthermore, Babu et al. have currently advanced these discoveries to engineer liquid materials capable of showcasing donor-acceptor type exciplex emission and room-temperature phosphorescence. [49]. Isoda et al. have reported on a homogeneous solution prepared by dissolving naphthalene in **PMDI**, generating exciplex emission [11]. They also found that naphthalene molecules in **PMDI** solution sublimate or evaporate with a smaller activation energy than naphthalene molecules alone. This suggests that naphthalene molecules are homogeneously dissolved in **PMDI**. In the field of Circularly Polarized Light luminescent (CPL) materials, which have much attentions in contemporary research, their advancements have been made in liquid materials. Bai et al. have notably achieved CPL expression through the incorporation of an optically active alkyl chain into naphthalene (**NA2**) [50]. Further, Takaishi et al. have reported on expressing CPL based on liquid BINAP (**BNOL**) with axial chirality as the backbone skeleton [51]. Other liquid materials based on tetraphenylethene with aggregate-induced emission [52] and liquid thiaziazole derivatives [53] have also been reported.

Furthermore, in recent years, stimulus-responsive liquid materials that change their liquid properties upon external stimuli have begun to be reported in Fig. 13.

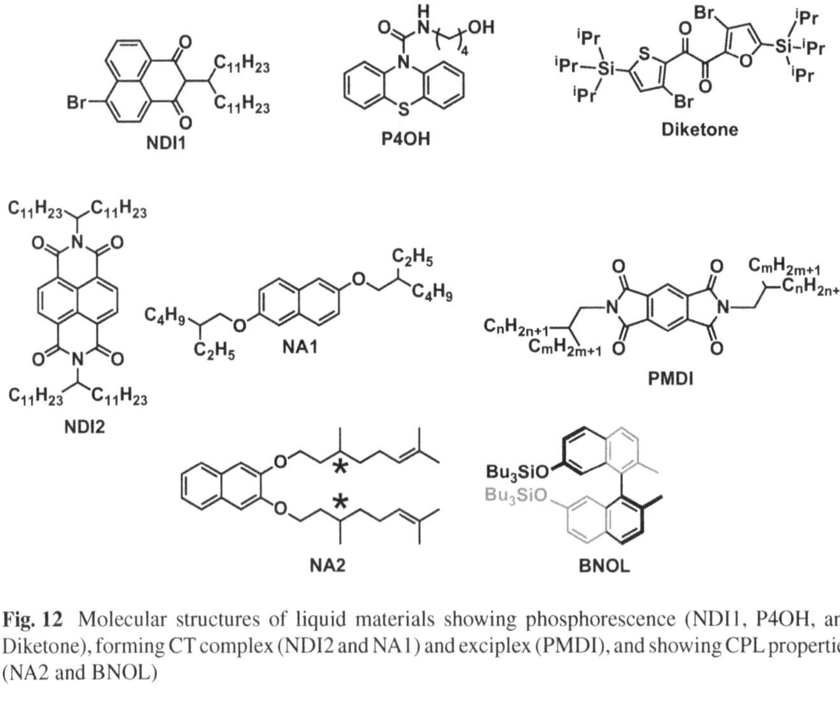

Fig. 12 Molecular structures of liquid materials showing phosphorescence (NDI1, P4OH, and Diketone), forming CT complex (NDI2 and NA1) and exciplex (PMDI), and showing CPL properties (NA2 and BNOL)

Kim et al. reported on molecular design of a thermally stable supercooled liquid based on diketopyrrolopyrrole (**DPP**), and shear-triggered crystallization showing dramatic optical property changes [54]. Upon shearing, **DPP** transforms to its crystal phase with a 25-fold increase in photoluminescence quantum efficiency and luminescent color change from red to yellow. Also, they elucidated shear-triggered crystallization by living-cell attachment. The obtained results reveal that the threshold shear stress value of **DPP** (0.90 kPa) with crystallization is comparable to the traction force range exerted by fibroblast cells per unit area (1–5 kPa). This observation suggests that the shear-triggered crystallization phenomenon exhibits a sensitivity approximately six orders of magnitude higher than the typical mechanochromic organic materials. Koga et al. have reported on supercooled liquid-to-crystal transformation induced by shearing, based on *N, N*-R-phenyl-7-amino-2,4-trifluoromethylquinoline derivatives (**TFMQ**). Isoda et al. reported a supercooled liquid based on thiophene-appended quinoxaline (**ThioQ**) and the possibility of its application in rewritable media. **ThioQ** showed a dramatic change in its photoluminescent color upon the transformation from liquid (yellow emission) to solid (green emission) state, of which phenomena were investigated by photoluminescent spectra as well as lifetime decay profiles [55].

Bardeen et al. reported on supercooled liquid-to-solid transformation of stable organic radical 2,2,6,6-tetramethyl-1-piperidine-1-oxyl (**TEMPO**) by continuous-wave (cw) and pulsed lasers at 532 nm [57]. Morita et al. reported on the sodium salt with polycyclic trioxotriangulene (**TOT**) anion liquid which shows halochromism upon exposure to HCl vapor in Fig. 14 [58].

Isoda et al. reported on liquid materials based on quinoxaline frameworks in Fig. 15 [55, 59–67]. Thiophene-appended liquid (**monoThio**) can change photoluminescent colors dependent upon the p*K*a of the Brønsted acid dissolved in **monoThio** [60]. Also, both asymmetric-structured liquid (**ThioBz**) tethering both thiophene and benzene rings can function as a site-selective acid sensing liquid [64]. For **ThioBz**, HCl molecule dominantly interacts with imino N atoms neighboring to phenyl ring, which is evaluated by single crystal *X*-ray analysis for protonated **ThioBz'** (**ThioBz' + H**$^+$). Single crystal *X*-ray analysis of **ThioBz' + H**$^+$ revealed that S atom in thiophene ring intramolecularly interacts with imino-N atom via through-space interaction, resulting in **ThioBz** is capable of the site-selective acid sensing. Also, quinoxaline-based liquids with two phenyl rings (**BzL**) and two pyridine rings (**PyL**) show similar photoluminescent spectra and change emission colors from light blue to green upon the exposure to HCl vapor [63]. Further, protonated **BzL** (**BzL + H**$^+$) can recover the initial emission color via deprotonation over time under ambient atmosphere, whereas the protonated state of **PyL** (**PyL + H**$^+$) can be maintained. This is due to the difference in p*K*a of conjugated acids between protonated **BzL + H**$^+$ and **PyL + H**$^+$.

Phenanthrene-fused quinoxaline liquid (***rac*-Phen**) can vary the luminescent color, morphology, and rheological property via a stimulus such as HCl vapor, reported by Isoda et al. [62] in Fig. 16. Also, the control of exposure time to HCl vapor can form wrinkle structures to aggregated structures with birefringence in Fig. 16b. Therefore, Ishiyama and Isoda et al. carried out the MD simulation to investigate how molecules aligned at the molecular level. ***Rac*-Phen** attains the disordered state

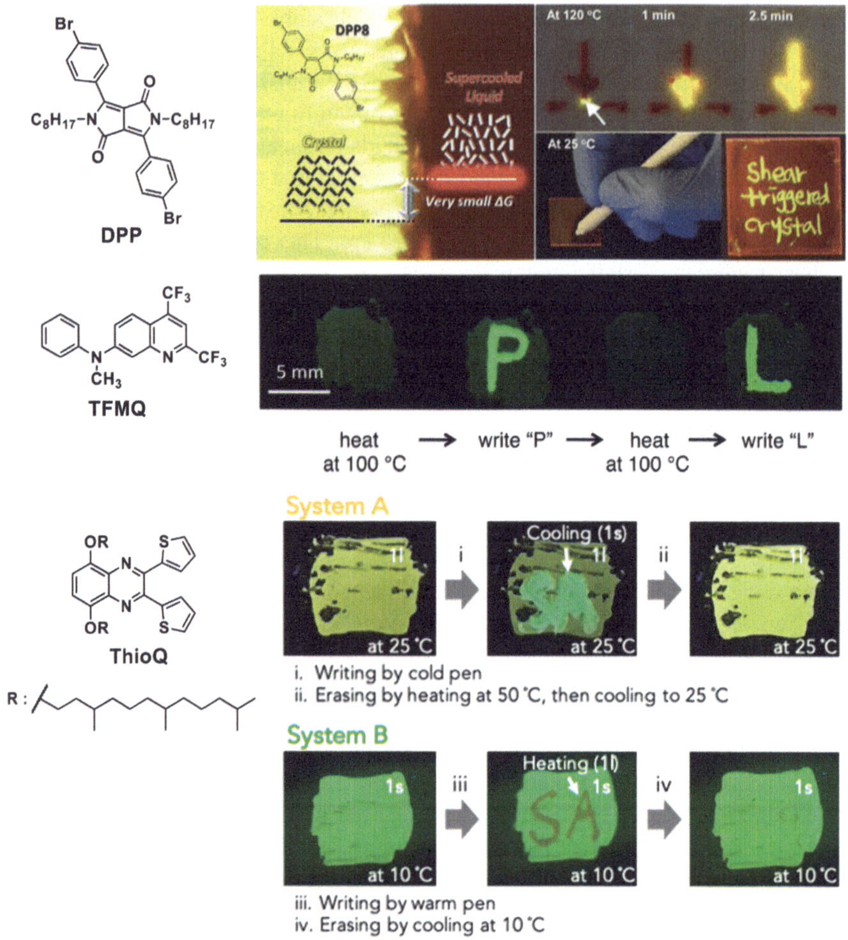

Fig. 13 Molecular structures of liquids DPP, TFMQ, and ThioQ and their stimuli-responsive behavior. Reproduced with permission from ref. [54–56]. Copyright 2014, 2015, and 2021 American Chemical Society

after 1000 ns, whereas protonated ***rac*-Phen** (***rac*-Phen-H⁺**) self-organized into the ordered structures in Fig. 16c. In the ordered structures, nanosegregation occurs between hydrophilic π-conjugate framework with HCl and hydrophobic alkyl chains. Based on these results, Ishiyama and Isoda et al. designed and developed ***R*-Phen** and ***S*-Phen** having optically active moieties in Fig. 17. For a theoretical investigation, the MD simulations revealed that ***R*-Phen** and ***S*-Phen** molecules were randomly dispersed, but protonated ***R*-Phen + H^+** and ***S*-Phen + H^+** exposed to HCl vapor were induced to self-organize, leading to supramolecular chirality through the influence of the chiral alkyl chains in Fig. 17a, b, c. On the other hand, the experimental investigations showed the same behavior as the above results in Fig. 17. ***R*-Phen**

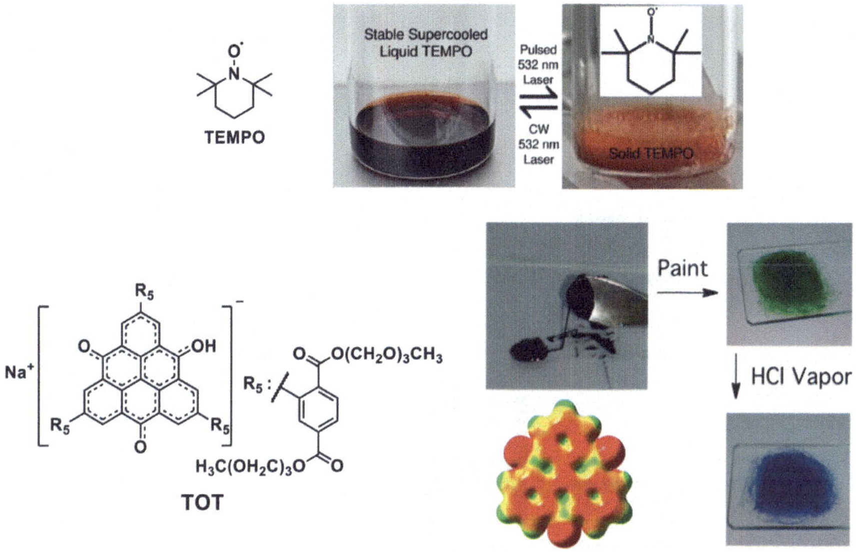

Fig. 14 Molecular structures of liquid TEMPO and TOT and their stimuli-responsive behavior. Reproduced with permission from ref. [57, 58]. Copyright 2019 and 2024 American Chemical Society

and ***S*-Phen** molecules adopt isotropic states. ***R*-Phen + H^+** and ***S*-Phen + H^+** are induced to show circular dichroism (CD) in Fig. 17d and CPL spectra in Fig. 17e with the mirror image between ***R*-Phen + H^+** and ***S*-Phen + H^+**.

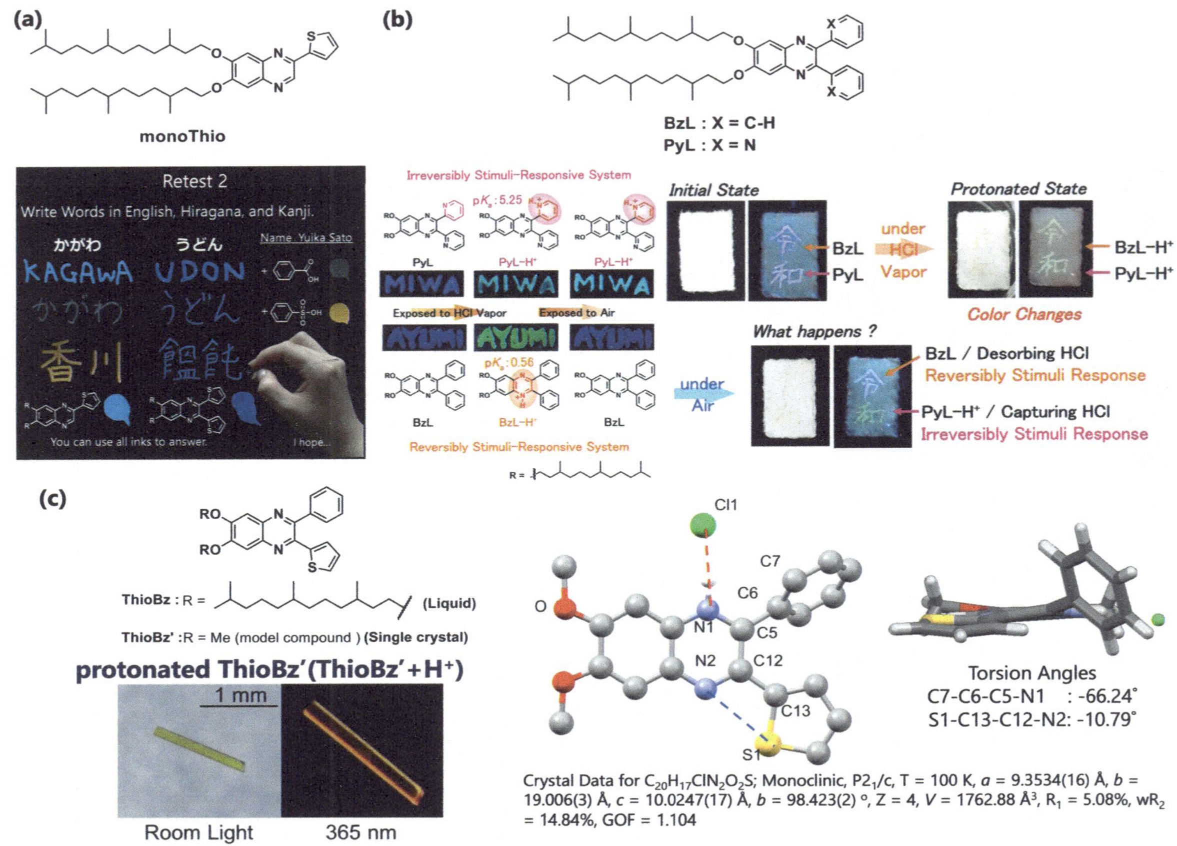

Fig. 15 Molecular structures and their application for **a** monoThiol and **b** BzL and PyL. Reproduced with permission from ref. [63]. Copyright 2019 The Royal Chemical Society. **c** Molecular structure of ThioBz+H$^+$ as model compound, photographs of single crystal under room and black light, and the result of single-crystal *X*-ray analysis. Reproduced with permission from ref. [64]. Copyright 2021 Elsevier

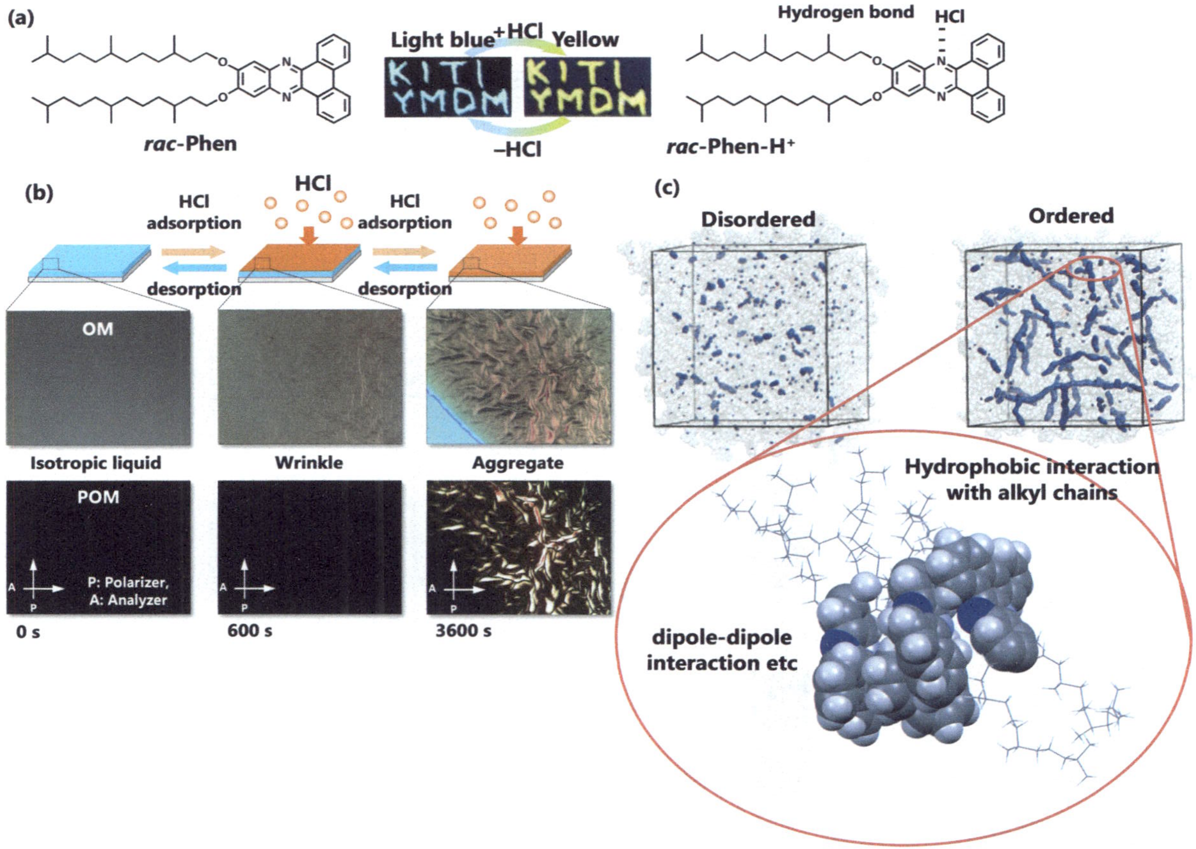

Fig. 16 **a** Molecular structure of *rac*-Phen and emission color changes upon exposure to HCl vapor. **b** OM and POM observations of *rac*-Phen after exposure to HCl vapor. **c** MD simulations of *rac*-Phen (left) and protonated *rac*-Phen-H^+ (light). Reproduced with permission from ref. [62]. Copyright 2019 American Chemical Society

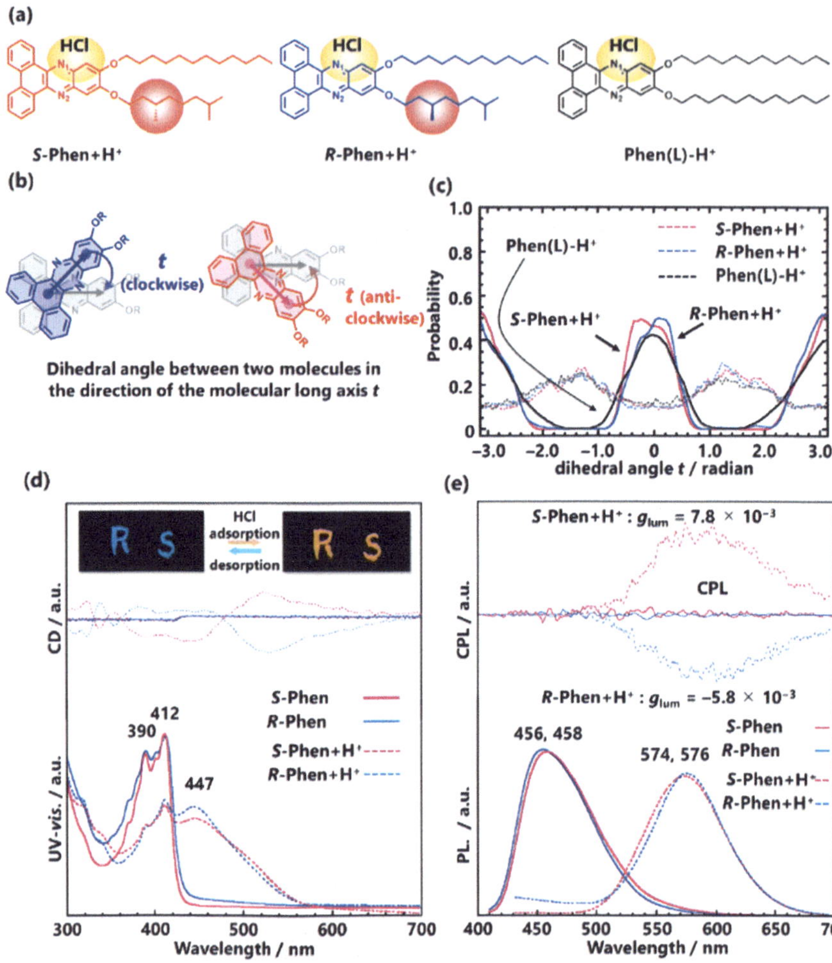

Fig. 17 **a** Molecular structures of protonated *S*-Phen+H$^+$ and *R*-Phen+H$^+$, **b** the torsion angle τ between two vectors for the dimer, **c** probability distributions of the stacking structures, **d** photographs under black light and UV-*vis* absorption and CD spectra, and **e** photoluminescent and CPL spectra of *S*-Phen, *R*-Phen, *S*-Phen+H$^+$, and *R*-Phen+H$^+$. Reproduced with permission from ref. [65]. Copyright 2019 American Chemical Society

References

1. T. Nakanishi, *Functional Organic Liquids* (Wiely-VCH Verlag Gmbh & Co, KGaA, 2019)
2. A. Ghosh, T. Nakanishi, Chem. Commun. **53**, 10344–10357 (2017)
3. F. Lu, T. Nakanishi, Adv. Opt. Mater. **7**, 1900176 (2019)
4. V.C. Wakchaure, G. Channareddy, S.S. Babu, Acc. Chem. Res. **57**, 670–684 (2024)
5. T. Machida, T. Nakanishi, J. Mater. Chem. C **9**, 10661–10667 (2021)
6. A. Tateyama, T. Nakanishi, Responsive Mater. **1**, e20230001 (2023)
7. M. Armand, F. Endres, D.R. Mac Farlane, H. Ohno, B. Scrosati, Nat. Mater. **8**, 621–629 (2009)

8. C.J. Clarke, W.-C. Tu, O. Levers, A. Bröhl, J.P. Hallett, Chem. Rev. **118**, 747–800 (2018)
9. I.T. Horváth, Chem. Rev. **118**, 369–371 (2018)
10. A. Ghosh, M. Yoshida, K. Suemori, H. Isago, N. Kobayashi, Y. Mizutani, Y. Kurashige, I. Kawamura, M. Nirei, O. Yamamuro, T. Takaya, K. Iwata, A. Saeki, K. Nagura, S. Ishihara, T. Nakanishi, Nat. Commun. **10**, 4210 (2019)
11. Y. Tanabe, H. Tsutsui, S. Matsuda, S. Shikita, T. Yasuda, K. Isoda, Chem Photo Chem **7**, e202200287 (2023)
12. E. Hendrickx, B.D. Guenther, Y. Zhang, J.F. Wang, K. Staub, Q. Zhang, S.R. Marder, B. Kippelen, N. Peyghambarian, Chem. Phys. **245**, 407–415 (1999)
13. J.C. Ribierre, T. Aoyama, T. Muto, Y. Imase, T. Wada, Org. Electron. **9**, 396–400 (2008)
14. D. Xu, C. Adachi, Appl. Phys. Lett., 95 (2009)
15. S. Hirata, K. Kubota, H.H. Jung, O. Hirata, K. Goushi, M. Yahiro, C. Adachi, Adv. Mater. **23**, 889–893 (2011)
16. T. Iwanaga, R. Tanaka, S. Toyota, Chem. Lett. **43**, 105–107 (2014)
17. J.-Y. Lee, T. Aoyama, M. Uchiyama, S. Matsumoto, Dyes Pigments **174**, 108030 (2020)
18. S.M. Tuladhar, D. Poplavskyy, S.A. Choulis, J.R. Durrant, D.D.C. Bradley, J. Nelson, Adv. Funct. Mater. **15**, 1171–1182 (2005)
19. M. Funahashi, J.-I. Hanna, Adv. Mater. **17**, 594–598 (2005)
20. T. Michinobu, T. Nakanishi, J.P. Hill, M. Funahashi, K. Ariga, J. Am. Chem. Soc. **128**, 10384–10385 (2006)
21. A. Shimizu, K. Takenaka, N. Handa, T. Nokami, T. Itoh, J.-I. Yoshida, Adv. Mater. **29**, 1606592 (2017)
22. S. Maruyama, K. Sato, H. Iwahashi, Chem. Lett. **39**, 714–716 (2010)
23. M. Morisue, G. Saito, D. Sasada, T. Umeyama, H. Imahori, K. Mitamura, H. Masunaga, T. Hoshino, S. Sakurai, S. Sasaki, Langmuir **36**, 13583–13590 (2020)
24. M. Morisue, M. Nakamura, Y. Miyake, Y. Kashiwagi, S. Watase, T. Hoshino, H. Masunaga, S. Sakurai, S. Sasaki, Cryst. Growth Des. **23**, 6747–6755 (2023)
25. A. Shinohara, C. Pan, Z. Guo, L. Zhou, Z. Liu, L. Du, Z. Yan, F.J. Stadler, L. Wang, T. Nakanishi, Angew. Chem. Int. Ed. **58**, 9581–9585 (2019)
26. A. Shinohara, M. Yoshida, C. Pan, T. Nakanishi, Polym. J. **55**, 529–535 (2023)
27. P. A. Wright, *Microporous Framework Solids* (The Royal Society of Chemistry, 2007)
28. A.K. Cheetham, G. Férey, T. Loiseau, Angew. Chem. Int. Ed. Engl. **38**, 3268–3292 (1999)
29. S. Kitagawa, R. Kitaura, S.-i. Noro, Angew. Chem. Int. Ed. **43**, 2334–2375 (2004)
30. A.P. Côté, A.I. Benin, N.W. Ockwig, M. O'Keeffe, A.J. Matzger, O.M. Yaghi, Science **310**, 1166–1170 (2005)
31. H.M. El-Kaderi, J.R. Hunt, J.L. Mendoza-Cortés, A.P. Côté, R.E. Taylor, M. O'Keeffe, O.M. Yaghi, Science **316**, 268–272 (2007)
32. N. O'Reilly, N. Giri, S.L. James, Chemistry—A Eur. J. **13**, 3020–3025 (2007)
33. T. Ogoshi, T. Aoki, R. Shiga, R. Iizuka, S. Ueda, K. Demachi, D. Yamafuji, H. Kayama, T.-a. Yamagishi, J. Am. Chem. Soc. **134**, 20322–20325 (2012)
34. T. Ogoshi, K. Maruyama, Y. Sakatsume, T. Kakuta, T.-a. Yamagishi, T. Ichikawa, M. Mizuno, J. Am. Chem. Soc. **141**, 785–789 (2019)
35. N. Giri, C.E. Davidson, G. Melaugh, M.G. Del Pópolo, J.T.A. Jones, T. Hasell, A.I. Cooper, P.N. Horton, M.B. Hursthouse, S.L. James, Chem. Sci. **3**, 2153–2157 (2012)
36. N. Giri, M.G. Del Pópolo, G. Melaugh, R.L. Greenaway, K. Rätzke, T. Koschine, L. Pison, M.F.C. Gomes, A.I. Cooper, S.L. James, Nature **527**, 216–220 (2015)
37. D. Umeyama, S. Horike, M. Inukai, T. Itakura, S. Kitagawa, J. Am. Chem. Soc. **137**, 864–870 (2015)
38. L. He, J. Chen, B. Cui, H. Yuan, F. Wang, Y. Wang, Y. Yu, T. Lee, Li, J. Am. Chem. Soc. **141**, 19708–19714 (2019)
39. A. Koutsianos, R. Pallach, L. Frentzel-Beyme, C. Das, M. Paulus, C. Sternemann, S. Henke, Nat. Commun. **14**, 4200 (2023)
40. C. He, Y.-H. Zou, D.-H. Si, Z.-A. Chen, T.-F. Liu, R. Cao, Y.-B. Huang, Nat. Commun. **14**, 3317 (2023)

41. T. Granier, B. Langlois d'Estaintot, B. Gallois, J.-M. Chevalier, G. Précigoux, P. Santambrogio, P. Arosio, JBIC J. Biol. Inorg. Chem. **8**, 105–111 (2003)
42. S. Santhosh Babu, J. Aimi, H. Ozawa, N. Shirahata, A. Saeki, S. Seki, A. Ajayaghosh, H. Möhwald, T. Nakanishi, Angew. Chem. Int. Ed. **51**, 3391–3395 (2012)
43. S.S. Babu, M.J. Hollamby, J. Aimi, H. Ozawa, A. Saeki, S. Seki, K. Kobayashi, K. Hagiwara, M. Yoshizawa, H. Möhwald, T. Nakanishi, Nat. Commun. **2013**, 4 (1969)
44. P. Duan, N. Yanai, N. Kimizuka, J. Am. Chem. Soc. **135**, 19056–19059 (2013)
45. A. Goudappagouda, V.C. Manthanath, K.C. Wakchaure, T. Ranjeesh, K. Das, T. Vanka, S.S. Nakanishi, Babu, Angew. Chem. Int. Ed. **58**, 2284–2288 (2019)
46. Z. Xu, Z. Wang, W. Yao, Y. Gao, Y. Li, H. Shi, W. Huang, Z. An, Angew. Chem. Int. Ed. **62**, e202301564 (2023)
47. M. Komura, T. Ogawa, Y. Tani, Chem. Sci. **12**, 14363–14368 (2021)
48. V.C. Wakchaure, L.V. Pillai, K.C. Goudappagouda, S. Ranjeesh, S. Chakrabarty, P.R. Ravindranathan, S.S. Rajamohanan, s, Chem. Commun. **55**, 9371–9374 (2019)
49. V.C. Wakchaure, S.D. Veer, A.D. Nidhankar, R. Goudappagouda, K. Nayak, S. Asokan, S.S. Ravindranathan, Babu, Chem. Commun. **58**, 1998–2001 (2022)
50. X. Bai, Y. Sun, Y. Jiang, G. Zhao, J. Jiang, C. Yuan, M. Liu, Angew. Chem. Int. Ed. **60**, 3745–3751 (2021)
51. K. Takaishi, T. Matsumoto, M. Kawataka, T. Ema, Angew. Chem. Int. Ed. **60**, 9968–9972 (2021)
52. T. Machida, R. Taniguchi, T. Oura, K. Sada, K. Kokado, Chem. Commun. **53**, 2378–2381 (2017)
53. V. Dyadishchev, D.O. Balakirev, N.K. Kalinichenko, E.A. Svidchenko, N.M. Surin, S.M. Peregudova, V.G. Vasilev, O.Y. Shashkanova, A.V. Bakirov, S.A. Ponomarenko, Y.N. Luponosov, Dyes Pigments **224**, 112003 (2024)
54. M.S. Chung, B.M. Kwon, A.G. Leung, M.S. Wong-Foy, J. Kim, S. Kim, J. Takayama, A.J. Gierschner, J.K. Matzger, A.C.S. Central, Science **1**, 94–102 (2015)
55. Y. Sato, S. Mutoh, N. Morishita, K. Tsurumachi, Isoda, J. Phys. Chem. Lett. **12**, 3014–3018 (2021)
56. R. Karasawa, Y. Hagihara, N. Abe, J.-I. Harada, N.K. Todo, Cryst. Growth Des. **14**, 2468–2478 (2014)
57. B. Rodriguez, K. Lam, T.B. Anwar, C.J. Bardeen, ACS Omega **9**, 11266–11272 (2024)
58. S. Enozawa, H. Ukai, T. Ito, Y. Murata, Morita, Org. Lett. **21**, 2161–2165 (2019)
59. Y. Isoda, D. Sato, Matsukuma, Chemistry Select **2**, 7222–7226 (2017)
60. Y. Sato, D. Mutoh, M. Matsukuma, T. Nakagawa, K.I. Kawai, Chem.—Asian J. **13**, 2619–2625 (2018)
61. A. Isoda, Ikenaga, Crystals **9**, 51 (2019)
62. T. Isoda, Y. Ishiyama, D. Mutoh, Matsukuma, ACS Appl. Mater. Interfaces **11**, 12053–12062 (2019)
63. M. Isoda, A. Matsubara, Y. Ikenaga, Y. Akiyama, Mutoh, J. Mater. Chem. C **7**, 14075–14079 (2019)
64. Y. Akiyama, K. Mutoh, Isoda, J. Photochem. Photobiol. A Chem. **407**, 113036 (2021)
65. Y. Ikenaga, T. Akiyama, M. Ishiyama, K. Gon, Y. Tanaka, K. Chujo, Isoda, ACS Appl. Mater. Interfaces **13**, 47127–47133 (2021)
66. Y. Isoda, Orita, Anal. Sci. **37**, 469–473 (2021)
67. Isoda, J. Synth. Org. Chem., Japan **81**, 105–114 (2023)

Functional Gels

Masafumi Yoshio

1 Introduction

Gels are a unique class of soft materials, consisting of a continuous solid framework dispersed within a liquid medium. This unique combination results in a semi-solid, jelly-like substance that exhibits characteristics bridging the properties of both solids and liquids. With their distinctive mechanical, rheological, and stimuli-responsive properties, gels are widely utilized across various fields, including drug delivery, tissue engineering, cosmetics, food products, and materials science. A broad spectrum of fluids can be immobilized within gels, ranging from common solvents like water and organic solvents to mineral oil, silicone oil, fluorinated solvents, metal alkoxides, ionic liquids, and liquid crystals.

This chapter focuses on supramolecular gels, which are formed through the self-assembly of low-molecular-weight organic molecules known as gelators [1–13]. These gelator molecules typically self-assemble into nanofibers with diameters ranging from tens to hundreds of nanometers, driven by reversible non-covalent intermolecular interactions, such as hydrogen bonding, halogen bonding, van der Waals forces, π-π interactions, charge-transfer interactions, and electrostatic interactions. The entanglement of these nanofibers creates a three-dimensional network within solvents, effectively entrapping the liquids within this network (Fig. 1). These physically crosslinked gels exhibit thermoreversible behavior, wherein changes in temperature can disrupt or restore the gel structure. They transition from a gel state to a more fluid-like sol state upon heating and revert to a gel state upon cooling. The topic of supramolecular gels has been extensively covered in numerous outstanding books and reviews [1–13]. This chapter explores the fundamental design of gelators and the development of electronic and optical functional gels.

M. Yoshio (✉)
Research Center for Macromolecules and Biomaterials, National Institute for Materials Science, Tsukuba, Ibaraki, Japan
e-mail: yoshio.masafumi@nims.go.jp

M. Funahashi (ed.), *Functional Soft Matter II*,
https://doi.org/10.1007/978-981-92-0100-6_15

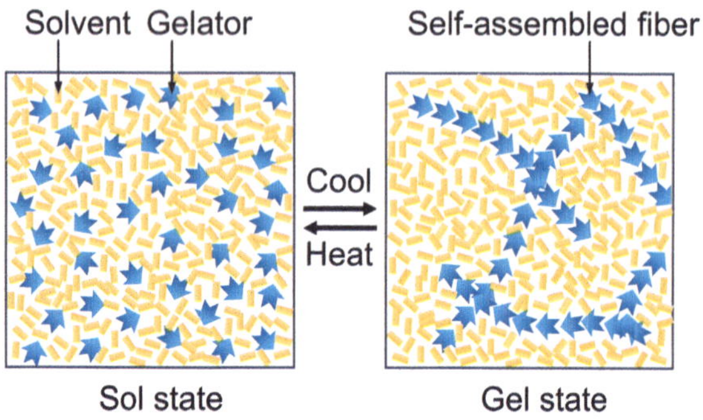

Fig. 1 Illustration of the self-assembly of low-molecular-weight gelators into an entangled fiber network through non-covalent interactions

2 Design of Gelators

A typical low-molecular-weight gelator incorporates structural motifs such as cholesterol, peptides, sugar, cyclohexyl amine, cyanurate, and gallate moieties. These motifs are connected by long hydrocarbon chains through amide and urea bonds (Fig. 2). Additionally, halogen bonds—formed between halogen atoms and other nucleophilic atoms with lone electron pairs, like the nitrogen in pyridine—can be utilized in gelator design [13]. Aromatic cores that promote π-π stacking can impart properties like luminescence, conductivity, and redox-activity to the gels. Furthermore, integrating metal-coordinating pyridyl ligands, photoreactive moieties—such as photocrosslinkable groups like acrylate, diacetylene, coumarin, or photochromic units like azobenzene—can enhance the catalytic capabilities of gelators [14] and enable the tuning of gel mechanical properties. Notably, some aromatic molecules such as 2,3-bis-*n*-decyloxyanthracene [15] and amphiphilic π-conjugated molecules containing oligoether chains [16, 17] can form π-stacked gels even without hydrogen-bonding linkers. The balance of molecular interactions between gelators and solvents, along with solvent characteristics like polarity and viscosity, is a crucial factor in the gelation process.

3 Conductive Gels

Discrete conductive molecular wires and networks have garnered significant attention due to their potential applications in electronic devices, sensors, and neural signal harvesting [6, 10]. In this context, π-conjugated gelators have emerged as promising candidates. A range of π-conjugated molecules, including perylene bisimides, oligothiophenes, oligophenylene vinylenes, triphenylenes, triphenylamines, porphyrins, hexabenzocoronenes, and fullerenes, have been incorporated into gelator designs. Their π-stacking interactions facilitate the formation of self-assembled nanofibers, nanoribbons, and nanotubes, which are beneficial for efficient charge transport

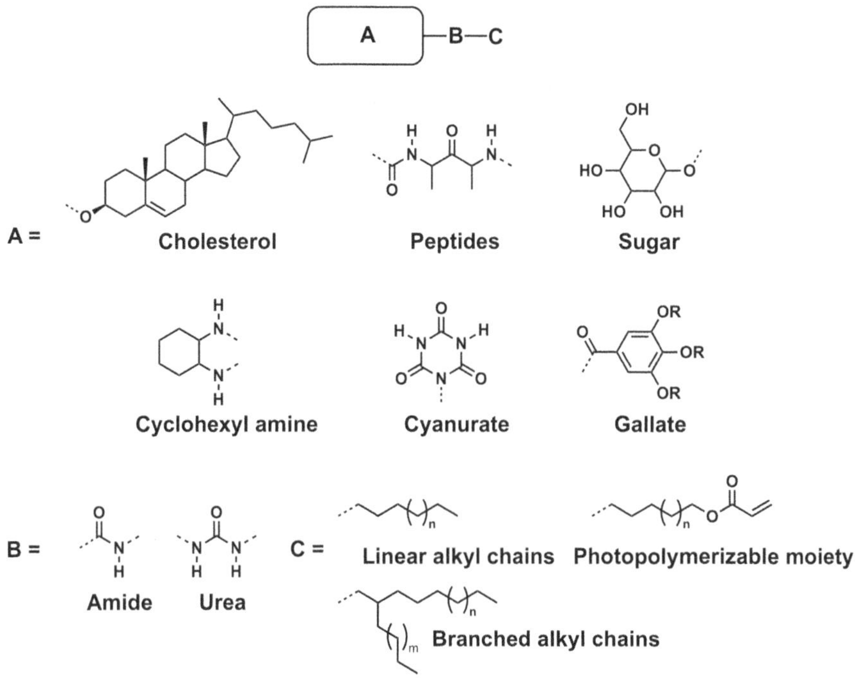

Fig. 2 Design of gelators. **a** represents structural motifs responsible for gelation. **b** indicates the linker connecting these motifs. **c** denotes the functional groups that enhance miscibility with solvents or provide photopolymerization capabilities

through a hopping mechanism. For example, the self-assembly, alignment, and conductive properties of π-conjugated gelators (Fig. 3) have been examined [17–23].

Fukushima, Aida, and their colleagues reported the formation of graphitic nanotubes through the self-assembly of amphiphilic hexabenzocoronene **1** in tetrahydrofuran (THF) [17]. Transmission electron microscopy revealed that the nanotube had a uniform diameter of 20 nm with an internal hollow diameter of 14 nm, indicating that formed by the tight rolling-up of bilayer hexabenzocoronenes through π-π stacking, with interdigitating alkyl chains and segregated hydrophilic triethylene glycol chains. In THF containing water, a helical coil structure was observed, resulting from the loose rolling-up of the bilayer graphitic tape of **1**. These nanotubes were subsequently oxidized with $NOBF_4$ to form the radical cations. A single oxidized nanotube, bridging a 180 nm gap between platinum electrodes, exhibited the resistivity of $2.5 \times 10^6\ \Omega$ at 285 K, comparable to that of an inorganic semiconductor nanotube made of gallium nitride.

Oxidation with iodine is a common method for doping organic conductors. Stupp and colleagues demonstrated that iodine-doped cast films made from entangled fibers, prepared from a 1 wt% THF gel of oligothiophene gelator **2**, achieved a conductivity of 7.9×10^{-3} S m^{-1} [18]. In contrast, cast films from THF solutions where self-assembly does not occur exhibit conductivities three orders of

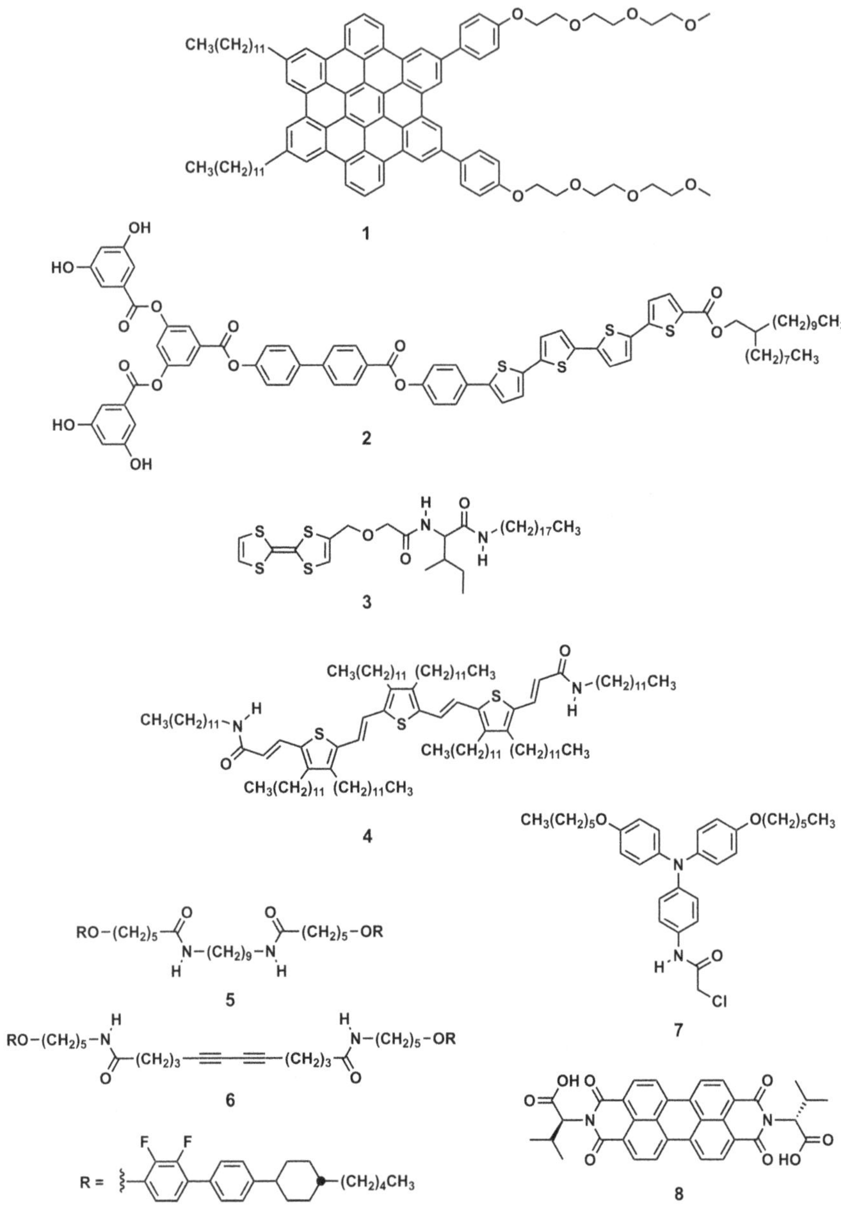

Fig. 3 Molecular structures of π-conjugated gelators

magnitude lower. Fiber alignment was achieved by casting the solution in an alternating current electric field, utilizing electrophoresis of the charged molecule. Kato and co-workers also demonstrated conductivities for randomly dispersed fibers of hydrogen-bonded tetrathiafulvalene gelator **3** doped with iodine and tetracyanoquinodimethane, reaching levels on the order of 10^{-3} S m^{-1} [19]. Uniaxial alignment of these fibers was achieved by gelation in an oriented nematic liquid crystal on a rubbed polyimide surface. Additionally, Ajayaghosh and colleagues reported aligned fibers of trithienylenevinylene-based gelator **4**, prepared by drop-casting a decane solution on mica [20]. The charge carrier transport properties of the gelator doped with a perylenedicarboxyimide derivative as an electron acceptor were evaluated using electrodeless flash-photolysis time-resolved microwave conductivity measurements, revealing a mobility of 10^{-2} cm^2 V^{-1} s^{-1}. Yoshio, Kato, and their colleagues demonstrated the electric field-induced alignment of self-assembled fibers formed by hydrogen-bonded gelators featuring fluoro-substituted mesogens with negative dielectric anisotropy (**5, 6**) in dodecylbenzene [21]. After this alignment, photopolymerization of a diacetylene derivative **6** produced potentially conductive polyacetylene coils. Conductive molecular coils are anticipated to function as molecular solenoids capable of generating magnetic force.

Additionally, a triarylamine derivative **7** can form light-induced self-assembled nanowires between two metallic electrodes in an electric field [22]. This assembly process is facilitated by *N*-cation radical formation and planarization of the triarylamine core. The aligned nanowires exhibit metallic behavior, achieving high electrical conductivity values ($> 5 \times 10^3$ S m^{-1}). The photoreduction of a perylene bisimide gelator **8**, resulting in the formation of a stable radical anion, has been reported to enhance the rheological properties of the gels [23]. The radical anion remains stable for several hours in air before decaying. This photoreduction process can be repeated multiple times. It was also demonstrated that the mechanical properties differed between the irradiated and non-irradiated regions of a patterned gel.

Reversible patterning of soft conductive materials can be achieved through a combination of peptide-based block copolymer hydrogelator **9** and photo-thermally active single-walled carbon nanotubes (Fig. 4) [24]. Unlike conventional physical gels, this gelator undergoes a transition from sol to gel upon heating, driven by the formation of ß-sheets and dehydration of poly(ethylene glycol) chains. The peptide derivative functions as both a dispersant for carbon nanotubes and a stimuli-responsive component. Notably, irradiation of the composite sol with light (532 nm or 785 nm laser, 250 mW) induces gelation through the transfer of heat generated by carbon nanotubes absorbing light to the thermo-responsive gelators. The photo-induced gels revert to sol states upon sonication. The dried photopatterned composite films exhibit high electrical conductivity ($> 10^2$ S m^{-1}).

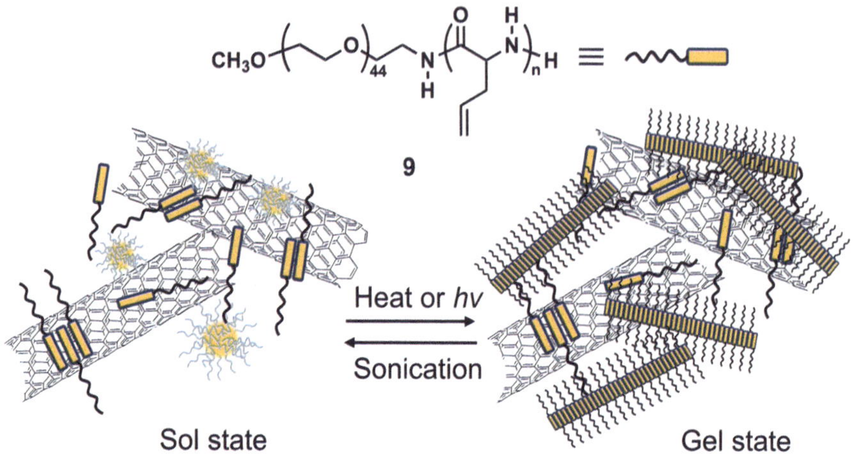

Fig. 4 Schematic illustration of the reversible transition between the sol state of peptide gelator **9**, which is dispersed with single-walled carbon nanotube in water, and the gel state is induced by heating or light irradiation. The gel reverts to the sol state upon sonication

4 Luminescent Gels

The molecular self-assembly approach has been recognized as an effective strategy for amplifying chirality and facilitating energy transfer [25]. The self-assembly of chiral gelators containing aromatic π-conjugated chromophores in achiral solvents, or achiral gelators with chromophore moieties in chiral solvents, offers a powerful method for generating π-conjugated helical assemblies.

For instance, Würthner and co-workers demonstrated helicity control in self-assembled fibers of perylene bisimide-based gelators **10**, **11** [26] (Fig. 5). Atomic force microscopy revealed that the chiral derivative **10** forms single-handed helical nanofibers in toluene, attributed to the helical face-to-face π-π stacking of perylene bisimide cores with a rotational displacement of about 30 °. Moreover, the single helicity of fibers from the achiral derivative **11** was achieved by employing a chiral solvent, such as (*R*)- or (*S*)-limonene. These assemblies exhibit intriguing optical properties, such as circularly polarized luminescence. Another key approach to developing circularly polarized luminescent-active gels involves the supramolecular chiral transfer from chiral gelators to achiral luminophores. Liu and co-workers demonstrated full-color tunable circularly polarized luminescence from nanotube structures formed by the C_3 symmetric chiral gelator **12** [27], achieved by incorporating achiral luminophores exhibiting aggregation-induced emission (Fig. 5). The co-assembly of **12** with Hexaphenylsilole (HPS) or a styrylbenzene derivative (MeCNS) in hybrid solvents of dimethyl sulfoxide and water resulted in gels that exhibit sky-blue and yellow circularly polarized fluorescence under UV light irradiation, respectively.

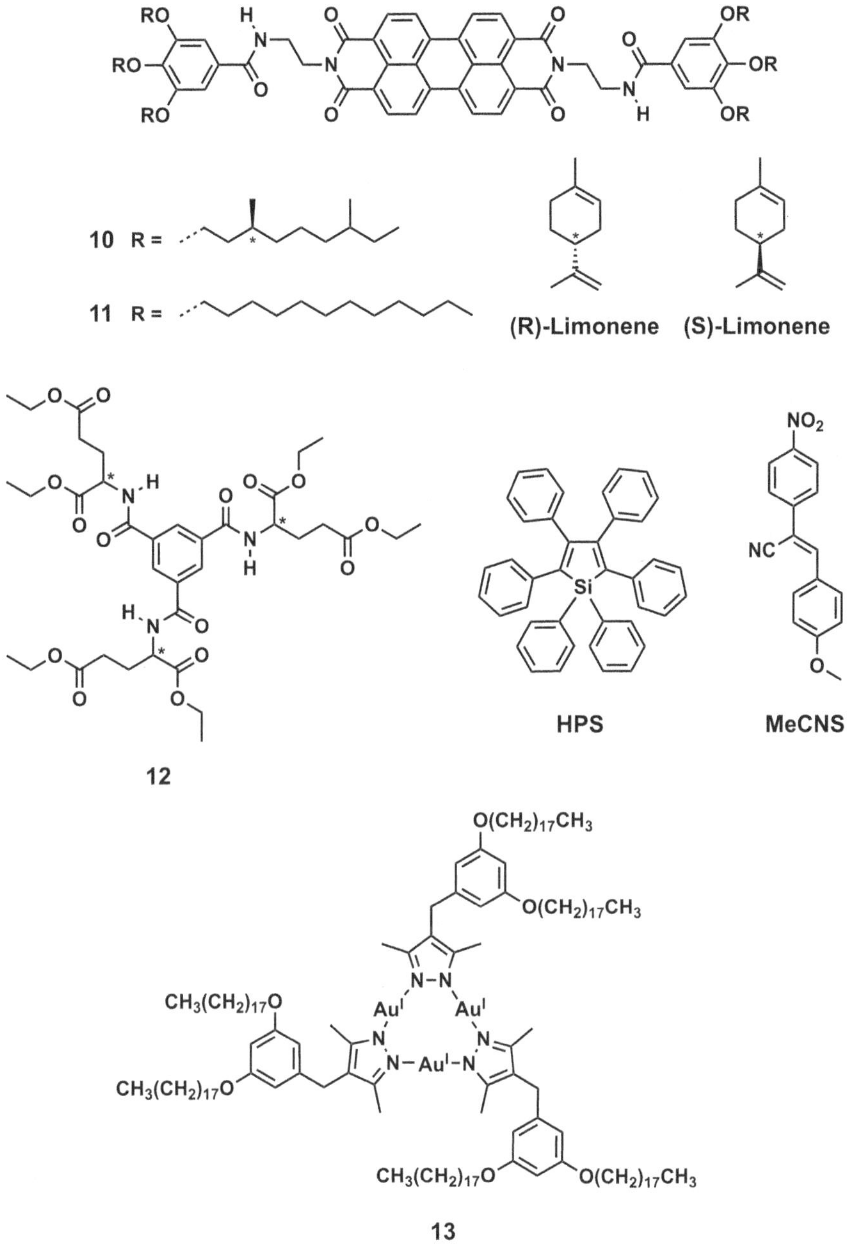

Fig. 5 Perylene bisimide gelator **10**, which is chiral, and achiral gelator **11**, forming helical fibers in the presence of limonene as a chiral solvent. Benzene triamide gelator **12**, capable of co-assembling with chromophores (HPS and MeCNS), exhibiting aggregation-induced emission. Phosphorescent gelator **13**, utilizing metallophilic interactions

Aida and co-workers reported the formation of a phosphorescent organogel using a trinuclear Au(I) pyrazolate complex **13** [28] via Au(I)–Au(I) metallophilic interactions in hexane (Fig. 5). The luminescent color of the gel shifted from red to blue upon doping with a small amount of silver. When heated, the blue luminescence of the silver ion-doped gel turned green in the sol state. These observed RGB luminescences are all long-lived (3–6 μm), indicating potential for use in novel imaging media.

5 Summary

This chapter provides an overview of the design principles behind gelators and the self-assembly into nanofibers through non-covalent interactions, such as π-π stacking and hydrogen bonding, in both organic solvents and water. The incorporation of π-conjugated molecules in gelator designs has led to the development of electro-active and optically active supramolecular gels, which can alter their conductivity and luminescence in response to external stimuli. These gels hold significant potential for advanced applications in devices like transistors, photodetectors, photovoltaics, and chiral sensors.

The introduction of new π-conjugated cores into gelator designs, coupled with the co-assembly of gelators with other functional molecules, has propelled the creation of innovative soft materials. For instance, Nakanishi and colleagues recently demonstrated that alkyl π-liquids—ambient-temperature liquids composed of π-conjugated molecules with branched alkyl chains—can function as liquid electrets, stabilizing static electricity [29]. When these liquids are gelled by hydrogen-bonding gelators, they enable the development of highly sensitive sensors capable of detecting subtle forces, such as the pulse of breathing.

Moreover, magneto-active gelators incorporating stable organic radical moieties [30] and metal-coordinating groups [31, 32] have been developed, showing promise in flexible spintronic materials and photon upconversion materials that convert lower-energy light into higher-energy light [32]. Supramolecular gels are not only promising for optoelectronic devices but also serve as versatile templates for the preparation of inorganic materials [33, 34].

The potential applications of supramolecular gels appear limitless. Their reversible nature and sensitivity to various stimuli make them excellent candidates for novel responsive systems.

References

1. F. Fages, Low molecular mass Gelators: Design, self-assembly, function, in *Topics in Current Chemistry*, vol. 256, (Springer, Berlin, 2005)

2. R.G. Weiss, P. Terech, *Molecular Gels: Materials with Self-Assembled Fibrillar Networks* (Springer, Netherlands, 2006)
3. J.H. van Esch, B.L. Feringa, New functional materials based on self-assembling organogels: From serendipity towards design. Angew. Chem. Int. Ed. **39**, 2263–2266 (2000)
4. D.J. Abdallah, R.G. Weiss, Organogels and low molecular mass organic gelators. Adv. Mater. **12**, 1237–1247 (2000)
5. N.M. Sangeetha, U. Maitra, Supramolecular gels: Functions and uses. Chem. Soc. Rev. **34**, 821–836 (2005)
6. S.S. Babu, V.K. Praveen, A. Ajayaghosh, Functional π-gelators and their applications. Chem. Rev. **114**, 1973–2129 (2014)
7. M. Shuzuki, K. Hanabusa, Polymer organogelators that make supramolecular organogels through physical cross-linking and self-assembly. Chem. Soc. Rev. **39**, 455–463 (2010)
8. K. Hanabusa, M. Suzuki, Development of low-molecular-weight gelators and polymer-based gelators. Polym. J. **46**, 776–782 (2014)
9. M. Liu, G. Ouyang, D. Niu, Y. Sang, Supramolecuar gelatons: Towards the design of molecular gels. Org. Chem. Front. **5**, 2885–2900 (2018)
10. O. Dumele, J. Chen, J.V. Passarelli, S.I. Stupp, Supramolecular energy materials. Adv. Mater. **32**, 1907247 (2020)
11. D.K. Smith, Supramolecular gels—A panorama of low-molecular-weight gelators from ancient origins to next-generation technologies. Soft Matter **20**, 10–70 (2024)
12. T. Kato, Y. Hirai, S. Nakaso, M. Moriyama, Liquid-crystalline physical gels. Chem. Soc. Rev. **36**, 1857–1867 (2007)
13. Y. Zhang, J. Wang, Y. Liao, X. Xie, A review on low-molecular-weight gels driven by halogen-effect. Chem. An Asian J. **18**, e202300097 (2023)
14. J.F. Miravet, B. Escuder, Pyridine-functionalised ambidextrous gelators: Towards catalytic gels. Chem. Commun., 5796–5798 (2005)
15. T. Brotin, R. Utermöhlen, F. Fages, G. Bouas-Laurent, J.-P. Desvergne, A novel small molecular luminescent gelling agent for alcohols. J. Chem. Soc. Chem. Commun., 416–418 (1991)
16. Z. Huang, H. Lee, E. Lee, S.-K. Kang, J.-M. Nam, M. Lee, Responsive nematic gels from the self-assembly of aqueous nanofibres. Nat. Commun. **2**, 459 (2011)
17. J.P. Hill, W. Jin, A. Kosaka, T. Fukushima, H. Ichihara, T. Shimomura, K. Ito, T. Hashizume, N. Ishii, T. Aida, Self-assembled hexa-*peri*-hexabenzocoronene graphitic nanotube. Science **304**, 1481–1483 (2004)
18. B.W. Messmore, J.F. Hulvat, E.D. Sone, S.I. Stupp, Synthesis, self-assembly, and characterization of supramolecular polymers from electroactive dendron rodcoil molecules. J. Am. Chem. Soc. **126**, 14452–14458 (2004)
19. T. Kitamura, S. Nakaso, N. Mizoshita, Y. Tochigi, T. Shimomura, M. Moriyama, K. Ito, T. Kato, Electroactive supramolecular self-assembled fibers comprised of doped tetrathiafulvalene-based gelators. J. Am. Chem. Soc. **127**, 14769–14775 (2005)
20. S. Prasanthkumar, A. Saeki, S. Seki, A. Ajayaghosh, Solution phase epitaxial self-assembly and high charge-carrier mobility nanofibers of semiconducting molecular gelators. J. Am. Chem. Soc. **132**, 8866–8867 (2010)
21. M. Yoshio, Y. Shoji, Y. Tochigi, Y. Nishikawa, T. Kato, Electric field-assisted alignment of self-assembled fibers composed of hydrogen-bonded molecules having laterally fluorinated mesogens. J. Am. Chem. Soc. **131**, 6763–6767 (2009)
22. E. Moulin, F. Niess, M. Maaloum, E. Buhler, I. Nyrkova, N. Giuseppone, The hierarchical self-assembly of charge nanocarrier: A highly cooperative process promoted by visible light. Angew. Chem. Int. Ed. **49**, 6974–6978 (2010)
23. E.R. Draper, R. Schweins, R. Akhtar, P. Groves, V. Chechik, M.A. Zwijnenburg, D.J. Adams, Reversible photoreduction as a trigger for photoresponsive gels. Chem. Mater. **28**, 6336–6341 (2016)
24. Z. He, J. Fan, J. Zou, K.L. Wooley, Reversible photo-patterning of soft conductive materials *via* spatially-defined supramolecular assembly. Chem. Commun. **52**, 8455–8458 (2016)

25. T. Zhao, P. Duan, M. Liu, Circularly Polarized Luminescence from Gelator Molecules: From Isolated Molecules to Assemblies, Chapter 11, 249–272, T. Mori Ed.; Circularly Polarized Luminescence of Isolated Small Organic Molecules, Springer Singapore (2020)
26. V. Stepanenko, Z.-Q. Li, J. Gershberg, F. Würthner, Evidence for kinetic nucleation in helical nanofiber formation directed by chiral solvent for a perylene bisimide organogelator. Chem. Eur. J. **19**, 4176–4183 (2013)
27. J. Han, J. You, X. Li, P. Duan, M. Liu, Full-color tunable circularly polarized luminescent nanoassemblies of achiral AIEgens in confined chiral nanotubes. Adv. Mater. **29**, 1606503 (2017)
28. A. Kishimura, T. Yamashita, T. Aida, Phosphorescent organogels via "Metallophilic" interactions for reversible RGB–color switching. J. Am. Chem. Soc. **127**, 179–183 (2005)
29. A. Tateyama, K. Nagura, M. Yamanaka, T. Nakanishi, Alkyl-π functional molecular gels: Control of elastic modulus and improvement of electret performance. Angew. Chem. Int. Ed. **63**, e202402874 (2024)
30. Y. Wu, Y. Hirai, Y. Tsunobuchi, H. Tokoro, H. Eimura, M. Yoshio, S. Ohkoshi, T. Kato, Supramolecular approach to the formation of magneto-active physical gels. Chem. Sci. **3**, 3007–3010 (2012)
31. K. Kuroiwa, N. Kimizuka, Self-assembly and functionalization of lipophilic metal-triazole complexes in various media. Polym. J. **45**, 384–390 (2013)
32. M. Hosoyamada, N. Yanai, K. Okumura, T. Uchihashi, N. Kimizuka, Translating MOF chemistry into supramolecular chemistry: Soluble coordination nanofibers showing efficient photon upconversion. Chem. Commun. **54**, 6828–6831 (2018)
33. Y. Ono, K. Nakashima, M. Sano, Y. Kanekiyo, K. Inoue, J. Hojo, S. Shinkai, Organic gels are useful as a template for the preparation of hollow fiber silica. Chem. Commun. 1477–1478 (1998)
34. H. Sukegawa, T. Nishimura, M. Yoshio, S. Kajiyama, T. Kato, One-dimensional supramolecular hybrids: Self-assembled nanofibrous materials based on a sugar gelator and calcite developed along an unusual axis. CrystEngComm **19**, 1580–1584 (2017)

Mechanochromic Luminescent Soft Materials

Yoshimitsu Sagara

1 Introduction

Force is one of the most common external stimuli, along with others like heat or light. Force is applied to a material through grinding, scratching, crushing, or stretching. When force is applied, many crystals fracture into fine particles while maintaining their molecular assembly. Some crystals become amorphous or undergo a phase transition to another crystalline phase (1–7). In addition, several other crystals, instead of breaking, change shape in response to external forces, exhibiting plastic, elastic, and superelastic behavior (8–11). In such crystalline materials, the molecular arrangement is changed by the application of force. Liquid crystals have also been reported to exhibit liquid crystal phase transitions when shear is applied (12, 13). When polymers are stretched, the internal arrangement of the polymer chains changes (14). If photoluminescent groups are introduced into materials whose molecular arrangement changes in response to forces, they can exhibit "mechanochromic luminescence" or "mechanoresponsive luminescence" properties. In general, when photoluminescent groups are assembled in an ordered fashion, excimer formation (15, 16), exciton coupling (17), or conformational changes in the luminophores occur, resulting in photoluminescence properties that are significantly different from those of the monomer state (18). In addition, when the assembly structure is changed, the interactions between the photoluminescent groups change; thus, the photoluminescence properties change. Therefore, the manipulation of luminophore arrangements through force enables the achievement of force-induced changes in photoluminescence properties (Fig. 1) (1–7).

The term "mechanochromic luminescence" is used to describe the phenomenon in which the color of luminescence changes in response to mechanical stimuli. When the photoluminescence intensity undergoes a significant change from the off state to

Y. Sagara (✉)
Department of Materials Science and Engineering, Institute of Science Tokyo, Tokyo, Japan
e-mail: sagara@mct.isct.ac.jp

M. Funahashi (ed.), *Functional Soft Matter II*,
https://doi.org/10.1007/978-981-92-0100-6_16

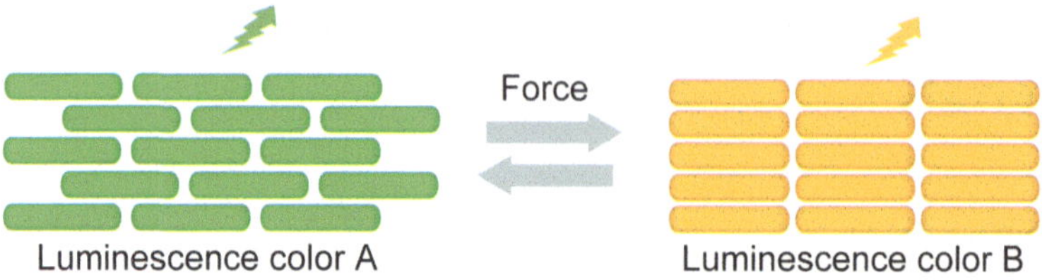

Fig. 1 Schematic illustration of mechanochromic luminescent materials based on molecular assemblies

the on state, the change can be considered as from colorless to colored luminescence. This phenomenon is known as mechanochromic luminescence. On the other hand, if the change in photoluminescence intensity is small, and thus the photoluminescence color remains the same, it is not classified as mechanochromic luminescence but as "mechanoresponsive luminescence."

This chapter presents examples of crystals, liquid crystals, and dye-doped polymers exhibiting mechanochromic luminescence. Subsequently, the discussion is extended to micelles and supramolecular mechanophores, which are geared toward developing more practical mechanosensors.

2 Mechanochromic Luminescent Crystals

The number of crystals exhibiting mechanochromic luminescence has now exceeded 2000. The mechanisms of color change are diverse and include excimer formation or inhibition, changes in exciton coupling, and conformational changes. Only a few representative examples are presented herein. However, review articles provide an idea of the diversity of their molecular structures (1–7).

The phenomenon of changing fluorescence properties under force was first reported in 1993 using pyridinium salt (19). Subsequently, the tetraphenylethylene derivative shown in Fig. 2 was found to undergo reversible changes in fluorescence color under mechanical and thermal stimuli, leading to a rapid increase in the number of mechanochromic luminescent compounds (20). The tetraphenylethylene derivative **C6TPPy**, which has four hexylamide groups, forms white crystals that exhibit blue fluorescence ascribed to the *H*-aggregation of the luminophores. The amide groups form uniform intermolecular hydrogen bonds in the initial crystals, resulting in a columnar structure for **C6TPPy**. When a blue-emitting crystal is ground, a crystalline-to-amorphous phase transition occurs, resulting in a change in the photoluminescence color from blue to blue-green. In the yellow amorphous phase, the initial uniform hydrogen bonds are disturbed, preventing the *H*-aggregation of the luminophore. After thermal treatment of the ground amorphous phase, the initial blue-emitting state is recovered because the amorphous state was thermodynamically metastable. This mechanochromic luminescence is caused by a change in

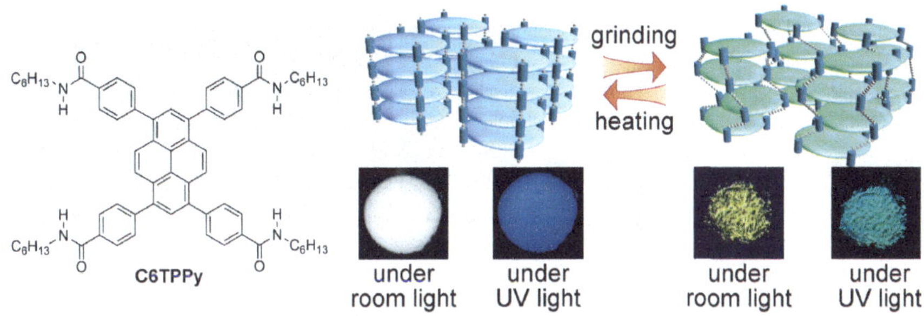

Fig. 2 Mechanochromic luminescence of crystalline tetraphenylpyrene derivative **C6TPPy**. Adapted with permission from (2). Copyright 2009, Nature Publishing Group

molecular arrangement and does not involve covalent bond breakage. Therefore, the fluorescence color could change repeatedly.

Organometallic complexes can also exhibit mechanochromic luminescence (21). A typical example is shown in Fig. 3. The gold complex is initially crystalline and exhibits blue luminescence under UV light irradiation. However, when ground, it changes from crystalline to amorphous, and the blue luminescence changes to yellow. When exposed to dichloromethane vapor, the original crystalline phase is restored, and the blue luminescence returns. This color change is due to a change in the aurophilic interactions. In the initial crystal, the gold atoms are sufficiently separated between adjacent complexes so that no aurophilic interactions occur. However, when they become amorphous upon grinding, the gold atoms move closer together and aurophilic interactions occur, resulting in a yellow luminescence color.

In the two aforementioned examples, the crystals became amorphous upon mechanical stimulation. The following example shows the crystal-crystal phase transition. The gold complex shown in Fig. 4 forms crystals that emit blue light and crystals that emit yellow light, depending on the recrystallization method (22). The former is in a kinetically isolated metastable state, whereas the latter is thermodynamically stable. When the former is ground, it changes to the latter without becoming amorphous. Furthermore, when the former crystal is touched with a needle, the blue-emitting crystalline phase transforms into a yellow-emitting crystal, and the yellow

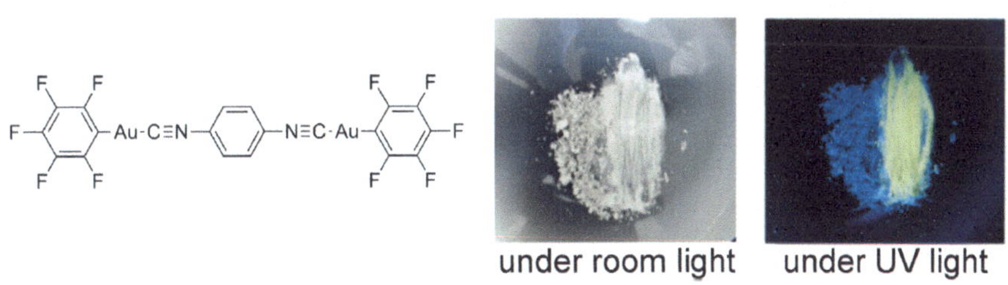

Fig. 3 Mechanochromic luminescence of crystalline Au(I) complex. The right part of the sample was ground. Adapted with permission from (21). Copyright 2008, American Chemical Society

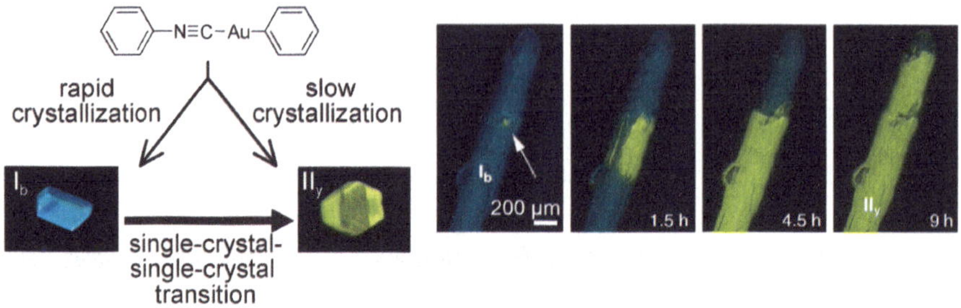

Fig. 4 Mechanochromic luminescence of Au(I) complex showing single-crystal-to-single-crystal transition. Adapted with permission from (22). Copyright 2012, Nature Publishing Group

crystalline phase spreads into the blue-emitting crystalline phase. This phenomenon is similar to that of a domino. In addition, the yellow crystalline phase spreads out when the blue crystals come into contact with the yellow crystals.

Superelastic crystals exhibiting mechanochromic luminescence have also been reported (23). Superelastic molecular crystals exhibit elastic bending characteristics when subjected to mechanical stress that are often accompanied by phase transitions. After the removal of the force, the initial crystal shape recovers. 7-Chloro-2-(2′-hydroxyphenyl)imidazo[1,2-*a*]pyridine forms superelastic crystals that exhibit green-yellow fluorescence (Fig. 5). When a shear force is applied to the crystals, another crystalline phase with orange fluorescence appears at the center. When the shear force is removed, the crystal returns to its original state, and only yellow-green fluorescence is observed.

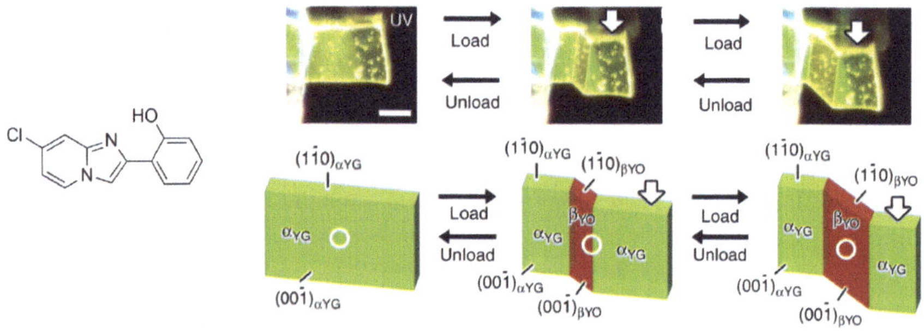

Fig. 5 Superelastic crystal exhibiting mechanochromic luminescence. Adapted with permission from (23). Copyright 2020, Nature Publishing Group

3 Mechanochromic Luminescent Liquid Crystals

Liquid crystals are soft materials that combine ordered molecular assemblies with mobility (24–26). When luminophores are introduced into Liquid-crystalline (LC) molecules to form ordered assemblies, they are properly aligned. In addition, in response to external stimuli such as heat, light, and force, liquid crystals undergo a phase transition to another LC phase with distinct molecular assemblies. Therefore, liquid crystals are ideal soft materials for the development of mechanochromic luminescent materials (2, 27). However, compared to crystals, there are far fewer reported examples of LC compounds that exhibit mechanochromic luminescence (28–33). Herein, the first mechanochromic luminescent liquid crystal (28) is described.

Liquid crystal **PyAm** has a 1,6-bis(phenylethynyl)pyrene group as the luminophore and two dendritic structures are introduced into the luminophore via the amide groups (Fig. 6). **PyAm** forms a cubic LC phase upon cooling from the isotropic phase. The π-stacked luminophores of **PyAm** in the cubic phase form static excimers to exhibit yellow fluorescence. When **PyAm** in the cubic phase is sandwiched between two glass substrates and sheared at 160 °C, a phase transition to the columnar LC phase occurs, and the emission color changes from yellow to blue-green. In the shear-induced columnar phase, uniform hydrogen bonds form between amide groups, preventing π-stacked structures that can form static excimers. Because the distance between amide groups forming hydrogen bonds is longer than that between π-stacked aromatic groups, uniform hydrogen bonds cannot form in the cubic phase. The cubic phase is thermodynamically metastable, whereas the shear-induced columnar phase is thermodynamically stable.

4 Mechanochromic Luminescence of Dye-Doped Polymers

Dye-doped polymers can be prepared by dipping polymer specimens in dye solutions or by mixing polymers and dyes without solvents at high temperatures (14). Appropriate conditions allow access to dye-doped polymers, in which dyes form nano/micro aggregates (34–37). Once the aggregation of dyes that can form static excimers occurs in the host polymers, the photoluminescence properties become excimer dominant. When mechanical forces such as tensile deformation are applied to polymer films containing dyes, the polymer chains undergo reorganization through slipping and tend to align in the direction of the force. The resulting stress exceeds the threshold for breaking the intermolecular interactions within the dye aggregates, causing the dye molecules to disperse molecularly within the polymer matrix as monomers. Therefore, the photoluminescence changes from excimer- to monomer-dominant. Because the introduced dyes need to form aggregates, they must be phase-separated from the host polymer to some extent. In addition, semi-crystalline polymers are desirable because their molecular assembly structures can be disrupted.

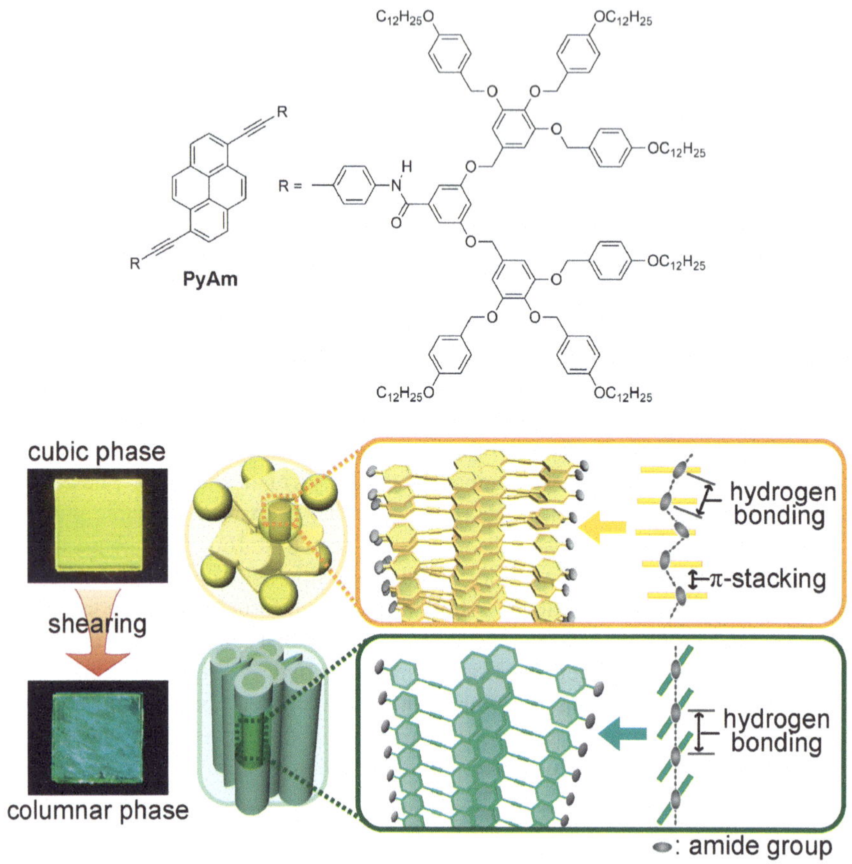

Fig. 6 Mechanochromic luminescence of LC compound **PyAm**. Adapted with permission from (28). Copyright 2008, Wiley-VCH

The most popular dyes used in mechanochromic luminescent dye-doped polymers are cyano-substituted oligo(*p*-phenylene vinylene)s (cyano-OPVs) (38–40). For example, cyano-OPVs **C1-RG** and **C1-YB** (Fig. 7) show green and blue monomer fluorescence, respectively, in chloroform. The introduction of two methoxy groups at the central phenyl group results in a red shift in the emissions of **C1-RG**. The crystals **C1-RG** and **C1-YB** exhibit red and yellow excimer emissions, respectively. The electron-withdrawing cyano groups overlapped the electron-rich central ring with the two methoxy groups of the adjacent molecules in the **C1-RG** crystal. The π-stacked structures result in excimer fluorescence. Dye-doped polymers containing these cyano-OPVs were prepared by melt mixing the dyes with linear low-density poly(ethylene) (LLDPE). Each pristine composite film shows excimer-dominated photoluminescence owing to the aggregation of **C1-RG** or **C1-YB**. When the films

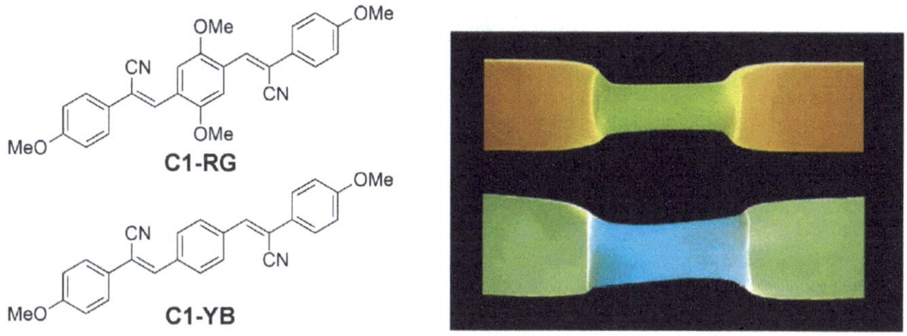

Fig. 7 Dye-doped polymers that show mechanochromic luminescence. Adapted with permission from (39). Copyright 2003, American Chemical Society

are stretched, the aggregates break and monomer-dominant fluorescence is observed for both films.

5 Mechanochromic Luminescent Micelles

The materials exhibiting mechanochromic luminescence described above are difficult to use as practical mechanosensors. Practical mechanosensors must exhibit quantitative properties. However, crystals, liquid crystals, and aggregates in dye-doped polymers cannot be uniformly sized. A small force is sufficient to break small crystals and induce mechanochromic luminescence, whereas a large force is required to break large crystals and completely change their luminescence color. Therefore, the development of practical mechanosensors that can detect and visualize small forces using molecular assemblies, constructing uniform molecular assemblies with a limited number of molecules, is a promising approach.

Therefore, mechanochromic luminescent micelles that function in water have been developed (Fig. 8) (41). The micelles consist of approximately 10–15 dumbbell-shaped amphiphiles containing a 1,6-bis(phenylethynyl)pyrene group as the hydrophobic fluorophore and two hydrophilic dendritic groups. Several hydroxyl groups at the peripheral positions are replaced by amine groups, allowing the micelles to be covalently attached to the amino group-modified glass and polymer beads using commercially available crosslinkers. The micelles show excimer fluorescence because the luminophores form π-stacked structures in the micelles. When a sufficient force is applied to the micelles on the substrates, the arrangement of the luminophore changes to exhibit green fluorescence. When the micelles are introduced onto the surface of glass beads and the glass beads are shaken in a glass vial filled with water, the latter collide with each other, applying force to the micelles, and resulting in a change in fluorescence color. However, when micelles are introduced into the polylactic acid beads and shaken under the same conditions, no fluorescence color change is observed. This is because the glass beads are harder than the polylactic acid

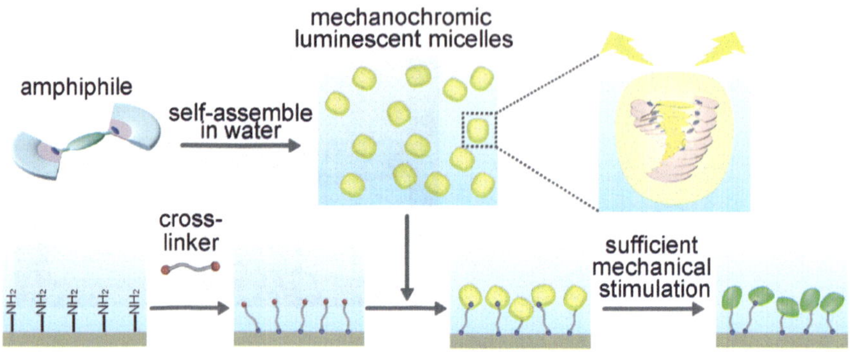

Fig. 8 Mechanochromic luminescent micelles that function in water. Adapted with permission from (41). Copyright 2014, American Chemical Society

beads. When shaken under the same conditions, insufficient force was applied to the micelles to induce emission color changes. When shaken with the glass beads, the emission color of the polylactide beads changed. These results indicate that a certain force threshold is required to induce fluorescence color changes of the micelles.

6 Supramolecular Mechanophores

Finally, supramolecular mechanophores are introduced. Mechanophores are molecular skeletons that exhibit various outputs in response to mechanical stimuli, such as the release of small molecules, initiation of catalytic reactions, and color changes (34, 42–44). Many mechanophores involve cleavage of covalent bonds, whereas supramolecular mechanophores do not. Therefore, instantaneous and reversible changes in luminescence properties occur (45–50).

Here, rotaxane-based supramolecular mechanophores (45–47) are described (Fig. 9). Rotaxane is an interlocked molecule that has been extensively investigated for decades and is regarded as a molecular assembly constructed with the smallest number of molecules. Rotaxane is composed of a ring with a fluorophore, a rod-shaped molecule with a matching quencher, and two stoppers that prevent the ring from dethreading. The quencher is intercalated by the ring through charge-transfer interactions in a force-free state, resulting in no fluorescence. After the mechanophore is covalently introduced into the segmented polyurethane elastomer, a dog-bone-shaped sample is prepared. When the film is stretched, the macroscopic force applied to it is transferred to the mechanophores. The ring then slides along the axis to move away from the quencher, resulting in strong photoluminescence.

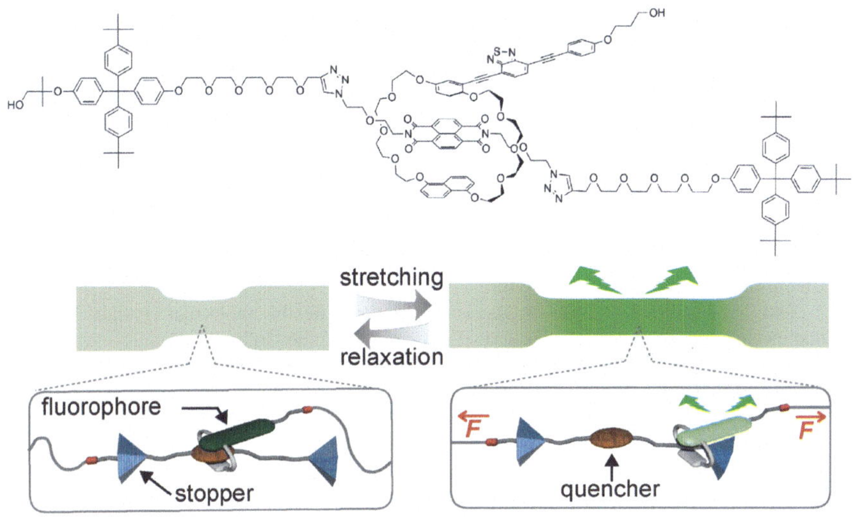

Fig. 9 Prototype of rotaxane-based supramolecular mechanophores and the mechanism of the mechanochromic luminescence.

7 Summary

This chapter briefly reviews mechanochromic luminescent materials based on molecular assemblies. The basic concept is simple: the arrangement of luminophores and quenchers is altered by force, resulting in mechanochromic luminescence. Many examples of crystalline materials have been reported, and further intriguing mechanochromic phenomena are expected to emerge. Liquid crystals and dye-doped polymers are not as common as crystalline materials, although these two types of mechanochromic luminescent materials are considered soft. Furthermore, to realize practical force-sensing applications, it is necessary to accumulate reports on low-dimensional mechanochromic luminescent materials such as micelles, 1D supramolecular fibers, and supramolecular mechanophores.

References

1. Y. Sagara, S. Yamane, M. Mitani, C. Weder, T. Kato, Mechanoresponsive luminescent molecular assemblies: An emerging class of materials. Adv. Mater. **28**, 1073–1095 (2016)
2. Y. Sagara, T. Kato, Mechanically induced luminescence changes in molecular assemblies. Nat. Chem. **1**, 605–610 (2009)
3. X. Zhang, Z. Chi, Y. Zhang, S. Liu, J. Xu, Recent advances in mechanochromic luminescent metal complexes. J. Mater. Chem. C **1**, 3376–3390 (2013)
4. Z. Chi, X. Zhang, B. Xu, X. Zhou, C. Ma, Y. Zhang, S. Liu, J. Xu, Recent advances in organic mechanofluorochromic materials. Chem. Soc. Rev. **41**, 3878–3896 (2012)

5. Z. Ma, Z. Wang, M. Teng, Z. Xu, X. Jia, Mechanically induced multicolor change of luminescent materials. ChemPhysChem **16**, 1811–1828 (2015)
6. S. Ito, Mechanochromic luminescence of soft crystals: Recent systematic studies in controlling the molecular packing and mechanoresponsive properties. J. Photochem. Photobiol. C Photchem. Rev. **51**, 100481 (2022)
7. S. Ito, Recent advances in Mechanochromic luminescence of organic crystalline compounds. Chem. Lett. **50**, 649–660 (2021)
8. S. Takamizawa, Y. Miyamoto, Superelastic organic crystals. Angew. Chem. Int. Ed. **53**, 6970–6973 (2014)
9. S. Hayashi, T. Koizumi, Elastic organic crystals of a fluorescent π-conjugated molecule. Angew. Chem. Int. Ed. **55**, 2701–2704 (2016)
10. T. Seki, N. Hoshino, Y. Suzuki, S. Hayashi, Functional flexible molecular crystals: Intrinsic and mechanoresponsive properties. CrystEngComm **23**, 5686–5696 (2021)
11. Z. Wang, W. Han, R. Shi, X. Han, Y. Zheng, J. Xu, X.-H. Bu, Mechanoresponsive Flexible Crystals. JACS Au **4**, 279–300 (2024)
12. L. Noirez, Shear induced Smectic-A-Smectic-C transition in side-chain liquid-crystalline polymers. Phys. Rev. Lett. **84**, 2164–2167 (2000)
13. C.R. Safinya, E.B. Sirota, R.J. Plano, Nematic to smectic-a phase transition under shear flow: A nonequilibrium synchrotron x-ray study. Phys. Rev. Lett. **66**, 1986–1989 (1991)
14. A.P. Haehnel, Y. Sagara, Y.C. Simon, C. Weder, Mechanochemistry in polymers with supramolecular Mechanophores. Top. Curr. Chem. **369**, 345–375 (2015)
15. J.B. Birks, Excimers. Rep. Prog. Phys. **38**, 903–974 (1975)
16. F.M. Winnik, Photophysics of preassociated pyrenes in aqueous polymer solutions and in other organized media. Chem. Rev. **93**, 587–614 (1993)
17. F. Würthner, T.E. Kaiser, C.R. Saha-Möller, *J*-aggregates: From serendipitous discovery to supramolecular engineering of functional dye materials. Angew. Chem. Int. Ed. **50**, 3376–3410 (2011)
18. K. Araki, T. Mutai, Packing-directed tuning and switching of organic solid-state luminescence. Photochemistry **43**, 191–225 (2016)
19. R. Gawinecki, G. Viscardi, E. Barni, M.A. Hanna, 4-*Tert*-butyl-1-(4′-dimethylamino-benzylideneamino)pyridinium perchlorate (BDPP): A novel fluorescent dye. Dyes Pigments **23**, 73–78 (1993)
20. Y. Sagara, T. Mutai, I. Yoshikawa, K. Araki, Material design for piezochromic luminescence: Hydrogen-bond-directed assemblies of a pyrene derivative. J. Am. Chem. Soc. **129**, 1520–1521 (2007)
21. H. Ito, T. Saito, N. Oshima, N. Kitamura, S. Ishizaka, Y. Hinatsu, M. Wakeshima, M. Kato, K. Tsuge, M. Sawamura, Reversible Mechanochromic luminescence of [$(C_6F_5Au)_2(\mu$-1,4-Diisocyanobenzene)]. J. Am. Chem. Soc. **130**, 10044–10045 (2008)
22. H. Ito, M. Muromoto, S. Kurenuma, S. Ishizaka, N. Kitamura, H. Sato, T. Seki, Mechanical stimulation and solid seeding trigger single-crystal-to-single-crystal molecular domino transformations. Nat. Commun. **4**, 2009 (2013)
23. T. Mutai, T. Sasaki, S. Sakamoto, I. Yoshikawa, H. Houjou, S. Takamizawa, A superelastochromic crystal. Nat. Commun. **11**, 1824 (2020)
24. T. Kato, N. Mizoshita, K. Kishimoto, Functional liquid-crystalline assemblies: Self-organized soft materials. Angew. Chem. Int. Ed. **45**, 38–68 (2006)
25. Handbook of Liquid Crystals. P. J. Collings. J. W. Goodby, H. Gleeson, P. Raynes, T. Kato, C. Tschierske, Ed., (WileyVCH, Weinheim, ed. 2nd, 2014)
26. T. Kato, J. Uchida, T. Ichikawa, T. Sakamoto, Functional liquid crystals towards the next generation of materials. Angew. Chem. Int. Ed. **57**, 4355–4371 (2018)
27. S. Yamane, K. Tanabe, Y. Sagara, T. Kato, Stimuli-responsive photoluminescent liquid crystals. Top. Curr. Chem. **318**, 395–405 (2012)
28. Y. Sagara, T. Kato, Stimuli-responsive luminescent liquid crystals: Change of photoluminescent colors triggered by a shear-induced phase transition. Angew. Chem. Int. Ed. **47**, 5175–5178 (2008)

29. V.N. Kozhevnikov, B. Donnio, D.W. Bruce, Phosphorescent, terdentate, liquid-crystalline complexes of platinum(II): Stimulus-dependent emission. Angew. Chem. Int. Ed. **47**, 6286–6289 (2008)
30. Y. Sagara, S. Yamane, T. Mutai, K. Araki, T. Kato, A. Stimuli-Responsive, Photoluminescent, anthracene-based liquid crystal: Emission color determined by thermal and mechanical processes. Adv. Funct. Mater. **19**, 1869–1875 (2009)
31. Y. Sagara, T. Kato, Brightly tricolored mechanochromic luminescence from a single-luminophore liquid crystal: Reversible writing and erasing of images. Angew. Chem. Int. Ed. **50**, 9128–9132 (2011)
32. M. Mitani, S. Ogata, S. Yamane, M. Yoshio, M. Hasegawa, T. Kato, Mechanoresponsive liquid crystals exhibiting reversible luminescent color changes at ambient temperature. J. Mater. Chem. C **4**, 2752–2760 (2016)
33. A. Seki, M. Yoshio, Multi-color photoluminescence based on mechanically and thermally induced liquid-crystalline phase transitions of a hydrogen-bonded Benzodithiophene derivative. ChemPhysChem **21**, 328–334 (2020)
34. H. Traeger, D.J. Kiebala, C. Weder, S. Schrettl, From molecules to polymers—Harnessing inter- and intramolecular interactions to create Mechanochromic materials. Macromol. Rapid Commun. **2000573** (2021)
35. C. Céline, G. Anirvan, W. Christoph, S. Stephen, Self-calibrating Mechanochromic fluorescent polymers based on encapsulated excimer-forming dyes. Adv. Mater. **30**, 1704603 (2018)
36. A. Pucci, R. Bizzarri, G. Ruggeri, Polymer composites with smart optical properties. Soft Matter **7**, 3689–3700 (2011)
37. A. Pucci, G. Ruggeri, Mechanochromic polymer blends. J. Mater. Chem. **21**, 8282 (2011)
38. C. Löwe, C. Weder, Oligo (*p*-phenylene vinylene) excimers as molecular probes: Deformation-induced color changes in Photoluminescent polymer blends. Adv. Mater. **14**, 1625–1629 (2002)
39. B.R. Crenshaw, C. Weder, Deformation-induced color changes in melt-processed Photoluminescent polymer blends. Chem. Mater. **15**, 4717–4724 (2003)
40. C. Calvino, Y. Sagara, V. Buclin, A.P. Haehnel, A. Del Prado, C. Aeby, Y.C. Simon, S. Schrettl, C. Weder, Mechanoresponsive, luminescent polymer blends based on an excimer-forming Telechelic macromolecule. Macromol. Rapid Commun. **40**, e1800705 (2019)
41. Y. Sagara, T. Komatsu, T. Ueno, K. Hanaoka, T. Kato, T. Nagano, Covalent attachment of mechanoresponsive luminescent micelles to glasses and polymers in aqueous conditions. J. Am. Chem. Soc. **136**, 4273–4280 (2014)
42. J. Li, C. Nagamani, J.S. Moore, Polymer Mechanochemistry: From destructive to productive. Acc. Chem. Res. **48**, 2181–2190 (2015)
43. D.A. Davis, A. Hamilton, J. Yang, L.D. Cremar, D. Van Gough, S.L. Potisek, M.T. Ong, P.V. Braun, T.J. Martinez, S.R. White, J.S. Moore, N.R. Sottos, Force-induced activation of covalent bonds in mechanoresponsive polymeric materials. Nature **459**, 68–72 (2009)
44. Y. Chen, G. Mellot, D. van Luijk, C. Creton, R.P. Sijbesma, Mechanochemical tools for polymer materials. Chem. Soc. Rev. **50**, 4100–4140 (2021)
45. Y. Sagara, M. Karman, E. Verde-Sesto, K. Matsuo, Y. Kim, N. Tamaoki, C. Weder, Rotaxanes as Mechanochromic fluorescent force transducers in polymers. J. Am. Chem. Soc. **140**, 1584–1587 (2018)
46. Y. Sagara, M. Karman, A. Seki, M. Pannipara, N. Tamaoki, C. Weder, Rotaxane-based Mechanophores enable polymers with mechanically switchable White photoluminescence. ACS Cent. Sci. **5**, 874–881 (2019)
47. T. Muramatsu, Y. Okado, H. Traeger, S. Schrettl, N. Tamaoki, C. Weder, Y. Sagara, Rotaxane-based dual function Mechanophores exhibiting reversible and irreversible responses. J. Am. Chem. Soc. **143**, 9884–9892 (2021)
48. Y. Sagara, H. Traeger, J. Li, Y. Okado, S. Schrettl, N. Tamaoki, C. Weder, Mechanically Responsive luminescent polymers based on supramolecular Cyclophane Mechanophores. J. Am. Chem. Soc. **143**, 5519–5525 (2021)
49. H. Traeger, Y. Sagara, D.J. Kiebala, S. Schrettl, C. Weder, Folded Perylene Diimide loops as Mechanoresponsive motifs. Angew. Chem. Int. Ed. **60**, 16191–16199 (2021)

50. S. Thazhathethil, T. Muramatsu, N. Tamaoki, C. Weder, Y. Sagara, Excited state charge-transfer complexes enable fluorescence color changes in a supramolecular Cyclophane Mechanophore. Angew. Chem. Int. Ed. **61**, e202209225 (2022)

Actuators Based on Soft Matter

Hirosato Monobe

1 Introduction

Actuators using soft matter (soft actuators) are actuators made of materials with excellent flexibility and deformability. Various drive power sources are available, including electric, magnetic, optical, thermal, humid, and so on. Soft actuators are being applied to medical and welfare equipment [1], robotics [2], and many other fields, taking advantage of their flexibility, safety, and lightweight. In the future, they are expected to be used in a wide variety of application fields. The details of actuators using soft matter can be covered by referring to the textbook [3]. In this section, we will focus on soft actuators using electrical stimulation, which are considered to be the easiest to apply and use.

The inverse piezoelectric phenomenon of ferroelectric polymers such as Polyvinylidene Fluoride (PVDF) and its copolymers has been known as a material for electric field-driven polymer actuators. However, in the 1990s, the electrostriction phenomenon with a large amount of expansion and contraction was found in dielectric elastomers and ferroelectric polymers in which elastic electrodes are bonded to elastomers, and their actuator applications are advancing [4, 5]. The performance of these Electroactive Polymer (EAP) actuators differs depending on the material, but in general, they are stretchable, have a large power density, and have excellent frequency characteristics. Electric current-driven (ionic EAP) actuators are driven at a low voltage (a few V or less), and although the expansion and contraction rate is also inferior to that of electronic field-driven (electronic EAP) actuators, they are large (several %) compared to piezoelectric ceramics. Ionic EAP actuators are low-voltage, high-deformation actuators. Because they are soft polymers, they can be easily made into thin films, and they are expected to be applied to thin devices such

H. Monobe (✉)
National Institute of Advanced Industrial Science and Technology (AIST), Research Institute of Core Technology for Materials Innovation, Osaka, Japan
e-mail: monobe-hirosato@aist.go.jp

M. Funahashi (ed.), *Functional Soft Matter II*,
https://doi.org/10.1007/978-981-92-0100-6_17

as haptic devices. Since the response is limited by ion migration, it depends on the size of the device and is about 100 Hz under optimal conditions.

2 Electroactive Polymer (EAP) Actuator

The main EAP actuators currently under research are shown in Table 1 [3]. Although there are a variety of materials, actuators mainly consist of a structure of soft-matter polymer materials laminated with stretchable electrodes (metal and carbon), which mechanically deform by various principles when voltage is applied. The actuator is made of a material that is mechanically deformed by various principles when a voltage is applied. All of these materials, being polymers, have excellent workability, are lightweight, soft, and easy to miniaturize, and have the potential to be integrated with sensors and structural materials. These materials can be roughly classified into electric field-driven (electronic) type and current-driven (ionic) type based on the driving principle as shown in Table 1. The former is a dielectric-based device whose driving principle is based on the electrostriction phenomenon or electrostatic effect. The latter is based on conductive materials and is driven by the principle of soft-matter polymer deformation, which also becomes ionic conduction based on redox currents, or charge/discharge currents in the electric double layer. As shown in Fig. 1, comparing the electronic and ionic types, the former is superior in all aspects, including expansion and contraction rate, work density, and efficiency. However, the drive voltage is extremely high, on the order of kV, and there are various problems in practical application. The latter has problems in high-speed response, etc., as its responsiveness is limited by the movement of ions, but the drive voltage is only a few volts, which, combined with the workability, lightweight, and small size of the polymer actuator, gives it an advantage over other actuators in its use.

Table 1 List of soft EAP actuators and their materials currently under research

Type	Materials	E/I
Ionic conductive polymer	Nafion/Gold, Flemion/Gold, Nafion/Metal, oxides Nafion/ PEDOT:PSS	I
Conductive polymer	Polypyrrol, Polyaniline, Polythiophene	I
Carbon nanotube (CNT)	Single wall CNT, Multi wall CNT	I
Electrostrictive polymer	Polyvinylidene fluoride, trifluoroethylene copolymer	E
Dielectric elastomer	Silicone elastomer, Acrylic elastomer	E
Dielectric gel	Polyvinyl chloride, PVA/DMSO	E

E: Electronic type, I: Ionic type

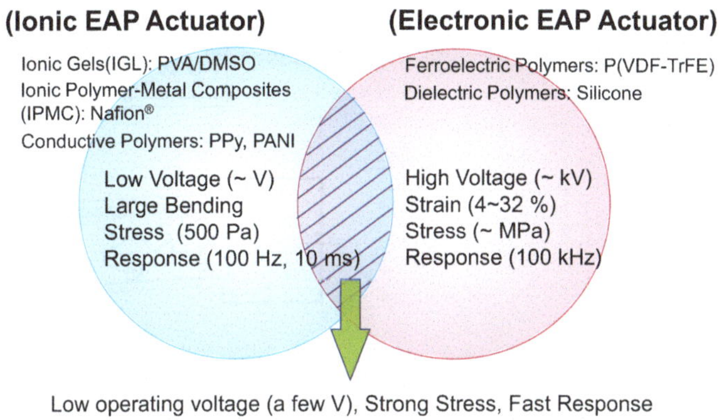

Fig. 1 Features of ionic and electronic EAP actuators

3 Ionic EAP Actuator

3.1 Ionic Conductive Polymer Actuators

IPMC (Ionic Polymer-Metal Composite) is a kind of polymer electrolyte gel actuator. In recent years, it is also called piezo ionic actuator because it is a deformation caused by the movement of ions, in contrast to piezo which is an electronic piezoelectric element [5]. As shown in Fig. 2, a structure in which electrodes are bonded to an ion exchange film such as Nafion, which is a polymer electrolyte gel, is used. When voltage is applied between the electrodes, volume change due to the electroosmotic flow of ions, hydrated water to the electrodes, and osmotic pressure and electrostatic force generated by ions adsorbed on the electrodes become driving forces, causing bending deformation. As electrodes, a method of electroless plating gold or platinum or a method of forming stretchable electrodes by coating metal oxides or carbon conductive particles are used. As ion exchange membranes, fluorine-based ion exchange membranes such as Nafion and Flemion are commonly used. The response characteristics are low-voltage driving (3 V or less), the response speed is relatively high (up to 100 Hz), and the deformation is large. The driving characteristics largely depend on the solvent and ion species contained in the ion exchange membrane. By using an ionic liquid, which is a salt in a liquid state at room temperature and is hardly volatile, as a solvent, it is possible to use an IPMC actuator in the air without coating. The basic deformation is bending, and its response, generating force, displacement, etc., depends on the thickness of the element. In recent years, research and development aiming at high-speed response have been advanced, for example, by using an ionic liquid crystal as an ion electrolyte membrane [6].

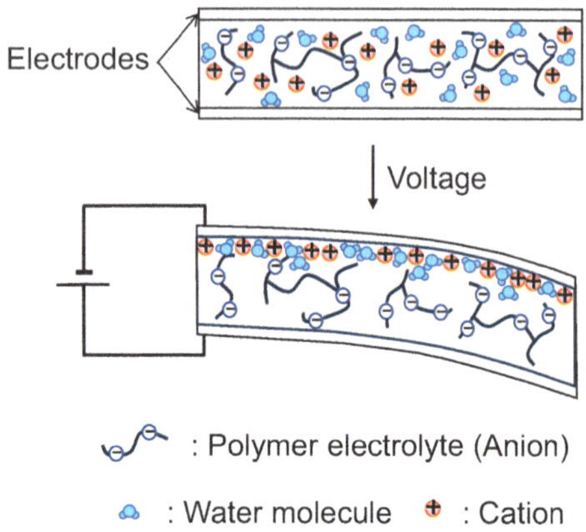

Fig. 2 Schematics of ionic EAP actuators; IPMC actuators

3.2 Conductive Polymer Actuator

Conducting polymers are polymers that exhibit electronic conductivity when conducting carriers such as electrons or holes flow through a conjugated system of double bonds. Various types of conducting polymers have been synthesized and applied to electrolytic capacitors, secondary batteries, organic EL, etc. Among them, several polymers have been studied as actuators, in which carriers are injected into a conductive polymer by causing an oxidation-reduction reaction in an electrolyte, and counter ions are doped into the polymer at that time to cause electric-field expansion and contraction in which the polymer swells and contracts (Fig. 3) [7]. The main conducting polymer actuator materials are polyaniline, polypyrrole, and polythiophene. These polymers exhibit excellent actuator performance, low-voltage drive (<2 V), high generating power (5 MPa), and high strain rate (10%). Usually, a conducting polymer and a counter electrode (and a reference electrode) are immersed in an electrolyte, and the conducting polymer is driven as an actuator by passing a current through the counter electrode. Bimorph-type devices are developed by gelling the electrolyte and bonding the conducting polymer as an electrode in order to make devices.

3.3 Carbon Nanotube (CNT)

It is known that a Carbon Nanotube (CNT) electrode expands and contracts when a voltage is applied between the electrode and a counter electrode immersed in an electrolyte [8]. This mechanism is different from that of a conductive polymer. The expansion and contraction of a CNT electrode is caused by the movement of ions during electric double-layer charge and discharge. As is well known, CNTs are either single-walled or multi-walled. Various forms of CNTs, such as single molecules, films, and fibers, have been synthesized [9–11]. The expansion and contraction phenomenon

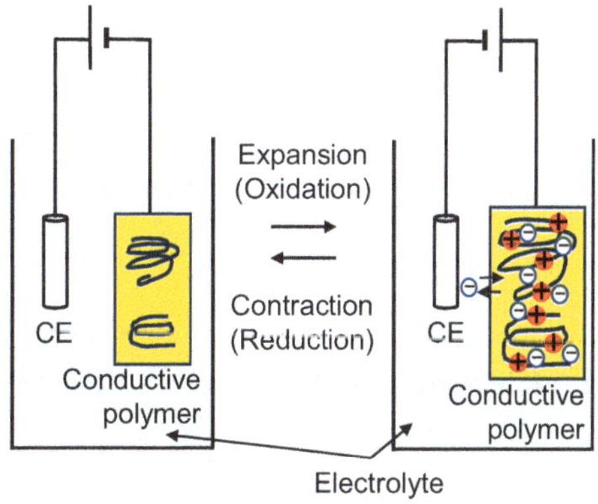

Fig. 3 Schematics of ionic EAP actuators; conductive polymer actuators

has been investigated in all of them. Furthermore, a bimorph-type device consisting of an electrode composed of a CNT, an ionic liquid, and a polymer, and an ionic liquid gel has been developed (Fig. 4) [12, 13]. This device can be driven in the air and exhibits low-voltage drive (<3 V), high generating power (10 MPa), and a large strain (2%). The principle of the response is considered to be the expansion and contraction of the electrode layer due to the volume effect of ions moved to the electrode layer. It can be molded by casting or printing, and it is easily being developed as a practical device.

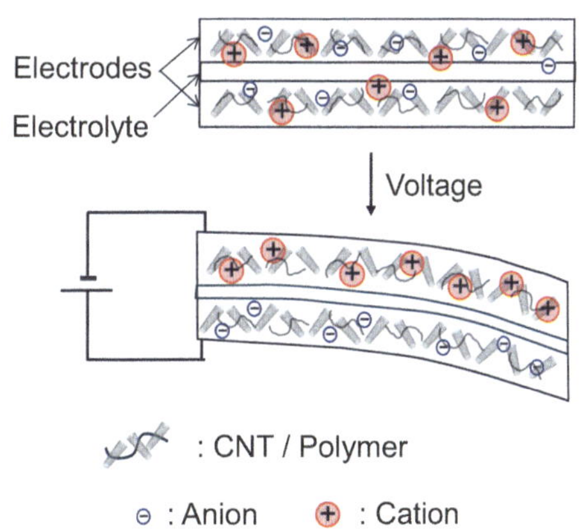

Fig. 4 Schematics of ionic EAP actuators; carbon nanotube actuators

Fig. 5 Schematics of electronic EAP actuators; electrostrictive polymer actuators with Maxwell stress

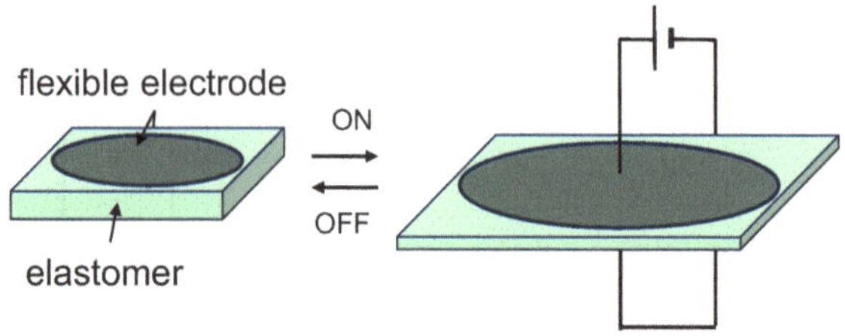

4 Electric Field-Driven Actuator

4.1 Electrostrictive Polymer Actuator

Ferroelectric polymers such as vinylidene fluoride copolymers (PVDF) exhibit inverse piezoelectricity. In other words, deformation is caused by the inversion of oriented molecular dipoles when a voltage is applied. However, the excellent electrostrictive properties of ferroelectric polymers are achieved under high applied electric fields, which not only reduce their lifetime and reliability, but also limit their application in wearable electronics and sensors. As is well known, this deformation phenomenon is a nonlinear phenomenon, and the amount of deformation is also very small. It has been found that a linear and large electromechanical response can be obtained by irradiating this material with an electron beam or the like to introduce defects into the crystal structure of oriented molecular dipoles, and research is progressing [14]. As shown in Fig. 5, the driving principle is based on the Maxwell stress between electrodes and the electrostriction effect, which is similarly proportional to the second power of the voltage, as in the dielectric elastomers described in the next section. The response of shrinkage is in the same direction regardless of whether the voltage is positive or negative. A driving voltage of 50 V/μm or more is required, and a high voltage is required, for example, a driving voltage of 1 kV is required even for a 20 μm-thick element. In addition, it has also been reported that deformation utilizing the electroclinic effect of a ferroelectric liquid crystal is reported [15, 16].

4.2 Dielectric Elastomer/Gel Actuator

As shown in Fig. 5, the actuator has a structure in which electrodes such as carbon are coated on a general-purpose polymer elastomer such as acrylic, silicone, polyurethane, etc. When a voltage of 1 kV or more is applied between the electrodes, deformation occurs due to the Maxwell stress between the electrodes and the elastic force of the elastomer. An actuator with characteristics far superior to those of a muscle in terms of stretch ratio, response speed, and energy density has

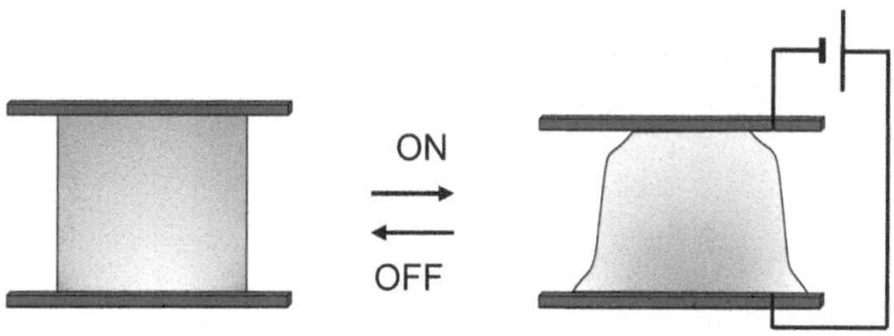

Fig. 6 Schematics of electronic EAP actuators; dielectric gel actuators

been developed by a technique in which electrodes are coated in a shape that maintains elasticity and conductivity, and a technique in which the elastomer is thinned by stretching the elastomer in advance and the voltage resistance of the elastomer is increased (pre-strain, pre-stretch). However, these are values obtained when the pre-strain is applied, and the flexibility of the original shape of the polymer is lost by the frame of the pre-strain. In addition, a booster circuit is required because a high voltage on the order of kV is required, and the system itself becomes large because a structure with consideration for safety is required. In other words, considering the whole system using a dielectric elastomer actuator, the work density, efficiency, etc., must be greatly reduced at present. Therefore, research on an elastomer that does not require pre-strain or a device that can be driven at a lower voltage is being advanced. In addition, development of an actuator using the deformation at the electrode interface by applying a voltage to a gel, as shown in Fig. 6, instead of an elastomer is also progressing [17]. Moreover, it has been reported that a liquid crystal elastomer, which is difficult to drive directly by an electric field, is wetted with a liquid crystal solvent to form a liquid crystal gel, and the liquid crystal gel is driven by an electric field [18].

5 Applications

In order to apply an actuator using soft matter to a new field as a soft actuator, it is necessary to develop appropriate device processing technology and control technology. For this purpose, a simulation using the finite element method, etc., is carried out to design the necessary actuator. According to the simulation, polymer molding, electrode formation, and patterning are carried out, and flexible electric wiring is used to integrate with a sensor, control circuit, etc., to fabricate a device. Various studies on the control method of soft actuators are also in progress. Polymer processing and electrode formation processes developed in the fields of polymer molding, batteries, etc., can be applied, and compared with other actuators, there is a possibility that various shapes and dimensions can be mass-produced at low cost.

As a specific application example, a braille display, as a tactile device, is being developed by taking advantage of the lightweight and thin characteristics of soft actuators [13]. Compared with off-the-shelf products using piezo bimorph elements, the actuator for driving braille pins can be made smaller and thinner. Therefore, there

(a)
(b)
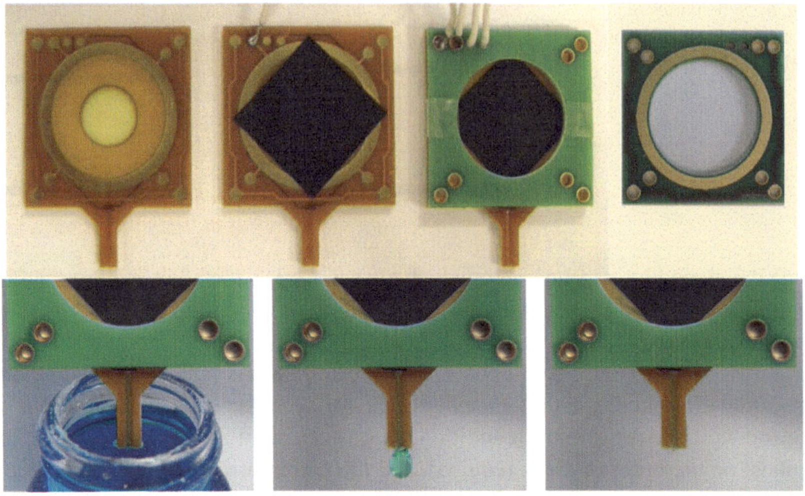

Fig. 7 Images of the silent mini pump unit using a CNT actuator. **a** Photos of parts unit of the device and **b** photos of suction and dispensing of liquid by a silent mini pump

is a possibility that a multi-line page display can be produced. In addition, a tactile page display, such as tactile drawings, can be developed as a thin device.

In the medical field, a device for controlling the tip of a catheter with an electric signal has been researched and developed since the early stages of the development of soft actuators. Since these are used in the body, research and development using a low-voltage, current-driven ion-conductive polymer and a conductive polymer actuator are being promoted. In addition, a silent mini pump that can be applied to medical treatment is also being developed as shown in Fig. 7 [19]. In addition, research on a wearable robot for the purpose of power assistance for rehabilitation is being promoted using a dielectric elastomer gel actuator.

In the robotics field, research and development of a soft robot using a soft actuator as a driving source is being promoted [20]. Research on a biomimetic robot that imitates the movement of various organisms has been promoted so far. In particular, research on an underwater microrobot is being promoted using an ion-conductive polymer actuator and a conductive polymer actuator.

In addition to the response as an actuator, it is known that an electric signal is generated by deformation not only in piezoelectric polymers, but also in ionic actuator materials as a reverse response [21]. By using these as mechanical sensors, research on a sensor system made of only soft materials without a hard substrate like conventional MEMS is being promoted. In addition, a dielectric elastomer has been put into practical use as a stretch sensor based on a ring-motion gel using the capacitance change due to its large stretch. In addition, research and development on power generation processes are being promoted, and applications such as wave power generation and human-body power generation are expected as new environmental power generation.

6 Summary

Among soft actuators using soft matter, electrically driven soft actuators, which are close to practical use, have been introduced in this section. Research on soft actuators using various materials and driving principles is in progress. Thin, lightweight, and quiet actuator elements utilizing the characteristics of soft matter have been realized and are expected to be applied in various fields. However, in terms of operating speed, power generation, and operating voltage, further research and development toward practical use are still necessary, and will be further developed.

References

1. M. Li, A. Pal, A. Aghakhani, A. Pena-Francesch, M. Sitti, Nat. Rev. Mater. **7**, 235 (2022)
2. N. El-Atab, R.B. Mishra, F. Al-Modaf, L. Johariji, A.A. Alsharif, H. Alamoudi, M. Diaz, N. Qaiser, M.M. Hussain, Adv. Intell. Syst. **2**, 2000128 (2020)
3. K. Asaka, H. Okuzaki, *Soft Actuators*, Second edn. (Springer Nature Singapore Pte Ltd., 2019)
4. C.M. Costa, V.F. Cardoso, P. Martins, D.M. Correia, R. Gonçalves, P. Costa, V. Correia, C. Ribeiro, M.M. Fernandes, P.M. Martins, S. Lanceros-Méndez, Chem. Rev. **123**, 11392 (2023)
5. K. Chen, D. Ho, Aggregate **5**, e425 (2024)
6. C. Liu, M. Yoshio, ACS Appl. Mater. Interfaces **16**, 27750 (2024)
7. K.J. Kim, S. Tadokoro (eds.), *Electroactive Polymers for Robotics Applications* (Springer Londan, 2007)
8. R.H. Baughman, C. Cui, A.A. Zakhidov, Z. Iqbal, J.N. Barisci, G.M. Spinks, G.G. Wallace, A. Mazzoldi, D. De Rossi, A.G. Rinzler, O. Jaschinski, S. Roth, M. Kertesz, Science **284**, 1340 (1999)
9. H. Mohammad, I.K. Haque, U. Vohrer, T. WAllmersperger, M. D'Ottavio, B. Kroplin, Proc. SPIE **5759**, 203 (2005)
10. T. Mirfakhrai, O. Jiyoung, M. Kozlov, E.C.W. Fok, M. Zhang, S. Fang, R.H. Baughman, J.D.W. Madden, Smart Mater. Struct. **16**, S243 (2007)
11. J. Foroughi, G.M. Spinks, G.G. Wallance, O. Jiyoung, M.E. Kozlov, S. Fang, T. Mirfakhrai, J.D.W. Madden, M.K. Shin, S.J. Kim, R.H. Baughman, Science **334**, 494 (2011)
12. T. Fukushima, K. Asaka, A. Kosaka, T. Aida, Angew. Chem. Int. Ed. **44**, 2410 (2005)
13. K. Asaka, K. Mukai, T. Sugino, K. Kiyohara, Polym. Int. **62**, 1263 (2013)
14. Federico Carpi and Elisabeth Smela (Eds.), Biomedical Applications of Electroactive Polymer Actuators, John Willey & Sons, Ltd., 2009
15. K. Hiraoka, M. Kobayasi, R. Kazama, H. Finkelmann, Macromolecules **42**, 5600 (2009)
16. C. Ohm, M. Brehmer, R. Zentel, Adv. Mater. **22**, 3366 (2010)
17. T. Hirai, T. Okinaka, S. Hayashi, Y. Amemiya, K. Kobayashi, M. Hirai, Angew. Macromol. Chem. **240**, 213 (1996)
18. T. Okamoto, K. Urayama, T. Takigawa, Soft Matter **7**, 10585 (2011)
19. Raphael Addinall, Takushi Sugino, Raphael Neuhaus, Urszula Kosidlo, Friedemann Tonner, Carsten Glanz, Ivica Kolaric, Thomas Bauernhansl, and Kinji Asaka, 2014 IEEE/ASME international conference on advanced intelligent mechatronics, Besacon, France, 2014, 1436
20. K. Suzumori, K. Fukuda, in *Ryuma Niiyama*, ed. by K. Nakajima, (The Science of Soft Robots, Springer Singapore, 2023)
21. M. Yamakita, N. Kamamichi, Z. Luo, K. Asaka, Robotic application of IPMC actuators with Redoping capability, in *Electroactive Polymers for Robotic Applications*, ed. by K.J. Kim, S. Tadokoro, (Springer, London, 2007), p. 199

Real-Sample Analysis Using Organic FET-Based Chemical Sensors

Yui Sasaki and Tsuyoshi Minami

1 Introduction

In our human bodies, sophisticated sensors (i.e., eyes, ears, nose, tongue, and skin) are incorporated to capture physical and chemical stimuli. Indeed, our behavior is regulated unconsciously or modified consciously in response to stimuli captured by these sensors [1]. To date, inspired by biological sensing systems, various artificial sensors integrated with optical, electrochemical, mass-sensitive detection mechanisms, etc., have been vigorously developed to visualize physical and chemical information. Chemical sensors consist of recognition elements that selectively interact with analytes and transducers that convert these interactions into measurable signals [2]. Sensor performance depends on the design and integration of the recognition element and transducer and is evaluated in terms of selectivity, sensitivity, response time, and other relevant metrics. In contrast to well-established physical sensors, the practical implementation of chemical sensors remains challenging, largely because of the difficulty in designing selective receptors [3]. For example, biogenic recognition materials such as enzymes and antibodies have been used in conventional sensor devices because of their specificity based on the lock-and-key model [4]. Indeed, a biosensor for glucose using an enzymatic electrode is a representative commercial device for diabetic patients. Immunoassay-based virus test kits are also used as self-testing devices. However, biogenic materials can be susceptible to denaturation under

Y. Sasaki
Research Center for Advanced Science and Technology, The University of Tokyo, Meguro-ku, Tokyo, Japan
e-mail: sayui@iis.u-tokyo.ac.jp

Y. Sasaki · T. Minami (✉)
Institute of Industrial Science, The University of Tokyo, Meguro-ku, Tokyo, Japan
e-mail: tminami@g.ecc.u-tokyo.ac.jp

Y. Sasaki
JST, PRESTO, Kawaguchi, Saitama, Japan

M. Funahashi (ed.), *Functional Soft Matter II*,
https://doi.org/10.1007/978-981-92-0100-6_18

nonphysiological conditions. In addition, the number of detectable analytes is limited by the library of biogenic recognition materials. Considering the above, the authors focus on synthetic molecular recognition materials that can be designed considering analyte sizes, charges, and molecular geometries [5–7].

Molecularly imprinted polymers (MIPs) are robust artificial recognition materials that can be prepared by imprinting specific analytes [8]. Functional monomers are employed considering intermolecular interactions with targets. Since the targets are surrounded by the functional monomers through molecular self-assembly, polymerization is carried out while preserving the complementary arrangement between the analytes and binding sites of functional monomers. Thus, cavities inside the polymers after extraction of template analytes provide three-dimensional recognition spaces (Fig. 1a). In MIP designs, appropriate ratios between functional monomers and analytes are determined by employing density functional theory calculations [5–7]. Such MIPs formed on solid-support materials can be introduced into transducer units, allowing chemical sensing at solid–water interfaces.

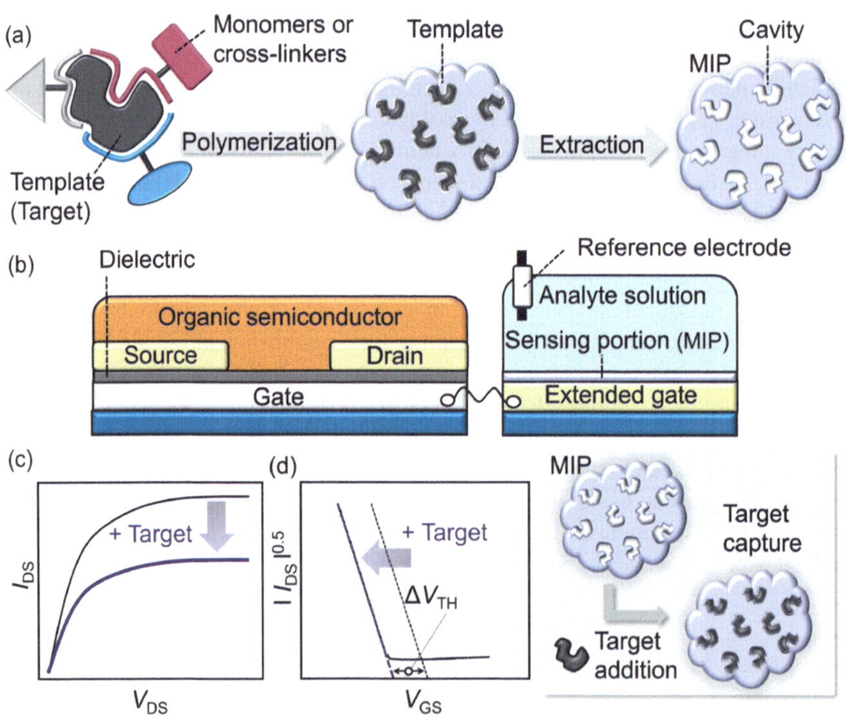

Fig. 1 **a** Conceptual figure of a molecularly imprinted polymer. **b** Schematic illustration of an extended-gate-type OFET chemical sensor. Changes in **c** output and **d** transfer characteristics upon target capture

Organic field-effect transistors (OFETs) are representative solution-processable electronic devices. Their applicability to printing technologies allows the fabrication of organic devices on flexible plastic films, which broadens the possibilities of OFETs for chemical sensing applications [9–11]. An OFET consists of gate, source, and drain electrodes, a dielectric layer, and an organic semiconductor layer [12]. The dielectric layer is sandwiched between the gate electrode and the semiconductor, thereby forming a capacitor-like structure. Owing to the amplification ability, OFETs functionalized with molecular recognition materials can visualize molecular recognition information sensitively through changes in drain currents (I_{DS}) or threshold voltages (V_{TH}) [13–16]. However, the instability of organic electronic devices is a bottleneck in chemical sensing in aqueous media. Therefore, the authors employed an extended-gate structure as a potent device platform to achieve reproducible sensing results using the OFET. The extended-gate electrode functionalized with molecular recognition materials is used as a sensing unit isolated from an operation unit (i.e., OFET) (Fig. 1b) [17, 18]. This chapter introduces an approach for real-sample analysis using extended-gate-type OFETs functionalized with MIPs (MIP-OFETs).

2 Detection Principle of an Extended-Gate-Type OFET

This section focuses on the principle of OFET operation and chemical sensing using an extended-gate electrode. A schematic illustration of an extended-gate-type OFET is summarized in Fig. 1b. OFETs operate according to the same fundamental field-effect principle as metal–oxide–semiconductor field-effect transistors. When a gate-to-source voltage (V_{GS}) is applied, an electric field is established across the dielectric layer, inducing charge carriers at the semiconductor/dielectric interface. When the magnitude of the gate bias exceeds the threshold voltage, a conducting channel forms, and a drain current (I_{DS}) flows upon application of a drain-to-source voltage (V_{DS}). The charge carriers are holes in p-type semiconductors and electrons in n-type semiconductors. V_{TH} is the threshold gate voltage at which the conducting channel forms between the source and drain electrodes [12]. Since transistor characteristics are manipulated by gate voltages, the gate electrode can be used as a sensing unit to detect surface potentials by chemical sensing at the solid–water interface [17, 18]. In this design, the extended-gate electrode is functionalized with molecular recognition materials. Molecular recognition at the functionalized extended-gate electrode alters the interfacial potential and/or capacitance, thereby modulating the effective gate bias and the conductance of the OFET. Consequently, molecular recognition events are transduced into changes in transistor characteristics, such as ΔV_{TH} or ΔI_{DS} (Fig. 1c, d).

3 Real-Sample Analysis Using MIP-OFETs

Herein, three applications of MIP-OFETs to real-sample analysis are introduced. The first example is the chemical sensing of cortisol in human saliva using an electrochemically deposited MIP electrode [5].

Cortisol has attracted considerable attention as a biomarker of psychological and physiological stress and is frequently investigated in studies of stress-related conditions [19]. Because salivary cortisol is typically present at nanomolar concentrations, analytical methods with nanomolar sensitivity are required [19]. Conventionally, reliable analytical methods such as high-performance liquid chromatography–tandem mass spectrometry [20] and immunoassays [21] have been applied to detect salivary cortisol with high accuracy. However, these methods often require specialized equipment, multistep procedures, and relatively long analysis times, limiting their suitability for on-site analysis. Meanwhile, chemical sensors using macrocyclic receptors (e.g., cucurbit[*n*]urils) have been applied to detect cortisol based on optical changes [22]. These solution-phase optical sensors transduce cortisol binding into changes in fluorescence. However, accurate discrimination of cortisol from structurally similar steroid hormones remains challenging because of the highly symmetric structures of the macrocyclic receptors.

In this study, the authors designed an MIP for cortisol using 1,2-diaminobenzene as a functional monomer (Fig. 2a). The employed monomer not only forms multiple hydrogen bonds with the target cortisol but also allows electrochemical polymerization for the fabrication of the MIP on the surface of the extended-gate Au electrode. The appropriate ratio between cortisol and the functional monomer was optimized using density functional theory (DFT) calculations. The DFT calculations were performed for cortisol/1,2-diaminobenzene molar ratios ranging from 1:1 to 1:5. With an increase in the number of 1,2-diaminobenzene molecules, the calculated binding became more favorable. According to the evaluation, the appropriate molar ratio between cortisol and the functional monomer was determined to be 1:5 (Fig. 2a). The MIP electrode was fabricated by electrochemical deposition in a mixture of cortisol and 1,2-diaminobenzene at the 1:5 molar ratio. The template was extracted from the polymer by cyclic voltammetry in a NaOH solution. The fabricated MIP electrode was integrated with the OFET device and applied to cortisol detection. The OFET device was manufactured by using a solution-processable π-conjugated polymer semiconductor, poly{2,5-bis(3-tetradecylthiophen-2-yl)thieno[3,2-*b*]thiophene} (PBTTT-C14) [23]. The OFET allowed device operation at voltages below 3 V in magnitude by employment of two dielectric layers made of metal oxide and a self-assembled monolayer with long alkyl chains (i.e., tetradecylphosphonic acid) [24]. A hydrophobic material (i.e., CYTOP™) was used as a passivation layer to cover the entire surface of the OFET device for stable operation under ambient conditions.

The MIP-OFET showed the highest response to cortisol among all tested analytes, including salivary biomarkers (i.e., uric acid, creatinine, glucose, and lactic acid) and structurally similar steroid hormones (i.e., prednisolone and progesterone). Cortisol

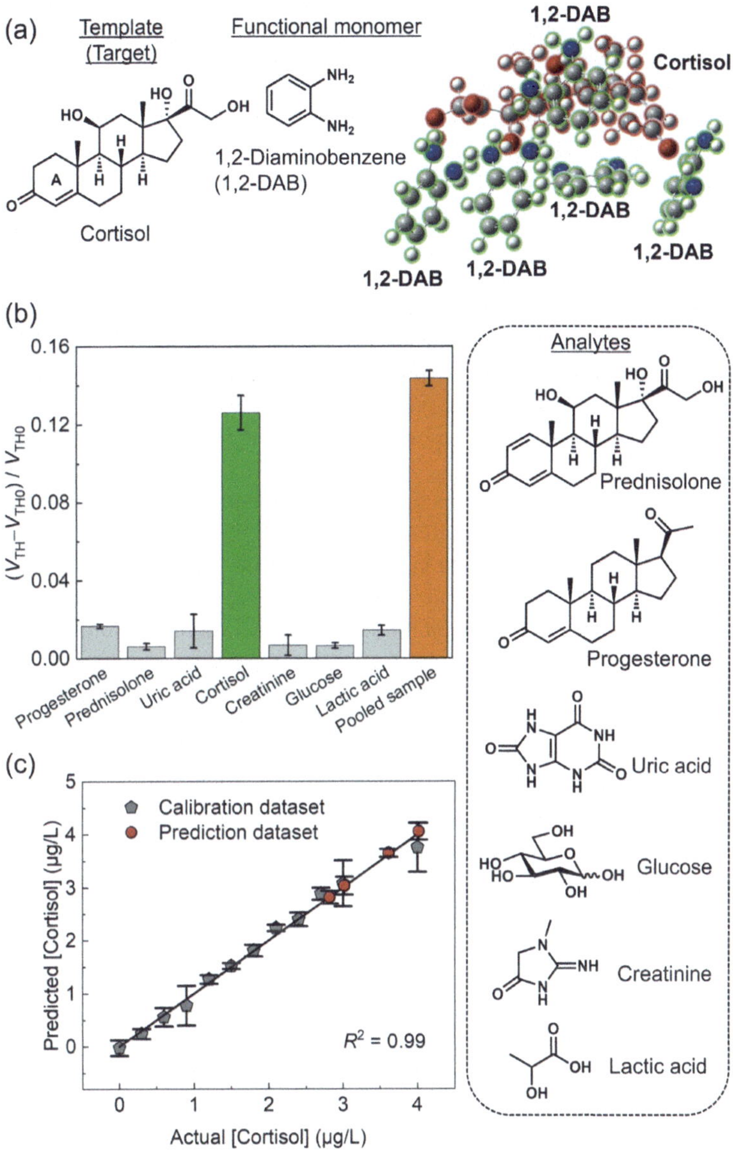

Fig. 2 **a** Chemical structures of cortisol and 1,2-diaminobenzene and the pre-organized structure optimized by DFT calculations. **b** Selectivity test results for cortisol (10 μg/L), prednisolone (10 μg/L), progesterone (10 μg/L), uric acid (8 mg/L), glucose (10 mg/L), creatinine (0.3 mg/L), and lactic acid (18 mg/L). The pooled sample contained cortisol and the above-mentioned analytes in a solution. The terms V_{TH0} and V_{TH} mean threshold voltages before and after adding analytes, respectively. **c** Spike-and-recovery test results for cortisol in human saliva

and prednisolone differ only in the presence of an additional $C_1 = C_2$ double bond in ring A of prednisolone [25, 26]. This subtle structural difference alters the rigidity and geometry of ring A and consequently affects molecular fitting within the imprinted cavity. Furthermore, a pooled sample containing cortisol and the above-mentioned analytes was used to evaluate the discrimination ability of the MIP-OFET. As shown in Fig. 2b, the change in the MIP-OFET sensor response to the pooled sample was similar to the response to cortisol in a buffer solution. The MIP-OFET exhibited a limit of detection of 2.0 nM, which is well below representative salivary cortisol concentrations reported in the literature [19]. The MIP-OFET was further applied to the quantification of cortisol in undiluted human saliva. A calibration line for a spike-and-recovery test was established by using collected transistor characteristics at different concentrations of cortisol. Spike-and-recovery test results suggested highly accurate recovery rates for cortisol (i.e., 102–110%), which implied the feasibility of an easy-to-use cortisol sensor based on the integration of molecular recognition chemistry and organic electronics.

The next demonstration concerns the detection of taurine in a food sample using an MIP-OFET. Taurine (Tau) is a sulfur-containing amino sulfonic acid with antioxidant properties. Taurine is involved in brain development and function, particularly in the hippocampus, cerebellum, and hypothalamus. Taurine is commonly added to energy drinks as a dietary supplement. Reliable quantification is therefore relevant to product quality control and to the assessment of beverages containing taurine together with caffeine and other ingredients [27]. An MIP for Tau was designed by using dopamine and 1,2-diaminobenzene as functional monomers (Fig. 3a) [6]. DFT calculation results indicated that an optimal pre-organized structure was obtained through hydrogen bonds at the 3:1:1 molar ratio (= dopamine: 1,2-diaminobenzene: Tau). In this regard, the independent gradient model was applied to visualize molecular interactions between the template Tau and the functional monomers. Figure 3a shows non-covalent interactions (i.e., hydrogen bonds) that stabilize the pre-organized structure. At this molar ratio, the MIP was obtained by electrochemical deposition on the extended-gate Au electrode. The template Tau was extracted using an aqueous HCl solution and deionized water. The fabricated MIP electrode and the OFET were integrated to construct the chemical sensor device. The MIP-OFET discriminated Tau from alanine (Ala), arginine (Arg), aspartic acid (Asp), cysteine (Cys), glutamic acid (Glu), and malonic acid (Mal) (Fig. 3b). Moreover, the reusability of the MIP was evaluated by repeated responses during alternating Tau addition and washing. The changes in V_{TH} corresponded to the addition of Tau and the washing process (Fig. 3c). Furthermore, the applicability of the MIP-OFET to real-sample analysis was evaluated by quantifying Tau in a diluted energy drink. The MIP-OFET achieved accurate Tau detection in the drink sample, which suggested the potential of the chemical sensor device for food analysis.

Next, the molecular imprinting technology was extended to drug analysis. For example, atropine is a clinically used tropane alkaloid that can cause anticholinergic adverse effects, such as memory impairment, hallucinations, and blurred vision [28]. Atropine is a racemic mixture, whereas (*S*)-hyoscyamine is the more pharmacologically active enantiomer [28]. In this study, an MIP electrode for (*S*)-hyoscyamine

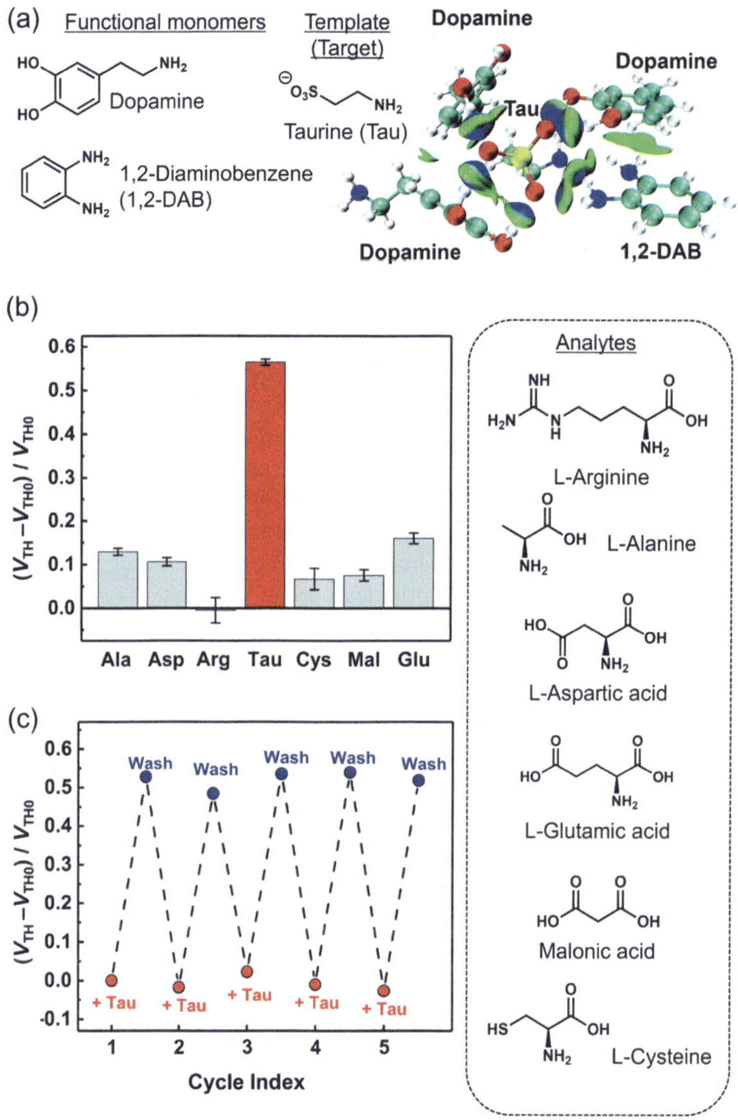

Fig. 3 **a** Chemical structures of dopamine, 1,2-diaminobenzene, and taurine, and the pre-organized structure optimized by DFT calculations. **b** Selectivity test results for Tau and structurally related compounds. [Analyte] = 10 μM. The terms V_{TH0} and V_{TH} mean threshold voltages before and after adding analytes, respectively. **c** Repeated responses of the MIP-OFET upon Tau addition (10 μM) and a washing process (with an aqueous HCl solution and deionized water)

was designed and synthesized using *N*-isopropylacrylamide (NIPAM) [29] and 2,2-dimethyl-4-pentenoic acid (DMPA) (Fig. 4a). These functional monomers were selected to form hydrogen bonds with the target (*S*)-hyoscyamine during molecular recognition. In addition, the two methyl substituents of DMPA impart steric hindrance, which plays a significant role in the recognition of sterically bulky drug molecules. DFT calculation results indicated an optimal molar ratio to form a pre-organized structure among NIPAM, DMPA, and (*S*)-hyoscyamine (= 4:1:1) through six hydrogen bonds (Fig. 4a). The MIP was prepared by polymerization on the Au electrode in the presence of a cross-linking agent (i.e., *N*,*N′*-methylenebisacrylamide). The template molecules were removed from the polymer on the extended-gate Au electrode by treatment with an HCl solution and a phosphate buffer solution. The synthesized MIP electrode was connected to the OFET device and was used for drug analysis. The fabricated MIP-OFET responded to (*S*)-hyoscyamine with high selectivity over metabolites of atropine (i.e., tropic acid and tropine) [30] and muscarinic-blocking drugs (i.e., scopolamine, tropicamide, solifenacin, and oxybutynin) [28] (Fig. 4b). The lowest detectable concentration of (*S*)-hyoscyamine using the MIP-OFET was estimated to be 1 μM. The MIP-OFET exhibited a larger response to tropine than to tropic acid, suggesting that the 8-azabicyclo[3.2.1]octane moiety contributes more strongly to the recognition of (*S*)-hyoscyamine than the 3-hydroxy-2-phenylpropanoate moiety. The MIP-OFET also discriminated (*S*)-hyoscyamine from scopolamine. This selectivity was attributed to a conformational difference arising from an intramolecular hydrogen bond involving the epoxide oxygen of scopolamine [31]. Finally, the determination of enantiomeric excess (*ee*) of (*S*)-hyoscyamine was performed by using the MIP-OFET. The *ee* determination was demonstrated over the range of 0.2–90.2% *ee* (Fig. 4c). The threshold voltage varied linearly with the enantiomeric excess of (*S*)-hyoscyamine in atropine. The correlation between V_{TH} and % *ee* of (*S*)-hyoscyamine in atropine implied the potential of the MIP-OFET for pharmaceutical analysis.

4 Summary

Real samples contain various chemical species at different concentrations. Therefore, selective and sensitive chemical sensors are required in practical real-sample analysis. Chemical sensors have been vigorously developed at the laboratory scale, but their practical implementation remains limited. A major challenge in chemical sensor development lies in receptor design. To this end, the authors focused on organic field-effect transistors functionalized with molecularly imprinted polymers (MIP-OFETs). MIPs are robust artificial molecular recognition materials with three-dimensional spaces for selective analyte capture. OFETs are switching devices possessing amplification abilities. Thus, the integration of appropriate recognition layers with the extended-gate electrodes of OFETs allows selective and sensitive chemical sensing in the presence of excess interferents. To achieve reproducible

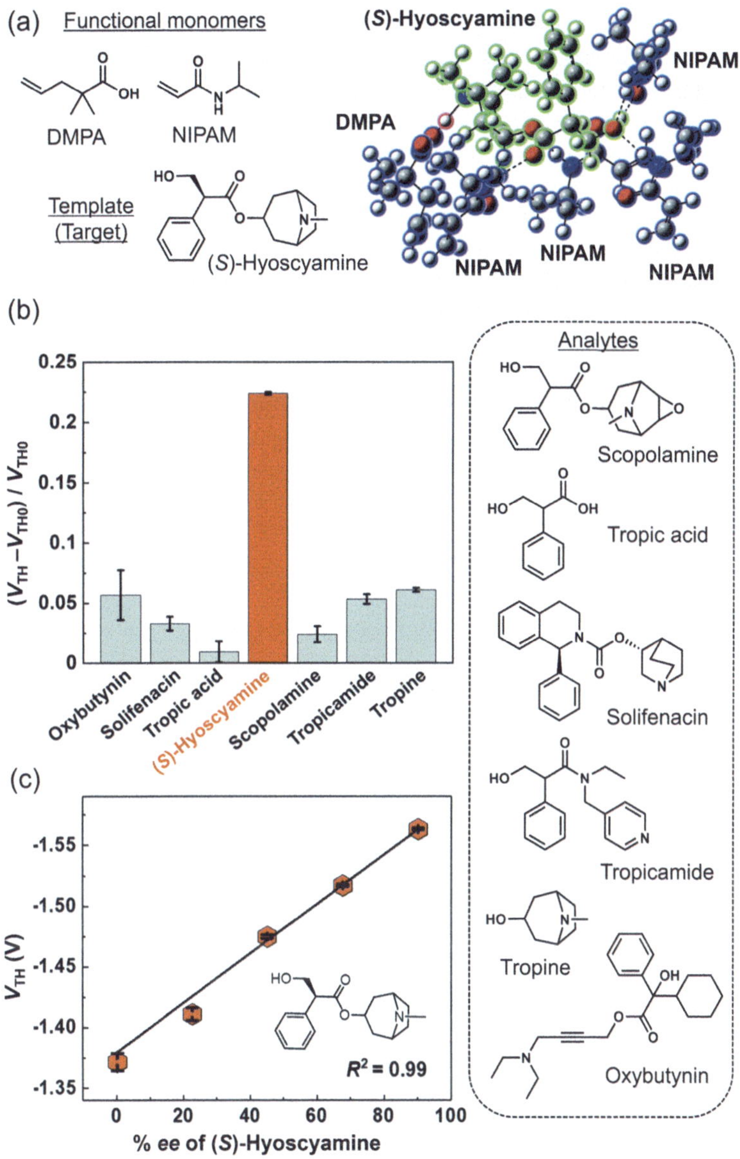

Fig. 4 **a** Chemical structures of 2,2-dimethyl-4-pentenoic acid (DMPA), *N*-isopropylacrylamide (NIPAM), and (*S*)-hyoscyamine, and the pre-organized structure optimized by DFT calculations. **b** Selectivity test results for (*S*)-hyoscyamine, metabolites of atropine (i.e., tropic acid and tropine) and muscarinic-blocking drugs (i.e., scopolamine, tropicamide, solifenacin, and oxybutynin). [Analyte] = 5 μM. The terms V_{TH0} and V_{TH} mean threshold voltages before and after adding analytes, respectively. **c** Correlation between V_{TH} and % *ee* values of (*S*)-hyoscyamine in atropine. The % *ee* determination was demonstrated at a total hyoscyamine concentration of 5 μM

chemical sensing, an extended-gate structure was selected as a sensor device platform. The target analytes comprised a stress biomarker (cortisol), a food-related nutrient (taurine), and a tropane alkaloid drug ((*S*)-hyoscyamine). Those analytes were challenging species in molecular recognition chemistry owing to their unique chemical structures. Complementarity between the analyte and the binding sites of the receptor is a key factor in the design of selective molecular recognition materials. In the molecular imprinting approach, density functional theory (DFT) calculations were employed to determine optimal molar ratios between target molecules (i.e., templates) and functional monomers. The combination of the optimized MIP electrode and the OFET was applied to chemical sensing at solid–water interfaces. The fabricated chemical sensors showed a favorable ability to discriminate specific analytes from structurally similar interferents. The demonstration in real samples revealed the potential of the combination of the MIP and the OFET as a novel chemical sensor device. The authors believe that the integration of molecular recognition chemistry and organic electronics represents a promising strategy for the realization of chemical sensor devices in real-world scenarios.

References

1. Y.H. Jung, B. Park, J.U. Kim, T.-I. Kim, Bioinspired electronics for artificial sensory systems. Adv. Mater. **31**, 1803637 (2019)
2. A. Hulanicki, S. Glab, F. Ingman, Chemical sensors: Definitions and classification. Pure Appl. Chem. **63**, 1247–1250 (1991)
3. D.J. Cram, J.M. Cram, Design of complexes between synthetic hosts and organic guests. Acc. Chem. Res. **11**, 8–14 (1978)
4. J.A. Goode, J.V.H. Rushworth, P.A. Millner, Biosensor regeneration: A review of common techniques and outcomes. Langmuir **31**, 6267–6276 (2015)
5. Y. Sasaki, Y. Zhang, H. Fan, K. Ohshiro, Q. Zhou, W. Tang, X. Lyu, T. Minami, Accurate cortisol detection in human saliva by an extended-gate-type organic transistor functionalized with a molecularly imprinted polymer. Sens. Actuators B Chem. **382**, 133458 (2023)
6. Q. Zhou, M. Wang, S. Yagi, T. Minami, Extended gate-type organic transistor functionalized by molecularly imprinted polymer for taurine detection. Nanoscale **13**, 100–107 (2021)
7. Q. Zhou, Y. Sasaki, K. Ohshiro, H. Fan, V. Montagna, C. Gonzato, K. Haupt, T. Minami, An organic transistor for the selective detection of a tropane alkaloid utilizing a molecularly imprinted polymer. J. Mater. Chem. B **10**, 6808–6815 (2022)
8. K. Haupt, P.X. Medina Rangel, B.T.S. Bui, Molecularly imprinted polymers: Antibody mimics for bioimaging and therapy. Chem. Rev. **120**, 9554–9582 (2020)
9. S. Yuvaraja, A. Nawaz, Q. Liu, D. Dubal, S.G. Surya, K.N. Salama, P. Sonar, Organic field-effect transistor-based flexible sensors. Chem. Soc. Rev. **49**, 3423–3460 (2020)
10. T. Minamiki, T. Minami, Y.P. Chen, T. Mano, Y. Takeda, K. Fukuda, S. Tokito, Flexible organic thin-film transistor immunosensor printed on a one-micron-thick film. Commun. Mater. **2**, 8 (2021)
11. K. Fukuda, T. Minamiki, T. Minami, M. Watanabe, T. Fukuda, D. Kumaki, S. Tokito, Printed organic transistors with uniform electrical performance and their application to amplifiers in biosensors. Adv. Electron. Mater. **1**, 1400052 (2015)
12. H. Sirringhaus, Device physics of solution-processed organic field-effect transistors. Adv. Mater. **17**, 2411–2425 (2005)

13. L. Torsi, M. Magliulo, K. Manoli, G. Palazzo, Organic field-effect transistor sensors: A tutorial review. Chem. Soc. Rev. **42**, 8612–8628 (2013)
14. B. King, B.H. Lessard, Review of recent advances and sensing mechanisms in solid-state organic thin-film transistor (OTFT) sensors. J. Mater. Chem. C **12**, 5654–5683 (2024)
15. R. Kubota, Y. Sasaki, T. Minamiki, T. Minami, Chemical Sensing Platforms Based on Organic Thin-Film Transistors Functionalized with Artificial Receptors. ACS Sens. **4**, 2571–2587 (2019)
16. Y. Sasaki, T. Minami, Organic transistor-based chemical sensors for real-sample analysis. Phys. Status Solidi A **220**, 2300469 (2023)
17. J. van der Spiegel, I. Lauks, P. Chan, D. Babic, The extended gate chemically sensitive field effect transistor as multi-species microprobe. Sens. Actuators **4**, 291–298 (1983)
18. T. Minamiki, T. Minami, R. Kurita, O. Niwa, S. Wakida, K. Fukuda, D. Kumaki, S. Tokito, Accurate and reproducible detection of proteins in water using an extended-gate type organic transistor biosensor. Appl. Phys. Lett. **104**, 243703 (2014)
19. C. Obel, M. Hedegaard, T.B. Henriksen, N.J. Secher, J. Olsen, S. Levine, Stress and salivary cortisol during pregnancy. Psychoneuroendocrinology **30**, 647–656 (2005)
20. W. Gao, T. Stalder, C. Kirschbaum, Quantitative analysis of estradiol and six other steroid hormones in human saliva using a high throughput liquid chromatography–tandem mass spectrometry assay. Talanta **143**, 353–358 (2015)
21. A.M. Sesay, L. Micheli, P. Tervo, G. Palleschi, V. Virtanen, Development of a competitive immunoassay for the determination of cortisol in human saliva. Anal. Biochem. **434**, 308–314 (2013)
22. A.I. Lazar, F. Biedermann, K.R. Mustafina, K.I. Assaf, A. Hennig, W.M. Nau, Nanomolar binding of steroids to cucurbit[n]urils: Selectivity and applications. J. Am. Chem. Soc. **138**, 13022–13029 (2016)
23. I. McCulloch, M. Heeney, C. Bailey, K. Genevicius, I. MacDonald, M. Shkunov, D. Sparrowe, S. Tierney, R. Wagner, W. Zhang, M.L. Chabinyc, R.J. Kline, M.D. McGehee, M.F. Toney, Liquid-crystalline semiconducting polymers with high charge-carrier mobility. Nat. Mater. **5**, 328–333 (2006)
24. H. Klauk, U. Zschieschang, J. Pflaum, M. Halik, Ultralow-power organic complementary circuits. Nature **445**, 745–748 (2007)
25. P. Manickam, F. Arizaleta, M. Gurusamy, S. Bhansali, Theoretical studies of cortisol-imprinted prepolymerization mixtures: Structural insights into improving the selectivity of affinity sensors. J. Electrochem. Soc. **164**, B3077–B3080 (2017)
26. O. Ramström, L. Ye, K. Mosbach, Artificial antibodies to corticosteroids prepared by molecular imprinting. Chem. Biol. **3**, 471–477 (1996)
27. C.P. Curran, C.A. Marczinski, Taurine, caffeine, and energy drinks: reviewing the risks to the adolescent brain. Birth Defects Res. **109**, 1640–1648 (2017)
28. A.M. Lakstygal, T.O. Kolesnikova, S.L. Khatsko, K.N. Zabegalov, A.D. Volgin, K.A. Demin, V.A. Shevyrin, E.A. Wappler-Guzzetta, A.V. Kalueff, DARK classics in chemical neuroscience: Atropine, scopolamine, and other anticholinergic deliriant hallucinogens. ACS Chem. Neurosci. **10**, 2144–2159 (2019)
29. S. Beyazit, B.T.S. Bui, K. Haupt, C. Gonzato, Molecularly imprinted polymer nanomaterials and nanocomposites by controlled/living radical polymerization. Prog. Polym. Sci. **62**, 1–21 (2016)
30. M.J. Van der Meer, H.K.L. Hundt, F.O. Muller, The metabolism of atropine in man. J. Pharm. Pharmacol. **38**, 781–784 (1986)
31. H. Weinstein, S. Srebrenik, S. Maayani, M. Sokolovsky, A theoretical model study of the comparative effectiveness of atropine and scopolamine action in the central nervous system. J. Theor. Biol. **64**, 295–309 (1977)

Printing Technology for Device Integration

Hiroaki Iino

1 Introduction

Organic materials having charge carrier transport properties can be used as organic semiconductors. The use of organic compounds as organic semiconductor materials enables high-productivity Roll-to-Roll manufacturing on flexible substrates using solution and printing processes. Therefore, new applications are expected, which are different from those of electronic devices using conventional inorganic semiconductors.

Amorphous and crystalline organic semiconductor materials have been studied for a long time [1, 2], and since the 1990s, liquid crystals have also been recognized as an organic semiconductor material [3, 4]. Mobility, which is important parameter for semiconductor materials, depends on the degree of molecular alignment and ordering. Amorphous materials exhibit low mobility of 10^{-6} cm^2/Vs to 10^{-3} cm^2/Vs, while crystalline materials exhibit high mobility up to 10 cm^2/Vs. The mobility in liquid crystalline materials exhibits between 10^{-4} cm^2/Vs and 10^{-1} cm^2/Vs, which is intermediate between those in amorphous and crystalline materials. Amorphous materials such as molecularly doped polymer and low-molecular-weight glass materials are used for practical applications such as photoconductor in xerographic applications and charge carrier transport layer in organic light emitting diodes applications, respectively. For further electronic device applications, crystalline materials with even higher mobility are required, but they have not been put to practical use because it is difficult to fabricate crystalline thin films with uniform properties by conventional solution processes.

In general, high crystalline materials have low solubility in common organic solvents. Anthracene, pentacene, and phthalocyanine are the examples of general

H. Iino (✉)
Laboratory for Future Interdisciplinary Research of Science and Technology, Institute of Integrated Research, Institute of Science Tokyo, Nagatuta, Midori-Ku, Yokohama, Japan
e-mail: iino@first.iir.isct.ac.jp

M. Funahashi (ed.), *Functional Soft Matter II*,
https://doi.org/10.1007/978-981-92-0100-6_19

crystalline materials. Nowadays, crystalline materials having high solubility have been developed, most of which have alkyl chains at both ends of the long axis of the rod-like conjugated system such as anthracene molecule. However, the polycrystalline thin films fabricated by conventional solution processes have poor uniformity because when soluble crystalline materials are formed in conventional solution process such as spin-coating method, recrystallization always occurs when organic solvents are evaporated on substrates, crystal nucleus randomly occurs, and crystal grains randomly grow up. Furthermore, since the orientation of the crystal grains cannot be controlled, the surface flatness of thin films is poor, and non-uniform crystal thin films are formed. Recently, it has been reported that crystalline thin films can be fabricated by crystal growth directly on a substrate via a solution process in order to achieve high crystallinity and uniformity [5, 6], but it is not an industrially usable thin film coating method because crystal growth is extremely slow, at several tens μm/s. This is because the crystals grow from solution, where molecules are completely random orientation and ordering, and the film must be formed at an extremely slow-speed to align at the molecular level as shown in the upper arrow of Fig. 1.

The nature of liquid crystals has a potential to dissolve the critical issue for fabrication of soluble crystalline materials. Liquid crystalline materials are used not only liquid crystalline thin films but also crystalline thin films. In here, liquid crystalline materials that exhibit liquid crystal phase at 100 °C or higher and exhibit a crystal phase at 100 °C or lower, which means that the liquid crystal material is used as a crystalline thin film at room temperature, are selected. When thin films are fabricated at the temperature in liquid crystal phase, liquid crystalline molecules are roughly oriented and ordered in an instant by liquid crystalline state, and then crystallize to form a crystalline thin film, so it is expected that polycrystalline thin

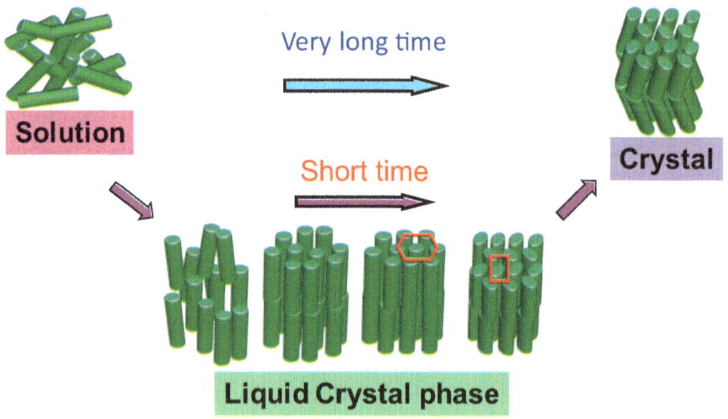

Fig. 1 Crystal thin film formation from solution. (Upper arrow) Normal materials: A long time is required to fabricate crystalline thin film with high uniformity and flatness. (Lower arrow) exhibiting liquid crystalline phase: It takes very short time to aggregate the molecules by self-assembling nature of liquid crystal phases. Crystalline thin films with high uniformity can be formed from uniform liquid crystalline films

films with excellent flatness and uniformity can be obtained even in industrially useful high-speed wet processes as shown in the lower arrow of Fig. 1. Since the fabricated thin films are crystalline state, it is expected to have a high mobility of 1–10 cm^2/Vs, which is similar to that of common crystalline materials.

In this chapter, the benefit of liquid crystallinity for the fabrication of uniform and high crystalline thin films by conventional solution process such as spin-coating, dip-coating, and blade-coating methods is described. It is focused on rod-like liquid crystalline organic semiconductor materials, which are suitable for organic field effect transistors (OFETs).

2 Fabrication of Thin Film by Solution Process Using Liquid Crystallinity

Dioctyl-terthiophene (8-TTP-8), which is a conventional liquid crystalline organic semiconductor, exhibits liquid crystal phases from 64 °C to 90 °C as shown in Fig. 2a [7]. 8-TTP-8 exhibits crystalline phase at room temperature and has three liquid crystal phases, i.e., smectic C (SmC), smectic F (SmF), and smectic G (SmG) phases. SmC phase is low-ordered liquid crystal phase, on the other hand, SmG phase is highly ordered liquid crystal phase. 8-TTP-8 is a suitable material to evaluate thin film fabrication at temperatures corresponding to the crystalline, liquid crystalline, and liquid phases. The polycrystalline thin films of 8-TTP-8 were fabricated by using a spin-coater in an oven to fabricate a uniform liquid crystal thin film using p-xylene as a solvent, and then cooling to room temperature.

Crystalline thin films fabricated at the temperature in crystal phase (below 50 °C) did not cover the whole surface of the substrate, and the aggregation of microcrystalline grains could be observed as shown in Fig. 2b. When the thin films were fabricated at higher temperature of phase transition between liquid crystal and liquid phases (> 90 °C), the film could not be kept because of the low viscosity of liquid state, and uniform thin films could not be produced as shown in Fig. 2c. On the other hand, uniform polycrystalline thin films with a thickness of 50 nm can be fabricated at the temperature in liquid crystal phase (from 60 to 90 °C) as shown in Fig. 2d, and the surface morphology observed by atomic force microscopy (AFM) shows only a 3 nm step corresponding to a molecular step and a flat film at the molecular level over 10 μm as shown in Fig. 2e [8, 9]. In conventional soluble crystalline materials such as dialkyl-quarterthiophene and dialkyl-benzothienobenzothiophene (BTBT) derivatives, the polycrystalline thin film fabricated at the temperature in liquid crystal phase shows uniform, flat, and high mobility in OFETs [8, 10]. From these results, it became clear that uniform and highly crystalline polycrystalline thin films can be easily fabricated by forming uniform liquid crystal thin films using the liquid crystal nature that has moderate viscosity, followed by cooling and crystallization.

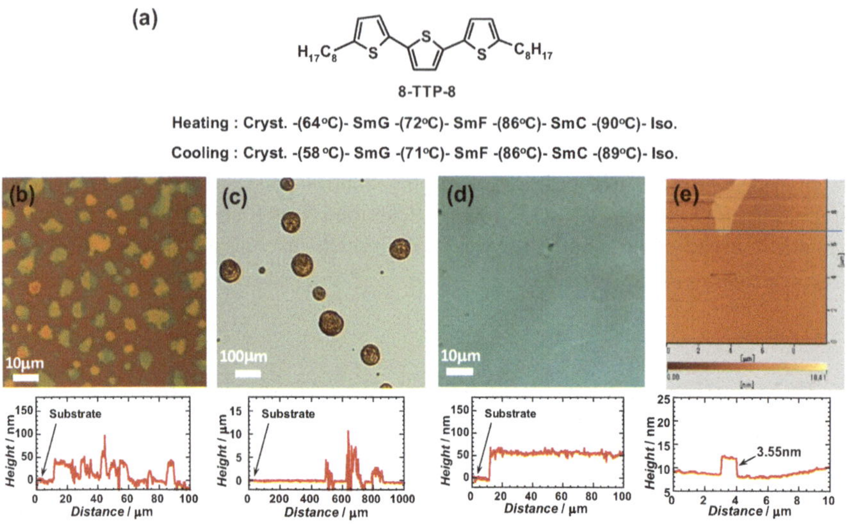

Fig. 2 Polycrystalline thin films of 8-TTP-8 spin-coated in crystal, liquid, and liquid crystal phases. **a** Chemical structure and phase transition temperature of 8-TTP-8. Optical microscopy textures of polycrystalline thin film fabricated **b** in crystal phase at 38 °C, **c** in isotropic phase at 102 °C, and **d** in smectic phase at 88 °C. The below of textures shows the cross-section profiles of the polycrystalline thin films measured by confocal laser scanning microscopy. **e** The topographical images of polycrystalline thin films spin-coated in smectic phase at 88 °C observed by atomic force microscopy

3 Liquid Crystalline Organic Semiconductors for OFETs

In order to fabricate flat and high crystalline thin films using liquid crystallinity, it is required that highly ordered liquid crystal phase such as smectic E (SmE) phase as shown in Fig. 3a. A highly ordered liquid crystal phase is structurally close to crystal phase. Therefore, it is more effective to design molecules by introducing crystallization-inhibiting fluctuations, to highly crystalline structures, rather than trying to impart order to liquids [11, 12]. Based on this idea, combining a large conjugated core favorable for crystallization, with a small aromatic ring core via single bonding is expected to induce fluctuations by providing degrees of freedom around the molecular axis, thereby inhibiting crystallization. In fact, derivatives with BTBT as the large core and thiophene rings as the small core linked by a single bond exhibit SmE phase [11, 12].

As a liquid crystalline material for OFETs, we designed a molecule consisting of BTBT as the large core and phenyl (Ph) ring as the small core to prevent crystallization. Additionally, flexible alkyl chains were attached to provide molecular anisotropy and to induce a smectic layer structure. The structure of decyl-phenyl-BTBT (Ph-BTBT-10: Fig. 3b) was designed. Ph-BTBT-10 exhibits two liquid crystal phases as shown in Fig. 3c. Polarized optical microscopy observation exhibit a typical fan-like texture in the liquid crystal phase from 210 °C to 223 °C, and disclination

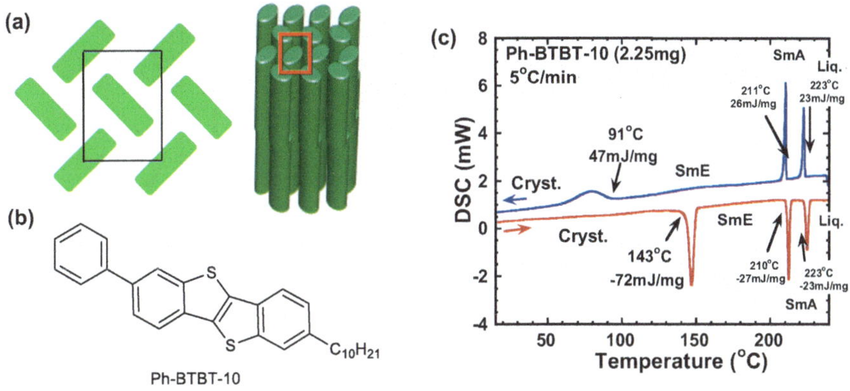

Fig. 3 Phase transition temperature of liquid crystalline Ph-BTBT-10. **a** Structure of SmE phase. **b** Chemical structure of Ph-BTBT-10. **c** Chart of differential scanning calorimetry of Ph-BTBT-10 [13]

lines within the fan-like domain in the liquid crystal phase from 143 °C to 210 °C. From the results of the polarized optical microscopy and XRD measurements, two liquid crystal phases were assigned as SmA and SmE phases [13].

4 Polycrystalline Thin Films Fabricated by Spin-Coating

In the polycrystalline thin film of Ph-BTBT-10 prepared by the spin-coating method at the temperature in liquid crystal phase, there is no aggregation of microcrystals associated with crystallization as shown in Fig. 4a, although there are some areas with different film thickness. The cross-section profile shows a step formation of 2.8 nm, which corresponds to the length of Ph-BTBT-10 molecular long axis as shown in Fig. 4b and Fig. 4c. Out-of-plane XRD measurement of this polycrystalline thin film shows only peaks corresponding to the molecular long axis in the small angle region and no peaks corresponding to π-π stacking in wide angle region as shown in Fig. 4d. This means that Ph-BTBT-10 molecules are oriented perpendicular to the substrate. In-plane XRD measurements of the thin film also show (110), (020), and (120) peaks in wide-angle region as shown in Fig. 4e, confirming that Ph-BTBT-10 molecules are oriented perpendicular to the substrate.

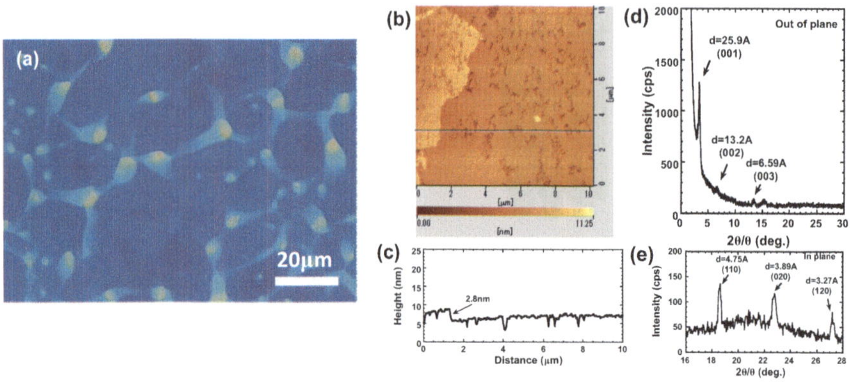

Fig. 4 Ph-BTBT-10 polycrystalline thin films fabricated at the temperature in liquid crystal phase. **a** Texture observed by optical microscope, **b** topo image observed by AFM, **c** cross-section profile of topo image observed by AFM, **d** out-of-plane XRD measurement, **e** in-plane XRD measurement [13]

5 Crystal Structure and OFET Performance of Ph-BTBT-10 Polycrystalline Thin Films

Silicon wafer with thermal oxide film was used as substrate with gate electrode and gate insulator, Ph-BTBT-10 polycrystalline thin film was formed on the substrate, and source drain electrodes are fabricated by the vacuum deposition of gold. The bottom-gate top-contact device was fabricated. The device exhibits p-channel operation, and the mobility is obtained to be 2.1 cm^2/Vs from the transfer characteristics in the saturation region. Interestingly, after a short thermal annealing at 120 °C for 5 min, the on-current increased by nearly one order of magnitude, reaching up to $5 \times 10^{-4} cm^2/Vs$ as shown in Fig. 5a. Concurrently, the mobility was greatly enhanced to 14.7 cm^2/Vs as shown in Fig. 5b.

It is interesting to know why the mobility increased significantly after thermal annealing at 120 °C for 5 min. The possible reasons for the increased mobility after annealing include an increase in grain size and a reduction in contact resistance. However, polarized optical microscopy observations revealed that the grain size remained around 10 μm both before and after annealing. Futhermore, the contact resistance unexpectedly increased from 1.2 kΩcm to 2.1 kΩcm upon annealing. The higher mobility of Ph-BTBT-10 due to thermal annealing is neither an increase in grain size nor an improvement in contact characteristics [13].

Although the wide angle region of in-plane XRD corresponding to the inter layer structure of Ph-BTBT-10 polycrystalline thin film is unchanged before and after annealing, the peak corresponding to the inter layer structure that appears in the small angle region of out-of-plane XRD is changed. The peak corresponding to one molecular length, 26 Å, decreased after thermal annealing and new peak corresponding to two molecular length, 52 Å, appears as shown in Fig. 5c. Comparing the topo images

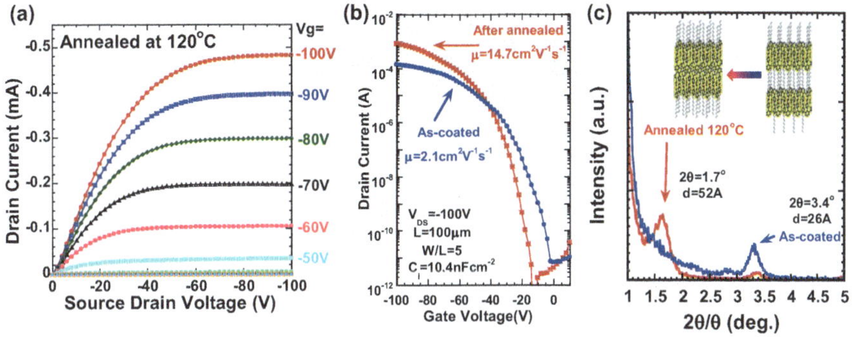

Fig. 5 Ph-BTBT-10 OFET characteristics. **a** Output characteristic of Ph-BTBT-10 OFET after thermal anneal at 120 °C for 5 min. **b** Transfer characteristics of Ph-BTBT-10 OFET before and after thermal anneal in saturation region. **c** Out-of-plane XRD of Ph-BTBT-10 polycrystalline thin films before and after thermal anneal [13]

observed by AFM before and after thermal annealing, only step corresponding to one molecular length is observed before annealing, but after thermal annealing at 120 °C, there is a 5.7 nm step corresponding to a length of approximately two molecules. This indicates that thermal annealing has changed the crystal structure to one unit of two molecules. Taken together, the large increase in mobility from 2.1 cm^2/Vs to 14.7 cm^2/Vs due to thermal annealing can be attributed to the change of crystal structure from monolayer crystal structure to bilayer crystal structure as shown in the inset of Fig. 5c. This bilayer crystal structure is the same structure of single crystal of Ph-BTBT-10 prepared from solution. Monolayer crystal structure comes from the structure of liquid crystal phase because the thin film fabrication using the precursor of liquid crystalline films. The thermal anneal increases the crystallinity of polycrystalline thin films from disordered structure like liquid crystal phase to ordered crystals such as single crystals prepared from solution.

6 Polycrystalline Thin Film Fabricated by High-Speed Dip-Coating Method

In order to realize to fabricate organic devices by the Roll-to-Roll process, the fabrication of Ph-BTBT-10 crystalline thin films by high-speed dip-coating technique using liquid crystallinity is demonstrated [14].

In the dip-coating technique, the uniform and flat polycrystalline thin films can be fabricated using the precursor of uniform liquid crystalline thin films by solution heated on the hot plate at 130 °C. Even at high dip-coating speed of 40 mm/s, uniform and flat crystalline thin films can be realized on the whole surface of about 1-inch square substrate. In the low dip-coating speed of 5 mm/s, the polycrystalline thin films with large crystal grains oriented in coating direction are formed

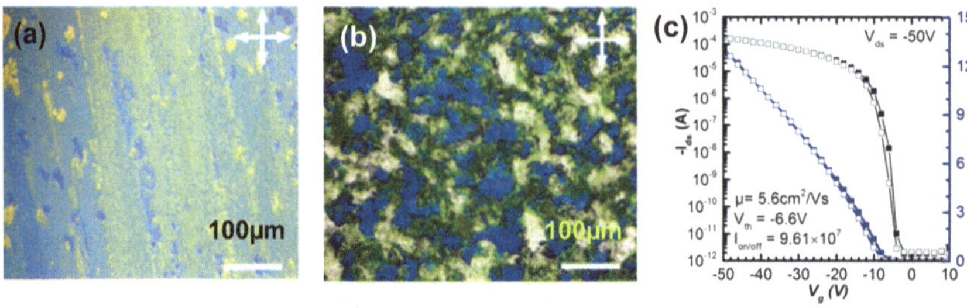

Fig. 6 Ph-BTBT-10 polycrystalline thin films fabricated by dip-coating. Optical microscopy textures of polycrystalline thin film fabricated by dip-coating speed of **a** 5 mm/s and **b** 40 mm/s. **c** Transfer characteristics of OFETs with Ph-BTBT-10 polycrystalline thin film fabricated by dip-coating speed of 40 mm/s

as shown in Fig. 6a. On the other hand, a mosaic-like texture is observed for the thin films fabricated at a highdip-coating speed of 40 mm/s, which indicates a polycrystalline structure with non-oriented grains as shown in Fig. 6b. Polycrystalline thin films with no crystal grain orientation have big benefit for practical applications because the OFET performance is independent of coating direction, and the design for integrated circuit is not limitation for coating directions. In fact, when bottom-gate bottom-contact FETs were fabricated using these polycrystalline thin films and the electrical characteristics were compared, the polycrystalline film with mosaic-like texture fabricated by high dip-coating speed of 40 mm/s exhibits high mobility of 5.6 cm^2/Vs as shown in Fig. 6c, furthermore, the standard deviation of the mobility of more than 25 OFETs divided by the average mobility is less than 20%, which indicates that the polycrystalline thin films are suitable for integrate circuit.

In order to investigate the scalability of high-speed dip-coating technique for large-area substrates, polycrystalline thin films are fabricated not only on 1-inch substrates, but also on 2-inch and even 4-inch substrates. Uniform and flat polycrystalline thin films were achieved regardless of substrate size, and the polycrystalline thin films are consisted of the mosaic-like crystal structure. Bottom-gate bottom-contact FETs (more than 200 transistors) were fabricated with the polycrystalline thin films on large 4-inch square substrate, and the OFETs exhibited high average mobility about 4 cm^2/Vs and small variation of mobility. These results of dip-coating technique indicate that the dip-coating technique using liquid crystallinity is scalable and applicable to the fabrication of high-quality organic semiconductor thin films.

7 Polycrystalline Thin Film Fabricated by High-Speed Blade-Coating Method

The high-speed thin film fabrication using the liquid crystallinity is not limited to the dip-coating technique but the blade-coating technique, which is often used in the Roll-to-Roll process for printed technology. The film fabrication speed of the blade-coating technique is 140 mm/s, which is faster than that of the dip-coating technique. In the film fabrication at room temperature of 30 °C, which is the temperature of the crystal phase, crystal nuclei and crystal growth are randomly generated and occurred, respectively, and the resulting crystalline thin film did not cover the whole surface of substrate and had big roughness as shown in Fig. 7a. This is due to recrystallization on the substrate when organic solvent is evaporated. On the other hand, when the crystalline thin films were formed at 80 °C, which is the temperature in supercooled liquid crystal phase, uniform and flat crystalline thin films were fabricated as shown in Fig. 7b [15]. This is because no recrystallization occurred when the thin films were fabricated at the temperature in supercooled liquid crystal phase, and crystallization occurred during the cooling process to room temperature after the uniform and flat liquid crystal thin films were fabricated. When OFETs with bottom-gate bottom-contact structure were fabricated with these polycrystalline thin films, the average mobility was as low as 0.24 cm^2/Vs due to the inhomogeneity of the polycrystalline thin films fabricated at room temperature, and the variation of transfer characteristics was extremely large as shown in Fig. 7c. On the other hand, the OFET of polycrystalline thin films fabricated at the temperature of liquid crystal phase showed an average mobility of 4.8 cm^2/Vs and a small coefficient of variation of 7.3%, which is standard deviation of mobility divided by average mobility as shown in Fig. 7d. The threshold voltage was also within the range of 1 V or less, suggesting that the OFETs can be used in transistor arrays for active matrix display and logic integrated circuits.

8 Summary

Liquid crystal materials have the benefit as organic semiconductor for printed technology such as blade-coating and dip-coating. Conventional soluble crystalline materials have an issue about the less uniformity and flatness crystalline films by conventional solution coating method. In organic semiconductor having liquid crystal phase, it is easy to fabricate the uniform crystalline thin films using the precursor of uniform liquid crystalline thin films by solution process. The uniform polycrystalline thin films show small variation of device performance and high mobility. Furthermore, high-quality crystalline thin films can be fabricated by even high-speed dip-coating and blade coating such as 140 mm/s. Thus, liquid crystalline materials are good candidate as organic semiconductor for printing technology.

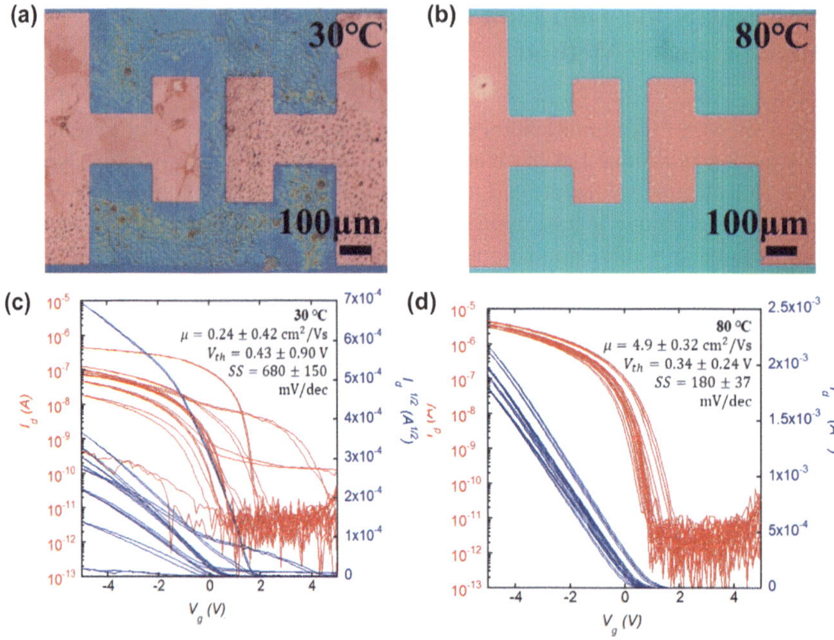

Fig. 7 Ph-BTBT-10 polycrystalline thin films fabricated by high-speed blade-coating. Optical microscopy textures of Ph-BTBT-10 polycrystalline thin film fabricated by blade-coating speed of 140 mm/s **a** at 30 °C in crystal phase and **b** at 80 °C in super cooled liquid crystal phase. Transfer characteristics of OFETs with Ph-BTBT-10 polycrystalline thin films **c** at 30 °C in crystal phase and **d** at 80 °C in super cooled liquid crystal phase

References

1. P.M. Bosenberger, D.S. Weiss, *(Organic Photoreceptors for Xerography* (Marcel Decker, Inc., 1998)
2. R.G. Kepler, Charge carrier production and mobility in Anthracene crystals. Phys. Rev. **119**, 1266–1229 (1960)
3. D. Adam, F. Closs, T. Frey, D. Funhoff, D. Haarer, P. Schuhmacher, K. Siemensmeyer, Transient photoconductivity in a discotic liquid crystal. Phys. Rev. Lett. **70**, 457–460 (1993)
4. M. Funahashi, J. Hanna, Fast hole transport in a new calamitic liquid crystal of 2-(4′-heptyloxyphenyl)-6-dodecylthiobenzothiazole. Phys. Rev. Lett. **78**, 2184–2187 (1997)
5. A. Yamamura, S. Watanabe, M. Uno, M. Mitani, C. Mitsui, J. Tsurumi, N. Isahaya, Y. Kanaoka, T. Okamoto, J. Takeya, Wafer-scale, layer-controlled organic single crystals for high-speed circuit operation. Sci. Adv. **4**, eaao5758 (2018)
6. S. Arai, S. Inoue, T. Hamai, R. Kumai, T. Hasegawa, Semiconductive single molecular bilayers realized using geometrical frustration. Adv. Mater. **30**, 1707256 (2018)
7. M. Funahashi, J. Hanna, High ambipolar carrier mobility in self-organizing terthiophene derivative. Appl. Phys. Lett. **76**, 2574–2576 (2000)
8. H. Iino, J. Hanna, Availability of liquid crystallinity in solution processing for polycrystalline thin films. Adv. Mater. **23**, 1748–1751 (2011)
9. H. Iino, J. Hanna, Liquid crystalline thin films as a precursor for polycrystalline thin films aimed at field effect transistors. J. Appl. Phys. **109**, 074505 (2011)

10. H. Iino, J. Hanna, Liquid crystalline thin films as a precursor for polycrystalline thin films aimed at field effect transistors. Mol. Cryst. Liq. Cryst. **510**, 259–267 (2009)
11. H. Iino, T. Kobori, J. Hanna, Improved thermal stability in organic FET fabricated with a soluble BTBT derivative. J. Non Cryst. Solids **358**, 2516–2519 (2012)
12. H. Iino, T. Kobori, J. Hanna, High uniformity and high thermal stability of solution-processed polycrystalline thin films by utilizing highly ordered smectic liquid crystals. Jpn. J. Appl. Phys. **51**, 11PD02 (2012)
13. H. Iino, T. Usui, J. Hanna, Liquid crystals for organic thin-film transistors. Nat. Commun.Commun. **6**, 6828 (2015)
14. H. Wu, H. Iino, J. Hanna, Scalable ultrahigh-speed fabrication of uniform polycrystalline thin films for organic transistors. ACS Appl. Mater. Interfaces **12**, 29497–29504 (2020)
15. I. Suzuki, J. Hanna, H. Iino, High-speed blade-coating using liquid crystalline organic semiconductor Ph-BTBT-10. Appl. Phys. Exp. **17**, 051007 (2024)

Liquid-Crystalline Metal Complexes

Shin-ichiro Kawano

1 Introduction

Metal complexes exhibiting liquid-crystalline behavior are known as metal-containing liquid crystals or metallomesogens. Following Vorländer's early work in 1923, which identified a smectic phase in diaryl mercury (II) complex with imino groups [1], various types of metal complexes have been reported, including those containing light metals, such as alkali and alkaline earth metal ions [2], as well as high-period elements, including lanthanoid ions [3]. Metallomesogens, particularly those with d- and/or f-orbitals hold promise for exerting advantageous properties, such as redox, magnetic, and photochemical attributes, distinguishing them from nonmetal mesogens composed of lighter elements [4, 5].

One of the most distinctive features of metal complexes lies in their coordination bonding and related specific chemical structures, such as square planar, tetrahedral, octahedral, and metallocene frameworks (Fig. 1). Furthermore, certain transition metal ions exhibit a relatively high number of chemical bonds per metal ion (as seen in ferrocene with a coordination number of 10 around Fe atoms) [6]. Additionally, metallomesogens can form multiple intra- and inter-molecular bonds between two metal ions [7, 8]. In such metal-to-metal bonds, reversible bonding often influences both the structure and electronic interactions of the self-assembled structure through labile or inert intermolecular interactions. Although not inherently planar structures conducive to the anisotropic direction orientation of molecules, metallomesogens impart propensity to orient themselves through aliphatic side chains.

S. Kawano (✉)
Department of Chemistry and Biomolecular Science, Faculty of Engineering, Gifu University, Yanagido, Gifu, Japan
e-mail: kawano.shinichiro.y0@f.gifu-u.ac.jp

M. Funahashi (ed.), *Functional Soft Matter II*,
https://doi.org/10.1007/978-981-92-0100-6_20

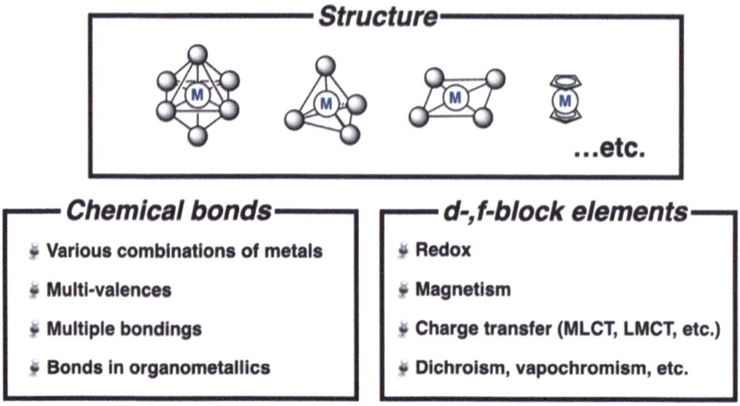

Fig. 1 Characteristics of metallomesogens

Moreover, compared to conventional organic compounds primarily composed of C, H, N, and O, metal complexes allow access to a wide range of element combinations. For instance, chemists can employ not only common Werner-type coordination bonds but also metal-chalcogens [9], metal-halogens [10], and metal–metal bonds [11]. Furthermore, certain metal ions can alter their valence around the metal center, thereby altering the redox-active behavior and chemical properties of the metallomesogens. Particularly, carbon forms relatively stable direct bonds with metal ions. These metal complexes, also referred to as organometallic compounds, have been reported to act as metallomesogens with distinct chemical properties, including luminescence [12].

In terms of thermal properties, metallomesogens often exhibit elevated melting temperatures (solid-to-mesophase) and clearing temperatures (mesophase-to-isotropic liquid), occasionally decomposing before transitioning into isotropic liquids. Nevertheless, they often offer potential advantages owing to their electronic and structural uniqueness [5, 13].

Furthermore, efficient synthetic methodologies in coordination chemistry enable the construction of various molecular module combinations in supramolecular architectures. Specifically, fluid materials composed of molecules with discrete nanospaces can be synthesized. Unlike porous "solid" materials, these fluid-porous materials allow deformation, resulting in thin films over a wide surface area owing to their low viscoelastic character.

This chapter primarily focuses on discussing the unique liquid-crystalline properties and functions attributed to metal ions, rather than examining each metallomesogen categorized by its elements. More detailed or comprehensive content can be found in the referenced books and reviews [4, 5].

2 Metallomesogens Used as Rigid Scaffolds for Liquid Crystals

As early examples, systematic examples of liquid-crystalline metal complexes include Ni(II), Pd(II), and Pt(II) complexes with dithiolene ligands (Fig. 2) [9]. Giroud reported that these compounds exhibited thermotropic liquid-crystalline behavior with a smectic structure in a temperature range from 109 to 184 °C. A crucial aspect of the emergence of thermotropic liquid crystals is the ability of transition-metal complexes with square-planar structures to act as planar mesogen structures when they possess an appropriate number of side chains. Based on their molecular design, these metallomesogens often exhibit smectic phases. Since then, many examples have been reported, including transition metal complexes with square-planar structures having versatile ligands, such as *N*,*N*-bis(salicylidene)-ethylenediamine (salen) [14], β-diketonate [15], and square-planar 1,3-diketonate Schiff base-type ligands [16]. These versatile ligands allow for the investigation of the metal dependency on liquid-crystalline properties within the same coordination geometry. In a related family of compounds, a higher clearing temperature was observed for larger atoms, such as Pd, compared to Ni complexes, possibly due to the greater association between the atoms.

Ohta et al. reported on a metal complex with a bis(diphenylglyoximato) ligand exhibiting thermochromic behavior in cast films. As the metal complex undergoes a phase transition, the metal–metal distance is altered by changing the temperature

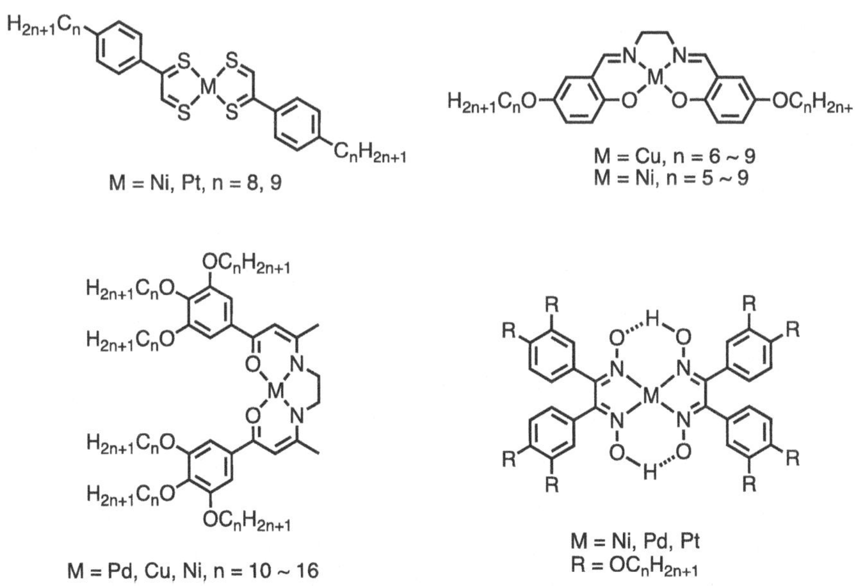

Fig. 2 Thermotropic liquid crystals composed of metallomesogens

[17]. The chromatic changes are critically dependent on the type of metal (d^8 metal ions such as Ni, Pd, and Pt) and the distance between the metal ions.

3 Magnetic Properties in Liquid Crystals

Compared to ordinary organic liquid-crystalline molecules, metal-containing complexes exert distinct control over the spin state of the d-electrons (Fig. 3). Hayami et al. presented a metallomesogen-containing Fe^{2+} that exhibited thermotropic liquid-crystalline behavior. Fe^{2+} ion complexes are endowed with a unique spin crossover phenomenon, where the molecule changes the high-spin state ($S = 2$) and the low-spin state ($S = 0$) upon photoirradiation [18]. This phenomenon known as "light-induced excited spin-state trapping (LIESST)" was observed alongside the thermotropic liquid-crystalline behavior. Below 100 K, the metal complex exhibited mainly a low-spin state (< 15% of the high-pin state), whereas above 217 K, the high-spin state was dominant. Upon UV irradiation at 5 K, LIESST behavior was observed, transitioning from a low-spin state to a high-spin state.

An example of electrochemically active liquid–crystal (LC) materials involves a calamitic mesogen tethering ferrocene moieties, as reported by Levelut [19]. The ferrocene-containing molecule exhibited no mesomorphism as its neutral form, while, at its one electron-oxidized state, the molecule exhibited a smectic A phase below 83 °C. Additionally, a Pt-complex with dioxylene ligands reversibly changes its redox state, acting as a "non-innocent ligand" [20]. This liquid-crystalline metallomesogen exhibited electroactive properties on an indium tin oxide (ITO) electrode during cyclic voltammetry.

(a)

(b)

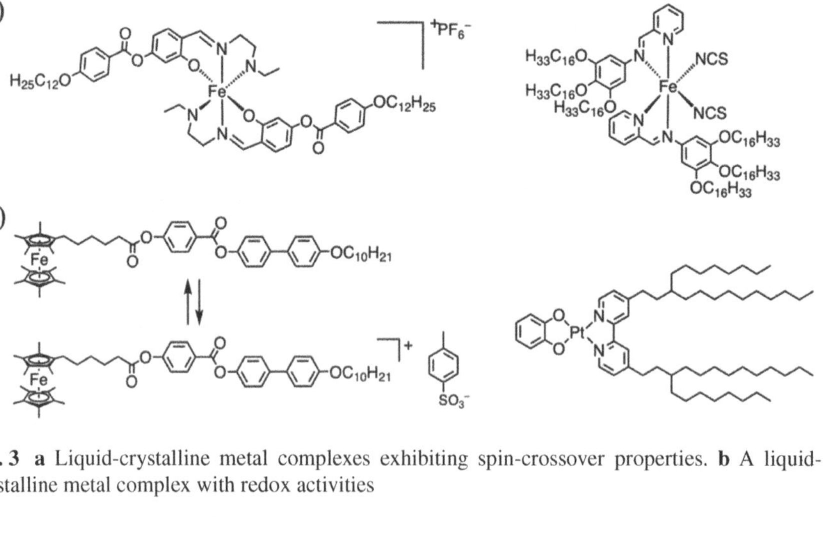

Fig. 3 **a** Liquid-crystalline metal complexes exhibiting spin-crossover properties. **b** A liquid-crystalline metal complex with redox activities

4 Lanthanidomesogens

Lanthanide metal ions are employed as molecular building blocks owing to their excellent properties such as magnetism and luminescence [21, 22]. Spherical and bulky trivalent oxidation lanthanide ions must be encapsulated by the ligand and typically require high coordination numbers.

The coordination geometries and nuclearities of metal complexes significantly influence supramolecular assembly in the mesophase. Different types of supramolecular structures in the mesophase, such as lamellar, columnar, and cubic structures, can be induced by selecting the metal center. For instance, Piguet et al. reported a series of lanthanide complexes with 2,6-bis(benzimidazole)pyridine ligands (Fig. 4) [23]. When larger lanthanide ions (Ln = La–Nd) coordinated by a tridentate ligand are utilized, a lamellar mesophase is observed. By contrast, the middle range of the metal centers (Ln = Sm–Ho) induced cubic phases. In addition, a hexagonal phase was observed for the smaller metal Ln = Er–Lu, where the metal center was effectively encapsulated by the ligand, as revealed by the single crystallography of the reference compound.

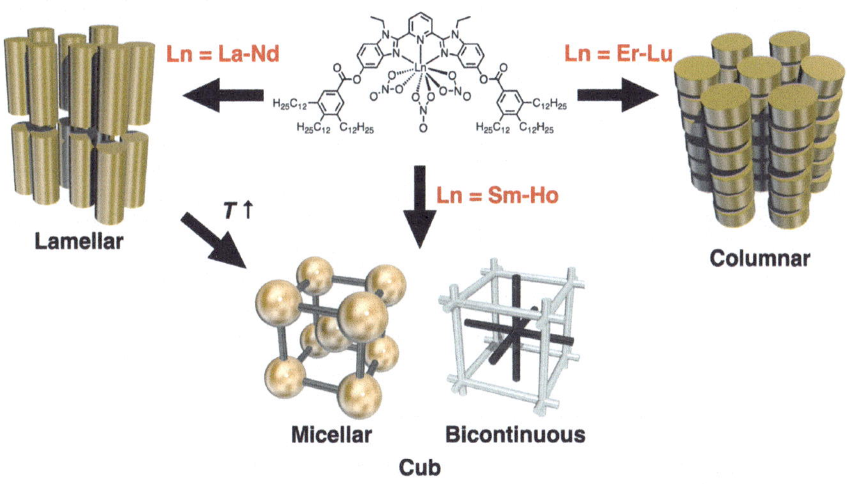

Fig. 4 Lanthanide complexes using 2,6-bis(benzimidazole)pyridine ligands exhibit diverse self-assembled structures dependent on the metal center

5 Induction of Ferroelectric Liquid-Crystalline Phases and Cholesteric Phases Using Chiral Metal Complexes

Serrano et al. reported ferroelectric liquid crystals composed of metallomesogens (Fig. 5) [24]. A metal complex with binuclear palladium ions coordinated by a 4,4'-bis(alkyloxybenzylidene)azine ligand, along with a bis μ-acetato ligand with chiral carbons, exhibited chiral smectic phases (Sc). This is the first organometallic ferroelectric LC displaying an enantiotropic Sc phase (102–119 °C). Ferroelectric properties can be observed in discotic LC compounds. Serrano et al. also reported bis(beta-diketo) metal complexes (M = VO, Cu, or Pd) with 10 chiral side chains [25]. These complexes exhibited rectangular columnar phases, demonstrating an electro-optic response under an alternating electric field.

Doping nematic liquid crystals with enantiomeric organic compounds often leads to the formation of chiral nematic (or cholesteric) phases [26]. Investigating the unique behavior of self-assembled structures owing to the chirality of chiral dopants at the molecular level can induce the transfer and amplification of chirality to a helical structure on a micrometer scale. However, compared to chiral organic molecules, little attention has been given to the Δ or λ isomers of metal complexes as chiral dopants.

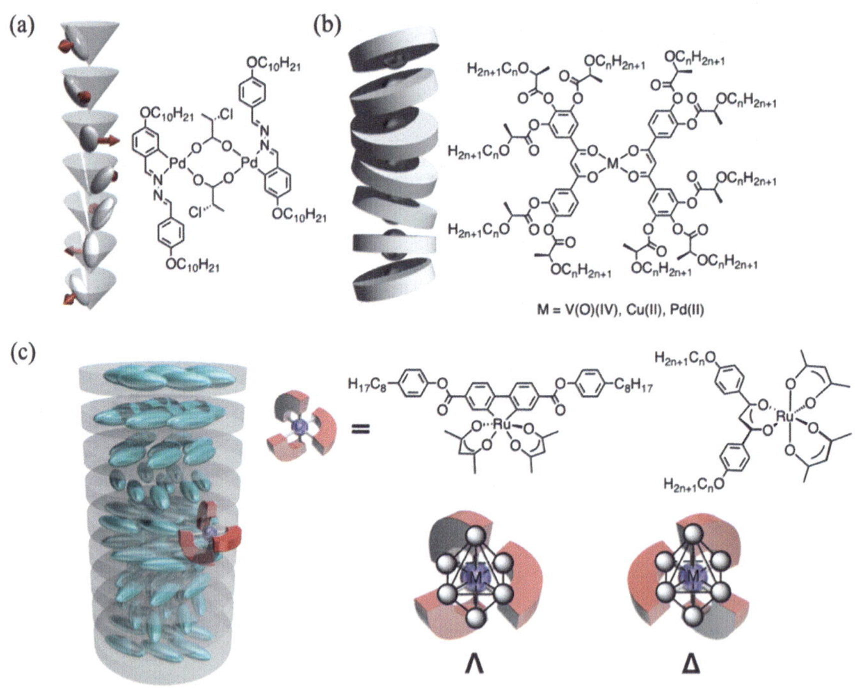

Fig. 5 **a** Metallomesogen displaying ferroelectric behavior. **b** Ferroelectric columnar liquid crystal. **c** Cholesteric liquid crystal having the chiral dopant of either Δ or λ enantiomers

Yamagishi et al. reported a series of chiral dopants comprising chiral Ru(II) complexes with octahedral coordination structures [27, 28]. The Δ- or λ-isomers of these metal complexes induced chiral nematic phases when mixed with nematic liquid crystals. The unique chirality of metal complexes with propeller-like structures sometimes results in relatively strong helical twisting power.

6 Liquid Crystals Containing Multimetal Complexes

Paddle wheel metal complexes consist of dinuclear metal complexes bridged by carboxylic acids (Fig. 6) [29, 30]. Core metal ions such as Cr, Mo, Ru, and Rh, with multiple bonds between the metals, are utilized, and when attached to aliphatic carboxylic acids, certain paddlewheel metal complexes exhibit thermotropic liquid-crystalline behaviors. Chisholm et al. presented a series of paddlewheel metallo-complexes exhibiting unique temperature-dependent thermochromic behaviors [31]. These chromatic changes are strongly related to the intermolecular interactions between neighboring molecules through their axial coordination. Furthermore, it has been reported that axial coordination by halide ions can be utilized to bridge the molecules, thereby tuning their electronic properties.

Multiple transition metals can form discrete metal clusters with octahedrally coordinated structures. These large metalloclusters also form mesophases when branched lipophilic chains are introduced at their peripheries. Since the first report

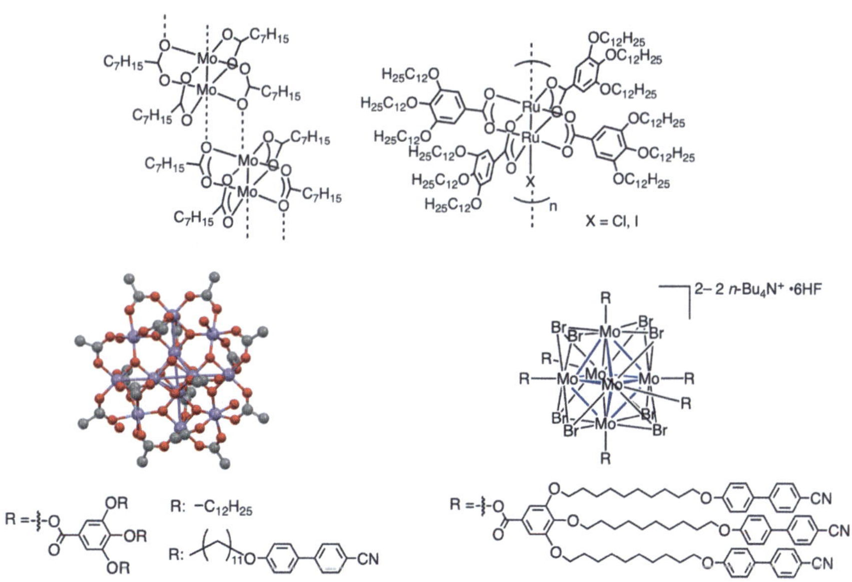

Fig. 6 Liquid-crystalline metal complexes having multinuclear complexes

in 1993, the discrete Mn cluster $[Mn_{12}O_{12}(OAc)_{16}(H_2O)_4]$, known as Mn_{12}, has attracted significant attention owing to its prominent bistability in magnetism at the single molecular level [32]. Mn_{12} clusters are envisioned as qubits for quantum computers, deemed as the ultimate storage bits of molecular magnetic memory. By replacing the peripherally coordinating acetates with gallic carboxylates containing long aliphatic chains, Mn_{12} clusters are endowed with the liquid-crystalline properties [33]. These LC Mn_{12} clusters exhibit either (1D) smectic or (3D) body-centered cubic mesophases depending on side chains such as cyanobiphenyl-containing chains or dodecyl chains, respectively. Furthermore, they show hysteresis loop characteristics typical of single-molecule magnets.

A metallocluster known as "clustomesogens," composed of six molybdenum ions has been synthesized by attaching branched aliphatic mesogens containing cyanobiphenyl groups [34]. Clustomesogens with a dendritic-like configuration exhibited a smectic phase, wherein the metallocluster formed a layered structure in the mesogens. What makes the LC molecules intriguing is that the cluster exhibits luminescence properties attributed to metal–metal charge transfer in the red-NIR region, with a maximum peak at 735–745 nm.

7 Nanoporous Soft Materials

7.1 Nanoporous Liquids and Lyotropic Liquid Crystals

In 2007, James et al. proposed the concept of microporous liquids, suggesting that molecules with well-defined micropores could form such liquids [35]. He categorized these liquids into three types: *Type I porous liquid,* a neat liquid composed of a molecule with a rigid internal nanospace, potentially resulting in a porous liquid with permanent porosity. In the liquid phase, molecules with inherent nanospaces must exhibit low melting points. In addition, such molecules must be rigid, excluding flexible crown ethers and related flexible host molecules. *Type II porous liquid*, composed of a molecule with an empty cavity dissolved in sterically hindered solvents, and *Type III porous liquid*, particles with microporous frameworks dissolved in a solvent forming a fluid phase with porosity. The dispersion of coordination frameworks and zeolites functionalized with solubilizing groups falls under this category.

Cooper et al. reported an imine-bridged organic cage functionalized with six 15-crown-5 moieties at the periphery (Fig. 7) [36]. The cage molecule has a cavity of 5 Å, which is large enough to capture gaseous molecules such as CH_4 and Xe. This cage molecule can be dissolved in a 15-crown-5 solvent.

The composite exhibits fluid properties in the liquid phase and has permanent porosity; thus, it is recognized as a Type II porous liquid. This porous liquid is capable of "dissolving" gaseous molecules by capturing them into the nanocavity of the cage. A similar approach was reported by Nitschke et al. for constructing discrete nanospaces in ionic liquids [37].

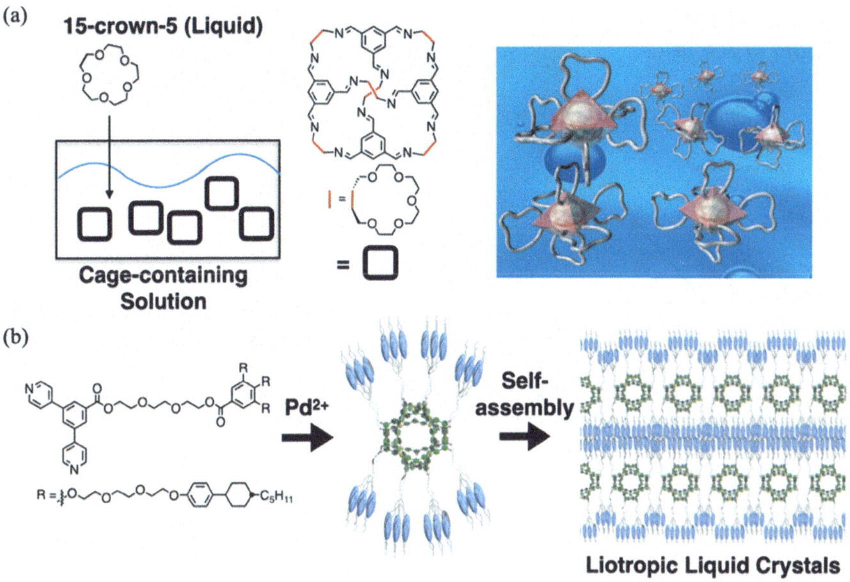

Fig. 7 **a** Porous liquid composed of a liquid 15-crown-5 and a cage molecule functionalized with crown ether at its periphery. **b** Giant spherical metal–organic cages displaying liquid-crystalline side chains connected with mesogens at the periphery, exhibiting lyotropic liquid-crystalline behavior

Recently, Kato et al. reported a giant coordination sphere with a diameter of 3.5 nm, constructed by the coordination-based self-assembly of 24 mesogenic ligands and 12 Pd(II) ions (Fig. 7b) [38]. Owing to the peripheral mesogenic folk-like dendrons connected through triethylene glycol moieties, the giant $M_{12}L_{24}$ spherical core exhibited lyotropic LC properties with layered ordering. Furthermore, they fabricated freestanding supramolecular LC gels composed of an LC coordination sphere, with 24 aldehyde groups on the surface, and polyethylene glycol (PEG) bishydrazide. The resulting LC gel exhibited self-healing properties after physical scission, attributed to the dynamic covalent bonding of the formed acylhydrazone.

7.2 *Nanoporous Liquid Crystals Composed of Shape-Persistent Macrocycles*

Discrete nanochannels or nanospaces can be constructed as columnar liquid crystals through the one-dimensional stacking of shape-persistent macrocycles with discrete nanospaces. These liquid-crystalline materials offer various advantages including optically anisotropic properties, unidirectional molecular orientations, and excellent stimuli responsiveness (Fig. 8) [39, 40].

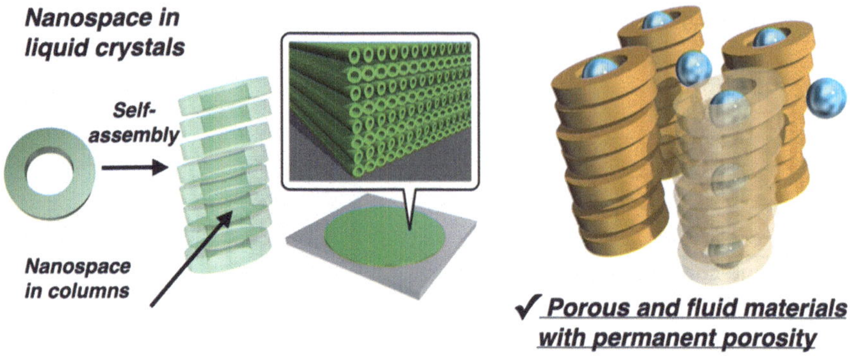

Fig. 8 Molecular design of columnar liquid-crystalline macrocycles. The shape-persistent macrocycles self-assemble to form columnar phases with a discrete nanochannel

The first example of a columnar phase composed of a macrocycle was reported with *N*-acylated hexamine, a liquid-crystalline macrocycle exhibiting a columnar LC phase [41]. However, the authors reported that the inner cavity might have been compressed in the condensed columnar assembly, attributed to the flexible molecular framework consisting of the sp^3 framework of the cyclic polyamines. Moore et al. reported a molecule with a phenylacetylene-based macrocyclic framework [42]. Shape-persistent macrocycles with aliphatic side chains at the periphery exhibit thermotropic columnar LC phases. The rigidity and planarity of the shape-persistent macrocycles are beneficial for constructing discrete nanochannels in columnar LC assemblies.

One of the most efficient strategies for accessing shape-persistent macrocycles at the nanometer scale is through self-assembly via coordination between transition metal ions and ligands with a precise angular design of the direction of the coordination bonds. In line with this strategy, a series of metallomacrocycles with different peripheral alkyl chains were synthesized through a 3:3 self-assembly based on metal complexation between naphthalene-2,7-dihydroxamic acid and 3-hydroxy-4-pyridones. These ligands have been utilized as versatile ligands with various divalent and trivalent transition metal ions (Fig. 9) [43–45]. A liquid crystal of the metallomacrocycle with appropriate side chains was successfully developed.

Furthermore, a large π-conjugated ligand, 9,10-diphenylanthracene (DPA) bearing two β-diketones was chosen as the building block for the macrocycle [46, 47]. A series of side chains were incorporated into DPA to impart fluid properties to the metallomacrocycles. The bis(β-diketone) ligands on DPA allow for the efficient formation of triangular 3:3 metallomacrocycles in the presence of square planar Cu^{2+}.

Cu-containing 3:3 metallomacrocycles with appropriate side chains exhibited thermotropic columnar LC phases, wherein the columns were arranged in rectangular arrays over a wide temperature range.

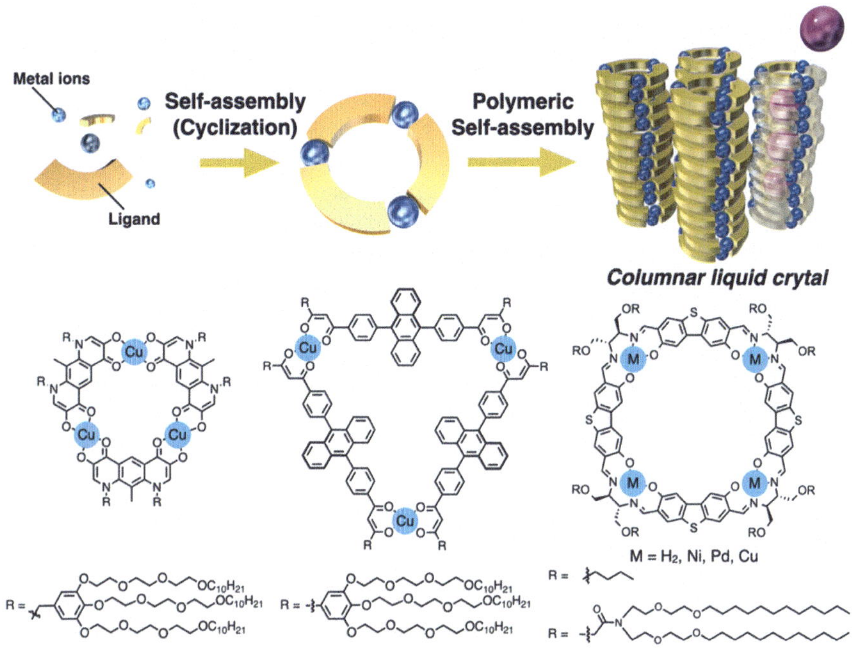

Fig. 9 Series of macrocycles composed of transition metal ions and ligands having branched alkyl chains. The mesomorphic behaviors of the macrocycles and the metallomacrocycles depend on the ligand moieties and the metal ions

Imine formation is one of the most efficient reactions for constructing supramolecular architectures [48, 49]. As a macrocyclic mesogen of columnar liquid crystals, a new series of shape-persistent imine-bridged macrocycles were synthesized using dynamic covalent chemistry [50]. The resulting macrocycle incorporates four *N'*-bis(salicylidene)-ethylenediamine (salen) ligands in the macrocyclic framework allowing tetranuclear metallomacrocycle preparation. The liquid-crystalline properties of these metallomacrocycles depend significantly on the characteristics of the coordinated metal ions. Moreover, the macroscopic organization and thermal properties of the corresponding liquid-crystalline metallomacrocycles were significantly affected by the subtle structural differences among the planar macrocycles, as revealed by single-crystal X-ray crystallographic analysis of macrocycles with shorter alkyl chains.

7.3 Exploration of Nanoporous Liquid Crystals

A series of imine-bridged, shape-persistent macrocycles was synthesized by the condensation of carbazole with two salicylaldehyde moieties and phenylenediamine with branched aliphatic side chains (Fig. 10) [51]. The obtained macrocycles contained alternating sequences of carbazole and salphen. Single-crystal X-ray crystallographic analysis of the model compound with short alkyl chains revealed a π-conjugated macrocycle with a planar structure and a discrete inner cavity measuring 1.4 nm diagonally. The size of the nanospace was sufficiently large to accommodate organic molecules and capture gaseous molecules such as Xe in the nanochannel structure [52]. In addition, solid-state ^{13}C NMR measurements provided local information about the LC macrocycle [53]. The mobility of the mesogenic macrocycle was enhanced after transitioning to higher-temperature LC phases. Tetranuclear metallomacrocycles of Ni(II) and Cu(II) complexes significantly stabilized the LC phases and exhibited a wide temperature range.

Expanding the nanospace defined by the macrocyclic framework involves rational designs for shape-persistent macrocycles, such as using π-extended building blocks [54]. In this context, the diindolocarbazole derivative served as a symmetrical extension of the carbazole precursor in a 4:4 macrocycle. The structure of the giant macrocycle was confirmed using scanning tunneling microscopy (STM). STM imaging of the giant macrocycle adsorbed on the Au(111) surface revealed the planar giant macrocycles corresponding to the 84-membered π-conjugated framework, observed as single molecules and uniformly distributed throughout the 50×50 nm^2 scanning range (Fig. 10a). The macrocycle appeared as a molecular square with sides of 2.0–2.2 nm and a square hollow in the middle. Introducing longer side chains to the macrocycle resulted in the giant macrocycle exhibiting a thermotropic liquid-crystalline phase from 16 to 213 °C. The expanded nanospace of columnar liquid-crystalline macrocycles offers the opportunity to entrap functional dyes of nanometer size and biomolecules such as proteins [55–57].

8 Conclusion

In this chapter, we have discussed metal complexes as versatile platforms for functionalizing liquid crystals. Owing to the intrinsic properties of metal centers, such as redox activity, magnetism, and the ability to form multiple bonds, a wide variety of liquid-crystalline materials can be created. Furthermore, large supramolecular architectures, including macrocycles and cages, were accessible using a supramolecular approach, offering new porous soft materials with promising functionalities such as adsorption–desorption of gases, ion transport, and molecular recognition.

Recently, liquid-crystalline metallacycles constructed by coordination-driven self-assembly were reported by Yang et al. [58]. These macrocycles, containing

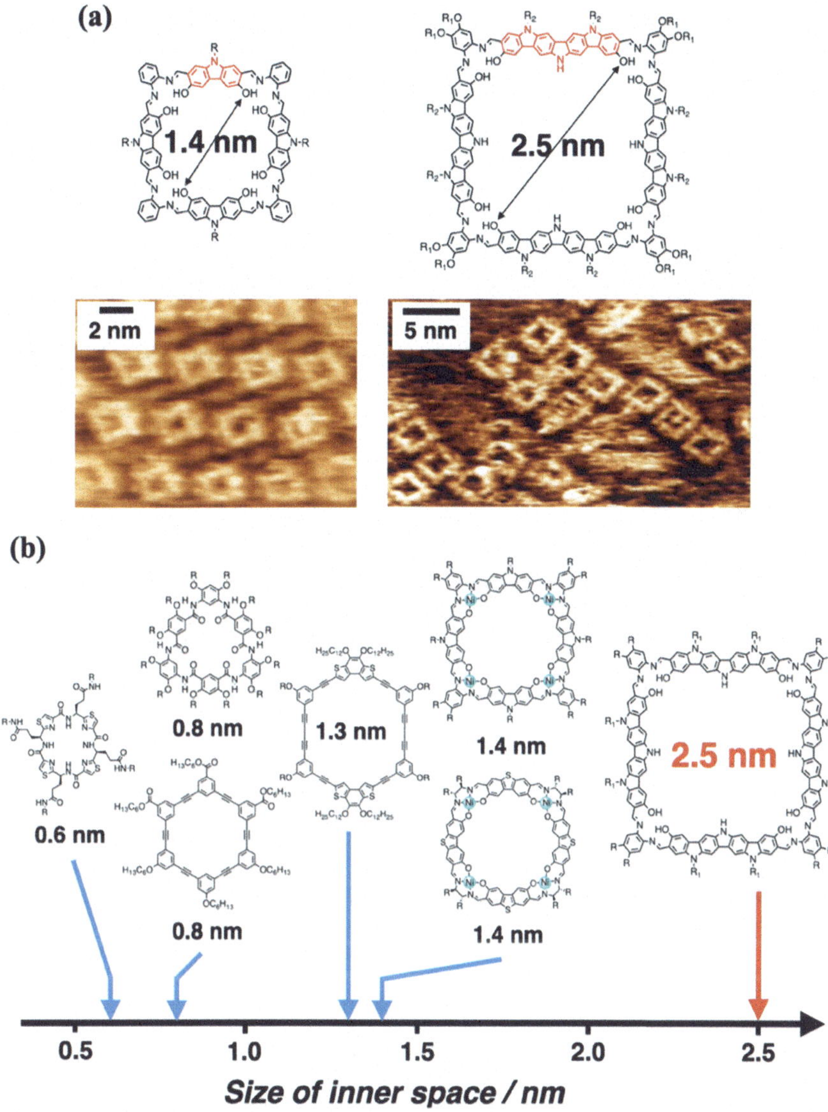

Fig. 10 **a** Imine-bridged macrocycles were visualized using scanning tunneling microscopy (STM) on the Au(111) surface, revealing its molecular structure with high resolution. **b** A series of shape-persistent macrocycles exhibiting thermotropic columnar liquid-crystalline phases are compared based on the size of their inner space. Reprinted with permission from J. Am. Chem. Soc., **144**, 6749. Copyright 2022 American Chemical Society

a tetraphenylethene moiety within their framework, exhibit unique fluorescent properties owing to the aggregation-induced emission (AIE) phenomena. This enables the preparation of films with unique holographic properties. Although this research is relatively incipient and requires further investigation, the fundamental research, into the fluidity of porous materials will pave the way for advancing printing technologies on various surfaces and to introduce new methodologies both in academic and industrial applications.

Acknowledgements This study was supported through a Grant-in-Aid for Scientific Research on Innovative Areas (Research in a Proposed Area) (22H04535) awarded to SK. SK acknowledges the Ogasawara Foundation for the Promotion of Science and Engineering, Asahi Glass Foundation, Iketani Science and Technology Foundation, Naito Research Grant, Tokuyama Science Foundation, and Fujimori Science and Technology Foundation.

References

1. D. Vorländer, Die Erforschung der molekularen Gestalt mit Hilfe der kristallinischen Flüssigkeiten. Z. Phys. Chem. **105**, 211–254 (1923)
2. D. Vorländer, Verhalten der Salze oorganischer Säuren beim Schmeizen. Ber. Dtsch. Chem. Ges. **43**, 3120–3135 (1910)
3. K. Binnemans, C. Görller-Walrand, Lanthanide-containing liquid crystals and surfactants. Chem. Rev. **102**, 2303–2345 (2002)
4. A.-M. Giroud-Godquin, P. Maitlis, Metallomesogens: metal complexes in organized fluid phases. Angew. Chem. Int. Ed. Engl. **30**, 375–402 (1991)
5. J.W. Goodby, P.J. Collings, T. Kato, C. Tschierske, H.F. Gleeson, P. Raynes, Handbook of Liquid Crystals, 2nd edn, vol. 5. (Wiley-VCH, 2014), pp. 175–241
6. J. Malthête, J. Billard, Mesomorphic derivatives of ferrocence. Mol. Cryst. Liq. Cryst. **34**, 117–121 (1976)
7. R.F. Grant, Proton magnetic resonance absorption in copper (II) stearate. Can. J. Chem. **42**, 951–953 (1964)
8. M. Takeyoshi, N. Watanabe, B. Tamamusbi, Copper stearate as a thermotropic liquid crystal. Colloid Polym. Sci. **256**, 588–590 (1978)
9. M. Giroud, U.T. Mueller-Westerhoff, Mesomorphic transition metal complexes. Mol. Cryst. Liq. Cryst. **41**, 11–13 (1977)
10. H. Adams, N.A. Bailey, D.W. Bruce, D.A. Dunmur, E. Lalinde, M. Marcos, C. Ridgway, A.J. Smiths, P. Styring, P.M. Maitlis, Mesogenic transition metal complexes liquid crystal phase behaviour and crystal and molecular structure of some nitrile complexes of the platinum metals. Liq. Cryst. **2**, 381–393 (1987)
11. D.V. Baxter, M.H. Chisholm, M.A. Lynn, E.F. Putilina, S.T. Trzaska, T. Swager, Studies of thermotropic properties and the mesophase of mixtures of n-alkanoates and perfluoro-n-alkanoates of dimolybdenum (M-4-M). Chem. Mater. **10**, 1758–1763 (1998)
12. R. Bayón, S. Coco, P. Espinet, Gold liquid crystals displaying luminescence in the mesophase and short F···F interactions in the solid state. Chem. Eur. J. **11**, 1079–1085 (2005)
13. K. Hatsusaka, K. Ohta, I. Yamamoto, H. Shirai, Discotic liquid crystals of transition metal complexes, part 30: spontaneous uniform homeotropic alignment of octakis(dialkoxyphenoxy)phthalocyaninatocopper(II) complexes. J. Mater. Chem. **11**, 423–433 (2001)
14. T.D. Shafffer, K.A. Sheth, Mesomorphic transition metal N_2O_2 chelate. Mol. Cryst. Liq. Cryst. **172**, 27–39 (1989)

15. B.J. Bulkin, R.K. Rose, A. Santoro, Synthesis and thermal properties of a possibly mesomorphic palladium chelate. Mol. Cryst. Liq. Cryst. **43**, 53–58 (1976)
16. H. Zheng, C.K. Lai, T.M. Swager, Transition metals in highly correlated discotic phases: designing metallomesogens with selected intermolecular organizations. Chem. Mater. **6**, 101–103 (1994)
17. K. Ohta, M. Ikejima, M. Moriya, H. Hasebe, I. Yamamoto, Disk-like liquid crystals of transition metal complexes. Part 20. pursuit of chemistry to directly visualize van der waals interactions. J. Mater. Chem. **8**, 1971–1977 (1998)
18. S. Hayami, K. Danjobara, K. Inoue, Y. Ogawa, N. Matsumoto, Y. Maeda, A photoinduced spin transition iron(II) complex with liquid-crystal properties. Adv. Mater. **16**, 869–872 (2004)
19. R. Deschenaux, M. Schweissguth, A.-M. Levelut, Electron-transfer induced mesomorphism in the ferrocene-ferrocenium redox system: first ferrocenium-containing thermotropic liquid crystal. Chem. Commun. 1275–1276 (1996)
20. H.-C. Chang, T. Shiozaki, A. Kamata, K. Kishida, T. Ohmori, D. Kiriya, T. Yamauchi, H. Furukawa, S. Kitagawa, A redox-active columnar metallomesogen and its cyclic voltammetric responses. J. Mater. Chem. **17**, 4136–4138 (2007)
21. C. Piguet, J.-C.G. Bünzli, B. Donnio, D. Guillon, Thermotropic lanthanidomesogens. Chem. Commun. 3755–3768 (2006)
22. Y. Galyametdinov, M.A. Athanassopoulou, K. Griesar, O. Kharitonova, E.A.S. Bustamante, L. Tinchurina, I. Ovchinnikov, W. Haase, Synthesis and magnetic investigations on rare-earth-containing liquid crystals with large magnetic anisotropy. Chem. Mater. **8**, 922–926 (1996)
23. E. Terazzi, S. Torelli, G. Bernardinelli, J.-P. Rivera, J.-M. Bénech, C. Bourgogne, B. Donnio, D. Guillon, D. Imbert, J.-C.G. Bünzli, A. Pinto, D. Jeannerat, C. Piguet, Molecular control of macroscopic cubic, columnar, and lamellar organizations in luminescent lanthanide-containing thermotropic liquid crystals. J. Am. Chem. Soc. **127**, 888–903 (2005)
24. P. Espinet, J. Etxebarria, M. Marcos, J. Perez, A. Remdn, J.L. Serrano, A novel approach to ferroelectric liquid crystals: the first organotransition metal compound displaying this behavior. Angew. Chem. Int. Ed. Engl. **28**, 1065–1066 (1989)
25. J. Barberá, R. Iglesias, J.L. Serrano, T. Sierra, M.R. de la Fuente, B. Palacios, M.A. Pérez-Jubindo, J.T. Vázquez, Switchable columnar metallomesogens. New helical self-assembling systems. J. Am. Chem. Soc. **120**, 2908–2918 (1998)
26. Y. Wang, Q. Li, Light-driven chiral molecular switches or motors in liquid crystals. Adv. Mater. **24**, 1926–1945 (2012)
27. N. Hoshino, Y. Matsuoka, K. Okamoto, A. Yamagishi, Δ-$[Ru(acac)_2L]$ (L = a mesogenic derivative of bpy) as a novel chiral dopant for nematic liquid crystals with large helical twisting power. J. Am. Chem. Soc. **125**, 1718–1719 (2003)
28. J. Yoshida, H. Sato, A. Yamagishi, N. Hoshino, On the parity in helical twisting power of Ru(III) 1,3-diketonates of C symmetry in nematic liquid crystals. J. Am. Chem. Soc. **127**, 8453–8456 (2005)
29. M.H. Chishokm, One-dimensional polymers and mesogens incorporating multiple bonds between metal atoms. Acc. Chem. Res. **33**, 53–61 (2000)
30. R. Kuwahara, S. Fujikawa, K. Kuroiwa, N. Kimizuka, Controlled polymerization and self-assembly of halogen-bridged diruthenium complexes in organic media and their dielectrophoretic alignment. J. Am. Chem. Soc. **134**, 1192–1199 (2012)
31. D.V. Baxter, R.H. Cayton, M.H. Chisholm, J.C. Huffman, E.F. Putilina, S.L. Tagg, J.L. Wesemann, J.W. Zwanziger, F.D. Darrington, Multiple bonds between metal atoms in ordered assemblies. 2. Quadruple bonds in the mesomorphic state. J. Am. Chem. Soc. **116**, 4551–4566 (1994)
32. R. Sessoll, D. Gatteschi, A. Caneschi, M.A. Novak, Magnetic bistability in a metal-ion cluster. Nature **365**, 141–143 (1993)
33. E. Terazzi, C. Bourgogne, R. Welter, J.-L. Gallani, D. Guillon, G. Rogez, B. Donnio, Single-molecule magnets with mesomorphic lamellar ordering. Angew. Chem. Int. Ed. **47**, 490–495 (2008)

34. Y. Molard, F. Dorson, V. Cîrcu, T. Roisnel, F. Artzner, S. Cordier, Clustomesogens: liquid crystal materials containing transition-metal clusters. Angew. Chem. Int. Ed. **49**, 3351–3355 (2010)
35. N. O'Reilly, N. Giri, S.L. James, Porous liquids. Chem. Eur. J. **13**, 3020–3025 (2007)
36. N. Giri, M.G.D. Pópolo, G. Melaugh, R.L. Greenaway, K. Rätzke, T. Koschine, L. Pison, M.F.C. Gomes, A.I. Cooper, S.L. James, Liquids with permanent porosity. Nature **527**, 216–220 (2015)
37. L. Ma, C.J.E. Haynes, A.B. Grommet, A. Walczak, C.C. Parkins, C.M. Doherty, L. Longley, A. Tron, A.R. Stefankiewicz, T.D. Bennett, J.R. Nitschke, Coordination cages as permanently porous ionic liquids. Nat. Chem. **12**, 270–275 (2020)
38. J. Uchida, M. Yoshio, H. Sota Sato, M. Yokoyama, T.K. Fujita, Self-assembly of giant spherical liquid-crystalline complexes and formation of nanostructured dynamic gels that exhibit self-healing properties. Angew. Chem. Int. Ed. **56**, 140851–214089 (2017)
39. T. Kato, J. Uchida, T. Ichikawa, T. Sakamoto, Functional liquid crystals towards the next generation of materials. Angew. Chem. Int. Ed. **57**, 4355–4571 (2018)
40. T. Kato, M. Yoshio, T. Ichikawa, B. Soberats, H. Ohno, M. Funahashi, Transport of ions and electrons in nanostructured liquid crystals. Nat. Rev. Mater. **2**, 17001 (2017)
41. J.-M. Lehn, J. Malthête, A.-M. Levelut, Tubular mesophases: liquid crystals consisting of macrocyclic molecules. J. Chem. Soc., Chem. Commun. 1794–1796 (1985)
42. J. Zhang, J.S. Moore, Nanoarchitectures. 6. Liquid crystals based on shape-persistent macrocyclic mesogens. J. Am. Chem. Soc. **116**, 2655–2656 (1994)
43. S. Kawano, K. Ideta, T. Banno, K. Tanaka, Liquid-crystalline metallomacrocycles composed of bis(hydroxypyridono)toluene ligand and Cu^{2+} ions. Chem. Asian J. **14**, 4415–4419 (2019)
44. S. Kawano, H. Inada, K. Tanaka, Self-assembled synthesis and liquid crystalline property of metallomacrocycle composed of naphthalenedihydroxamate and Cu^{2+} ions. Chem. Lett. **45**, 1105–1107 (2016)
45. S. Kawano, A. Sasaki, K. Tanaka, Arraying and guest inclusion of soluble metal-organic nanotubes composed of macrocyclic paddle-wheel metal complexes. Angew. Chem. Int. Ed. 64, e202422546 (2025)
46. S. Kawano, T. Murai, T. Harada, K. Tanaka, Columnar liquid-crystalline macrocycles synthesized via metal ion-assisted self-assembly. Inorg. Chem. **57**, 3913–3919 (2018)
47. S. Kawano, T. Harada, A. Sasaki, K. Tanaka, Kinetically-locked metallomacrocycle for host-guest chemistry with bulky anions. Eur. J. Inorg. Chem. **26**, e202200592 (2018)
48. S. Kawano, T. Fukushima, K. Tanaka, Specific and oriented encapsulation of fullerene C_{60} molecular double-decker cage composed of shape-persistent macrocycles. Angew. Chem. Int. Ed. **57**, 14827–14831 (2018)
49. S. Kawano, M. Nakaya, M. Saitow, A. Ishiguro, T. Yanai, J. Onoe, K. Tanaka, Thermally stable array of discrete C_{60} s on a two-dimensional crystalline adlayer of macrocycles both in vacuo and under ambient pressure. J. Am. Chem. Soc. **144**, 6749–6758 (2022)
50. S. Kawano, T. Hamazaki, A. Suzuki, K. Kurahashi, K. Tanaka, Metal-ion-induced switch of liquid-crystalline orientation of metallomacrocycles. Chem. Eur. J. **22**, 15674–15683 (2016)
51. S. Kawano, Y. Ishida, K. Tanaka, Columnar liquid-crystalline metallomacrocycles. J. Am. Chem. Soc. **137**, 2295–2302 (2015)
52. S. Kawano, H. Yoshimizu, K. Tanaka, Continuous nanospace in nanoporous liquid crystal investigated by ^{129}Xe NMR spectroscopy. Angew. Chem. Int. Ed. **63**, e202316523 (2024)
53. S. Kawano, K. Tanaka, Solid-State NMR study of dynamic properties of a columnar liquid crystalline macrocycle. Bull. Chem. Soc. Jpn **90**, 826–830 (2017)
54. S. Kawano, M. Kato, S. Soumiya, M. Nakaya, J. Onoe, K. Tanaka, Columnar liquid crystals from a giant macrocycle mesogen. Angew. Chem. Int. Ed. **57**, 167–171 (2018)
55. K. Sato, Y. Itoh, T. Aida, Columnarly assembled liquid-crystalline peptidic macrocycles unidirectionally orientable over a large area by an electric field. J. Am. Chem. Soc. **133**, 13767–13769 (2011)
56. M. Fritzsche, A. Bohle, D. Dudenko, U. Baumeister, D. Sebastiani, G. Richardt, H.W. Spiess, M.R. Hansen, S. Höger, Empty helical nanochannels with adjustable order from low-symmetry macrocycles. Angew. Chem. Int. Ed. **50**, 3030–3033 (2011)

57. X. Li, B. Li, L. Chen, J. Hu, C. Wen, Q. Zheng, L. Wu, H. Zeng, B. Gong, L. Yuan, Liquid-crystalline mesogens based on cyclo[6]aramides: distinctive phase transitions in response to macrocyclic host-guest interactions. Angew. Chem. Int. Ed. **54**, 11147–11152 (2015)
58. Y.-X. Hu, X. Hao, L. Xu, X. Xie, B. Xiong, Z. Hu, H. Sun, G.-Q. Yin, X. Li, H. Peng, H.-B. Yang, Construction of supramolecular liquid-crystalline metallacycles for holographic storage of colored images. J. Am. Chem. Soc. **142**, 6285–6294 (2020)

Generation and Control of Circularly Polarized Luminescence in Organic Materials

Yoshitane Imai

1 What is Circularly Polarized Luminescence (CPL)?

1.1 CPL

When light is considered as a wave, light is said to be linearly polarized if the oscillating planes of the light wave are aligned in a straight line. Conversely, when the oscillating plane of the light wave rotates as it propagates, it is said to be circularly polarized. Thus, a light wave is circularly polarized when the oscillating plane rotates with the frequency of the light wave in a plane perpendicular to the direction of light propagation, and the amplitude is constant and independent of the direction. The circular polarization of light occurs when the orthogonal wave vectors are out of phase by a quarter wavelength. The emission of circularly polarized light from light-emitting materials is called circularly polarized luminescence (CPL).

1.2 CPL and Circular Dichroism (CD)

CPL represents the difference between the intensities of the left and right circularly polarized emitted light and serves as a spectroscopic tool to evaluate the optical activity of a chiral substance in its excited state. Thus, the chirality associated with transition from the excited state to ground state of the chiral substance can be discussed in terms of the CPL of the emitted light. Conversely, CD is a property by which a chiral substance differentially absorbs left and right circularly polarized light. CD can therefore be used as a spectroscopic tool to measure the optical activity of the ground state of a chiral substance, using which the chirality in the absorption

Y. Imai (✉)
Faculty of Science and Engineering, Kindai University, Higashiosaka, Osaka, Japan
e-mail: y-imai@apch.kindai.ac.jp

M. Funahashi (ed.), *Functional Soft Matter II*,
https://doi.org/10.1007/978-981-92-0100-6_21

transition from the ground state to excited state of the substance can be discussed. CD and CPL therefore provide very different information about the chirality of the material under investigation. In general, CPL and CD cannot be regarded simply as inverse processes of each other because the chirality of the excited state differs significantly from that of the ground state.

1.3 CPL Anisotropy Factor

The anisotropy factor provides a measure of the CPL performance and is expressed by the Kuhn asymmetry factor: $g_{lum} = 2 \times (IL - IR)/(IL + IR)$, where *IL* and *IR* are the intensities of the left and right circularly polarized components of the emitted light, respectively, and $(IL + IR)/2$ is the total emission intensity. The anisotropy factor, g_{lum}, represents the proportion of circularly polarized component in the total luminescence intensity. An anisotropy factor, $|g_{lum}|$, of 2 indicates that a substance has ideal circular polarization properties.

1.4 Design Guidelines for Circularly Polarized Luminescent Materials

Molecular design can, in principle, lead to a circularly polarized luminescent material with a theoretical maximum $|g_{lum}|$ value of 2. The anisotropy factor can also be expressed as $g_{lum} = 4(m{\cdot}\mu)(|m|^2 + |\mu|^2)^{-1} = 4|m||\mu|(|m|^2 + |\mu|^2)^{-1}\cos\theta$, where θ is the angle formed between the electric (μ) and magnetic (m) transition dipole moments. Therefore, luminescent materials do not exhibit circular polarization if the two dipole moments are orthogonal ($\theta = 90°$), even if μ or m is of high intensity. In addition to the intensity of μ and m, these dipole moments must be aligned appropriately in the same direction to ensure the emission of circularly polarized light.

The applications of circularly polarized light emitted from a luminescent material typically require a large anisotropy factor (large circular polarization: firm rotation) and high luminosity (brightness). Accordingly, a large $|g_{lum}|$ value and high-emission quantum efficiency (quantum yield) are required: however, a trade-off relationship exists between the $|g_{lum}|$ and quantum efficiency of organic circularly polarized light emitters. According to the definition of the Kuhn asymmetry factor, high quantum efficiency typically requires a large electric transition dipole moment, μ, in which case the value of $|g_{lum}|$ is small. Conversely, a large magnetic transition dipole moment, *m*, is required to obtain a large $|g_{lum}|$ value, which reduces the quantum efficiency. The molecular design of a circularly polarized luminescent material with both a large $|g_{lum}|$ value and high quantum efficiency must therefore consider the steric structure of the molecule and its transition dipole moment.

2 Flexible and Soft Pyrene-Based Luminophores

The rotational direction of the CPL (left- or right-handed CPL, denoted by the sign of the CPL) can be controlled using the same fluorophore attached to the optically active units with opposite chirality. In the case of flexible and soft luminophores, the sign of the CPL can be controlled by adjusting the structure of the linker between the fluorescent and chiral units with the same absolute configuration. Thus, changing the type of the linking chain can afford fluorescent units with clockwise or counterclockwise orientations in a predictable, and therefore controllable, manner.

Molecule (*R*)-**1** (Fig. 1), which contains long oligoether tethers with ten single bonds, and molecule (*R*)-**2**, which has shorter alkyl ether tethers, exhibited intramolecular CPL at similar wavelengths (480 nm) through the twisted orientation of the photoexcited pyrene units in $CHCl_3$. Interestingly, despite the (*R*)-binaphthyl unit in both molecules having the same chirality, (*R*)-**1** exhibits a negative CPL ($g_{CPL} = -0.46 \times 10^{-3}$ at 480 nm), whereas (*R*)-**2** exhibits a positive CPL ($g_{CPL} = +1.0 \times 10^{-3}$ at 480 nm). Conversely, both (*R*)-**1** and (*R*)-**2** show similar CD spectra in which the CD signals of the 0–0' band had the same negative Cotton sign. These results suggest that although the chirality-oriented conformations of the two pyrenes in (*R*)-**1** and (*R*)-**2** are identical in the ground state, their relative orientations may have opposite handedness in the photoexcited state [1].

Furthermore, the direction of CPL rotation can also be controlled by changing the distance between the fluorescent sites rather than by regulating chirality. The flexible peptide main chains of the peptide-pyrene luminophores (*L*)-**3** and (*L*)-**4** have the same absolute configuration; however, the distance between the fluorescent pyrene units differs owing to the different number of methylene groups in the chain between the pyrene moieties (Fig. 2).

Interestingly, in peptide-pyrene luminophores (*L*)-**3** and (*L*)-**4**, the signs of the CPL were reversed (Fig. 2, black and grey lines, respectively). In contrast, the signs

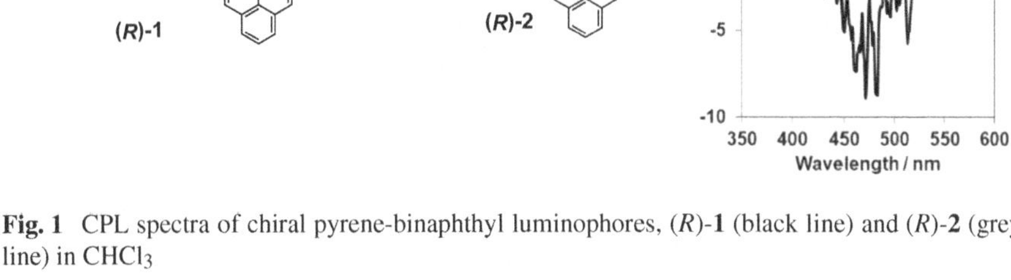

Fig. 1 CPL spectra of chiral pyrene-binaphthyl luminophores, (*R*)-**1** (black line) and (*R*)-**2** (grey line) in $CHCl_3$

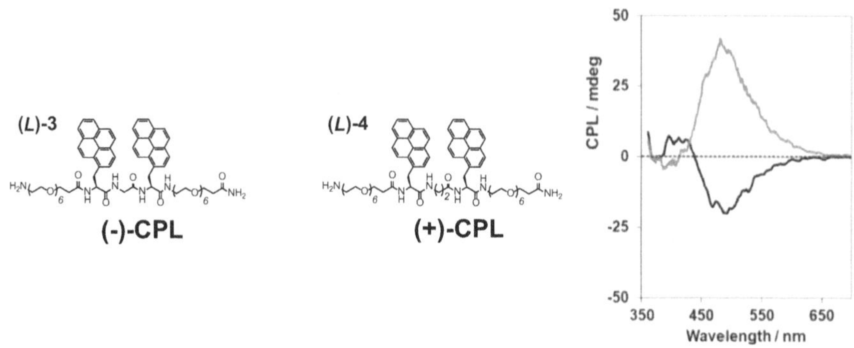

Fig. 2 CPL spectra of chiral pyrene-peptide luminophores (*L*)-**3** (black line) and (*L*)-**4** (grey line)

of the CD spectra of the (*L*)-forms of the peptide–pyrene luminophores were identical. Thus, it can be assumed that the relative chiral configuration of the intramolecular pyrene ring in two chiral peptide–pyrene luminophores with the same absolute configuration is identical in the ground state, but is reversed according to the distance between the pyrene units in the photoexcited state [2].

The distance between luminescent units can also be controlled by altering the hydrophilic/hydrophobic balance of luminophores containing stereocenters of the same chirality in water. Bipyrenyl oligopeptide luminophores ((*L*)-**5** and (*L*)-**6**) contain different numbers of hydrophilic arginine units in the oligopeptide skeleton to balance the hydrophilicity and hydrophobicity of the luminophore (Fig. 3).

(*L*)-**5** and (*L*)-**6** clearly exhibited strong excimer CPL in water with $|g_{CPL}|$ values of ~ 5.3×10^{-3} and 3.5×10^{-3}, respectively. As expected, (*L*)-**5** and (*L*)-**6** yielded excimer CPL spectra with opposite signs: (negative ($\lambda_{CPL} =$ 486 nm) and positive ($\lambda_{CPL} =$ 469 nm), respectively), despite having identical absolute configurations (Fig. 3, black and grey lines, respectively). However, (*L*)-**5** and (*L*)-**6** exhibited the

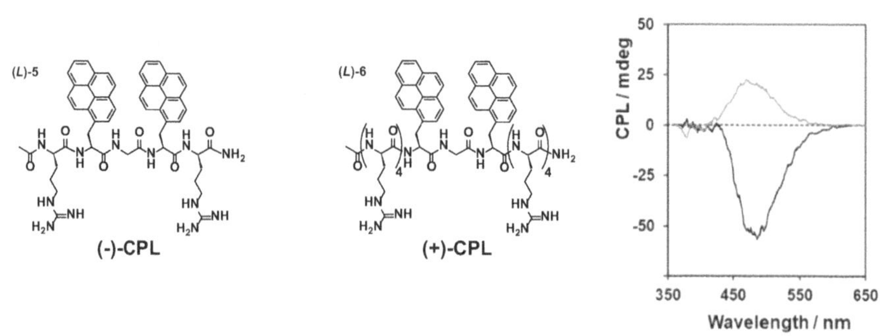

Fig. 3 CPL (upper) and PL (lower) spectra of chiral pyrene-peptide-arginine luminophores (L)-**5** (black lines) and (L)-**6** (grey lines) in water

same CD sign, demonstrating that the configurations of both excimers are identical in the ground state.

As the number of arginine units increases, the chain between the two pyrene moieties becomes more hydrophilic and the interpyrene distance extends, whereas the hydrophilicity and the interpyrene distance both decrease with the number of arginine units decreases. In this case, the chiral bipyrenyl motif was identical in the ground state and showed the same CD sign, but differed significantly in the photoexcited state, suggesting that the sign of CPL was reversed [3].

3 Flexible and Soft Pyrene-Based Cryptochiral Luminophores

The most common chiral CPL luminophores are both CD-active and CPL-active; however, cryptochiral CPL luminophores, a type of flexible and soft chiral luminophore, are CD-silent/CPL-active. Chiral binaphthyl luminophore **7** (Fig. 4) consists of a central chiral 2,2-dimethyl-1,3-dioxolane backbone and two fluorescent pyrene units connected by a flexible linking chain (Fig. 4). In the spectra of (*R,R*)- and (*S,S*)-**7**, the intensities of the CD signals corresponding to main π-π* vibronic transitions derived from the two pyrene moieties were extremely weak (Fig. 4, black and grey lines, respectively). In contrast, both luminophores exhibited strong CPL with positive and negative signs (g_{CPL} = ~ 8.9×10^{-4} at 460 nm), respectively (Fig. 4, black lines and grey lines, respectively).

Figure 5 shows this mechanism of cryptochirality. The pyrene units of **7** adopt an almost achiral T-shaped structure in the ground state (Fig. 5, left), resulting in a very weak CD spectrum. In contrast, in the photoexcited state, the pyrene units adopt a chiral π-stacked arrangement (Fig. 5, right), thereby giving rise to CD-silent/CPL-active cryptochiral behaviour. In this luminophore, both pyrene units are located far from the central chiral 1,3-dioxolane unit and act as a hidden chiral excimer pair prior to photoexcitation, which constitutes the core principle of cryptochirality [4].

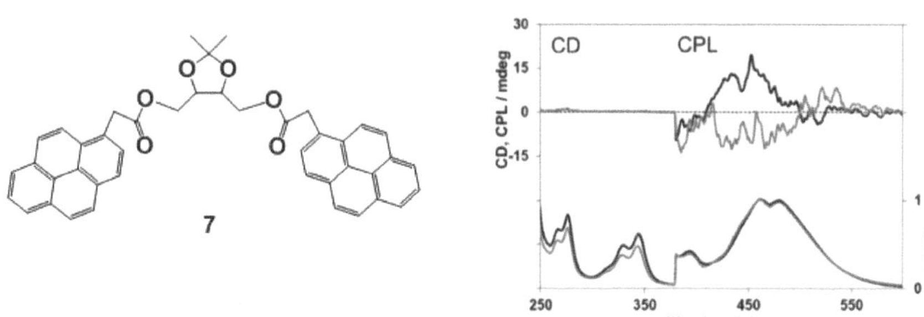

Fig. 4 CPL, PL, CD and absorption spectra of chiral pyrene-dioxolane luminophore **7**

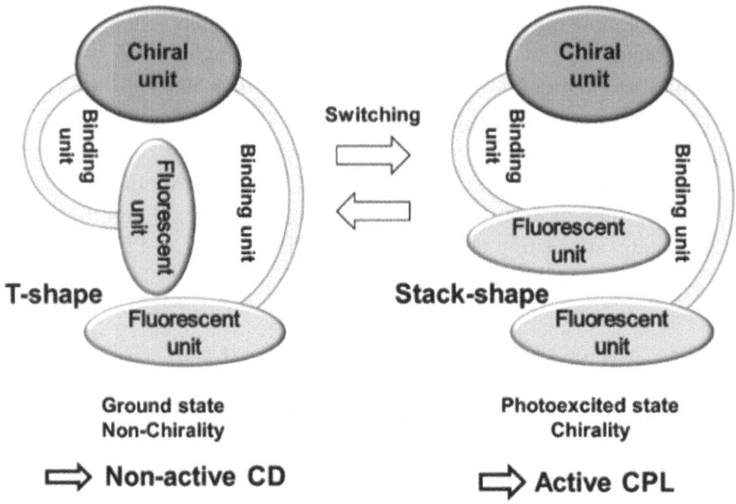

Fig. 5 Operational mechanism of the cryptochiral CPL system

4 Solvent-Dependent CPL in Flexible and Soft Pyrene-Based Luminophores

Differences in the external environment affect the structure of flexible and soft luminescent materials; thus, changes in the interaction between the external environment and functional groups such as the luminescent main chain and luminescent unit propagate to a change in the relative orientation of the luminescent units.

Intramolecular pyrene interactions in (*L*)-**8** resulted in an excimer CPL with a maximum wavelength (λ_{CPL}) and anisotropy factor, $|g_{CPL}|$ of 484 nm and 0.5 $\times$ 10^{-2}, respectively (Fig. 6). Interestingly, a positive CPL was observed in the relatively nonpolar $CHCl_3$, whereas a negative CPL was observed in planar *N*,*N*-dimethylformamide DMF. In DMF, λ_{CPL} and $|g_{CPL}|$ were 464 nm and 0.2 $\times$ 10^{-2}, respectively. In the CD spectra of the luminescent pyrene units in the ground state, the sign of the Cotton effect at the longest wavelength was identical in both solvents, suggesting that chirality inversion in different solvents occurs only in the photoexcited state [5].

5 CPL in Optically Inactive Pyrene

CPL can be induced even from achiral luminophores in the solution and solid states using magnetic fields. The unique characteristics of this magnetic CPL (MCPL) facilitate the development of versatile CPL systems using various optically inactive flexible and soft molecules.

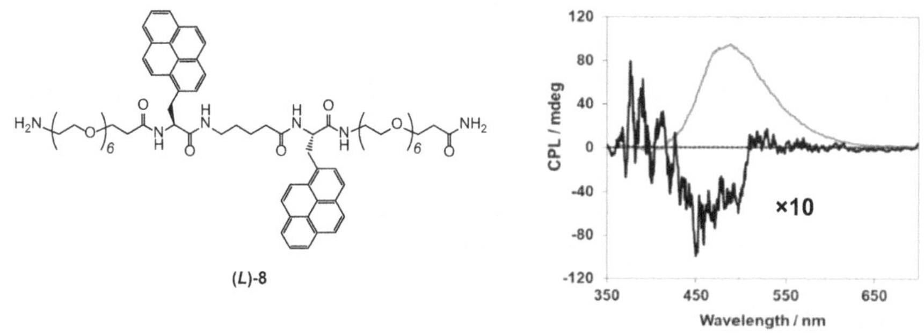

Fig. 6 CPL spectra of chiral pyrene-peptide luminophore (L)-**8** in **a** $CHCl_3$ (black lines) and **b** DMF (grey lines) solutions

Achiral pyrene, which does not possess any stereogenic centre or restricted stereogenic bond, clearly exhibits mirror-image MCPL spectra in $CHCl_3$ solutions under a magnetic field of 1.6 T with Faraday geometry N → S or S → N geometry (Fig. 7). MCPLs corresponding to the 0–0' (374 nm), 0–1' (393 nm) and 0–2' (416 nm) vibronic PL bands of the pyrene ring of the monomer species were observed, with a $|g_{MCPL}|$ of $0.82 \times 10^{-2}\ T^{-1}$ at 374 nm in $CHCl_3$. In contrast, pyrene excimers formed in the excited state do not contribute to MCPL signals [6].

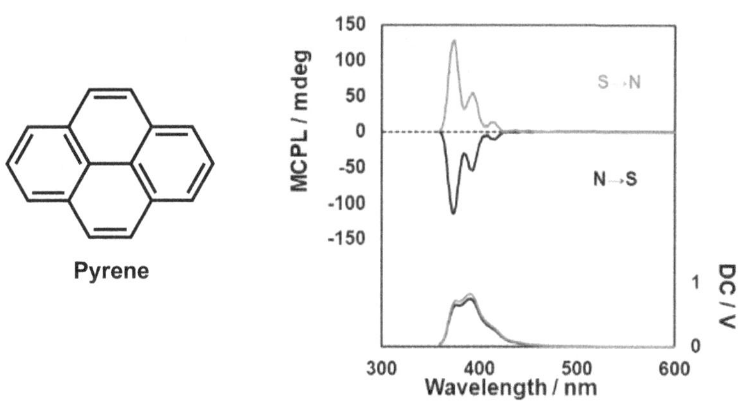

Fig. 7 MCPL (upper) and PL (lower) spectra of achiral pyrene in N → S (black lines) and S → N (grey lines) configurations under a magnetic field (1.6 T) in $CHCl_3$ (1.0×10^{-4} M)

6 CPL Switching of a Chiral Perylene Diimide-Doped Nematic Liquid Crystal

Achiral N-LC 4'-pentyl-4-biphenylcarbonitrile (5CB) was doped with chiral (*R,R*)/(*S,S*)-*N,N'*-bis(1-cyclohexylethyl)perylene-3,4,9,10-tetracarboxylic diimide [(*R,R*)/(*S,S*)-CPDI] to produce a flat, luminescent chiral nematic liquid crystal N*-LC-CPDI (Fig. 8). Subsequently, a reversible CPL switching system was obtained by applying a DC electric field to liquid crystal devices (LCDs) prepared on conductive indium tin oxide (ITO) glass substrates.

CPLs (547 nm and 583 nm (λ_{CPL})) derived from perylene monomers were observed from the liquid–crystal state of N*-LCs-CPDI. The spectrum of N*-LCs-(*R,R*)-CPDI shows a strong negative signal (Fig. 9a). The $|g_{CPL}|$ values of N*-LCs-(*R,R*)-CPDI in the liquid crystalline state were 4.0×10^{-1} (547 nm) and 4.7×10^{-1} (583 nm). Selective reflection is negligible in this system because the reflection band of the liquid crystal lies outside the CPL emission region.

Furthermore, the application of a 30 V DC electric field to N*-LCs-CPDI changed its POM texture, indicating a structural change. Interestingly, the CPL intensities of N*-LCs-CPDI were significantly reduced (Fig. 9b), and sign inversion of the CPL spectra was observed. The $|g_{CPL}|$ value under an applied DC electric field of 30 V was 6.9×10^{-3} at 594 nm (λ_{CPL}), which is remarkably lower than that at 0 V. Upon removing the applied DC field, the CPL intensities and anisotropy factors ($|g_{CPL}|$) of N*-LCs-CPDI reverted to their initial states (3.1×10^{-1} (547 nm) and 3.5×10^{-1} (583 nm), as shown in Fig. 8c, indicating that the CPL from the N*-LC can be turned on and off by switching the DC electric field on and off, which also regulates the direction of CPL rotation. The switching characteristics of the CPL are thought to arise from the alignment of the planar CPDI along the bar-shaped LC molecule (5CB), which is synchronized with the 5CB and reoriented by the applied voltage [7].

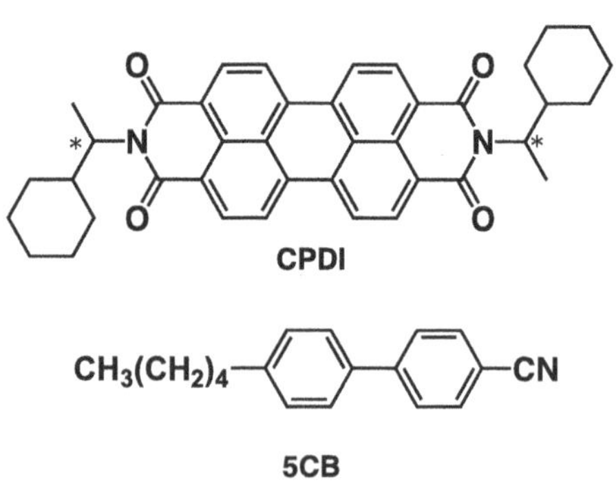

Fig. 8 Chiral perylene luminophore CPDI and nematic liquid crystal 5CB

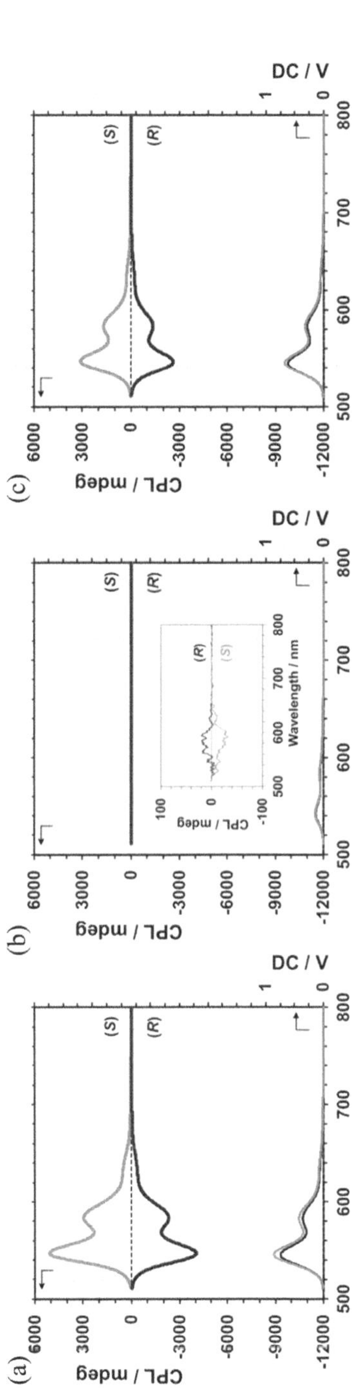

Fig. 9 CPL (upper) and PL (lower) spectra of N*-LCs-(*R*,*R*)-CPDI (black lines) and N*-LCs-(*S*,*S*)-**CPDI** (grey lines) in the liquid crystalline state at **a** 0 V before the application of a DC electric field, **b** 30 V (inset: magnified CPL spectrum) and **c** 0 V after the application of a DC electric field

7 Circularly Polarized Electroluminescence (CPEL) from Optically Inactive Devices

CPL can be extracted by applying an external magnetic field to an optically inactive luminophore. Similarly, circularly polarized electroluminescence (CPEL) can be extracted by applying an external magnetic field to an organic light-emitting diode (OLED) incorporating an optically inactive luminophore. Circularly polarized organic light-emitting diodes (CP-OLEDs) with a multi-stack structure were fabricated using achiral 4CzIPN, which exhibits thermally activated delayed fluorescence (TADF), as the emissive dopant (Fig. 10).

This device exhibited a maximum luminance (L_{max}) of 11,770 cd·m^{-2} at 7.4 V. Notably, the external quantum efficiency (EQE) of the device, which reached a maximum of 15.5% at 6.5 V, is considerably higher than the theoretical EQE limit of conventional fluorescent OLEDs (~5%), which is a characteristic of TADF-based OLEDs.

The MCPEL characteristics were then investigated at room temperature by applying a fixed voltage of 8 V to the device under a magnetic field of 1.7 T. As expected, a green MCPEL was obtained from the 4CzIPN-based device (Fig. 11).

As with CPL, the performance of the 4CzIPN-based device was quantitatively evaluated using the anisotropy factor given by $g_{MCPEL} = 2 \times (IL\text{-}IR)/(IL + IR)$. The anisotropy factor, $|g_{MCPEL}|$, of the 4CzIPN-based device was $3.7 \times 10 - 4$ (T^{-1}) at 537 nm. Furthermore, the sign of the MCPEL could be completely controlled by changing the Faraday configuration (N-up/S-up) of the applied external magnetic field [8].

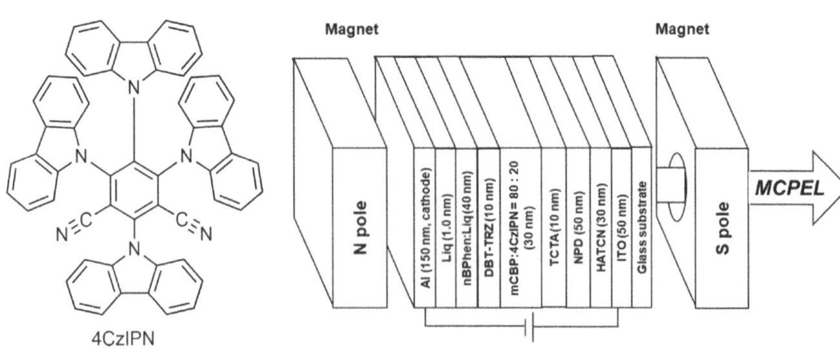

Fig. 10 Optically inactive TADF 4CzIPN and MCP-OLED system

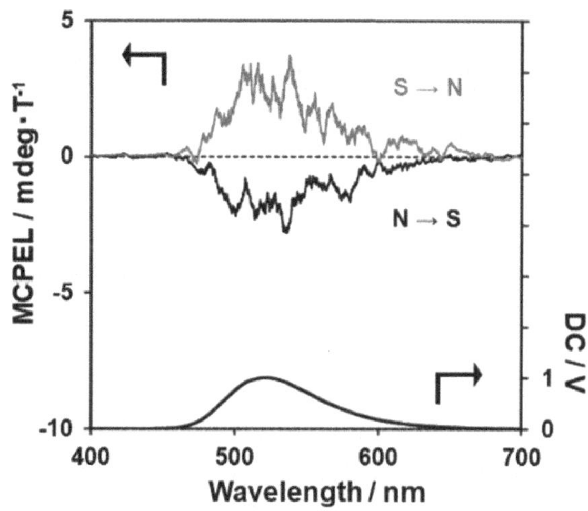

Fig. 11 MCPEL (upper) and MEL (lower) spectra of the 4CzIPN-based OLED under 1.7 T N-up (black lines) and S-up (grey lines) magnetic fields

8 Conclusion

The chirality of the excited and ground states of flexible and soft luminescent materials differs; thus, CPL and CD cannot be considered simple inverse processes of each other. Measuring the CPL from a flexible and soft luminescent material can reveal very interesting excited state structures and optical properties.

References

1. K. Nakabayashi, S. Kitamura, N. Suzuki, S. Guo, M. Fujiki, Y. Imai, Eur. J. Org. Chem. **1**, 64 (2016)
2. T. Nishikawa, S. Kitamura, M. Kitamatsu, M. Fujiki, Y. Imai, ChemistrySelect **1**, 831 (2016)
3. Y. Mimura, Y. Motomura, M. Kitamatsu, Y. Imai, Asian. J. Org. Chem. **10**, 14 (2021)
4. T. Amako, K. Nakabayashi, N. Suzuki, S. Guo, N.A.A. Rahim, T. Harada, M. Fujiki, Y. Imai, Chem. Commun. **51**, 8237 (2015)
5. Y. Mimura, S. Kitamura, M. Shizuma, M. Kitamatsu, M. Fujiki, Y. Imai, ChemistrySelect **2**, 7759 (2017)
6. D. Kaji, H. Okada, N. Hara, Y. Kondo, S. Suzuki, M. Miyasaka, M. Fujikiand, Y. Imai, Chem. Lett. **49**, 674 (2020)
7. S. Suzuki, K. Kaneko, T. Hanasaki, M. Shizuma, Y. Imai, ChemPhotoChem. **8**, e202300224 (2024)
8. T. Kuroda, M. Kitahara, S. Yagi, Y. Imai, Front. Chem Nanoscience. **11**, 1281168 (2023)

Colorless and Transparent Photoluminescence Materials (CLT-PLM)

Hiroki Iwanaga

1 Introduction

1.1 Advantages of Colorless and Transparent Photoluminescence Materials

Recently, colorless and transparent photoluminescence materials (CLT-PLMs) consisting of phosphors dissolved in polymers have been attracting increasing attention as a new type of functional soft matter in the fields of electronic devices, security applications, and sensing applications. They not only convert light from one wavelength to another but also have the remarkable properties of very high light absorption and low scattering.

For inorganic phosphors, particle size must be made small enough to prevent light scattering. However, the smaller the particle size, the lower the quantum yield. In recent years, much attention is focused on lanthanide complexes, organic phosphors, nanometer phosphors [1, 2], quantum dots [3–5], and the use of fluorescent sources in CLT-PLMs. Among these, lanthanide complexes have been extensively investigated as luminescent materials in a wide range of scientific and industrial applications [6, 7].

Figure 1 shows examples of CLT-PLMs involving lanthanide complexes. They are colorless and transparent under room light and emit pure red light when irradiated with ultraviolet (UV) light [8].

H. Iwanaga (✉)
Photoluminescence Material Project Team, Innovation Business Promotion Office, Next Business Development Div., Toshiba Corporation, 72-34 Horikawa-cho, Saiwai-ku, Kawasaki 212-8585, Japan
e-mail: hiroki.iwanaga.r66@mail.toshiba; 5691465@rikkyo.ac.jp

Toshiba Future Luminescence Molecule Project Laboratory, Graduate School of Science, Rikkyo University, 3-34-1 Nishi-Ikebukuro, Toshima-ku, Tokyo 171-8501, Japan

M. Funahashi (ed.), *Functional Soft Matter II*,
https://doi.org/10.1007/978-981-92-0100-6_22

Fig. 1 Colorless and transparent photoluminescence materials (CLT-PLMs) produced by dissolving Eu(III) and/or Tb(III) complexes in a polymer [8]

Lanthanide complexes are colorless because they absorb almost no visible light. At the same time, completely dissolving the lanthanide complexes in polymers makes them transparent. However, many lanthanide complexes have low solubility because of their large molecular weight, especially in hydrophobic polymers or solvents. High quantum yields are required in order to obtain high photoluminescence intensity under irradiation with UV light. To achieve both high solubilities and high quantum yields for lanthanide complexes, novel molecular design concepts are needed.

1.2 Fundamental Characteristics of Lanthanide Complexes

While absorption cross section is indeed low, the quantum yield of Eu(III) ion itself is extremely high, if no vibrational-rotational quenching modes are present. However, light emission from lanthanide ions can be enhanced through the antenna effect of ligands (Fig. 2). β-Diketonates efficiently absorb light and transfer energy to lanthanide ions [9, 10], and are known to be effective ligands for increasing the photoluminescence intensity of Eu(III) complexes. The photoluminescence intensity of Eu(III) complexes depends largely on the β-diketonate substituents because the triplet-state energy levels of β-diketonates are derived from the molecular structures of the substituents. However, emission intensity sufficient for use in many applications can be difficult to obtain simply by adjusting the of β-diketonates substituents.

There are two main types in ligands of lanthanide complexes: ionic and non-ionic. β-Diketonates are prominent ionic ligands and neutralize the charge of lanthanide

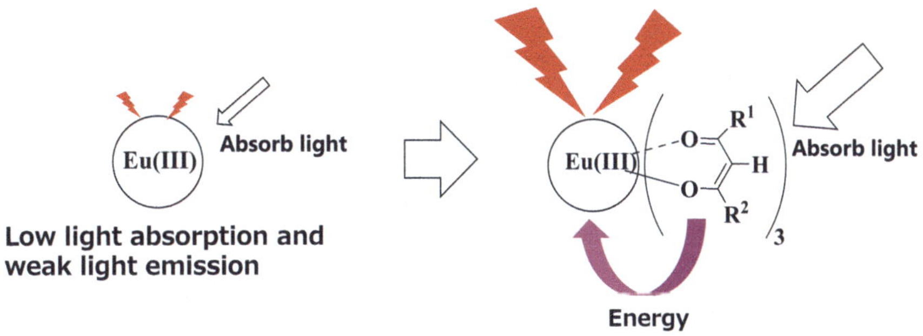

Fig. 2 Schematic drawings of Eu(III) complexes

ions. The use of non-ionic ligands in addition to β-diketonates is known to enhance photoluminescence intensities. Phosphine oxide compounds are strong Lewis bases and excellent non-ionic ligands for increasing photoluminescence intensity of Eu(III) complexes [11–13]. However, further improvements in solubility and photoluminescence intensity are required for lanthanide complexes to be used in larger-scale practical applications.

2 Development of Highly Soluble and Strongly Photo-Luminescent Lanthanide Complexes

2.1 *Novel Molecular Design of Lanthanide Complexes for Realizing High Solubility and Strong Photoluminescence*

For the application of functional organic compounds in many types of electronic devices, one of their most important properties is solubility in organic solvents or polymers. However, many functional compounds do not dissolve sufficiently because of their large π conjugation systems, which induce strong π-π stacking among molecules. Eu(III) complexes can be soluble in polar and hydrophilic solvents such as ethyl acetate or alcohol, but their solubility drastically decreases in hydrophobic solvents [14]. It was found that making the polar parts of dichroic dye asymmetric is effective for increasing their solubility [15, 16]. Based on these results, a novel molecular design concept for Eu(III) complexes to improve their solubility was established. Coordination of two different phosphine oxide structures together with β-diketonates to one Eu(III) or Tb(III) ion is effective for improving solubility [14, 17, 18]. However, ligand exchange of phosphine oxide was detected for Eu(III)-β-diketonates by NMR analysis [19] and is expected to have an undesirable effect on durability. To overcome this problem, asymmetric diphosphine dioxide ligands [20, 21] which have molecular structures consisting of two different phosphine oxide parts connected via

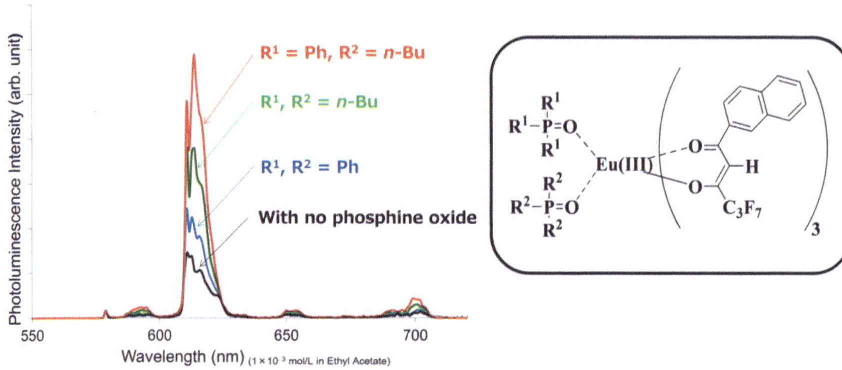

Fig. 3 Relationships between molecular structures and photoluminescence spectra of Eu(III)-β-diketonates [22]

a methylene unit were developed. They suppress ligand exchange via the chelate effect. Eu(III) complexes with an asymmetric diphosphine dioxide ligand are highly soluble, even in fluorinated solvents.

Next, correlation between the molecular structures of Eu(III) complexes and photoluminescence intensity are described. Figure 3 shows the effects of phosphine oxide ligands on the photoluminescence intensity of Eu(III) complexes [22]. Complexes without a phosphine oxide ligand showed relatively low photoluminescence intensity. When two triphenyl phosphine oxides (TPPOs) were coordinated, photoluminescence intensity increased, and when two tributyl phosphine oxides (TBPO) were coordinated, the enhancement of photoluminescence intensity was even greater. These results are attributable to TBPO being a stronger Lewis base than TPPO. Moreover, with coordination of both TPPO and TBPO (1 equiv each), the largest photoluminescence intensity was obtained.

2.2 *Asymmetric Diphosphine Dioxide Ligands for Lanthanide Complexes*

The molecular structures of diphosphine dioxide ligands and the properties of their Eu(III) complexes have been investigated in detail. The number of methylene units, n, between $P = O$ groups in a diphosphine dioxide ligand, which restrict the relative positions of the nearest oxygen atoms around the Eu(III) ion, influences the shape of the $^5D_0 \rightarrow {}^7F_2$ transition [21]. Stark splitting is considered to be influenced by n. A thienyl-substituted diphosphine dioxide ligand suppresses the concentration quenching of Eu(III) complexes [23], and bulky alkyl groups in diphosphine dioxide ligand increase the half-widths of the $^5D_0 \rightarrow {}^7F_2$ transition of Eu(III) complexes [24]. The mutual positions of the nearest oxygen atoms around the Eu(III) ion can be strained by bulky substituents, increasing the energy transfer efficiency (Φ_{ET}).

Fig. 4 Molecular structure of the optimal asymmetric diphosphine dioxide ligand for Eu(III) complexes to increase photoluminescence intensity (DPDO-mCF_3 ligand) [25]

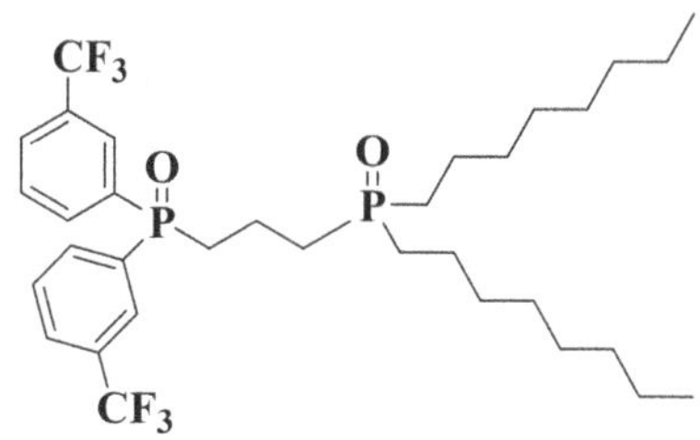

DPDO-mCF_3 ligand

Introducing CF_3 groups into aromatic groups in diphosphine dioxides effectively increases their solubility [21, 25]. The optimal molecular structure of an aromatic- and aliphatic-substituted diphosphine dioxide ligand is shown in Fig. 4. The number of methylene units between the two $P = O$ groups (n), the aliphatic groups, and the aromatic groups were optimized to increase the photoluminescence intensity and solubility of the Eu(III) complexes. Substitution of CF_3 groups at the meta positions of the phenyl groups is a characteristic of the optimal diphosphine dioxide ligand for maintaining solubility [25].

Figure 5a and b shows the relationships between excitation wavelength and absolute quantum yield for Eu(III)(hfnh)$_3$ **1** and Eu(III)(hfnh)$_3$(DPDO-mCF_3) **2** in both the solid and solution states. Compound **1** is a 6-coordinate Eu(III) complex and its quantum yield is relatively low over a wide wavelength range. The quantum yield is lower in the solid state than in solution due to the effects of concentration quenching. On the other hand, the quantum yield of Eu(III) complex **2** with a DPDO-mCF_3 ligand is markedly larger than that of complex **1**. Also, the quantum yield of complex **2** is larger in the solid state than in solution over a wide wavelength region, in contrast to the case of complex **1,** and reaches 0.82 at an excitation wavelength of 280 nm [25].

2.3 Mechanism of Quantum Yield Enhancement by Diphosphine Dioxide Ligands on Eu(III) Complexes

To elucidate the effects of phosphine oxide ligand coordination, a 6-coordinate Eu(III) complex was synthesized that was designed to have low luminescence and a large absorption coefficient (Tris{6,6,7,7,8,8,8-heptafluoro-1-[2-(9,9-dimethylfluorenyl)]-1,3-octanedionate} europium(III) (Eu(III)(hfod)$_3$ **3**, Fig. 6) [26].

Dimethylfluorenyl groups are bulky aromatic substituents with large absorption coefficients. Partially fluorinated alkyl groups in β-diketonates are also very bulky. Methylene units in partially fluorinated alkyl groups decrease the efficiency of energy transfer from the ligands to the Eu(III) ion.

Fig. 5 Action spectra (excitation wavelength vs. Φ_{TOT}) of **a** Eu(III)(hfnh)$_3$ **1** and **b** Eu(III)(hfnh)$_3$(DPDO-mCF$_3$) **2** in the solid and solution states [25]

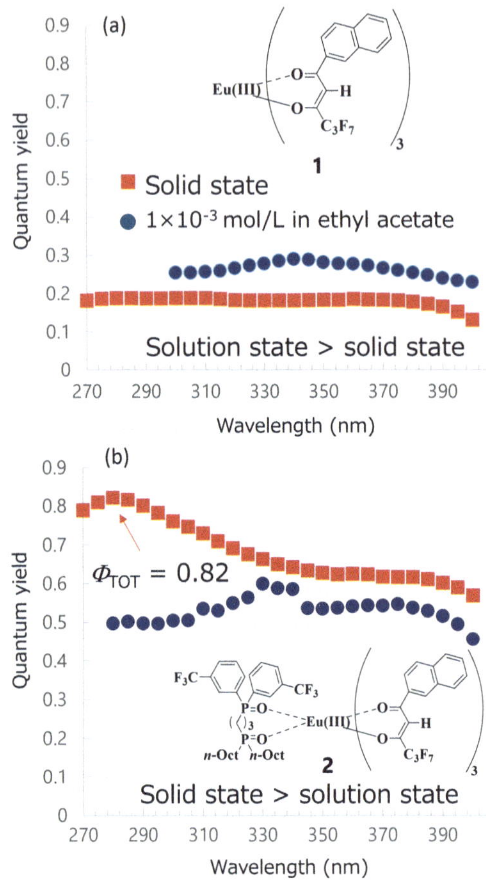

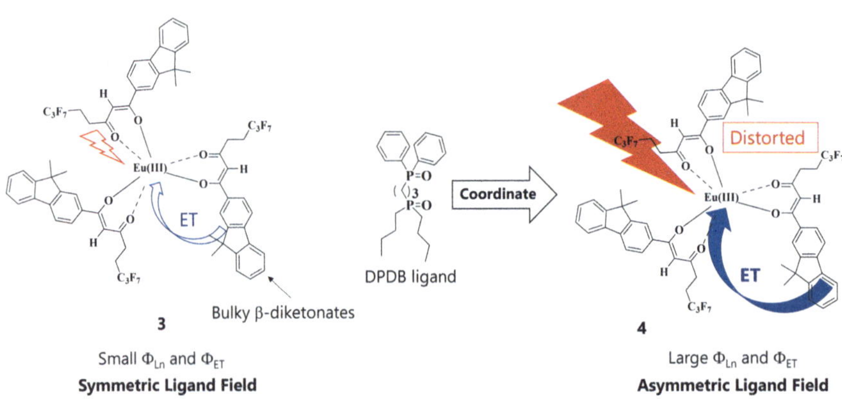

Fig. 6 Effects of coordination of a DPDB ligand to Eu(III)(hfod)$_3$ with bulky β-diketonates, hfod [26]

The photoluminescence intensity of Eu(III)(hfod)$_3$ **3** is dramatically enhanced by coordination of a diphenyl-4-(dibutylphosphinyl)butyl phosphine oxide (DPDB) ligand (generating Eu(III)(hfod)$_3$(DPDB) **4**) due to the increased Φ_{TOT}, Φ_{Ln}, and Φ_{ET} in both the solid and solution states.

Iwanaga et al. proposed the following hypothesis for the increase in Φ_{TOT} of the 6-coordinate Eu(III) complex by the effects of a DPDB ligand. When the DPDB ligand coordinates with the Eu(III) ion, the positions of the nearest oxygen atoms around the Eu(III) ion are shifted by steric repulsion and the relative positions of the nearest oxygen atoms become distorted. Asymmetry of the ligand field is induced by this distorted coordination environment, which increases Φ_{TOT}.

According to a theoretical analysis by Sims and co-workers, coordination of various non-ionic ligands to the Eu(III) ion effectively increases the quantum yield of Eu(III) complexes [27, 28]. In 2020, Iwanaga et al. proposed a new definition, effective dipole moment (μ) which is a metric of ligand field symmetry. It was shown that the ligand fields of Eu(III) ions become asymmetrical by coordination of two different phosphine oxides, thus increasing the quantum yield of the resulting Eu(III) complexes [26]. It was successfully shown that the efficiency of energy transfer from the lowest triplet state of the ligands to the $^5D_1$ level of the Eu(III) ion (Φ_{ET}) increases when ligand fields of Eu(III) ion become more asymmetric due to coordination of a DPDB ligand.

2.4 *Dinuclear Eu(III) Complexes with a Tetraphosphine Tetraoxide Ligand and Their Properties*

Dinuclear Eu(III) complexes with a macrocyclic TPTO ligand (1,5,9,13-tetraphosphacyclohexadecane-1,5,9,13-tetraphenyl-1,5,9,13-tetraoxide) have been developed. They have relatively high quantum yields even at high concentration in a solvent [29] (Fig. 7).

The TPTO ligand serves as a "separator among Eu(III) ions" and prevents concentration quenching caused by maintaining the distance among Eu(III) ions. However, in dinuclear Eu(III) complexes with TPTO ligand, two identical phosphine oxides coordinate to one Eu(III) ion because the four phosphine oxide structures of the macrocyclic TPTO ligand are the same. Also, the two Eu(III) ions in a dinuclear complex are located in equivalent coordination environments. Quantum yields of Eu(III) complex **5** is restricted by these characteristic molecular structures of the macrocyclic TPTO ligand. On the other hand, asymmetric diphosphine dioxide ligands enhance the quantum yields of Eu(III) complexes. By fusing these two molecular design concepts, the novel branched tetraphosphine tetraoxide ligands TDPBPO (tris(4-diphenylphosphinylbutyl)phosphine oxide) and TDPPPO (tris(5-diphenylphosphinylpentyl)phosphine oxide), which form dinuclear Eu(III) complexes, were developed (Fig. 8 [30, 31]). They are based on the following two concepts: separating the two Eu(III) as is done by the macrocyclic TPTO ligand,

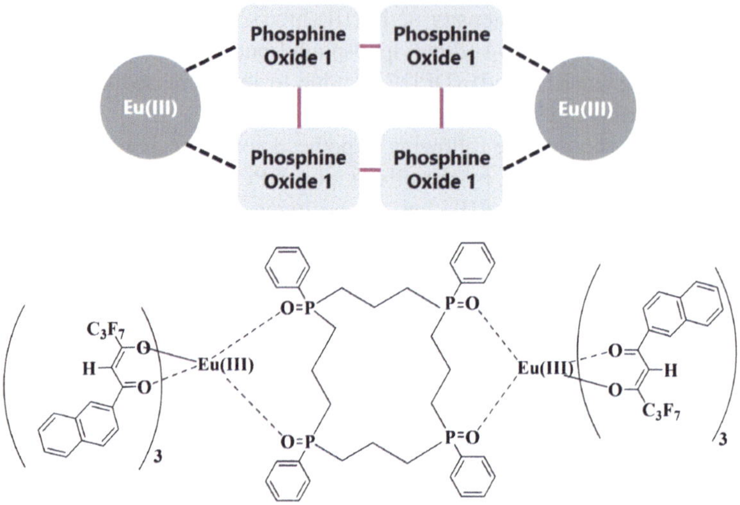

5

Fig. 7 Molecular structures of dinuclear Eu(III) complexes with a macrocyclic TPTO ligand. [29]

and coordinating two different phosphine oxides to one Eu(III) ion as is done by asymmetric diphosphine dioxide ligands (Fig. 8).

The phosphine oxide structures in the center of TDPBPO and TDPPPO have aliphatic substituents (type 1) and aromatic and aliphatic substituents (type 2), respectively. When TDPBPO or TDPPPO coordinate to Eu(III) ions, dinuclear Eu(III) complexes with different types of phosphine oxides are generated. Moreover, the two Eu(III) ions in a dinuclear complex are located in different coordination environments because one has phosphine oxide types 1 and 2 and the other have two of phosphine oxide type 2 (Fig. 9, Eu(III) complexes **6**, **7**).

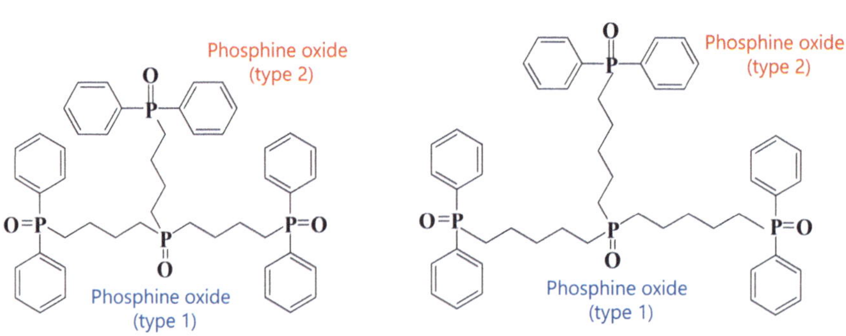

TDPBPO **TDPPPO**

Fig. 8 Molecular structures of branched tetraphosphine tetraoxide ligands for Eu(III) complexes [29]

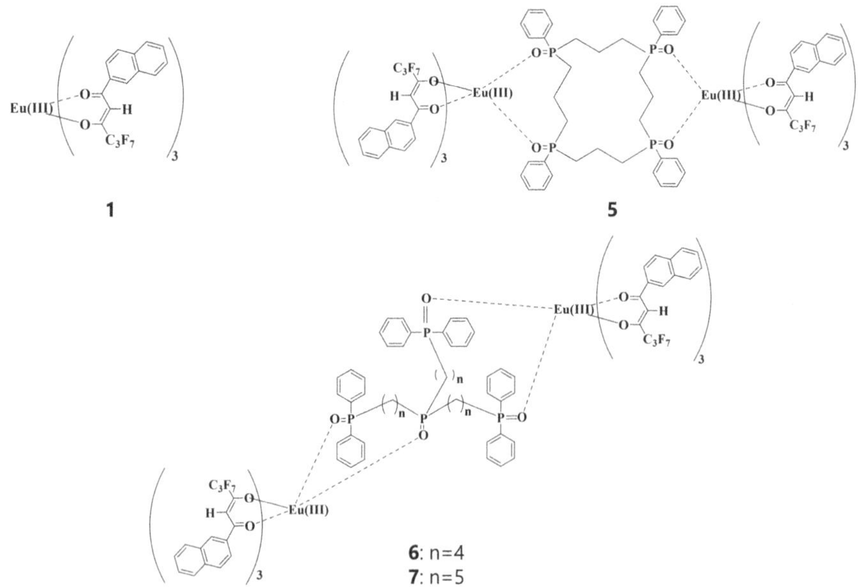

Fig. 9 Molecular structures of Eu(III) complexes with a branched tetraphosphine tetraoxide ligand 6, 7 and complexes 1 and 5 to compare [29]

The photoluminescence properties of Eu(III) complexes **6** and **7** are compared with those of Eu(III) complexes **5** and **1**. Eu(III) complex **5** has a macrocyclic tetraphosphine tetraoxide ligand which has the same phosphine oxide structure. Eu(III) complex **1** has no phosphine oxide ligand. The differences in photoluminescence properties of complexes **6** and **7** compared with **5** will reveal the effects of coordinating two different phosphine oxide structures to one Eu ion. The differences in these properties of complexes **6** and **7** compared with **1** will reveal the comprehensive effects of branched tetraphosphine tetraoxide ligands on photoluminescence properties.

Figure 10 shows the photoluminescence spectra of Eu(III) complexes **1**, **5**, **6**, and **7** in ethyl acetate at a concentration of 1×10^{-3} mol/L.

The photoluminescence intensity of Eu(III) complexes **6** and **7** is larger than that of complex **1** and the half-width of the $^5D_0 \rightarrow {}^7F_2$ transition of complexes **6** and **7** is also larger than that of complex **1**. These results indicate that the branched tetraphosphine tetraoxide ligand reduces the symmetry of the ligand field and increases the photoluminescence intensity. Notably, the photoluminescence intensity of complexes **6** and **7** is sixfold that of complex **1**. These results show that coordination of branched tetraphosphine tetraoxide ligands can increase the photoluminescence intensity of Eu(III) complexes to a greater extent than coordination of two different phosphine oxides with a Eu(III) ion, which results in the fourfold increase shown in Fig. 3 [22].

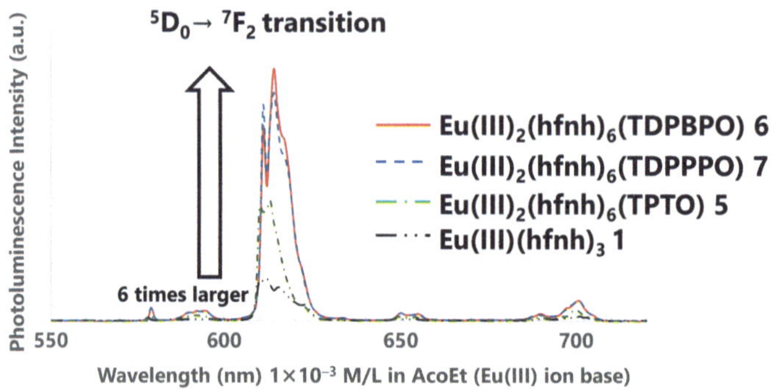

Fig. 10 Photoluminescence spectra of Eu(III) complexes **1**, **5**, **6**, and **7** in ethyl acetate at a concentration of 1×10^{-3} mol/L [31]

3 Applications

3.1 Target Applications of Transparent Photoluminescence Materials

Figure 11 shows target applications of CLT-PLMs. In the early stage, we are aiming at UV light sensing, especially deep UV light visualization, security printing, circadian rhythm lighting, and detection of harmful materials. In the next stage, our aim will be scaling production in order to commercialize these materials for micro-LED use.

The following is an example of detecting harmful materials. Dichlorvos is a general-purpose insecticide for agricultural, household, and animal use. However, it is harmful to humans, and a rapid, high-sensitivity detection method is needed. The photoluminescence intensity of Eu(III)(hfnh)$_3$ **1** is drastically and rapidly decreased

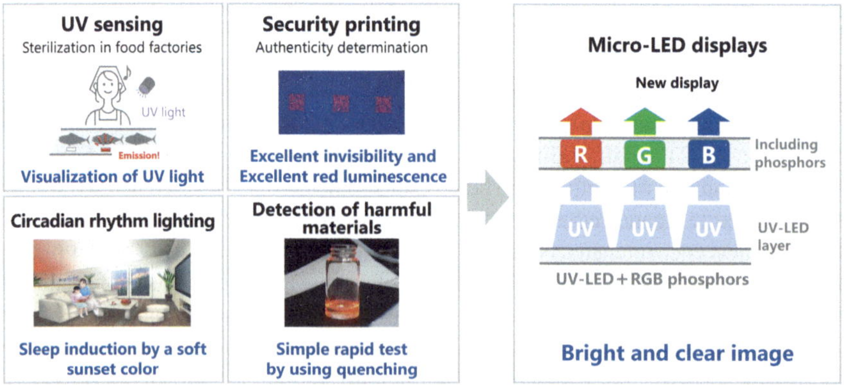

Fig. 11 Target applications of transparent photoluminescence materials

when a dilute solution of dichlorvos is added to a solution of Eu(III)(hfnh)$_3$ **1** [32]. Thus, Eu(III)(hfnh)$_3$ **1** is a highly sensitive luminescent probe for dichlorvos.

Table 1 shows the history of displays and future predictions. LCDs have been the most widely used display for a long time. Despite being a highly mature technology, general LCDs are not suitable for use in bright outdoor environments. Around 2020, organic LEDs (OLEDs) have begun to appear on the market, partially taking the place of LCDs. OLEDs have excellent visibility in bright outdoor environments, but their lifetime is not so long. Micro-LED displays are attracting considerable attention because they have excellent visibility in bright outdoor environments and long lifetime. However, there are problems for practical applications, particularly low production yields and insufficient photoluminescence intensity of red phosphors.

For display use, phosphors with large photoluminescence intensity and high color purity are required. In addition to this, for micro-LED displays consisting of UV or blue LED arrays and phosphors, where the chips used are very small, the particle size of phosphors must be small to suppress variation in hues for each pixel. There is an especially strong demand for a red phosphor that satisfies these requirements.

For inorganic phosphors, quantum yields decrease with decreasing particles size of phosphors [33] (Fig. 12). In addition, the half-widths of emission spectra of inorganic phosphors are large and the color purity is low.

For Eu(III) complexes, on the other hand, particle size is theoretically not related to quantum yield because each molecule of a Eu(III) complex absorbs and emits light. From this point of view, Eu(III) complexes are promising candidates as red phosphors for use in micro-LED displays [30, 34].

Eu(III) complexes with two different phosphine oxide structures are highly soluble in polymers and are colorless and transparent under visible light. Their sharp photoluminescence spectra are attractive for use in displays because they realize high color purity, which enables a wide range of color reproduction in displays. They also have large quantum yields and molar absorption coefficients.

LED devices consisting of a 402-nm LED chip and dimethyl silicone polymer including Eu(III) complexes with asymmetric diphosphine dioxide ligand or KSF

Table 1 History of displays and future predictions

	LCD → 2020	OLED → 2030	Micro-LED
Brightness	Low	Large (Self-emitting)	Very large (Self-emitting)
Color reproduction area	Middle	Large	Large
Lifetime	60000 h	30000 h	100000 h
Characteristic	Not suitable for outdoor use	Suitable for outdoor use Short lifetime	Suitable for outdoor use Long lifetime Low production yield Insufficient photoluminescence Intensity of red phosphor

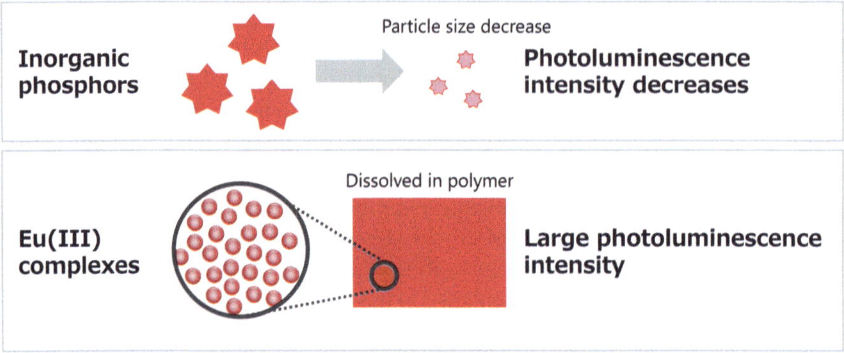

Fig. 12 Comparison of inorganic phosphors and Eu(III) complexes

phosphor have been prototyped and their photoluminescence intensities were measured (Fig. 13) [35].

With increasing Eu(III) complex concentration, the photoluminescence intensities increase sharply. With the KSF phosphor [36–38], a much greater concentration is required to obtain strong emission. Photoluminescence of a KSF phosphor was rarely observed in the low-concentration region between 0.1 wt% and 5 wt%, and the concentration of KSF phosphor were 5 wt%, 20 wt%, and 40 wt%. The photoluminescence intensity per unit concentration in LED devices employing a Eu(III) complex was notably larger than that in LED devices employing KSF phosphor, which indicates that this complex is a promising candidate for red phosphors in micro-LEDs [35]. Figure 14 shows the photoluminescence spectra of the mixture of the Eu(III)

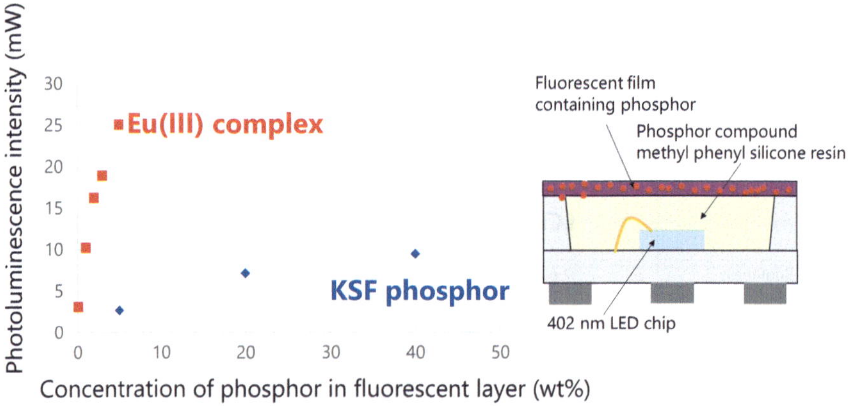

Fig. 13 Photoluminescence intensity of LED devices containing Eu(III) complexes with an asymmetric diphosphine dioxide ligand or KSF phosphor. Concentrations of Eu(III) complexes with two different phosphine oxides were between 0.1 and 5.0 wt%. Concentrations of KSF phosphor were between 5 and 40 wt% [35]

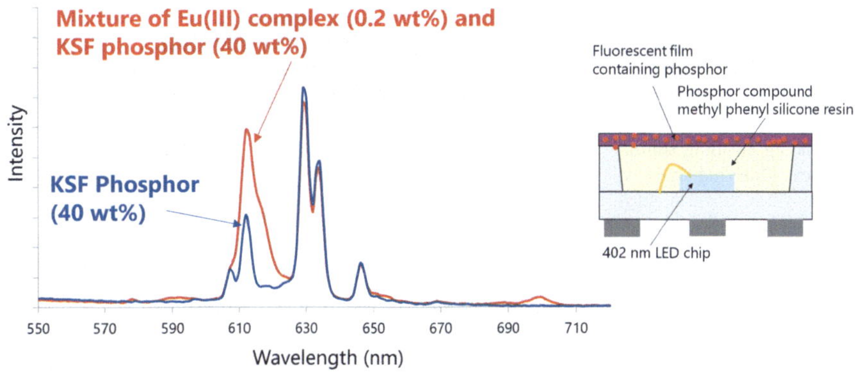

Fig. 14 Photoluminescence spectra of the mixture of the Eu(III) complex with two different phosphine oxide structures and a KSF phosphor in the fluorescence film of an LED [35]

complex with two different phosphine oxide structures and a KSF phosphor in the fluorescence film of an LED [35].

The dominant wavelength of photoluminescence spectra of the mixture is 619 nm, which is suitable for displays. When only 0.2 wt% of Eu(III) complex is mixed with 40 wt% of KSF phosphor, the total photoluminescence intensity is increased by 18%.

4 Summary

This chapter focused on highly soluble lanthanide complexes with large quantum yields for CLT-PLMs in view of their rapid progress. A novel molecular design concept, namely, coordination of two different phosphine oxide structures with one lanthanide ion together with β-diketonates, has been established for lanthanide complexes to improve their solubility and photoluminescence intensity. Moreover, the effective dipole moment (μ), which is a novel metric of ligand field symmetry, has been determined. The ligand fields of Eu(III) ions were shown to become asymmetrical due to the coordination of two different phosphine oxides, thus enhancing the quantum yields of these complexes. In line with this novel concept, many lanthanide complexes with two different phosphine oxide structures have been developed. CLT-PLMs with these lanthanide complexes are a promising new type of functional soft matter for micro-LED displays, security application, and sensing applications. Figure 15 shows that CLT-PLMs with novel Eu(III) complexes can find use in practical applications, shining like glittering stars.

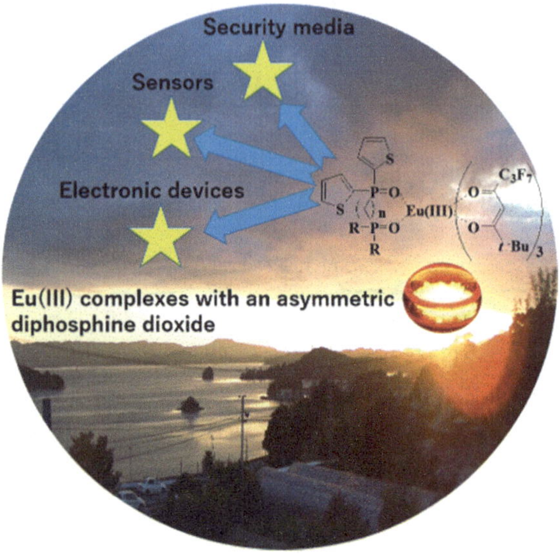

Fig. 15 Inside cover of the *bulletin of the chemical society of Japan* [23]. Eu(III) complexes with two different phosphine oxides used in practical application, shining like these glittering stars

References

1. T. Igarashi, M. Ihara, T. Kusunoki, K. Ohno, T. Isobe, M. Senna, Relationship between optical properties and crystallinity of nanometer phosphor. Appl. Phys. Lett. **76**, 1549–1551 (2000)
2. G. Wakefield, E. Holland, P.J. Dobson, J.L. Hutchison, Luminescence properties of nanocrystalline Y_2O_3:Eu. Adv. Mater. **13**(20), 1557–1560 (2001)
3. A.I. Ekimov, A.A. Onushchenko, Quantum size effect in three-dimensional microscopic semiconductor crystals. JETP Lett. **34**, 345–349 (1981)
4. R. Rossetti, S. Nakhara, L.E. Brus, Quantum size effects in the redox potentials, resonance Raman spectra, and electronic spectra of CdS crystallites in aqueous solution. J. Chem. Phys. **79**, 1086–1088 (1983)
5. R. Katakami, K. Sato, A. Ogura, K. Takao, Y. Iso, T. Isobe, Open system synthesis of narrow-bandwidth red-fluorescent carbon quantum dots with a function of multi-metal ion sensing. J. Mater. Chem. C **11**, 4143–4152 (2023)
6. K. Binnemans, Lanthanide-based luminescent hybrid materials. Chem. Rev. **109**, 4283–4374 (2009)
7. J.C.G. Bunzli, On the design of highly luminescent lanthanide complexes. Coordin Chem Rev. **293–294**, 19–47 (2015)
8. H. Iwanaga, Emission properties, solubility, thermodynamic analysis and NMR studies of rare-earth complexes with two different phosphine oxides. Materials **3**, 4080–4108 (2010)
9. S. Sato, M. Wada, T. Seki, Some properties of europium β-diketone chelates 1. (Synthesis and fluorescent properties), Jpn. J. Appl. Phys. **7**, 7–13 (1968)
10. S. Sato, M. Wada, Relations between intramolecular energy transfer efficiencies and triplet state energies in rare earth β-diketone chelates. Bull. Chem. Soc. Jpn **43**, 1955–1962 (1970)
11. W.G. Andrew, Platt, Lanthanide phosphine oxide complexes. Coordin Chem. Rev. **340**, 62–78 (2017)
12. J. Yuan, K. Matsumoto, Fluorescence enhancement by electron-withdrawing groups on β-diketones in Eu(III)-β-diketonato-topo ternary complexes. Anal. Sci. **12**, 31–36 (1996)
13. K. Nakamura, Y. Hasegawa, Y. Wada, S. Yanagida, Novel luminescent Eu(III) complex with remarkably narrow emission band. Chem. Phys. Lett. **398**, 500–504 (2004)

14. H. Iwanaga, A. Amano, F. Furuya, Y. Yamasaki, Solubility in fluorinated medium and thermal properties of europium(III) complexes with phosphine oxides. Jpn. J. Appl. Phys. **45**, 558–562 (2006)
15. H. Iwanaga, K. Naito, F. Effenberger, Oligothiophene dyes for guest host liquid crystal displays. Liq. Cryst. **27**, 115–123 (2000)
16. H. Iwanaga, F. Aiga, Correlation among molecular structures, solubilities in fluorinated media, thermal properties and absorption spectra of anthraquinone dichroic dyes with phenylthiol and/or anilino groups. Liq. Cryst. **38**, 135–148 (2011)
17. H. Iwanaga, A. Amano, F. Aiga, K. Harada, M. Oguchi, Development of ultraviolet LED devices containing europium (III) complexes in fluorescence layer. J. Alloys Compd. **408–412**, 921–925 (2006)
18. H. Iwanaga, F. Aiga, Novel Tb(III) complexes with two different structures of phosphine oxides and their properties. J. Lumin. **130**, 812–816 (2010)
19. H. Iwanaga, A. Amano, M. Oguchi, Study of molecular structures and properties of Europium(III) complexes with phosphine oxides by NMR analysis. Jpn. J. Appl. Phys. **443**, 3702–3705 (2005)
20. H. Iwanaga, F. Aiga, A. Amano, The molecular structures and properties of novel Eu(III) complexes with asymmetric bis-phosphine oxides. Mater. Res. Soc. Symp. Proc. **965**, 211–216 (2006)
21. H. Iwanaga, Relationships between molecular structures of aromatic- and aliphatic-substituted diphosphine dioxide ligands and properties of Eu(III) complexes. Opt. Mater. **85**, 418–424 (2018)
22. H. Iwanaga, Investigation of strong photoluminescence and highly soluble Eu(III) complexes with phosphine oxides and β−diketonates. J. Lumin. **200**, 233–239 (2018)
23. H. Iwanaga, Photoluminescence properties of Eu(III) complexes with thienyl-substituted diphosphine dioxide ligands. Bull. Chem. Soc. Jpn **92**, 1385–1393 (2019)
24. H. Iwanaga, Effects of alkyl groups in diphosphine dioxide-ligand for Eu(III)-β-diketonate complexes on photoluminescence properties. Chem. Phys. Lett. **736**, 136794 (2019)
25. H. Iwanaga, A CF_3-substituted diphosphine dioxide ligand that enhances both photoluminescence intensity and solubility of Eu(III) complexes. J. Alloys Compd. **790**, 296–304 (2019)
26. H. Iwanaga, F. Aiga, Quantum yield and photoluminescence intensity enhancement effects of diphosphine dioxide ligand on a 6-coordinate Eu(III)-β-diketonate complex with low luminescence. ACS Omega **6**, 416–424 (2021)
27. N.B.D. Lima, S.M.C. Gonçalves, S.A. Júnior, A.M. Simas, A comprehensive strategy to boost the quantum yield of luminescence of europium complexes. Sci. Rep. **3**, 02395 (2013)
28. N.B.D. Lima, J.D.L. Dutra, S.M.C. Gonçalves, R.O. Freire, A.M. Simas, Chemical partition of the radiative decay rate of luminescence of europium complexes. Sci. Rep. **6**, 21204 (2016)
29. H. Iwanaga, F. Aiga, Dinuclear Eu(III) complexes with macrocyclic tetraphosphine tetraoxide ligand and their photoluminescence properties. J. Lumin. **208**, 1–5 (2019)
30. H. Iwanaga, C. Huang, T. Hongo, A. Okuno, Dinuclear Eu(III)-β-diketonates with tetraphosphine tetraoxide ligand and their LED applications, *2024 International Conference on Electronics Packaging (ICEP)*, (Toyama, Japan, 2024), pp. 263–264, https://doi.org/10.23919/ICEP61562.2024.10535654
31. H. Iwanaga, F. Aiga, S. Sasaoka, T. Wazaki, Development and photoluminescence properties of dinuclear Eu(III)-β-diketonates with a branched tetraphosphine tetraoxide ligand for potential use in LEDs as red phosphors. IEICE Trans. **E107.C**, 34–41 (2024)
32. H. Iwanaga, F. Aiga, A simple and sensitive detection method for the pesticide dichlorvos in solution using Eu(III)-β-diketonate as a luminescent probe. Jpn. J. Appl. Phys. **59**, SDDF05 (2020)
33. C. Bois, P. Bodrogi, T.-Q. Khanh, H. Winkler, White LED light characteristics as a function of phosphor particle size. ECS J. Solid State Sci. Technol. **1**(5), R131–R135 (2012)
34. H. Iwanaga, Photoluminescence properties of Eu(III) complexes with an asymmetric diphosphine dioxide ligand for potential uses in LED, security, and sensing devices, *2022 International Conference on Electronics Packaging (ICEP)* (Sapporo, Japan, 2022), pp. 17–18 (Invited talk)

35. H. Iwanaga, Novel Eu(III) complexes with an asymmetric diphosphine dioxide ligand and their potential uses in micro-LED displays as red phosphors, *22nd International Symposium on Microelectronics and Packaging joined with the 18th International Conference Reliability and Stress-Related Phenomena in Nanoelectronics*, pp. I1–25 (Invited talk)
36. I. Hendy, J. Murphy, A. Setlur, Latest advances in narrow-band phosphors and their role in color management. Inf. Display **39**(3), 37–42 (2023)
37. A.G. Paulusz, Efficient Mn(IV) emission in fluorine coordination. J. Electrochem. Soc. **120**(7), 942–947 (1973)
38. T. Takahashi, S. Adachi, Mn4 + -activated red photoluminescence in K2SiF6 phosphor. J. Electrochem. Soc. **155**(12), E183–E188 (2008)

Subject Index

M. Funahashi (ed.), *Functional Soft Matter II*,
https://doi.org/10.1007/978-981-92-0100-6

The manufacturer's authorised representative in the EU is Springer Nature Customer Service Centre GmbH, Europaplatz 3, 69115 Heidelberg, Germany. If you have any concerns regarding our products, please contact ProductSafety@springernature.com

Printed and bound by CPI Group (UK) Ltd, Croydon, CR0 4YY
17/09/2026
02218430-0012